Annual Meeting of the Federation of Analytical Chemistry and Spectroscopy Societies (FACSS 2017)

The Great SCIentific eXchange SCIX2017

And the National Meeting of the Society for Applied Spectroscopy (SAS)

Reno, Nevada, USA
8 - 13 October 2017

ISBN: 978-1-5108-5214-3

Printed from e-media with permission by:

Curran Associates, Inc.
57 Morehouse Lane
Red Hook, NY 12571

Some format issues inherent in the e-media version may also appear in this print version.

Printed by Curran Associates, Inc. (2018)

Federation of Analytical Chemistry and Spectroscopy Societies (FACSS)
PO Box 24379
Santa Fe, NM 87502
USA

Phone: (505) 820-1648
Fax: (505) 989-1073

facss@facss.org

Additional copies of this publication are available from:

Curran Associates, Inc.
57 Morehouse Lane
Red Hook, NY 12571 USA
Phone: 845-758-0400
Fax: 845-758-2633
Email: curran@proceedings.com
Web: www.proceedings.com

TABLE OF CONTENTS

The Analytical and Economic Challenges of Maintaining Food Safety in a Global Supply Chain

Janie Dubois

Maintaining a safe and abundant food supply in the face of a rapidly increasing human population is a huge challenge facing governments and regulators around the globe. The problem is significantly complicated by the diverse geographic origins for those foods within a highly complicated supply chain, which often includes any number of additional unit operations and packaging for processed foods. For example, the US imports almost 600 million gallons of apple juice every year, mostly from China. As a result, the industry is becoming increasingly dependent on the application of novel technologies deployed in a preventive holistic approach to accurately enforce regulatory compliance and quickly identify any divergence from these requirements and its root cause. However, to complicate this further, and because of its historically 'slim' profit margins, the food sector has also not been perceived as a premium market for analytical technology, resulting in our analytical arsenal being somewhat neglected, and largely composed of instruments developed for other markets and adapted for food samples. But how similar are the requirements for food from the original targets for those technologies? How much is the population prepared to pay for "more safety"? Given the huge tonnage of imports into the US and Western Europe, and the enormous analytical testing burden that implies, can food safety be verified at source when derived from countries that have made little or no investment in analytical measurement technology? Does tested even mean safe, and can it be readily traced to improvements in health? The landmark Food Safety Modernization Act of 2011 brought to light the sophistication of the food supply chain and its control systems, and raised the profile of this market with technology developers. With 180,000 processors in the US alone, it was realized that the sheer volume could make up for price sensitivity in some segments; it also became clear that simply applying tools borrowed from the pharmaceutical, medical or chemical industries may not be the best approach anymore. In our world, measuring up to 300 different chemicals at trace levels in table grapes is considered easy, but the industry is expected to measure trace levels of pesticides, antibiotics and industrial chemicals in a complex fatty and sweet blueberry muffin matrix for the same cost! Is this practical or even possible? In 2017, and with the huge strides made in genomics, finding evidence of the presence of a toxin-producing E. coli in a sample made up of a hundred billions of cells, is now common practice, and for many laboratories, is often a far easier challenge than measuring trace chemical contaminants in the same sample. With the significantly growing requirement for fit-for-purpose tools to measure trace microbial and chemical contaminants, food testing looks to be the next high-growth market for analytical instrumentation and ideally will result in a win-win for the consumer, producer and testers alike. To achieve this however, we must also understand the regulatory, technical and societal challenges faced by the diverse group of stakeholders that we simply describe as the 'food industry'. In this presentation I will describe the challenges facing food safety analysts, paying special attention to the realities that may not be commonplace for many applications, such as the natural variability of the biomatrix and the high volume of testing needed. These challenges will be discussed in the context of trade requirements, economic impacts, health and societal norms.

The Small Matter of Bioanalysis: Adventures at Less Than 10 nm with Quantum Dots and/or FRET

Russ Algar

Bioanalysis encompasses a wide range of quantitative measurements for biomolecular targets and on biological systems, spanning from *in vitro* diagnostics to cellular imaging. Nanomaterials have had a significant impact in this field because their unique and often size-dependent properties offer new capabilities and new opportunities. This presentation will summarize some of our recent and ongoing research in this area, particularly focusing on brightly luminescent quantum dots (QDs), which are semiconductor nanocrystals smaller than 10 nm in size, and also Förster resonance energy transfer (FRET), which is generally limited to length scales less than 10 nm. The development and characterization of materials, fundamental spectroscopy, and proof-of-concept applications in bioanalysis will all be highlighted through examples such as smartphone-based assays, multifunctional concentric FRET probes, photonic logic probes, and more.

Tackling Global Health Challenges with Biosensor Technologies
Tony Cass

Despite the impressive advances in basic and applied biomedical sciences and engineering over the past five decades, humanity today is faced with many newly emerging challenges to global health. From diseases associated with contamination of drinking water, to antimicrobial resistance; from zoonotic infections to 'lifestyle choice' conditions such as diabetes or heart disease; from neurodegenerative diseases to cancers. Whilst they may have distinct geographical or socioeconomic dimensions they can all be said to be 'global' in their scope or potential impact. If we add to this the increasing pressure on healthcare budgets, whether in high or low and middle income countries, it's becoming apparent that the words of Mike Bloomberg "If you can't measure it you can't manage it and you can't fix it" can apply equally to global health as to New York City's climate policies. In the context of this lecture the measurement will be based on the use of biosensor technologies for both point of care and continuous measurements, applications for which they are uniquely well suited. The focus will be on the interface of biological molecules with micro- and nanostructured materials and how our ability to engineer the properties of both open the way to improving traditional devices such as lateral flow assays and enzyme electrodes as well as enabling new measurement methods that promise improved sensitivity and specificity.

Fine Tuning Model Updating for Multivariate Calibration Maintenance
Anit Gurung, John H. Kalivas, Erik Andries

In the field of chemometrics, an important issue in multivariate calibration is model updating. Model updating is the adaption process in which a model obtained for a given set of samples and measurement conditions (primary) is updated to predict the analyte in new samples and measurement conditions (secondary). Primary and secondary conditions can be different due to variations in the geographical situation, instrumentation, or environment. Model updating can be performed using labeled data sets containing samples with reference analyte values for both conditions. A common approach is performed by sample augmenting the larger primary labeled sample set with a small weighted secondary labeled sample set. In this situation, only one updating model is obtained to predict the analyte amount in both primary and secondary conditions. The proposed new approach is similar to this common approach, but instead of one updated model, two models are formed simultaneously. One model is used to only predict samples from the primary conditions and the second model is based on this primary model but modified relative to the weighted augmented secondary samples. This second model is used to predict samples from the secondary conditions. Both model updating methods require multiple tuning parameters (penalties). A new fusion approach of multiple model quality measures is presented for selection of tuning parameters. The two updating approaches are evaluated with near-infrared spectral data sets of soil, corn, and pharmaceutical tablets.

Leveraging Multiple Linear Regression for Wavelength Selection
Tony Lemos, John Kalivas

In multivariate calibration, wavelengths selection is often used to lower prediction errors of sample properties. As a result, many methods have been created to select wavelengths. Several of the wavelength selection methods involve many tuning parameters that are typically complex or difficult to work with. The purpose of this poster is to show an easy way to select wavelengths while using a few simple tuning parameters. The proposed method uses multiple linear regression (MLR) as an indicator to which wavelengths should be used to create a model. From a collection of random MLR models, those models with an acceptable bias/variance balance are evaluated to determine the wavelengths most frequently used. Portions of the most frequently used wavelengths are chosen for the final MLR selected wavelengths. These MLR selected wavelengths are then used to produce a calibration model by the method of partial least squares (PLS). This proposed wavelength selection method is compared to PLS models containing all wavelengths using several near infrared data sets. The PLS models with the selected wavelengths show an improvement in prediction error, suggesting this method as a simple way to select wavelengths.

Robustness across instruments: a case study on calibration transfer of NIR-based multivariate calibration models

Hua Yu

In the world of global pharmaceutical manufacturing, it is important that analytical chemists be deliberate and proactive in their planning for analytical method lifecycles. Spectroscopy-based analytical methods require special consideration as compared with their traditional lab-based counterpart methods over their respective method lifecycles. For example, spectral variation is expected with the deployment of NIR models across different sites, different instruments, applied to excipients and APIs from multiple sources. These same changes may have no impact to traditional lab techniques, but may cause complications for PAT-based methods. For multivariate calibration model based on NIR spectra, the idea of a universal model working seamlessly across instruments is always intriguing. However, making a universal model is challenging due to the fact a wide range of instrument-associated variation can register in spectra in variety of ways, which often makes calibration transfer complex. This work will present advantages and disadvantages of both approaches: building enough variability into the original model such that a single model applies to all field instruments, or building a master model on a single instrument and employing the use of calibration transfer to field instruments. To test the concept in pharmaceutical application, in this work, a set of spray-dried dispersion (SDD) samples were prepared with a range of moisture levels and measured using two separate NIR instruments. A Partial Least Square (PLS) model for moisture was developed using original spectra collected from the primary instrument. Instrument-associated variation on spectra was observed. The impact of such variation on transferability of calibration model was investigated. Spectra collected from the secondary instrument, in original or transferred form, were used to replace the original calibration spectra collected from the primary instrument. Those updated models were evaluated in terms of their ability to predict moisture regardless of instrument used.

Comparing NIR, MIR and Raman Spectroscopy for Classification of Edible Oils and Prediction of Edible Oil Peroxide Values

Joshua Ottaway, J. Chance Carter, Kristl Adams, Karl Booksh

Determining the correct spectral technique is the first step in any experimental design. The correct technique provides enough information to answer the question being asked, while also being time and cost efficient. This work will compare Near Infrared (NIR), Mid Infrared(MIR), and Raman spectroscopy for determining the identity of edible oils, as well as predicting the peroxide value of the edible oils. Determining the identity of edible oils is the first step in ensuring the purity and authenticity of the oil which is necessary to prevent fraud. The peroxide value of an oil is the primary measure of the freshness of an oil, so the ability to quickly determine peroxide value via spectroscopy, instead of the more standard titration method, is extremely useful for quality control. Partial least squares (PLS) and partial least squares discriminate analysis (PLS-DA) were used to build models on NIR, MIR, and Raman data that was collected on a single set of oil samples.

Robust Calibration Model Transfer Between Portable Near-Infrared Spectrometers to Predict API of Pharmaceutical Oral Dosage Form

Suyang Wu, Md Nayeem Hossain, Shikhar Mohan, Md Anik Alam, Dr. James Drennen, Dr. Carl Anderson

Near-infrared Spectroscopy (NIRs) is widely used in pharmaceutical industry for raw material characterization, in process monitoring and product evaluation because of its rapidness, non-destructive nature and portability. Multivariate calibration techniques are often required to extract the required information to develop a quantitative model. While developing method, it is necessary to ensure that model meets the performance requirements; along with accuracy and precision, robustness is critical to ensure long term model performance. However, a problem occurs when an existing model is applied to spectra that were collected in different instruments. To handle this, calibration transfer method is required. Moreover, in routine analysis, often incoming batches contain samples which are prepared in different seasons and from different raw materials. One possible solution is to add this newly generated sample to update the model. However, this process would be both costly and time consuming. An efficient way is to develop chemometric methods which are insensitive to these variances. Therefore, the objective of this study to see the effect of instrument variation and sample variation on long term model performance, and develop a robust efficient chemometric method to handle these variabilities. Model drug formulation containing acetaminophen, lactose, microcrystalline cellulose, hypromellose, and magnesium stearate were used to prepare all the samples. Partial Least Square was applied to preprocessed spectra to construct API quantitative model. During the course of study, different granule batches were manufactured over the years using a fluid bed granulator, and subsequently tablets were prepared from these granules, in lab and manufacturing scale. Due to seasonal differences, large variations existed in the particle-size distribution of the granules as well as in the moisture content of the excipients and granules. Three different portable near infrared spectrometers (coded as Port1, Port2 and Port3) were used to scan all the samples. These spectrometers were also manufactured over the years by the same manufacturer. It was observed that, for the samples prepared in same season, spectra collected with master and slave spectrometers performance was similar. However, samples prepared in different season, performance of calibration transfer degrades. Understanding the impact of these variances prior transfer obtained robust predictions.

Forensic IR Spectroscopy: A Look Back, at the Present and Ahead

Edward Bartick, Rohit Bhargava

Looking back, the use of dispersive infrared spectrometers was limited with forensic application. Among other drawbacks, low available radiant energy, requiring long sampling times, hindered workload throughput. However, Forensic IR analysis has come a long way since the advent of the first commercial FT-IR spectrometers in 1969 when infrared spectroscopy was revolutionized. Forensic analysis has benefited significantly by these instruments. In general, the initial use was hampered by marginally stable interferometers and the tedious machine language of the minicomputers. The applicability was enhanced by improved interferometers and microcomputers with user friendly software having advanced application algorithms such as spectral library searching and chemometric analysis. More sensitive detectors (mercury cadmium telluride) were also an enhancement. A very large impact was made by the advancement of accessories and their applications. Diffuse reflectance was applied to drugs and copy toners. attenuated total reflectance (ATR), particularly with single reflection, has been used for the analysis of a multitude of evidential materials. Microscopes attached to the spectrometers have revolutionized trace evidence analysis. At present, focal plane array detectors are having a strong impact in general IR analysis spectral imaging applications. While the reporting authors introduced forensic FT-IR imaging of latent fingerprints and trace evidence fifteen years ago, it has not come to fruition for common use as yet. With the evolving technology of quantum cascading laser (QCL) IR the potential exists for forensic imaging analysis. Because of the higher energy and faster samling times, we anticipate trace evidence spectral imaging analysis in the laboratory and in the field is very possible. The analysis of single fibers without flattening, identification of bi-component fibers and IR microscopy with and without ATR should also be available in the field. These and other advances should be available in the near future with this third generation IR spectroscopy.

Infrared Absorption Spectra of Gas Phase Uranium Oxides

Batikan Koroglu, Marco Mehl, Michael Armstrong, Jonathan Crowhurst, David Weisz, Joseph Zaug, Harry Radousky, Mark Cappelli, Timothy Rose

The high temperature (2000K < T < 5000K) infrared spectra of gas phase uranium oxides (e.g. UO2 and UO3) is being studied using an atmospheric pressure plasma flow reactor. As the reactant (uranyl nitrate) passes through the plasma torch, it is decomposed into its constituent elements and the subsequent cooling along the reactor results in the formation of oxide molecules. An Optical Parametric Amplifier / Difference Frequency Generator system (OPA/DFG) is used to record the spectra of molecules. In this work, changes in the relative intensities of absorption spectra for various species will be presented as a function of distance, and therefore temperature, along the reactor. Similar measurements will be shown for two different cooling rates to understand the role of kinetics of chemical reactions in the resulting speciation of molecules. The experimental data is useful for inferring the extent to which high temperature chemistry of uranium oxide molecules is controlled by quasi-equilibrium thermodynamics versus gas phase chemical kinetics.

Determination of residual organic compounds in uranium ore concentrates by GC/Q-TOF after monolithic material sorptive extraction: application to nuclear forensic investigations.

Maxime BRIDOUX, Françoise Leprince

Illicit trafficking of nuclear material poses a serious threat to the international community. Since the first event of illicit nuclear trafficking of nuclear and material about 50 years ago, the International Atomic Energy Agency (IAEA) reported over 200 cases of such illicit traffic. In this context, the field of nuclear forensic, whose global aim is to trace geographical origin and industrial history of unknown seized nuclear materials by searching for specific nuclear forensic markers, emerged over the last two decades. However, one of the analytical challenges of the field of nuclear forensics is to identify novel markers related to origin, intended use and production processes of the seized nuclear material . Several nuclear forensic signatures have been developed during the last years, including the concentration of major, minor and trace level constituents, isotopic composition of impurities or material morphology. Organic compounds such as extractants (organophosphorus compounds, alkylamines,...) and diluents (organic solvents, various mixtures saturated hydrocarbons,...) are widely used in the extraction process and could, in principle provide useful information about the origin and process of seized nuclear materials . Here, we report on the development of a novel method for the analysis of residual organic compounds at trace levels in uranium ore concentrates by using a Monolithic Material Sorptive Extraction. The MonoTrap RSC18 is placed in the headspace of the vial without any contact with the radioactive species in the UOC and the vial heated at about 175 C. The MonoTrap is then placed in the thermal desorption unit and analyzed by Gas Chromatography/ Quadrupole-Time of Flight Mass Spectrometry (GC/Q-TOF-MS) directly, without the need for radiological contamination controls. Using this method, the chemical composition of a series of UOC of known industrial history was investigated. We will show that in most cases, organic fingerprinting of UOC using MonoTrap and GC/Q-TOF-MS analysis can be used to facilitate sourcing of unknown UOC.

TNT Reduction in Clostridium acetobutylicum Affected by Oxidation State of Substrates

Sanchao Liu, Christian Sund, Elliot Gerlach, Matthew Servinsky

Abstract Bioremediation of soil contaminated by nitroaromatics is an approach that is more cost-efficient and less destructive to the environment. Biodegradation of TNT with Clostridium has been used as an effective method for anaerobic TNT reduction. In this presentation we studied the reduction of TNT by different oxidized hexose derivatives. The efficiency of TNT reduction with Clostridium acetobutylicum was examined to analyze the effects of feedstocks at different oxidation states. The metabolites of the fermentation products and the TNT reduction products were characterized by HPLC/MS and GC/MS. The results showed that when the substrate switched to a more oxidized state, i.e., from glucose to gluconate to galacturonate, the TNT reduction rate was decreased gradationally. When TNT was fermented with galacturonate, only minimum amount of TNT reduction could be observed.

Optimization of Single Hair Proteomics for Human Identification

Fanny Chu, Katelyn Mason, Deon Anex, A. Daniel Jones, Bradley Hart

DNA is considered the gold standard for forensic identification. However, in the absence of viable DNA for analysis, no other evidence type achieves a comparable level of statistical probability and power of discrimination. Proteins are an attractive alternative due to their persistence and derivation from DNA; non-synonymous single nucleotide polymorphisms (SNPs) in individuals are often conserved as single amino acid polymorphisms (SAPs). The peptides in which SAPs are found are known as genetically variant peptides (GVPs). Hair evidence is especially amenable for protein analysis as it is commonly found at crime scenes, and proteins in hair are notoriously robust. Single hair analysis, while ideally suited for forensic identifications, is challenging due to minute protein quantities. This research aims to optimize methods to maximize extraction of proteins from inch lengths of a single hair for human identification. To that end, protein concentration via acetone precipitation was investigated across a variety of sample preparation conditions. Extracts were digested with trypsin and then analyzed via nano-LC-Orbitrap-MS/MS. Extraction efficacy was compared across acetone precipitation conditions and to a bulk hair high volume method and another single hair sample preparation method using acidified ethyl acetate. Metrics to evaluate performance include the number of proteins and unique peptides identified, the number of GVPs detected, and sequence coverage. Comparison of experimental conditions using the four metrics showed that acetone precipitation and volume reduction effectively concentrated proteins; more than 530 proteins and 2500 peptides were identified, 14 GVPs were detected, and wide sequence coverage was achieved for many proteins with single hair analysis. These results are comparable to those obtained using the bulk hair high volume method and far surpass the protein and peptide identifications from single hair samples prepared using acidified ethyl acetate. Additionally, differences in protein and peptide identifications were observed between the individual hair samples; individuals were differentiated using these metrics. While further method optimization is required for implementation for forensic identification, this research demonstrates the potential of single hair proteomics. This work was performed under the auspices of the U.S. Department of Energy by Lawrence Livermore National Laboratory under Contract DE-AC52-07NA27344. IM #LLNL-ABS-735860

One-Class Classification Method for Pharmaceutical Product Authentication System using Spectroscopy

Md Nayeem Hossain, Carl Anderson, James Drennen

The threat of counterfeit medicines exists worldwide. According to the World Health Organization (WHO), 10%-30% of pharmaceuticals in the world are counterfeit/substandard medicines, with developing countries suffering the highest rates. Spectroscopic techniques, e.g., Near-Infrared (NIR) and Raman spectroscopic techniques offer rapid and non-destructive methods for screening suspected samples. Current miniaturized, low-cost, portable instruments create opportunities for field use in rural hospitals, clinics, drugstores, etc. To implement spectroscopic techniques, it is required to develop classification techniques. These classification techniques are broadly categorized by discriminate analysis and one-class modeling. However, it was revealed that the one-class modeling approach was more appropriate, as counterfeit containing new information can be erroneously assigned to an existing class by discriminant analysis. Here in this study, authors proposed the direction of the development of one- class classification method to detect counterfeit products. Also, a new one-class classification approach based on machine learning theory was proposed for spectroscopic based analyses. The performance of this method was compared with the traditional PCA based one-class modeling (SIMCA). Genuine samples were created with the model drug Acetaminophen and the following excipients: lactose, microcrystalline cellulose, starch, croscarmellose sodium and magnesium stearate. After defining an authentic formulation, simulated fake products were prepared at lab scale by varying excipient compositions of genuine sample. Mixture design was generated to vary lactose, microcrystalline cellulose and starch of the genuine sample. The manufacturing conditions for genuine and simulated fake products were kept consistent. Both Raman and NIR spectrometers were used to collect spectra. While developing this method, two important figures of merit were monitored: sensitivity and specificity. The ability of the method to authenticate genuine samples reflects the sensitivity of the model. The ability of the method to recognize counterfeit products and reject them represents specificity. It was observed that one-class classification method was suitable for counterfeit drug detection. High sensitivity and specificity were observed for the developed method.

Using LIBS in Hardness Measurements for Samples of Different Matrices

Ahmed Galmed, Mohamed Abdel Harith, Malik Maaza

Sample hardness measurement is one of the most essential tests in material science. The search for a fast and reliable as well as portable means of hardness measurement is becoming more important. Laser Induced Breakdown Spectroscopy (LIBS) have been used recently as a tool for hardness measurements. Many studies have been done proved that LIBS is a promising way for relative hardness measurement. The fastness, lake of sample preparation and the portability of such technique gave it the advantage over other techniques in measuring the material hardness. In all of the previous work samples used for hardness comparison are all have the same matrix but with different hardness value. In this work we are trying to measure compare the hardness of highly rigid materials of different matrices and study the matrix effect on the relation between the samples surface hardness and the ionic to atomic spectral lines' intensity ratio as well as the relation between the samples surface hardness and (Te). The results were compared with the conventional mechanical hardness measurement.

Time-Resolved Absorption Measurements in Transient Plasmas: Towards Single-Shot Optical Measurements of Uranium Isotope Ratios

Jonathan Merten, Bruce Johnson

Previous researchers have used LIBS for isotope ratio determinations. The measurement is complicated by the low entendue of the high resolution spectrographs required to resolve the isotopes' emission. Alternatively, tunable laser probes (absorption or fluorescence) of laser-ablation plasmas avoid the issue of entendue, but cannot measure absorption a multitude of wavelengths simultaneously. We present here an extension of continuum source atomic absorption spectroscopy that allows single picometer-resolved absorption measurements in transient plasmas without scanning the probe laser. The dual-beam measurement employs a novel, but simple spectrograph design and an optical parametric oscillator pseudocontinuum source whose width exceeds that of the atomic absorption profile by just a few orders of magnitude, but is not a true blackbody emitter. We discuss our proof of principle nanosecond-resolved absorption measurements of noble gas metastable and metal ground state absorption in pulsed see-through hollow cathode lamps. We also present our time-resolved measurements of neon and argon metastable in LIBS plasmas, as well as spatial mapping of the absorption in the plasma. We will discuss the factors affecting the background noise and the implications for measurements of isotope ratios in laser-induced plasmas via this new technique, which we call pseudocontinuum source atomic absorption spectroscopy. In addition to the potential for nuclear forensics, the technique can allow ground-state number density measurements in the plasma (though not of trace elements) and may permit improved understanding of departure from local thermodynamic equilibrium in LIBS plasmas.

Multivariate Analysis of Complex Samples Using Atomic and Molecular Emission Spectra from LIBS/SIBS

Sofia Pozsonyiova, Prasoon Diwakar, Nathalie Manzano, Theodore Caplow

Laser induced breakdown spectroscopy (LIBS) and spark induced breakdown spectroscopy (SIBS) are powerful multi-elemental detection techniques for detection of variety of samples including solid, liquid, gases and aerosol. Both the techniques utilize pulsed laser or high voltage source to generate plasma which in turn ablates minuscule amount of sample followed by its atomization, excitation and ionization resulting in emission spectra. LIBS and SIBS spectra are very rich and contains significant information about the sample in the form of atomic and molecular lines as well as background emission. For a complex sample such as soil, biological sample, forensics, the spectral information gets very complicated hindering the identification and classification of samples. In this study, LIBS and SIBS methodology is used to generate the spectra (atomic and molecular emission) using samples from day to day life and compared with standards containing known toxic elements. Using multivariate analysis approach and principal component analysis, samples are compared and identified for presence of toxic components in the sample. Ambient gas (Ar and He) is used to study the effect of gas on the improvement of identification and classification of samples based on this approach. A library of chemicals present in materials used in day to day life will be generated. Detailed discussion of the improvement in classification and identification of toxic elements in samples by using combination of atomic and molecular lines in presence of Ar and He as ambient gas are discussed and presented.

Classification of Milk Vetch Root by its geographical origins using Laser-Induced Breakdown Spectroscopy

ChangHwan Eum

As the global trading of agricultural products grows, the deception of geographical origin becomes increased. Therefore, fast and accurate identification of geographical origins of agricultural products become more demanding. Generally, near-infrared (NIR) and Raman spectroscopy have been widely used for this purpose because both methods are fast and non-destructive with no or minimal sample pretreatments. Molecular structural information of components in samples provided by these methods enabled the identification of geographical origins by combining with various chemometric methods such as linear discriminant analysis (LDA) and support vector machine (SVM). Alternatively, the composition of elements could be valuable information for the identification, so laser-induced breakdown spectroscopy (LIBS) able to provide information on diverse elements has drawn an attention. LIBS is also quick in data acquisition and quasi-nondestructive as comparable with NIR and Raman spectroscopy. In this study, we attempted LIBS-based identification of geographical origin of milk vetch roots. LIBS spectra of pellets made from domestic and imported milk vetch roots were collected by optimizing necessary experimental conditions such as delay time and number of averaging spots. The compositions of Na were relatively more different between two groups. To assess the accuracy, SVM was used for the discrimination. Also, NIR spectra of the same pellets were collected, and an effective chemometric strategy able to improve accuracy by cooperatively combining both LIBS and NIR data has been discussed.

Laser-induced breakdown spectroscopy for analysis of gold and silver in mineral synthetic samples

Daniel Diaz, David Hahn, Alejandro Molina

Laser-induced breakdown spectroscopy was tested for quantification of precious metals Au and Ag in synthetic mineral samples with the purpose of exploring its application as an in situ and online monitoring technique for the precious metal mining industry. Two sand matrices were doped with standard solutions of Au and Ag in concentrations from 10 ppm to 500 ppm to produce a set of synthetic samples that, after being prepared as pressed pellets, were analyzed with a LIBS setup that included a 1064 nm Nd:YAG laser, a Czerny-Turner spectrograph, an ICCD camera and an optical setup configured to collect the plasma light in backscatter mode. Parameters such gate and delay times, lens-to-sample distance and laser energy were optimized to obtain the highest signal for the emissions lines Au I 267.59 nm and Ag I 328.06 nm. The calibration plots obtained for both analytes showed a linear behavior with coefficients of determination above 0.966 and a dynamic range wider than 10 to 50 ppm for Au and from 125 ppm to 250 ppm for Ag. The limits of detection and quantification were, respectively, 0.8 ppm and 2.6 ppm for Au and 1.1 ppm and 3.6 ppm for Ag. These values illustrate the ability of LIBS to quantify and detect these elements in real time and with relatively simple sample preparation, and are particularly encouraging for Ag quantification given its relatively higher abundance. However, as the gold mining industry demands techniques that measure concentrations of the order of 1 ppm or below, quantification of Au would be challenging for the conventional LIBS technique.

Detection of impurities leached from sugarcane by LIBS after liquid-to-solid matrix conversion
Fabiola Pereira, Wesley Guedes, Daniel Andrade

The sugar cane sector in Brazil produces and processes million tons of sugar cane with the sucrose, one of the main agroindustrial product used in the production of sugar and ethanol. Vegetal impurities in the sugar cane compromise the useful life of machinery and mineral impurities (soil) can lead to browning of the final product, especially in the case of sugar. Although the impurities directly influence the quality of the sugar cane, there is no analytical method that allows the identification of these impurities inherent to the post-harvest process. In this context, the work main objective was to analyze residues extracted from sugar cane using deionized water by Laser induced breakdown spectroscopy (LIBS) after the immobilization of the solutions in a polyvinyl alcohol (PVA) polymer. A mass of 800 mg of 10% w/v PVA was accurately weighted and mixed with 200 mg of the sample. The mixture was homogenized, transferred to an aluminum holder positioned over a glass support and dried in an oven at 50 C for 2 hours. The solid polymer film formed with the immobilized sample was analyzed by LIBS. For the treatment of the data generated by LIBS, 12 different standardization strategies of the dependent variables were tested. In order to obtain a global vision of all analytes and their relations, principal component analysis (PCA) was calculated using entire spectra profiles. The analysis of the score plots shows a predominant segregation in the positive part of the PC2 relative to the experiments with the highest percentage of impurity in the extractions (8-10%) relative to standardization using mean, with ionic lines of C for 193.1 and 247.9 nm. In this case, a high correlation with the concentrations of C, Mg and Na in the samples was observed from loadings of PCA. Considering the advantages inherent in the analysis of solids by LIBS, the conversion of the analytes using only water from the liquid sample to a solid phase produced a useful system for the analysis of solutions in LIBS. The method provides speed, simplicity and sensitivity for detection of impurities of sugar cane.

Uncertainty Assessments and Plasma Modeling for Quantitative LIBS Applications
David Surmick, Noureddine Melikechi, Hacene Boukari, Jonathan Woodward, Ashley Stowe

Quantitative laser-induced breakdown spectroscopy (LIBS) is a collection of experimental and analytical methodologies that are meant to obtain information regarding sample composition. These methodologies are sought after for applications in forensics, industrial process monitoring, medicine, and isotopic identification. In this vein, we consider numerical approaches in LIBS data analysis for the purpose of isotopic identification and quantification. One aspect of this work addresses uncertainty assessments in quantitative LIBS analysis. We demonstrate a method for determining the uncertainty of sample constituent relative number densities following line profile fitting of multi-lined spectra. An uncertainty budget is defined by conserving the integral of a measured spectrum. The absolute difference between the integrals of a fit and spectrum describe the total uncertainty of the area. This is redistributed to each of the lines in the spectrum through a weighting factor determined from the absolute difference between the lines and fit over each line kernel. As a demonstration of the this method, we present relative number density ratios inferred from fits to aluminum and zinc multiplets measured from laser produced plasma on the surface of an Al alloy 7050 sample. The second aspect of this work regards the integration of plasma and spectral simulations into LIBS data analysis. To this end, hydrodynamic simulations of the plasma were completed followed by collisional radiative modeling to produce corresponding spectra. The simulations were performed in order to better quantify the results of time and space integrated LIBS measurements. Initial post process steps of the simulated spectra are targeted towards modeling temporally and spatially integrated spectra from time and space resolved calculations, which match the experimental capabilities of many spectroscopic instruments used in field and industrial plant applications. The numerical aspects of this work were supported and compared to spectral measurements of laser induced optical breakdown initiated on the aluminum alloy in air, as well as lithium hydride pellets under a flow of helium in air.

Challenges and strategies for the simultaneous determination of Cd, Cr, Ni and Pb in micronutrient fertilizers by laser-induced breakdown spectroscopy

Lidiane Cristina Nunes, Francisco José Krug

The analytical performances of laser-induced breakdown spectroscopy (LIBS) were evaluated aiming at the quantitative determination of Cd, Cr, Pb, and Ni in pressed pellets of micronutrient fertilizers. A set of 25 samples containing B, Cu, Fe, Mn, and Zn was ground and homogenized by using a planetary ball milling for 10 min. Pellets (test samples) were prepared by transferring 800 mg of the resulting laboratory samples to a 15 mm die set and applying 8.0 ton cm-2 for 3 min. Experiments were carried out with a Q-switched Nd:YAG laser at 1064 nm (5 ns, 360 mJ, 10 Hz), the laser pulse being focused on the pellet surface by a convergent lens; part of the plasma emitted radiation was collected by a telescope into an optical fiber coupled to the entrance slit of a model ESA 3000 spectrometer with Echelle optics and an ICCD detector (LLA Instruments GmbH, Germany). Instrument parameters were defined by varying the laser fluence from 10 to 50 J/cm2 , either in argon or air atmosphere, and the most appropriate the number of accumulated laser pulses per site. All measurements were made after 2.5 μs delay with 10.0 μs integration time gate. Each spectrum was obtained after the accumulation of 10 to 50 laser pulses per site. Thirty accumulated spectra were obtained in the analysis of 30 independent sampling sites on the pellet surface. Data processing was based on spectra averaging, where the average of ten accumulated spectra (data from 10 sites) was used as a representative spectrum of each replicate. All experiments were carried out with three replicates. In cases of ICCD detector signal saturation, e.g. Cr II 267.716 nm emission line, it will be demonstrated that it is possible to overcome this effect by choosing the most appropriate number of accumulated laser pulses per site, by keeping the same test portion mass. Calibration models were obtained by using laboratory samples previously analyzed by ICP OES after microwave-assisted acid digestion. Linear correlation coefficients (r) between the analyte mass fractions and their corresponding emission intensities ranged from 0.9830 (for Cd) to 0.9985 (for Cr). Acknowledgments: CAPES, and CNPq (Proc. 309679/2014-1) for fellowship, grant and financial support.

Spatio-temporal Evolution of Metal Oxide Diatomic Molecules in Tailored Ultrafast Laser-Induced Plasmas
Yonghoon Lee, Xianglei Mao, George Chan, Jhanis Gonzalez, Rick Russo, Vassilia Zorba

We investigate the distribution of atoms and metal oxide diatomic molecules in femtosecond-laser-induced plasmas generated at the focus and below the focus of the laser beam, where non-linear phenomena give rise to the formation of a weakly ionized plasma channel. We use direct plasma imaging and optical emission spectroscopy to study plasma expansion and physical and chemical separation processes. Plasma splitting takes place below the laser focus and is accompanied by a change in the spatio-temporal distribution of the molecules and atoms compared to sampling in focus. Furthermore molecular emission is favored at specific locations and is enhanced over atomic as we move away from the laser focus. These findings open up the possibility to enhance or impede molecular specie formation versus atomic at specific spatio-temporal locations in the laser induced plasma.

Water Content Determination for Bilayer Tablets Using Near Infrared Spectroscopy and Multivariate Statistical Analysis

Zhenyu Lu, Yanqiao Shawn Xiang, Daniel Pohlman, David Webster, Pete Mustonen, Mark D. Trone, Michael Palmieri, Jr.

The degradation kinetics of the drug substance in tablets in development has been shown to be correlated with the amount of free water in the drug product. The current technique for measuring total water content is volumetric Karl Fischer titration. Due to the hygroscopicity of these tablets, the measured moisture is sensitive to the environmental moisture which makes the sample preparation critical for accurate analysis. In order to minimize the exposure of the tablet to the ambient environment after removal from packaging, whole tablets have been used in lieu of composite powder samples (i.e. ground tablets) for the titration. However, the use of a whole tablet is challenging due to slow and incomplete dissolution of the tablets in the sample solvent. Near infrared (NIR) spectroscopy, which requires minimal sample preparation, is a non-destructive, fast, and solvent-free alternative for water content determination. In this work, NIR combined with multivariate statistical analysis was used to construct calibration models to measure water content of these tablets. The variations introduced by differences in sampling direction, coating, and dosage strength were investigated to improve the quality of the models. The constructed NIR calibration model was used for rapid and accurate determination of water content in the tablets for monitoring of tablet drying, packaged release testing, and stability testing.

Considerations of Model Validation of NIR Based Quantitative Methods

Dongsheng Bu, Jim Pratt, Gary McGeorge

We have developed and implemented a Real Time Release testing method for Assay determination (drug content within a tablet) of a pharmaceutical product. Quantitative multivariate models incorporating near infrared (NIR) spectroscopy with partial least squares (PLS) modeling were applied and model validation was carried out as an essential part of model development and implementation. This presentation will discuss a few selected key elements we think are important in general for multivariate quantitative method deployment. The presentation will focus on the method validation procedures/acceptance criteria and after deployment the continuous monitoring to ensure that the models are still fit for purpose and don't drift over time. Current analytical method validation guidelines, e.g. ICH Q2(R1), are usually applied to univariate models, and it is necessary to modify testing procedures, such as linearity and specificity, in order to be applicable to multivariate model validation. In addition to general rules and a statistical basis, we have considered model performance design when setting acceptance criteria. Then we will discuss the concept of continuous model verification over time throughout the lifecycle of the product that is somewhat specific to spectroscopic multivariate methods since their calibration set is not "re-built" at the time of use as in traditional HPLC Assay methods.

Triboluminescence Analysis of Pharmaceutical Formulations

Garth J. Simpson, Casey J. Smith, Scott R. Griffin, Gregory Eakins, Fengyuan Deng, Atanu Sangupta

Triboluminescence (TL) is demonstrated for selective detection of trace crystallinity within a model amorphous solid dispersion (ASD) containing trace quantities of a crystalline active pharmaceutical ingredient cast within an amorphous polymer. While ASDs are increasingly used for the preparation of pharmaceutical formulations, trace crystallinity during manufacturing or storage can negatively impact both shelf-life and efficacy. In the present study, TL measurements of a model ASD consisting of griseofulvin in polyethylene glycol produced limits of detection of 142 ppm. Separate studies of the particle size dependence of sucrose crystals and the dependence on polymorphism in clopidogrel bisulphate particles are both consistent with a mechanism for TL closely linked to the piezoelectric response of the crystalline fraction. Whereas disordered polymeric materials cannot support piezoelectric activity, molecular crystals produced from homochiral molecules adopt crystal structures that are overwhelmingly symmetry-allowed for piezoelectricity. Consequently, TL may provide a broadly applicable and simple experimental route for sensitive detection of trace crystallinity within nominally amorphous materials.

Real-Time Fundamental Calibrations of Process Instruments -Properties and Advantages

Robert Lascola, Patrick O'Rourke

We have developed a double beam diode array spectrometer which implements real time wavelength and periodic intensity/absorbance calibrations to provide stabile and precise measurements over a range of operating conditions. The instruments have been in use at our facilities for over a year and have demonstrated improved performance compared to "normal" diode array instruments. Examples include the reduction of instrument drift over a wide range of operating conditions, standardization between instruments which allows the use of chemometric models without maintenance of a transfer matrix, and operation as a low cost, high precision wavelength meter with several interesting applications. Here we will describe the construction of the spectrometer, the necessary properties of the components, and the implementation of the calibrations. Several practical examples, based on applications for nuclear materials monitoring, will be presented which illustrate the advantages and limitations of these instruments. These spectrometers, constructed from commercially available components, are an affordable option for high precision measurements.

Portable, Rapid Analysis of MEA-Triazine and Dithiazine via Raman Spectroscopy
Samuel Kleinman, Merwan Benhabib, Natalya Zherebnenko, Mark Peterman

Hydrogen sulfide is a naturally-occurring, poisonous, corrosive, and putrid component of natural gas and crude oil. To avoid corrosion and increase workplace and environmental safety, it is beneficial to remove H_2S as early as possible. MEA-triazine is a commonly used hydrogen sulfide scavenger for oil and gas applications. It is non-regenerative and reacts stoichiometrically during the sweetening process to produce dithiazine. We describe a new technique to quantify both MEA-triazine and dithiazine concentration using Raman spectroscopy. Rapid analysis of triazine and dithiazine concentration protects assets from corrosion and avoids work stoppages due to gas flow issues. It also provides a tool to optimize chemical usage by maximizing scrubbing capability, while avoiding negative issues of overspending MEA-triazine solutions. The method is robust and covers triazine concentrations ranging from 10 to 97% as well as dithiazine concentrations up to 25%. Additionally, correct quantification is minimally affected by the addition of alcohols to the triazine solution—a common industrial diluent. The test method employs a portable Raman spectrometer and simple user-friendly protocols to make the measurement as simple as possible. This is the first demonstration of MEA-triazine and dithiazine concentration analysis in industrially-relevant activity ranges employing an optical technique. In this talk we will detail method development as well as provide real-world examples and case studies.

In-Situ FT-IR Reaction Monitoring using Standard Detector
Michael Kleimann

Traditional in-situ FT-IR reaction monitoring is performed using ATR probes and requires a sensitive detector to compensate for the intrinsically weak signal associated with any kind of ATR measurement, combined with the additional loss of throughput due to the absorption of fibre optics in infrared range. This is normally done with MCT detectors (Mercury-Cadmium-Telluride) that require liquid nitrogen cooling. Those are very fast and sensitive detectors, however there are a number of issues associated with supply of liquid nitrogen in some industrial sites (not allowed in some areas), and also with the regular refilling of a dewar. Alternatively thermoelectrically-cooled MCT detectors ("Sterling cooling") can be used for reaction monitoring, but they are less sensitive and are known to be prone to electrical and mechanical interferences. They also require expensive refurbishment at regular intervals (typically every 2 years). We present here applications of FT-IR in-situ reaction monitoring using the ABB MB3000 FT-IR spectrometer with a standard DTGS detector. This detector is significantly easier to operate and maintain than an MCT detector, but it usually produces spectra with noise level 10 to 100 times higher than MCT detector.

However, in combination with an innovative design of FT interferometer providing state-of-the art signal-to-noise performance, and using a newly designed and optimised fibre optics interface, we demonstrate that the setup allows reaching limits of detection similar to MCT configurations, at significantly lower costs and using a maintenance-free configuration. This system opens a number of opportunities for scientists looking for rugged and simple to use FT-IR reaction monitoring equipment at research laboratories or pilot plants.

Principal Component Correlation Based Variable Selection to Improve Glucose Concentration Prediction in Mammalian Cell Cultivations by Raman Spectroscopy

Bence Kozma, Szilveszter Gergely, László Párta, András Salgó

Raman spectroscopy is applied frequently for bioprocess monitoring nowadays, because it is a very promising technique for the biopharmaceutical industry due to its insensitivity to water but high sensitivity to the media components. One of the frequently monitored media components is glucose, which is the primary carbon source of the cells thus on-line monitoring and controlling of its concentration is crucial for bioprocesses. In this study, four different Chinese hamster ovary (CHO) cell cultivations were followed by in situ, on-line Raman spectroscopy and the sensitivity of the Raman technique for glucose was demonstrated. First, the spectra (1890.0–200.0 cm?1) were analysed with Principal Component Analysis (PCA) and it was found that the score values of higher principal components (PC) correlates with glucose concentration. This was proven by the value of the calculated correlation coefficients between the off-line measured glucose concentration and the score values of the principal components. The first PCA model was built on the spectra of one run at a time. The highest correlation was experienced with the fourth, fifth or sixth principal component, depending on the dataset. The second PCA was calculated on the spectra of all four runs combined to one dataset and the highest correlation coefficients, which were varied between 0.71 and 0.97 for both PCA models, were observed with the eighth principal component for all four runs. The loadings of the two PCA models were also examined for the PCs with the highest correlation coefficients. The loadings were the highest for similar wavenumber regions in both PCA models, which is the verification that both PCA models captured the same, glucose related effect but with different PCs. This was further verified with the literature. The utilization of this variable selection technique could improve the accuracy and transferability of the Raman spectroscopy based glucose concentration determination model by selecting the most glucose specific variables for calibration.

Graphene Analyzer for Rapid On-line or At-line Graphene Quality Monitoring
Dawn Yang, Kristen Frano

Graphene is a 2D material made from carbon atoms. A single layer graphene has the crystal structure of one-layer graphite, where carbon atoms are neatly arranged in hexagonal pattern. Graphene's unique properties are due to the arrangement of carbon atoms. Its excellent electrical and thermal conductivity makes it a very attractive material as conductor for memory chips fabrications and battery industries. Its flexibility and durability finds potential applications in the construction materials. Although it may be decades before graphene potentials are fully realized, graphene manufacturing has been increasing largely due to various applications where graphene materials are incorporated into products that worth many billions of dollars per year. For graphene industry that is being developed for mid-to-large scale of manufacturing graphene in various formats at affordable price, there is an urgent need for simple and rapid graphene analysis tool that can easily and rapidly characterize graphene quality for effective quality monitoring leading to consistent yields. Over the years Raman spectroscopy has been used extensively by the research communities on graphene materials. In this study, a Raman Graphene Analyzer is introduced as rapid graphene analysis tool to conducted at-line or on-line process monitoring for graphene manufacturers. No microscope is needed which is ideal for graphene material in powder forms.

Feed-forward Process Control in Enzymatic Protein Hydrolysis of Byproducts: A spectroscopic approach

Sileshi Wubshet, Jens Petter Wold, Nils Kristian Afseth, Ulrike Böcker, Ingrid Måge

Enzymatic protein hydrolysis(EPH), i.e., protease catalyzed digestion of protein, is one of the well-recognized bioprocesses in valorization of byproducts and side-streams from food industries. Extensive heterogeneity of by-products from, for example, meat-processing pauses a major challenge in production of protein hydrolysates with stable quality attributes. Therefore, EPH of meat byproducts requires a strict control of the raw material composition and process settings. Here, we are presenting a new feed-forward process control strategy for EPH of poultry by-products based on rapid spectroscopic measurements. By-products from poultry processing span significant variation in major constituents such as protein, ash and fat. Therefore, production of protein hydrolysates with predefined quality such as molecular weight distribution and protein yield is a challenging task. In the present work, we investigated the potential of four different spectroscopic techniques, i.e., NIR scanner, MicroNIR, Flourescence and Raman, as tools for continuous on-line characterization of the raw materials. Partial least squares (PLS) prediction models for ash, protein and fat content of raw materials were developed based on Raman, Flourescence and MicroNIR measurements, respectively. In an effort to establish a feed-forward process control strategy, we have developed statistical models that enabled prediction of end-product characteristics, i.e., protein yield and average molecular weight of peptides (Mw), as a function of raw material quality and controllable process parameters. The Raman-based model gave a relatively moderate prediction model for Mw ($R2 = 0.56$ and $RMSE = 150$). The best model was developed for protein yield prediction based on combined use of both MicroNIR and Fluorescence ($R2 = 0.88$ and $RMSE = 4.8$). Such prediction models can be vital process control tools for EPH in a feed-forward manner.

Raman spectroscopy in polymer applications from laboratory discoveries to large-scale production
Karen Esmonde-White, Patrick Wiegand, Ian Lewis

Raman spectroscopy offers a non-invasive, non-destructive method to measure polymer chemical composition and molecular structure. These features of Raman have been exploited in many polymer applications ranging from commercial polymer manufacturing to laboratory synthesis of microgels to quality control of multilayer thin films. As a process analytical instrument, an on-stream Raman analyzer can be used in qualitative and quantitative analyses to improve process knowledge, enable in-process corrections and safely measure reactions. We present recent Raman applications in polymers which include polymerization blending, extrusion monitoring, density, crystallinity, and copolymer content. Raman spectroscopy can also be coupled to rheometry or calorimetry equipment, which enables simultaneous collection of composition, molecular structure and rheological or thermal properties. Biological polymers, plant-based polymers and extruded films containing an active pharmaceutical ingredient (API) are important applications within the polymer industry. We show examples where the use of Raman spectroscopy has been shown to improve understanding polymer chain entanglement and API distribution within the extruded film. Through these representative examples, we demonstrate Kaiser Raman spectroscopy as a versatile technique in measuring and understanding polymer processes from laboratory to manufacturing.

Raman spectroscopy technologies enabling in situ real-time bioprocess monitoring

Karen Esmonde-White, Maryann Cuellar, Alexander Pitters, Sean Gilliam, David Strachan, Herve Lucas, Bruno Lenain, Ian Lewis

In situ Raman spectroscopy enables simultaneous measurement of multiple bioprocess parameters within a cell culture or fermentation bioprocess. Raman spectroscopy provides detailed chemical and physical information that can be used to generate multi-component qualitative and quantitative predictive models. We describe technology developments in Raman spectroscopy for real-time understanding and monitoring bioprocesses. There are scientific and logistical needs of upstream bioprocessing, all of which affect technological, ergonomic, safety and logistical aspects of developing analyzers, sampling probes and software. In this paper, we describe how consideration of these aspects affected the development of Raman spectroscopy hardware and software for upstream bioprocessing applications. We show a representative example that compares sampling probe compatibility with stainless steel, glass or single use bioreactor environments. Specially-designed sampling probes address not only the needs of collecting in situ measurements but also compatibility with bioreactor cleaning protocols and reliable performance over the course of a multi-day bioprocess. Software which integrates into an advanced process control and integrates a general analytical model can provide quantitative information without significant method development. We show examples where Kaiser Raman successfully integrates a general model and where Kaiser Raman was used for rapid development of a control model in process development in order to improve culture duration, product quality and higher titer.

Photo oxidative stabillity study of active layers in bulk heterojunction organic photovoltaics by micro-RAMAN spectroscopy.

Vasilis Gregoriou, Christos Chochos, Michalis Spanos

Even though, a large number of new conjugated polymers have been synthesized, most of these publications have been focused solely on high efficiency while neglecting stability. The result is although interesting, many of the active materials based on these compounds, may not be useful for real world applications. Of various factors affecting to the stability of polymer:fullerene solar cells, among which light, heat, oxygen, and humidity are the essential considerations. While the effect of oxygen and humidity can be suppressed by proper encapsulation, light and accumulated heat from illumination are inherently unavoidable and must be addressed with the highest priority. The active layer component in the organic solar cell is the part of the device that is very sensitive to light and heat and since it is integral to the device functionality this translates directly into a degradation of the PCE. In this study, we will present that the main mechanism affecting the stability of the OPV devices is morphological variation and not changes in the chemical structures of the active layer components as demonstrated by micro-RAMAN studies on pristine active layers films and upon photo-oxidation

Maximizing performance and reducing variability for a SERS substrate self-assembled hot spot array developed by HP based on nanofingers

Milo Overbay, Michael Delos-Reyes, Steven Barcelo, Christopher N. Young, Anita Rogacs

Surface enhanced Raman spectroscopy (SERS) has been shown to be most effective when plasmonic hot spots are formed in narrow gaps of a nanostructure or aggregated colloids. The distribution and interaction of these hotspots affect the enhancement factor and variability during analysis. Further, adjustments need to be made in power density, integration time and total irradiated sample volume for optimum results. In addition, analyte concentration and adsorption affinity play a role in optimum analysis. Different SERS substrates designs also need different conditions to maximize performance. This presentation will explore these factors for a SERS substrate developed by HP based on gold nanofingers [1]. 1. Kim A, Ou F S, Ohlberg D A A, Hu M, Williams R S and Li Z 2011 Study of molecular trapping inside gold nanofinger arrays on 2 surface-enhanced Raman substrates. J. Am. Chem. Soc. 133 8234–9

Time Domain Diffuse Raman Instrumentation based on a TCSPC camera for depth analysis of diffusive media

Sanathana Konugolu Venkata Sekar, Sara Mosca, Gianluca Valentini, Werner Zuschratter, Rainer Erdmann, Antonio Pifferi

Raman spectroscopy of biological media most often suffers from background fluorescence that reduces the signal to noise ratio (SNR) of measurements. One possible way to overcome this limitation is by time gating the Raman signal which can effectively distinguish Raman from the delayed fluorescence signal. In this work, we present a time resolved Raman instrumentation based on a new time resolved single photon counting camera that enables simultaneous broadband acquisition of time resolved Raman spectra. A compact portable probe was built to perform Time resolved Raman measurements in diffusion regime. The instrument was tested on a bilayer tissue mimicking phantom. The limitation of fluorescence contaminating the signal was effectively removed by time gating the acquired signal which demonstrated the potential of time resolved Raman spectroscopy in the presence of fluorescence. Furthermore, time gating at different intervals allowed probing at different depths of diffuse media as late arriving photons carry depth information.

Spectrochemical detection of chemical analytes in microfluidic systems
Nalin Andersen, Kateryna Artyushkova, Ivana Matanovic, Plamen Atanassov

A surface-enhanced Raman scattering (SERS) platform has been developed for use with a porous media-based microfluidic device. A motorized scan stage combined with confocal Raman spectroscopy provides quasi *in situ* fundamental and dynamic chemical analyses of analytes for optimizing and quantifying catalytic systems. The SERS platforms have been used to characterize the analytes and analyte mixtures present in the oxidation of glycerol to CO_2, a cascade reaction involving 9 sequential steps. The fabrication method for the plasmonic detection platform including the synthesis of silver nanoparticles (AgNP) will be presented. The SERS enhancement factor has been calculated to be 10^7 and can be made in less than 10 minutes with facile techniques that are reproducible. The analytical device is designed for characterization of mechanistically distinct catalytic cycles and can be used across a wide range of scientific fields.

Through-skin analysis of quality parameters in intact fish using spatially offset Raman spectroscopy
Ulrike Böcker, Nils Kristian Afseth, Jens Petter Wold, Chris Welsby, Pavel Matousek

Recent instrumental developments within Raman spectroscopy have provided new possibilities for probing deeper layers of opaque biological materials. One of these Raman approaches, namely spatially offset Raman spectroscopy (SORS), is based on the acquisition of Raman spectra from regions spatially offset on the sample surface from the excitation laser beam. In this way, SORS allows molecules located deeper into biological tissue to be measured. As biological samples and foods are in general notoriously heterogeneous, SORS thus holds promise as a tool to improve representative sampling within the field of food analysis. This field, however, has so far not been extensively explored. The authors have previously shown the potential of employing SORS in the characterisation of quality parameters of salmon through the skin. Our studies showed that SORS applying 830 nm excitation provides specific and clear spectral signatures of fatty acids in salmon muscle both when analysed through light grey skin as well as through dark black skin. SORS also effectively suppressed the broad and dominating fluorescence background of the salmon skin. In the present study, we have utilized a commercial SORS-setup (i.e. the portable RapID system from Cobalt Light Systems) to perform depth penetration studies in model samples mimicking the tissue composition beneath fish skin. Depth penetration of the laser for different components could be estimated by adjusting sample heights with thin slices of ham (i.e. proteins), lard (i.e. fats) and carrot (i.e. pigments), respectively. The same SORS-setup was used to probe pigmentation and lipids in intact samples of Atlantic salmon and Arctic char. The results suggest that through-skin probing of pigmentation in intact is possible. In the future, it is expected that non-invasive tools like SORS for assessing quality parameters of live or slaughtered fish would be of major benefit for farming and breeding purposes, and for end-product quality documentation.

Fabrication and Evaluation of Surface Enhanced Raman Scattering Substrates of Gold Nanoparticles Embedded in a Polymer Matrix

Md Shah Alam, Mary Tecklenburg

The objective of this study is to fabricate a long term stable substrate for surface enhanced Raman scattering (SERS), which measures the increased Raman scattering of a molecule adsorbed on or in proximity to a nanostructured metal surface. A substrate of aqueous gold nanoparticles (AuNPs) embedded in a polymer matrix was designed to enhance the sensitivity of conventional Raman spectroscopy. To accomplish the purpose, preparation of reproducible AuNPs of similar size and shape as well as narrow size distribution were required. The structural features of AuNPs influence strongly the strength of enhancement of Raman scattering by a molecule of interest. Stability of the nanoparticles for SERS was accomplished by incorporating gold nanoparticles of desired geometry into a polymer matrix (branched polyethyleneimine PEI) which may prevent AuNPs from being oxidized over time. For preparing the SERS substrate, gold nanoparticles were synthesized in aqueous solution of a surfactant, cetyltrimethyl ammonium bromide (CTAB), via a seed mediated process and then the particles were centrifuged to remove excess CTAB before embedding in polyethyleneimine (PEI) and casting polymer films. The SERS films were prepared using an optimized amount of gold nanoparticles imbedded in the polymer and in the presence and absence of the cross-linker adipic acid. The reproducibility of the AuNP size, shape, and particle distribution was confirmed by TEM measurements. The surface plasmon resonance (SPR) of aqueous gold nanoparticles was evaluated by UV-Visible absorption spectroscopy. Raman spectra were collected from aqueous drops of thiosalicylic acid (TSA) on the PEI-AuNP film at the optimum pH of 7.0. The detection limit determined for TSA on as fabricated SERS films was 10-8 M. Initial long-term studies of SERS signal reproducibility and stability for TSA on the fabricated PEI-AuNP templates showed degradation over 3 weeks. The data suggests that alternative cross-linkers may provide better stability and homogeneity of the SERS signal.

Raman Spectroscopy for Undergraduate Laboratories

Alexander Osterbaan, Justin Shorb, Stacey Carrier, Giora Proskurowski

There are a variety of instruments that are used for quality control in industry. Each method has its own set of constraints such as long run times, extensive sample preparation, or destruction of samples. Recent advancements by MarqMetrix and Hellma, USA, have resulted in the design of the Reva Raman instrument that allows for simple, rapid, and nondestructive acquisition of Raman spectra. Raman spectroscopy can be used independently or in conjunction with other instrumental methods to analyze samples. Raman has a variety of applications such as identification, molecular feature analysis, and analytical analysis. Currently, Raman is rarely used in industry as very few people are familiar with this method. Most undergraduate chemistry major programs do not include an introduction to Raman spectroscopy throughout their four years. The objective of this project is to design labs using the Reva instrument for use at the undergraduate level.

Experimental artifacts influencing polarization sensitive magneto-Raman spectroscopy

Komalavalli Thirunavukkuarasu, Zhenguang Lu, Liqin Su, Yifei Yu, Linyou Cao, Mariana Ballotin, Peter Christianen, Yong Zhang, Dmitry Smirnov

Since the discovery of graphene, there has been an explosion of research on two-dimensional layered materials such as transition metal dichalcogenides (TMDs). Among several experimental techniques utilized for studying these materials, Raman spectroscopy has proven to be a very powerful tool due to it's sensitivity to layer numbers, interlayer coupling etc., [X. Zhang *et al.*, Nanoscale **8**, 6435 (2016)]. Layered MoS_2, belonging to the family of TMDs, is a typical example with promising applications in nano-optoelectronics [X. Xu *et al.*, Nature Physics **10**, 343 (2014), K.F. Mak and J. Shan, Nature Photonics **10**, 216 (2016)]. A recent magneto-Raman investigations on MoS_2 published by Ji *et al.*, reported an observation of giant magneto-optical effect [J. Ji *et al.*, PNAS **113**, 2349 (2016)]. In this work, the intensity of A_{1g} Raman mode was found to decrease in intensity with increasing magnetic field. This effect was attributed to field-induced symmetry breaking of the Raman scattering cross-section. However, the results are slightly unclear as to how this symmetry breaking may be induced. Due to the ambiguous nature of the interpretation presented in this publication, we performed further Raman studies on MoS_2 at high magnetic fields up to 14 T to illustrate the experimental factors overlooked by the previous study. Based on our results, we emphasize that it is highly important to consider the magnetic field-induced rotation of the polarization of the light and its effect on the Raman active phonon modes in anisotropic materials. A detailed report of our magneto-Raman experiments and their outcomes will be presented.

SEHRS of Rhodamine Derivatives as a Probe of Two-Photon Properties
Jake Olson, Jon Camden

Tunable dye molecules with strong two-photon properties show promise as imaging probes and non-linear optical switches. Surface-enhanced hyper-Raman scattering (SEHRS) is a non-linear optical process with the ability to probe the nature of two-photon accessible states. In this work, we utilize the SEHRS effect to systematically study the evolution of electronic properties of a series of rhodamine-like dye molecules. Specifically, we look at tetramethyl rosamine dyes (TMRs) with a variation of the substituted chalcogen atom. These conjugated dye molecules exhibit non-Condon effects in their vibronic transitions causing changes in the SEHRS spectrum. Through variation of the excitation wavelength, vibrational band enhancements on resonance with electronic states elucidate vibronic behavior. This behavior gives insights to symmetry changes of states during transitions, providing a better understanding of the two-photon accessible states. Two-photon properties of chemical systems can be applied to two-photon imaging, all optical switching, and optical data storage.

Hydration effect on lysozyme in gas phase studied by UV-photodissociation spectroscopy in combination with droplet-beam IR laser ablation
Norishi Kawauchi, Hiroya Asami, Jun-ya Kohno

Proteins are usually present in solution, interacting with ambient solvent molecules. Therefore, it is of importance to elucidate the hydration effect of the protein molecules. The hydration effect depend on a valance state of the protein molecules. However, valence selective measurements are impossible in solution. In contrast, a mass spectrometry in gas phase can separate both the hydration number and the valence state. Especially, an ion trap technique enables us to select the ions with the desired hydration and valence state and keep them in gas phase. The gas-phase protein molecules have been investigated by an electrospray ionization in combination with a photodissociation spectroscopy.

However, there studies have been limited to high valence ions, which is different from their state in am aqueous solution. In this study, we aim to establish a method for analyzing the gas-phase proteins and reveal the interaction between the protein and the solvent molecules under a valence selective condition using an electrodynamic ion trap technique. We observed a UV-photodissociation spectrum of the lysozyme (Lys) isolated in the gas phase by a droplet-beam IR laser ablation method. The valence selective trap of Lys is accomplished by optimizing the trap condition, such as radio frequency and amplitude of the RF field applied to the trap. We measure the UV-photodissociation spectrum of Lys^{2+} at deep UV region (215-275 nm). Ions, H^+ and H_3O^+, which are UV-photodissociation products, are observed by irradiation of the UV laser onto trapped Lys^{2+} ions. These ions are generated through a two-photon dissociation process. The intensities of the ions are measured as a function of the UV-laser wavelength, which gives the UV-photodissociation spectrum. In a conventional UV-absorption spectrum of the Lys in aqueous solution, the absorption band locates at around 220 nm. On the other hand, the UV-photodissociation spectrum of Lys^{2+} in the gas phase exhibits a peak at around 230 nm. The blue shift by the hydration is explained by a larger hydration stabilization in the ground state than that in the exited state.

Study for structure analysis of ions Deep Eutectic Solvent by ATR-Far UV

Kazutaka Nishikido, Yusuke Morisawa

Deep Eutectic Solvent (DES) is solution of electrolyte mixed with Lewis base in a molar ratio of 30-22%. DES has a similar property as ionic liquid (IL), such as high viscosity, low volatility, high conductivity and broad electrochemical window. For example, DES composed of $LiNO_3$ and Urea shows significant melting depression. As a result, DES is liquid at room temperature. These materials are cheaper than IL and easily preparable. Thus, DES is expected as low-cost solution alternative to IL. Allowed electronic transitions of molecules in condensed phase can be easily measured by using Attenuated Total Reflectance Far-UV spectroscopy (ATR-FUV) technique. If we observe the electronic transition of DES and know the behavior of electrons in solution, one can develop structure analysis of liquids in the high concentration state. In the present study, we used DES composed of Li salt as electrolyte and Urea and NMA as Lewis base. As to DES of $LiNO_3$ and Urea, there are two peaks in FUV region. The band at around180-190nm assigned to a transition due to ?-?* of Urea [1], and the other observed in 220-230nm which is assigned to be transition due to ?-?* of NO_3^- [2]. NO_3^- band (220-230nm) is observed as separated band in 2nd derivative. Molar ratio dependences were considered to assign bands in the spectra. Moreover, longer wavelength side is decreased of the intensity with the lapse of time after turning off and stirrer, but shorter wavelength side is increased of that. This means that structure of ions in the DES is changing. We examine the difference of electronic transition in various Lewis bases and anions of Li salt. [1]Morisawa et.al,J.chem.Phys,2013,139,154301,[2] Ikehata et.al,J.Phys.chem.A,2010,114,8319-8322

Accessing the Secondary Structure of Monoclonal Antibody (mAb) Pharmaceuticals under Stresses by Deep-UV Resonance Raman (DUVRR) Spectroscopy

Chen Qiu, Sergey Arzhantsev

Monoclonal antibody (mAb) therapeutic is playing an important role in the biopharmaceutical industry because of their target selectivity, minimal side effects and ability to treat rare diseases. In order to assure the safety and quality of therapeutic mAb products during manufacture and clinical application, it is important to develop a rapid analytical method for monitoring and screening the protein secondary structure change resulting from different types of stresses. Currently, several conventional analytical techniques have been applied to determine the secondary structure of protein drugs, including NMR, IR, and Circular Dichroism. However, due to the structural complexity of mAbs and the complicated nature in drug formulation, all the current techniques have their limitations. In our previous work, we reported the development of an alternative spectroscopic technique, deep-ultraviolet resonance Raman (DUVRR) spectroscopy for studying the protein secondary structure changes under thermal stress in the drug formulation. DUVRR has been proved to be capable for measuring the higher order structure changes in mAbs with high sensitivity and selectivity. In a recent study, we use this DUVRR technique to investigate the protein secondary structure change under a common type of chemical stress: SDS-protein interaction, together with chemometrics modeling and two-dimensional correlation analysis. This protein-SDS interaction has been visualized spectroscopically, with different types of folding and unfolding processes. This work illustrates the promise of DUVRR as a powerful tool for probing the mAb secondary structure change, as well as the potential of combining this novel spectroscopic technique and advanced data analysis for analyzing complex protein drug products in formulations directly.

The History of Glow Discharge Mass Spectrometers – A Manufacturer's Viewpoint

Peter Robinson

The first commercially available Glow Discharge Mass Spectrometer was introduced in the mid-1980's. It was a low power, "static" source coupled to a double focusing reverse geometry Nier-Johnson magnetic sector analyser. In the initial years its development was rapid, driven by market needs to analyse different material types to ever more demanding levels. GDMS is a semi-quantitative technique when used without standards for calibrations, and was therefore very powerful in the analysis of "novel materials" and could be applied to quality control of ultra-pure metals, semi-conductors and alloy materials. A version of the instrument with the same source was available using a quadrupole analyser for a short time, but that did not prove to be commercially acceptable; the high resolution associated with the magnetic sector instruments was necessary for splitting away multi-ionised and molecular species. Development of the technique slowed after the initial years until a research grant was awarded by the European Community in 2000, and a three-year project was undertaken by a group of participants from manufacturers, users and academic institutions to develop a high powered fast flow source coupled to a more modern mass spectrometer. This was accomplished and after some time became commercially available. In recent years development of GDMS has gathered pace again and there are now three magnetic sector instruments commercially available from three different manufacturers. There is also, in its early stages, an instrument available with a GD source coupled to a Time of Flight system. The three magnetic sector instruments all use double focusing analysers for high resolving power, but all three offer fundamentally different features. There are fast flow and static versions, rf and dc discharges, pin and flat sample geometries, and both bulk and depth profiling analysis. So after a quiet period, we can say that the present time is exciting for Glow Discharge Mass Spectrometry, and it looks as though the future will be as well.

Glow Discharge Mass Spectrometry: a superstar for fast depth profiling.
Jorge Pisonero Castro, Jonatan Fandino, Nerea Bordel

A plethora of innovative materials are produced by depositing thin and ultrathin coatings on different substrates. For instance, hard disks able to store terabytes of information are based on multiple magnetic and non-magnetic nano-layers. Moreover, protectives coatings might be used to enhance the mechanical properties of substrate materials; for instance Zn-based coatings are used in the Al industry. In this context, solid analytical techniques are required to provide fast multi-elemental chemical analyses with high depth resolution (to monitor the different layers) and high sensitivity (to detect major, minor and trace elements). Direct solid analytical techniques, including Secondary Ion Mass Spectrometry, Secondary Neutral Mass Spectrometry, Auger Electron Spectroscopy or X-Ray Photoelectron Spectroscopy provide very valuable information about the atomic chemical composition of the surfaces/coatings; however, they also have some major drawbacks, such as high operating costs, complex sample pre-treatment and handling, low sample throughput and/or severe matrix effects that result in difficult quantification procedures. In order to overcome some of these drawbacks Glow Discharge Mass Spectroscopy (GD-MS) is proposed as a complementary methodology that provides an ideal solution for fast and accurate bulk and layer analyses [1-3]. In this work, we evaluate the advantages and limitations of this technique and we discuss about recent progresses and new applications. [1] C. González-Gago, J. Pisonero, R. Sandín, J. F. Fuertes, A. Sanz-Medel, N. Bordel, Analytical potential of RF-PGD-TOFMS for depth profiling of oxidized thin film composite membranes, Journal of Analytical Atomic Spectrometry, 2016, Volume 31, Issue 1, January 2016, Pages 288-296, DOI: 10.1039/C5JA00104H [2] C. Venzago, J. Pisonero, Book Chapter: "Glow discharge mass spectrometry", Sector Field Mass Spectrometry for Elemental and Isotopic Analysis (editors T. Prohaska, J. Irrgeher, A. Zitek and N. Jakubowski), Series of New Developments in Mass Spectrometry, Volume 2015-January, Issue 3, 2015, Pages 319-380; DOI: 10.1039/9781849735407; ISBN: 978-1-84973-392-2 [3] J. Pisonero, N. Bordel, C. Gonzalez de Vega, B. Fernández, R. Pereiro, A. Sanz-Medel, "Critical evaluation of the potential of radiofrequency pulsed glow discharge time of flight mass spectrometry for depth profile analysis of innovative materials", Analytical and Bioanalytical Chemistry, 2013, 405(17), 5655-5662.DOI 10.1007/s00216-013-6914-1.

Unique features and recent developments for GD-OES open the way to new applications

Matthieu Chausseau, Philippe Hunault, Patrick Chapon, Sofia Gaiaschi, Kayvon Savadkouei

Glow Discharge Optical Emission Spectroscopy, GD-OES, is well known for its capability to perform depth profiles on a large variety of samples for metallurgical applications, corrosion studies or hard disks characterization. The use of GD-OES for more recent applications such as CIGS photovoltaic cells, Perovskite solar cells, Lithium batteries, which also benefit from depth profile characterization for the study of layers composition and thickness, interfaces and diffusion effects, was not obvious few years ago as for those samples several issues were to be overcome. Samples were possibly modified by heat dissipation happening during the sputtering, layers of interest were embedded below polymers layers exhibiting (too) slow sputtering rates or samples were air sensitive and should be handled in an air-free environment. Determining the thickness of layers on such materials was also not easy to derive from the standard calibration procedures as it is impossible to use sputtering rate tables. For most of these materials, these rates are unknown and their determination is not always possible due to strong variations notably in the density of the materials involved Significant improvements of the technique were needed to address such applications in an efficient way. These required improvements and what they are bringing for the analysis of PV materials such as CIGS and Perovskite cells or for Lithium batteries characterization will be presented. To provide a better understanding of the level of performance that GD-OES can offer for such applications, results obtained in application laboratories worldwide as well as results obtained and published by scientists using the GD Profiler 2 instrument will be shared.

Diffusion Study of Multilayer Film stack by pulsed RF Glow-Discharge Optical Emission Spectrometry (GD-OES)
Helia Jalili, Qi Zhang, Kayvon Savadkouei, Matthieu Chausseau

Molybdenum (Mo) thin films are commonly used as a cap/barrier layer for the Aluminum (Al) line, and is also being considered for the Copper (Cu) line, in thin film transistor (TFT) structures due to their thermal stability, matching thermal expansion to Silicon (Si), low resistivity and excellent diffusion barrier properties. In certain applications, the anti-corrosion properties of Molybdenum are a growing new requirement. Alloying the metallic material (i.e. Mo) with sacrificial elements to form a stable oxide is a common method to improve the corrosion resistance of the cap/barrier layers. The sacrificial elements have a high affinity for oxygen, which leads to improvement of stability of the material to corrosive environments. However, these cap/barrier layers are very thin (20-30 nm) and the diffusion of the alloying elements to the Aluminum line during the thermal process is a main concern. Pulsed RFGD-OES was used to study the annealing effect on diffusion of alloying elements on multilayered stacks of Mo-alloy (30 nm) / Al (300 nm) / Mo-alloy (30 nm) films. Even with the low concentration of alloying elements, non-conductive glass substrate and low thickness of the cap/barrier layer, diffusion was studied down to the ppm level and depth resolution of sub-nanometer scale. Simultaneous measurement of all the elements regardless of their atomic number and binding energy makes the technique a powerful one for the characterization of such samples, providing useful information in a short time compared to other depth profiling techniques normally used.

Comparison of "half" and "full" dielectric barrier discharges – LTP vs. DBDI

Pascal Vogel, Felix David Klute, Sebastian Brandt, Antje Michels, Charlotte Reininger, Daniel Thurston, Beatrix Biskup, Paul Farnsworth, Joachim Franzke

Dielectric barrier discharges (DBD) have become an integral part in analytical applications over the last decade. They have proven to be versatile tools due to their simple designs and ease of operation making them ideal candidates for standardized commercial equipment. This success has led to a variety of different discharge designs optimized to their respective analytical application making a complete overview over the field difficult. One reason for this variety is, that according to Kogelschatz et al., for dielectric barrier discharges, at least one dielectric layer has to be in between the electrodes [1]. This definition of a DBD does not differentiate between the case when the discharge gas is in contact with a conducting surface which can be called "half DBD" or not which can be referred to as the "full DBD". It will be shown that this differentiation is important contrary to the official definition and should be taken into consideration. A new discharge design, the variable dielectric barrier discharge (VDBD), was built to create a fair comparison basis between half and full DBD. The VDBD enables a quick switch between both discharge modes making a direct live comparison possible.
[1] Kogelschatz, U.; Eliasson, B.; Egli, W. J. Phys. IV 1997, 7, 47-66.

Investigations of plasmon-enhanced fluorescence and Förster Resonant Energy Transfer in composite multilayer nanoparticles

Denis Boudreau, Jérémie Asselin, Samuel Ouellet, Josée Richard-Daniel, Nicolas Fontaine

The use of fluorescent nanoparticles is a popular tool in bioanalytical chemistry and cellular imaging, thanks to their higher brightness as compared to molecular fluorophores and their facile coupling chemistry for grafting oligonucleotides, antibodies, etc. The inclusion of a metallic core within these nanostructures increases the local electric field intensity and the excitation/emission rates of surrounding fluorophores while decreasing the lifetime of excited states, with the overall effect being sizable improvements in brightness and robustness towards photobleaching. In recent years plasmon-enhanced core-shell nanostructures have been introduced, where a metal core is coated with multiple concentric silica shells, with some layers incorporating fluorophores while others act as spacers to position the dye-doped layers at precise distances form the core. The presence of the core in these multi-shell nanocomposites is also known to increase the Förster Resonant Energy Transfer (FRET) efficiency and range between fluorophore pairs immobilized within the nanostructures. In this presentation, we will discuss how geometry in these concentric nanocomposites impacts brightness, relaxation rates, photostability and FRET efficiency/range between donor and acceptor fluorophores, and we will show how these features can be implemented in the design of nanoprobes for various sensing applications.

Engineering and Imaging Excitons for Brain Imaging of Modulatory Neurotransmitters

Markita Landry, Abraham Beyene, Jackson Travis Del Bonis O'Donnell, Ralph Henry Page

For over 60 years, drugs that alter, mimic, or block modulatory neurotransmitters have formed the core arsenal for the treatment of neurological disorders such as depression, addiction, schizophrenia, anxiety, and Parkinson's disease. However, methods to diagnose and validate drug efficacy have remained largely the same: questionnaires and behavioral observations. The archaic nature of neurological disorder diagnosis results from the lack of tools to detect the molecular 'key players' of neuronal communication – the three primary modulatory neurotransmitters dopamine, serotonin, and norepinephrine. In this talk, we describe the design, characterization, and implementation of near-infrared optical sensors to image neurotransmitter dopamine. We describe how exciton engineering is achievable with molecular dynamics simulations, to create in vivo suitable sensors. Our nanosensors enable direct visualization of endogenous dopamine release over multiple rounds of acute brain slice stimulation, for over 80 minutes. We next introduce a new form of fluorescence microscopy for deep-brain neurotransmitter imaging: double infrared excitation-emission imaging. We characterize our findings in the context of their utility for high spatial and temporal neurotransmitter imaging in the brain, describe nanosensor exciton behavior from a molecular dynamics (MD) perspective, validate nanosensor use in vitro, and for nanosensor use in vivo, to correlate external stimuli (experiences, behavior) to chemical output (neurotransmission).

A Single Tb-to-Quantum Dot FRET Pair for Temporally Multiplexed Detection of Nucleic Acids.

Xue QIU, Jiajia GUO, Zongwen JIN, Igor L. Medintz, Niko Hildebrandt

Multiplexed photoluminescence (PL) detection plays an important role in chemical and biological sensing. Different to most frequently developed approaches, which use spectral separation (color multiplexing) of different fluorophores, we demonstrate the applicability of a single Förster resonance energy transfer (FRET) pair (Tb donor and quantum dot (QD) acceptor) for multiplexed biosensing using multiple time-gated (TG) PL intensity detection windows. We accomplished tunable Tb-QD distances by two strategies: 1) DNA-polyhistidine self-assembly and streptavidin-biotin-DNA interaction on the QD surfaces; 2) Specific location of the Tb donor by using different lenghs of DNA. Both systems used acceptor-sensitization and donor-quenching for quantifying the biomolecular recognition and modification of the donor-acceptor distance for tuning the PL decays. We also show that TG detection of a single Tb-to-QD pair can be used to selectively quantify low nanomolar concentrations of multiple microRNAs in a single sample. TG intensity detection provided extremely low background noise and a quick and simple one-step assay format. Single TG-FRET pair multiplexing can be combined with spectral (color) and spatial (mircoarrays) resolution, paving the way for biosensing with unprecedented high order multiplexing capabilities.

Fabrication of enzyme-powered 3D DNA nanomachine for rapid detection of nucleic acids and single nucleotide variants

Feng Li, Xiaolong Yang, Sean Mason, Zechen Yu, Yanan Tang

Single nucleotide mutations in DNA or RNA sequences can lead to important biological consequences and techniques capable of detecting such single nucleotide variants (SNVs) are hence of great value for biological research as well as accurate disease diagnosis and management. Here, we report a novel 3D DNA nanomachine technology that is able to detect SNVs with high sensitivity and specificity. The 3D DNA nanomachine is essentially a DNA walking device that moves a DNA walker along a three-dimensional (3-D) DNA functionalized gold nanoparticle (DNA-AuNP). The movement of the DNA walker is powered by a nicking endonuclease that cleaves specific DNA substrates on the track. During the movement, each DNA walker cleaves multiple fluorophore-labeled substrates, amplifying the fluorescence signal. To detect SNVs, we incorporated the target-specific sequence into the DNA walker and deactivated it using a protecting DNA. The target DNA can displace the protecting DNA and activate the 3D DNA nanomachine through a toehold exchange mechanism. By carefully tuning the lengths of reverse and forward toeholds, we were able to detect target nucleic acid with single nucleotide specificity. We further demonstrated that it was possible to further enhance the detection sensitivity and specificity by integrating the 3D DNA nanomachine with a non-covalent DNA catalysis. Overall, the 3D DNA nanomachine is a rapid and isothermal sensing platform for nucleic acid and SNVs with high sensitivity and specificity and therefore has the potential to be used in applications such as point-of-care disease diagnosis.

Focusing Light Energy with DNA Nanostructures
Igor Medintz

The maturity of DNA structural technology now provides the unparalleled ability to position attendant molecules and especially fluorophores within any designer 1-, 2-, and 3-dimensional nanoscale architecture with high precision and intimately controlled spacing relative to each other. Although many applications can be undertaken with this capability, we focus on light harvesting and exciton directing and delivery based on creating Förster resonance energy transfer networks and cascades. Our long-term goals in this area include understanding the design principles that can allow almost any fluorescent material, ranging from organic dyes to nanoparticles, to be incorporated into these types of structures while still functioning optimally. Structurally, we wish to determine the complexity of the structures that can be assembled along with their size and efficiency limitations. Several examples that we have studied in detail will be presented including linear photonic wires, star-shaped structures, dendrimers and photonic wires incorporating semiconductor quantum dots, and bioluminescent enzymes such as luciferase. The data will be presented in the context of detailed energy transfer analysis and modeling of the underlying processes. Each configuration brings with it both a set of functional and experimental benefits and liabilities that must be accommodated in order to facilitate these studies and this will also be highlighted.

A spectroscopic serum based blood test for brain tumours: Optimisation for high-throughput sampling and the health economic impacts
Holly Butler, Matthew Baker, Mark Hegarty, David Palmer, Ewan Gray, Duncan Finlayson

It has recently been demonstrated that a spectroscopic test using blood-serum is able to effectively identify brain tumours in patients with sensitivities and specificities as high as 91.5% and 83% respectively. The approach employs Attenuated Total Reflectance - Fourier Transform Infrared (ATR-FTIR) spectroscopy to derive a disease specific spectroscopic signature, that using pattern recognition and machine learning algorithms is able to identify brain tumour cases.The ability to identify brain cancer cases based on serum samples alone raises the possibility of systematic screening prior to investigation with more expensive (MRI/CT imaging) and invasive (biopsy) tests. However, current methodology is currently limited to single patient analysis due to limitations in standard instrumentation. Here, we investigate the performance of a novel ATR-FTIR spectroscopic approach that allows automated, disposable, and multi-patient sampling. Our results indicate that this optimised approach is able to stratify brain tumour patients at performances equivalent to previous approaches, with sensitivities and specificities as high as 94.56% and 87.65% respectively. In advance of any prospective clinical study results being available, a cost-effectiveness analysis (CEA) was conducted to calculate the effects on health outcomes and health service costs of introducing this technology as a triage tool in primary and secondary care. Results indicate that using a serum spectroscopy test for brain tumours in both scenarios may be considered highly cost-effective in an HTA agency decision making process. Incremental cost-effectiveness ratios were well below standard threshold values of £20,000 to £30,000 per quality adjusted life year (QALY) gained used in the UK, and similar thresholds used internationally. In primary care, the test has the potential to be both more effective and cost saving for the health service.

Raman Spectroscopy to aid diagnosis of Colorectal Cancer.
Cerys Jenkins, Peter Dunstan, Catherine Thornton, Dean Harris

Colorectal cancer (CRC) is the third largest cause of cancer related death worldwide [1]. Most cases present at a later stage leading to poorer prognosis so quick methods of diagnosis are essential for improved outcomes. The aim of this work is to investigate if there is a link between a Raman spectral signature and colorectal malignancy. It aims to see if serum Raman spectroscopy can be used as a potential tool to be combined with symptoms improve the rate of CRC detection and lower referral rates for colonoscopy and flexible sigmoidoscopy. The presentation will outline an optimized protocol that produces and processes spectra from two laser excitation sources for liquid samples. An

optimized high throughput dry serum protocol will also be described. Raman spectra have been collected from patients with varying stages of colorectal cancer and matched control patients with clear colonoscopy (n ? 60). Analysis of the spectral data has aided the development of a classification model to predict 'unknown' samples. Further to this, patients who have polyps greater than >1cm in size have also been compared to the control dataset to evaluate the ability of the Raman test to differentiate pre-cancerous lesions. Sample collection is ongoing and current results will be presented. A multi-centre, multi-phase trial (n ? 300) is underway to validate the high throughput Raman spectroscopy test. In the first stage of this study, general practitioners will recruit patients into the trial and the high throughput Raman test will be compared to clinical outcomes to compute the positive predictive value and adoption rates of the Raman test. In the second phase, it is hoped that provided with results from the Raman test the number of colonoscopy referrals from primary to secondary care will be significantly reduced. Acknowledgements: The authors would like to thank Cancer Research Wales for funding the pilot studies and ABMU health board for funding the trial study. We would also like to thank all patients who have consented to donating blood to our studies.
[1] Cancer Research UK Key-Facts, Accessed January 2017. http://www.cancerresearchuk.org/health-professional/cancer-statistics

Cardiovascular disease related micro-RNA detection using paper based devices and surface enhanced Raman scattering

Samuel Mabbott, Syrena Fernandes, Karen Faulds, Charles Mace, Duncan Graham

Cardiovascular disease (CVD) causes more than a quarter of all deaths in the UK, equating to an average of 435 people per day, or one death every 3 minutes. Not only is it the UK's biggest killer, but it is also the leading cause of death worldwide, with the majority of deaths being caused by heart attacks. Micro-RNA (miRNA) detection offers an alternative route for early detection of CVD, so that treatment can be applied much more rapidly and/or lifestyle changes can made. The small non-coding RNA molecule (~22 nts) is responsible for gene silencing through the formation of an RNA-induced silencing complex with a ribonuclear particle. The short nucleotide strands have been shown to be involved in a wide range of biological processes in healthy people; however, they have also been associated with a number of diseases including CVD. miRNAs provide a desirable target for the diagnosis of CVD, because there are many different variations relating to the disease, and circulating miRNA is also extremely stable. Here, SERS and paper based diagnostics will be utilised to detect CVD related miRNA strands with an aim of carrying out multiplexed detection of short non-coding RNA contained within biological fluids.

Point-of-care field trials in Laos and Papua New Guinea of ATR spectroscopy for malaria diagnosis and blood
Bayden Wood, Phiip Heraud, David Perez-Guiata, Anja Rüether, Moritoshi Iwagami, Paul Brey, Christian Doerig

In March 2017 we undertook two scientific expeditions into Laos and Papua New Guinea to deploy ATR technology for point-of-care diagnosis of malaria in remote villages. The aims are two-fold. 1) Determine if the ATR technology has the sensitivity to detect asymptomatic carriers of malaria in a field setting. 2) Ascertain the robustness of the technology for point-of-care field deployment in a harsh and tropical environment. In the Laos study pin prick blood samples from 595 patients were analysed with ATR spectroscopy. A blood spot was also be collected for qPCR analysis. Spectra of lysed red blood cells were dried onto a glass slide and recorded using a Bruker alpha portable spectrometer using 8 cm-1 spectral resolution and 64 co-added interferograms for the sample. In the PNG study whole blood was analyzed from 235 patients in both wet and dry state. The results of the Laotian trial will be presented at SciEx 2016.

Diffuse Resonance Raman Spectroscopic Study of Red Blood Cells Inside Polymer-bag
Rekha Gautam, Joo-Yeun Oh, Rakesh Patel, Richard Dluhy

Red blood cells (RBCs) continue to metabolize via the glycolytic pathway and degrade during storage at 4-6 C. The transfusion with older, degraded blood products underlies the lack of efficacy and could cause toxicity in critical patients. Current methods used for RBCs screening, e.g. histochemistry, biochemical and microarray based assays are invasive and time-consuming. Thus, there remains a clear need for non-invasive tools which could facilitate in-process quality control and inventory management. Resonance Raman spectroscopy is a label-free modality which provides enhanced Raman spectra of molecules comprise of electronic transitions in resonance with excitation wavelength. However, the use of excitation wavelength in UV-Vis region increases the risk of photodegradation of samples and limited by the depth of diffusion. In this study, we discuss our recent efforts in developing a novel strategy - Diffuse Resonance Raman Spectroscopy (DRRS) – to acquire depth sensitive Raman spectra. Two main landmarks of this methodology include (i) illumination by diffuse beam instead of a focused beam which decreases the photon density at the sample and penetrates farther and
(ii) inclusion of additional detector (CCD) pixels adjacent to the pixels aligned with the optical path of the micro-spectrometer to augment the detection of diffuse photons. Unlike the excitation with near-infrared wavelengths, DRRS requires ten-fold less power and acquisition time to get analogous signal-to-noise ratio. We employed this approach for identification of spectral features to characterize RBCs inside blood-segments beneath 1 mm polymer layer. Further, a multivariate statistical analysis is performed on Raman spectra using Partial Least Squares-Discriminant Analysis (PLS-DA) which classified the Young (6-8 days) and Old (35-42 days) stored RBCs with > 95% sensitivity and specificity. Based on our results, it is evident that DRRS is a rapid, non-invasive and non-destructive approach to attain subsurface information for the molecules in resonance with excitation wavelength and has the potential for in-vivo applications targeting molecules such as hemeproteins and carotenoids.

Multivariate Methods for Analyzing Designed Data
Federico Marini

Since its first systematization by Fisher, design of experiments plays a fundamental role in science and, in particular, in chemistry, especially when the aim is to evaluate the effect of possible sources of variation on univariate or multivariate responses by performing trials under controlled conditions. On the other hand, as the presence of multiple sources of variation can hinder the interpretation (exploratory approach) or introduce nonlinearities (predictive modeling), particular care should be taken when analyzing the multivariate responses coming from designed exeriments. In this context, whereas in the case of univariate or oligovariate responses, well established methods for the evaluation of the effect of the controlled factors and their interactions exist in the framework of the theory of linear models, the classical multivariate analysis of variance becomes inapplicable when the number of experiments is insufficient compared to the number of responses. Accordingly, different methods have been proposed so far in the chemometric literature to provide a reliable analysis of multivariate data coming from designed experiment and an accurate interpretation of the effects of the different factors and interactions, ANOVA-PCA [1] and ANOVA-Simultaneous component analysis (ASCA) [2] being the most frequently used. As a possible alternative to the aforementioned methods, another approach, based on the application of discriminant PLS on an opportunely coded dummy matrix followed by target projection after proper deflation of the experimental matrix (ANOVA-TP), has recently been introduced in the literature [3]. In the present communication, the main characteristics of the different methods proposed for the multivariate analysis of designed data will be discussed also by means of real world examples. [1] P.B. Harrington et al., Analysis of variance-principal component analysis: a soft tool for proteome discovery, Anal. Chim. Acta 544 (2005) 118–127. [2] A.K. Smilde et al., ANOVA-simultaneous component analysis (ASCA): a new tool for analyzing designed metabolomic data, Bioinformatics 21 (2005) 3043–3048. [3] F. Marini et al.,

Analysis of variance of designed chromatographic data sets: the analysis of variance-target projection approach, J. Chromatogr. A 1405 (2015) 94–102.

Enhancing Hyperspectral Image Exploration via Preprocessing and Targeted Anomaly Detection
Neal Gallagher

Hyperspectral images contain an enormous amount of information but extracting the desired information can seem daunting. For example, changes in sample, measurement conditions and how target signal of interest manifests in the sample matrix can lead to variability between images. This is true for raw material inspection in the chemical process industry and is prevalent in biological samples. The result is that between-image comparison to find anomalies of interest can be quite challenging. Fortunately, individual images contain thousands of spectra that vary spatially and these measurements can be used to characterize clutter (interferences and noise) in the image. Differences in spectral signatures also can be enhanced via data preprocessing to maximizing signal-to-clutter. Additionally, concepts related to targeted anomaly detection, image whitening and weighting can be used to suppress clutter signal and enhance detection of signal of interest. This talk will demonstrate these concepts with examples from different imaging techniques.

Exploring Hyperspectral Images with Topological Data Analysis
Ludovic Duponchel

Hyperspectral imaging has become an essential tool in many scientific domains for the exploration of complex and heterogeneous samples providing simultaneously spectral and spatial information from the acquired data cube. Current instrumental developments become faster and faster generating an always higher number of pixels. This trend is even more accentuated with the increasing importance accorded to multimodal spectroscopic experiments. K-means clustering is certainly one of the most widely used methods for the exploration of hyperspectral images. Because it clearly has its pros and cons, it could be a good opportunity to explore new concept of data analysis. Topology, a sub-field of pure mathematics, is the mathematical study of shape. Although topologists usually study abstract objects, they have developed recently what they call Topological Data Analysis (TDA) [1]. The idea is here to use topology in order visualize and explore high dimensional and complex real-world data sets. This concept has been successfully used in different topics like gene expression profiling on breast tumors [2,3], T-cell reactivity to antigens for different type of diabetes [4], viral evolution [5], population activity in visual cortex [6] but also on unexpected topic as 22 years of voting behavior of the members of the US House of Representatives [7], characteristics of NBA basketball players via their performance statistics [7]. Lately, we have demonstrated that TDA could be used to discriminate single bacteria with Raman spectroscopy [8]. The two main tasks of TDA is the measurement of shape and its representation. One fundamental idea of the concept is to consider a data set to be a sample or point cloud taken from a manifold in some high-dimensional space. The sample data are used to construct simplices, generalizations of intervals, which are, in turn, glued together to form a kind of wireframe approximation of the manifold. This manifold and the wireframe represent the shape of the data set i.e. the topological network. Our aim will be to demonstrate that TDA could be useful in partitioning hyperspectral data sets. A comparison with the well-known K-means algorithm will be also proposed.

Mass Informatics of Stable Isotope Assisted Metabolomics
Xiang Zhang

Stable isotope assisted metabolomics (SIAM) uses heavy isotope tracers (e.g., 13C, 18O, 15N) to identify and discern pathways involved in biochemical processes, by measuring the incorporation of heavy atoms into the metabolites produced downstream of the tracer(s). A significant feature of SIAM data is that one metabolite in an unlabeled sample is represented by multiple isotopologues in a labeled sample because different numbers of tracer atoms are incorporated into the metabolite through different biosynthesis pathways. I will introduce a suite of algorithms for automatic analysis of SIAM data acquired by LC-MS and GC-MS. To increase the accuracy of isotopologue assignment, metabolites detected in the unlabeled samples were used as reference metabolites to generate possible isotopologue candidates for analysis of isotopic peaks detected in the labeled samples. A mixture score function was also employed for cross-sample chromatographic peak list alignment to recognize the chromatographic peaks generated by the same metabolite in different samples, by simultaneously evaluating the similarity of retention time and mass spectra of two chromatographic peaks. An iterative linear regression model was used to deconvolute the intensity of overlapping isotopologues. The developed algorithms have been tested with mixtures of metabolite standards and biological samples.

Multiple Versus Single Set Validation to Avoid Erroneous Conclusions

Peter Harrington

Many scientists working in the fields of -omics and industries that require authentication of materials rely on multivariate chemometric methods for modeling, classification, and calibration. In the modern information age, scientists should break from the habit of using single measurements to evaluate the performance of their chemometric methods. Surely, with laboratory measurements, the measurement is repeated to characterize drift in the signal by measuring the same sample at different times. The irony is that when using multivariate modeling, classification, and calibration that may involve sophisticated calculations, measures of performance are based on a single measurement or figure of merit (FOM). The figure of merit often is a measurement of prediction accuracy or error for a single validation set of data. This approach is wrong, especially in the modern era. It does not make good use of the measurement data which for some cases may be obtained from precious samples that are difficult to obtain, measurements made on expensive instruments, or large complex experiments. In addition, basing a conclusion on a single set of validation data or data point can lead to false inferences about the accuracy of the chemometric method. A systematic approach is advocated to use bootstrapping that efficiently uses all the measurements for validation, provides generalized estimates and confidence bounds about the FOMs.[1] In addition, it provides a conservative measure of the accuracy of the classification or calibration algorithm. This approach also is useful for optimization and automation of chemometric methods. 1. P.B. Harrington, Statistical Validation of Classification and Calibration Models Using Bootstrapped Latin Partitions. Trends in Analytical Chemistry, 25:11 (2006) 1112-1124, DOI: 10.1016/j.trac.2006.10.010.

Structure and Morphology of Biosynthesized and Biodegradable Polymer Ultrathin Films and Single Crystals Using AFM-IR, XRD and Selected Area Electron Diffraction (SAED)
John Rabolt, Changhao Liu, Bruce Chase, Isao Noda

The combination of atomic force microscopy (AFM) and infrared (IR) spectroscopy is a powerful tool that provides chemical, conformational and morphological information at a spatial resolution of 50-100 nm. Using an AFM-IR instrument, we have explored ultrathin films and isothermally grown single crystals of poly[(R)3-hydroxybutyrate-co-(R)-3-hydroxyhexanoate] (PHBHx), which have been produced with a wide range of crystallinity and morphology. They have been characterized using XRD, AFM-IR spectroscopy and SAED in order to assess the role of processing on crystal structure and morphology. For example, AFM-IR allows direct investigation of both the crystalline and amorphous phases of ultrathin polymer single crystals and provides IR spectra and AFM images simultaneously of sub-100 nm features. The source is a tunable quantum cascade laser (QCL) whose wavelength can be swept through the infrared "fingerprint" region in less than 1 minute. If one of the wavelengths is absorbed by the sample, then expansion of the sample occurs on the ns time scale, which causes a modulation of the oscillating AFM cantilever. This creates a "ringdown" at that particular frequency which decays as the heat dissipates. The positive amplitude of the oscillation represents the IR band intensity and hence as the frequency is tuned through the IR region (900 cm-1 – 3600 cm-1), an IR spectrum is obtained at a spatial resolutions of 50 - 100 nm. More details of this instrumentation is reported elsewhere1. The homopolymer, poly(hydroxybutyrate) (PHB) is a biodegradable, aliphatic polyester that can be produced by chemical processes or bacterial fermentation2. The fact that it can be produced biologically with the properties of a thermoplastic and is naturally biodegradable, have made it a subject of research in many industrial and academic laboratories worldwide. However, bacterially produced PHB results in a 100% stereoregular, highly crystalline polymer that is brittle and lacks flexibility. In order to modify these properties 3-hydroxyhexanoate (3HHx) has been added as a co-monomer and these new materials, referred to as PHBHx, exhibit a significantly reduced crystalline content, resulting in improved mechanical properties and processability. 1. Dazzi, A.; Prater, C. C. B.; Hu, Q.; Chase, D. B.; Rabolt, J. F.; Marcott, C. Appl. Spectrosc. 2012, 66, 1365–1384. 2. Satkowski, M. M.; Melik, D. H.; Autran, J. P.; Green, P. R.; Noda, I.; Schechtman, L. A., Physical and Processing Properties of Polyhydroxyalkanoate Copolymers. Biopolymers 2001, 3B, 231-263

Investigation of Nanodomain Composition in Impact Copolymer Polypropylene by AFM-IR

Peite Bao, Fuguang Tang, Zhaohui Su

Chemical analysis of individual nanoscopic objects has always been a challenge. Molecular spectroscopy can provide rich chemical information, but with poor spatial resolution, whereas electron microscopy affords high spatial resolution, but produces little chemical information. The AFM-IR technique emerged recently combines AFM with IR and enables chemical analysis at sub-100 nm scale, which can provide structural information not assessable to other techniques. Here for the first time we establish a quantitative method by the AFM-IR technique, and show that the rigid cores (~100 nm size) of the rubber particles in a commercial grade impact copolymer polypropylene (ICP), an important material with complicated morphological structure, mainly compose of polypropylene rather than polyethylene, which contradicts the previous conclusion, established on the basis of electron microscopy and electron diffraction analyses, demonstrating the unique capability of the novel method.

Nanophotonic AFM Transducers Transform Chemical Composition and Thermal Conductivity Measurements at the nanoscale

Andrea Centrone, Georg Ramer, Jungseok Chae, Vladimir Aksyuk

The use of atomic force microscopy (AFM) and related techniques is pervasive in nanoscience and biology because, in addition to the sample topography, they provide maps of many other properties with high spatial resolution. For example, photothermal induced resonance (PTIR) is an emergent technique that combines the spatial resolution of AFM with the specificity of absorption spectroscopy, enabling material identification, biomolecule conformation analysis, mapping of composition and electronic bandgap at the nanoscale. In PTIR, the absorption of a laser pulse induces a rapid thermal expansion of the sample. Conventional cantilevers are too slow to track the thermal expansion dynamics of the sample but the fast sample expansion kicks the cantilever in oscillation, with amplitude proportional to the absorbed energy. In this work, we revolutionize AFM signal transduction by integrating cavity-optomechanics for sensing the motion of fast, nanosized/picogram scale AFM probes with unprecedented precision and bandwidth, thereby breaking the trade-off between AFM measurement precision and ability to capture transient events. Applied in PTIR, the probe near-field ultralow detection noise and wide bandwidth improves the time resolution, signal-to-noise ratio and throughput by a few orders of magnitude each. Remarkably, this synergy enables a new PTIR measurement modality: capturing the previously inaccessible fast thermal-expansion response of the sample to nanosecond laser pulses, thus allowing concurrent measurement of the chemical composition and thermal conductivity, at the nanoscale, for the first time. As first demonstration, we validate these new capabilities using polymer films and leverage them to measure the intrinsic thermal conductivity (?) of metal-organic framework

(MOF) individual microcrystals, a property not measurable by conventional techniques. MOFs are a class of nanoporous materials promising for catalysis, gas storage, sensing and thermoelectric applications where accurate knowledge of ? is critically important. We strongly believe that the radical AFM innovation enabled by nanofabrication, nanophotonics and cavity-optomechanics is broadly-applicable and will benefit a wide range of dynamic observations in nanoscience and biology and greatly improve the impact of the PTIR technique in those fields.

AFM-IR for the Examination of Coating Nanostructure and Environmental Damage

Suzanne Morsch, Stuart Lyon, Simon Gibbon

Network polymers are increasingly used in manufacturing for a variety of structural and functional applications. Epoxy resins in particular rank amongst the most widely used synthetic polymers, frequently comprising the matrix component of advanced composites (used in e.g. airframes, wind turbine blades, sports equipment); functional coatings (e.g. self-cleaning, self-healing, conductive, biocidal, anti-graffiti); protective and aesthetic coatings (e.g. decorative and corrosion protective paints, coatings for food containers and beverage cans) and adhesives. Many processes lead to the degradation and service failure of these materials including: photo-oxidation, environmental stress cracking, chemical attack and mechanical creep, and yet the molecular network structure that defines their performance remains poorly understood. Previously, investigators have been wholly reliant on topographic evidence to characterize resin nanostructures, leading to significant differences in the interpretation of microscopic evidence. Now, the advent of nanoscale analysis techniques such as AFM-IR offer unprecedented insights into polymer network nanostructure and degradation. For example the disputed nodular morphology found within epoxy resins has been shown to correspond to chemical heterogeneity associated with an inhomogeneous cure using AFM-IR characterization. A key application of epoxies is in protective organic coatings, which are widely employed for corrosion control due to their relatively low cost and proven efficacy. Nonetheless, the processes governing the long-term deterioration of intact organic coatings (in the absence of defects) are not fully understood. In-situ time-resolved FTIR has been applied to monitor both the chemical integrity of the coating and water uptake as a function of immersion time in water or electrolyte. Meanwhile, ex-situ AFM-IR experiments were used to characterize coating morphology and local moisture sorption under humid conditions. Epoxy coatings were found to deform locally as a result of water sorption during immersion in electrolyte or water, yielding nanoscopic regions of heterogeneous hydrophilicity, yet no change in functional group chemistry corresponding to network deterioration was found to occur.

Orientation and Disentanglement in Electrospun Fibers

Christian Pellerin, Marie Richard-Lacroix, Elise Siurdyban

Electrospun fibers show unusual properties that are still poorly understood, such as an exponential increase of elastic modulus with a decreasing diameter. Here, we combined standard IR, NanoIR and polarized Raman spectroscopy to probe orientation and molecular structure in individual fibers of polystyrene. We show by Raman that this increase in modulus is directly correlated with molecular orientation. Polystyrene fibers were initially selected as a model system due to its amorphous nature but results indicate a partial disentanglement of the polymer chains in fibers which also correlate with fiber diameter. The conditions promoting this incomplete reentanglement during the spinning process are studied by NanoIR. The results suggest a surface localization for the phenomenon and a much higher amplitude than observed on bundles by standard IR. Our results help understanding the properties of electrospun fibers.

Determination of rare earth elements in geological minerals by laser induced breakdown spectroscopy
Jinesh Jain, Chet Bhatt, Christian Goueguel, Christina Lopano, Dustin McIntyre

Laser induced breakdown spectroscopy (LIBS) has been extensively used for the determination of major elements in geological samples. However, its application for quantification of rare earth elements (REEs) is relatively new as REE spectral lines in complex matrix samples are generally interfered by other prominent elemental peaks. In this study, we identified low and high intensity spectral lines for various rare earth elements in REE-rich minerals [i.e., monazite, characterized by X-Ray Diffraction (XRD) analysis]. Spectra recorded from test samples were compared with those recorded from the REE standards and the NIST database was used as a reference. The calibration models were developed using partial least squares regression (PLS-R) for quantification of these elements. The detection limits were relatively higher but the predicted results using calibration models were found comparable to those obtained by ICP-MS analysis, indicating that LIBS has a potential to quantify REEs in geological minerals/ores.

The Versatility of LIBS -Industrial Applications
Dominik Schiller, Tino Seger, Lars Jacobsen

LIBS is still in its growing stages and continually has to demonstrate its strengths against other established analytical techniques such as RAMAN, XRF or ICP-MS. With some of its strongest advantages being that it is fast, contact free and almost non-destructive, LIBS is a deployable and reliable measuring method for mono & multi-elemental calibration and qualitative & quantitative analysis, all of which can be performed in real time. LTB's ability to combine mechanical, electrical and software engineering with our knowledge in spectroscopy and optical design, allows the swift in-house development of measuring systems. From high-end laboratory equipment -- for example, the LIBSlab as a compact and complete LIBS measurement system --to robust industrial on-line solutions – such as the in-line process monitoring of solar cells -- LIBS is an incredibly versatile analysis method. Sorting processes which utilize conveyor belts -- to sort coal, ore or scrap, for example – are quickly becoming a significant field for LIBS technologies. The continuing development of faster cameras and improved mathematical algorithms has enabled LIBS systems to perform both a high quantity of measurements as well as in-depth spectral analysis in a short time. The LTB LIBSorter is one such measurement solution used not only to sort non-ferrous scrap metal, but also to distinguish and sort aluminum alloys at a rate of up to 40 pieces per second. The implementation of LIBS in the recycling industry has vastly improved the efficiency of the recycling process, thereby saving resources and offering a rapid return on investment from its low operating cost. Furthermore, LTB is developing a test to identify jumbled components in the processing chain of steel production, thereby providing reliable quality control under harsh environmental conditions. LTB places heavy emphasis on partnering with other market leaders and research partners to expand the field of potential LIBS applications. Past collaborative research projects have included not only the LIBS-based analysis of molten minerals, but also the analysis of recyclables in sewage sludge ashes, and the in-situ monitoring of the chemical compositions of stainless steel during the welding process.

Commercial LIBS Instrument that combines surface preparation and elemental analysis for metal industries
Lutfu Ozcan, François Doucet, Altan Muftuoglu

The metal industry has been using Spark-AES (Atomic Emission Spectroscopy) for over 70 years for controlling the chemical composition of their alloy production. Spark-AES technology is now considered mature; nevertheless, it still requires mechanical preparation of sample surfaces prior to analysis. There is a need for an Automated Online metal Ana-lyzer to achieve accurate and fast elemental analysis of materials, as well as to lower human error during manipulations. However, direct analysis of the process stream is very challenging due to surface preparation and environmental conditions. Laser-Induced Breakdown Spectroscopy is an emerging technology that allows for control of the chemical composition of the metallurgical process stream. Several groups have developed solutions that enable the direct analysis of metal on the process using LIBS technology with encouraging performances. Even though LIBS technology needs little sample preparation, in some applications (especially in the metals industry), a mechanical preparation of the surface needs to be done in order to remove scale or rust that are not representative of the material to be analyzed. In this presentation, a novel commercial LIBS instrument that combines sample surface preparation and analysis will be presented. This novel LIBS analyzer has the capability to analyze a sample in a few seconds, without mechanical surface preparation. In addition, this instrument has demonstrated repeatability and reproducibility better than 1% and 3%, respectively, on the alloying elements in steel and aluminum. This new LIBS instrument has demonstrated multiple correlation coefficients (R^2) better than 0.9999 for the most common alloying elements in steel and aluminum. Finally, this new LIBS analyzer has demonstrated accuracy better than 4% on several external validations made on certified reference material.

Analysis of gold in rock samples using laser-induced breakdown spectroscopy: matrix and heterogeneity effects

Kheireddine Rifai, Marcel Laflamme, Marc Constantin, Mohamad Sabsabi, Alain Blouin, François Vidal, Paul Bouchard, Konstantinos Fytas

There is an increasingly need in the gold mining industry for evaluating the concentration of gold in ore samples in real time and in situ. X-ray fluorescence (XRF) spectroscopy, which has been used successfully for a number of metals such as copper, zinc and nickel, is inadequate for the desired gold ore application, which requires a detection limit of 50 to 100 times lower. Other existing technologies such as infra-red (IR) spectroscopy allows determining the mineralogical matrix but doesn't allow an elemental analysis. Laser-induced breakdown spectroscopy (LIBS) appears to be a potential candidate for the quantitative analysis of gold in rocks. Our aim is to develop the knowledge for building up a portable gold analyser based on LIBS technology. When analysing ore samples, one faces many challenges, including the variation of the mineralogical matrix and the heterogeneity of the gold grain distribution in the ore. We have established the LIBS calibration curve using 44 calibrated samples with gold concentrations ranging between 0 and 1000 ppm and different matrix compositions. Consistently with the chemometrics principal component analysis, which indicates that our samples fall into two categories, Si-rich and Fe-rich, the calibration curve splits into two branches showing detection limit of about 0.8 ppm, which is adequate for the desired application (~ 1 ppm). The analyses of the two halves of 3 different drill cores from a gold mine show discrepancies less than 50% between the LIBS results and those from a certified laboratory. This relative error is of the same order as the discrepancies observed in the same conditions between two certified laboratories, and is a consequence of the heterogeneity of the gold grain distribution in the samples. We have characterized further this distribution heterogeneity effect by cutting 32 rocks from 6 different mines into parallelepipeds of few centimeters in size and analysed each face of the parallelepipeds by using LIBS. Then, the same samples have been analysed by conventional techniques (atomic absorption or gravimetry) for comparison with the LIBS results. The detailed study will be presented in the talk.

Versatile applications of Laser Induced Breakdown Spectroscopy (LIBS): From nanocatalysts and semiconducting materials characterizations to biomedical applications.
Dibyendu Mukherjee

Laser Induced Breakdown Spectroscopy (LIBS) is a relatively non-destructive spectrochemical characterization technique that has increasingly gained prominence among both industrial and laboratory scale analytical tools due to its simplicity and robustness in carrying out quantitative chemical analysis with minimal sample preparations. Recent years have seen a surge in the development of calibration-free quantitative LIBS for chemical composition analysis of materials in aerosol, liquid and/or solid phase. To this end, recent works in our research group have successfully developed and implemented LIBS using background matrix as the internal calibration standards for high-accuracy quantitative elemental analyses of complex intermetallic nanomaterials, biomaterials and industrial grade semiconducting materials. The current presentation will focus on some of these works highlighting the versatility of LIBS for: 1) quantitative elemental analyses in binary/ternary nanoalloys (NAs) and nanocomposites (NCs) synthesized as advanced nanocatalysts (namely, PtNi, PdCo, PtCuCo (NAs) and PtCo (NCs) with varied elemental ratios) by our in-house built pulsed laser ablation technique; and 2) rapid and accurate measurement of thin films and dissolved oxygen in semiconducting materials for the rapidly growing semiconductor manufacturing sector in the US. LIBS results will be benchmarked against standard ICP-OES analytical measurements for the respective case studies presented here. Results will also be supported by transmission electron microscopy (TEM), atomic force microscopy (AFM), energy dispersive X-ray spectrometry (EDX), ellipsometry, and laser crater profilometry measurements to provide concurrent morphological, physical and qualitative analyses as and when required for the specific studies presented here. The highlights of the results presented here showcases the versatility of LIBS in demonstrating extremely good agreements with elemental analyses obtained from ICP-OES and other measurement techniques while being devoid of cumbersome sample preparations and external calibration standards. The talk will conclude by briefly touching upon our recent works involving biomedical applications of LIBS for early stage in-vitro detection of calcium depositions in valvular interstitial cells (VICs) within aortic valves.

Recent Advances is Analyzing Single-Walled Carbon Nanotube Samples by Optical Spectroscopy

R. Bruce Weisman

Single-walled carbon nanotubes (SWCNTs) are a family of artificial nanostructures of great interest in basic and applied research. A major challenge in such studies is characterizing the composition of structurally mixed SWCNT samples, which are typically heterogeneous in aggregation state; nanotube diameter, roll-up angle, and length; and electronic and optical properties. Recent progress will be presented in the use of optical fluorescence and absorption spectroscopy for such characterization. Results include measurements of structure-specific absorption cross sections, fluorescence quantum yields, fluorescence side bands, and fluorescence excitation spectra. The use of the new method of variance spectroscopy for SWCNT analysis will also be described. Finally, an integrated scheme will be presented that combines fluorescence and absorption spectral analysis to deduce absolute concentrations of specific semiconducting SWCNT species in mixed samples.

Differentiating Left- and Right-handed Carbon Nanotubes by DNA
Ming Zheng

New structural characteristics emerge when solid-state crystals are constructed in lower dimensions. This is exemplified by single-wall carbon nanotubes, which exhibit a degree of freedom in handedness, and a multitude of helicity that gives rise to three distinct types of electronic structures -metals, quasi-metals, and semiconductors. Here, we report the use of intrinsically chiral single-stranded DNA to achieve simultaneous handedness and helicity control for all three types of nanotubes. We apply polymer aqueous two-phase systems to select special DNA-wrapped carbon nanotubes, each of which we argue must have an ordered DNA structure bound to a nanotube of defined handedness and helicity, resembling a well-folded biomacromolecule with innate stereo-selectivity. We have screened over 300 short single-stranded DNA sequences with palindrome symmetry, leading to the selection of more than 20 distinct carbon nanotube structures that have defined helicity and handedness and cover the entire chiral angle range and all three electronic types. The mechanism of handedness selection is illustrated by a DNA sequence that adopts two distinct folds on a pair of (6,5) nanotube enantiomers, respectively, rendering them large differences in fluorescence intensity and chemical reactivity. This result establishes a first example of functionally distinguishable left- and right-handed carbon nanotubes. Taken together, our work demonstrates highly efficient enantiomer differentiation by DNA, and offers a first comprehensive solution to achieve simultaneous handedness and helicity control for all three electronic types of carbon nanotubes.

Quantum Defects of Carbon Nanotubes: Room Temperature, 1.5 μm Single Photon Emitters for Quantum Information Technology
Han Htoon

Quantum light sources capable of supplying a regular stream of single photon hold the key for realization of novel quantum information technologies. For practical implementation, these quantum light sources are required to operate at room temperature and telecommunication wavelengths. Despite intensive research effort, an ideal quantum material system capable of meeting both of these requirements has not emerged as of today. Last year, we have demonstrated that room T single photon emission at wavelength approaching 1.3 μm can be achieved via introduction of quantum defect states by covalent oxygen functionalization of (6,5) single wall carbon nanotubes (SWCNTs). Because these quantum defect states are created relative to the band edge of the SWCNT, it becomes possible to tune this emission to 1.3 and 1.55 μm via functionalizing larger diameter nanotubes that have smaller ban gap. However, oxygen functionalization becomes difficult to control with the increase of tube diameter. Furthermore, the single photon emission of these oxygen dopant states also becomes unstable at room temperature. As a result, long-desired goal of establishing room-T single photon emission in telecommunication wavelength remain unfulfilled. Here in this work, we utilize highly controllable aryl diazonium reaction chemistry to create quantum defect states on large diameter, (7,5) and (10,3) SWCNTs and successfully demonstrate room-T single photon emission at wavelengths extending from 1.1 μm to the deep center of the telecom C band, 1.55 ?m. We achieved 99% single photon purity, a critical milestone for applications such as quantum key distribution, logic gates, and memory. This figure exceeds the best single photon purity reported in other RT single photon emitters such as NV centers (70%-90%). The single photon emission is also characterized with exceptional, shot-noise limited count rate stability. These results establish quantum defects states of diazonium functionalized SWCNTs as near perfect single photon sources meeting almost all the key technological requirements.

2D Nanomaterials for Membrane-based Water Purification
Baoxia Mi, Sunxiang Zheng, Casey Finnerty, Zhongying Wang

Recent advances in two-dimensional (2D) nanomaterials offer unprecedented opportunities to fabricate a new class of filtration membranes that can potentially revolutionize the membrane-based water purification. In this talk, I discuss the promise of 2D nanomaterials as building blocks to ingeniously make new nanostructured membranes with exceptional mechanical, electrical, chemical, and biological properties, thereby significantly improving the separation efficiency of water purification membranes. In particular, I will present some of the exciting results from our ongoing research on graphene oxide (GO) and MoS2 membranes made by layer-stacking and bonding of appropriately spaced 2D nanosheets. Compared with today's widely used polymeric membranes, 2D-structured membranes are fabricated by facile methods using inexpensive bulk materials and have a fundamentally different separation mechanism. Most important, 2D-structured membranes exhibit many appealing properties such as high water permeability, excellent selectivity, multifunctionality, and superior fouling resistance. Additionally, I will point out several outstanding challenges in manipulating the nanostructure of 2D-strucutured membranes to maximize their separation capability.

Tunable Single-Photon Emission at Telecom Wavelengths from Carbon Nanotube Quantum Defects
Xiaowei He, Nicolai F. Hartmann, Xuedan Ma, Jeffrey L. Blackburn, Weilu Gao, Junichiro Kono, Han Htoon, Stephen K. Doorn

The intentional incorporation of impurities and defects can serve as a powerful tool for modification of the electronic and optical properties of host nanomaterials and enabling of new functionalities. Introduction of Sp3 defects by covalent functionalization to single-walled carbon nanotubes (SWCNTs) generates new emission states with energies lower than the optical bandgap (E11) of SWCNTs by 100 to 300 meV. With controllable emission wavelengths and high quantum yield, these defect states open up new opportunities for the applications of SWCNTs in photonics such as lasing, photon upconversion and single photon emission. Recently, it has been shown that such doping sites are capable of emitting single photons at room-T, which opens the possibility of building room-T electrically-driven single photon emitters based on doped-SWCNTs. The realization of these devices needs the fine control of doping densities, to the point of introduction of solitary dopants on individual SWCNTs. It also requires the high emission quality from doping states, such as sharp linewidth of the emission spectra and non-blinking of the emission intensity. In this work, we showed a chemical approach capable of creating solitary dopants on SWCNTs of different diameters. Our method was based on the fine control of the doping of surfactant-suspended and poly[9,9-dioctylfluorenyl-2,7-diyl] (PFO) wrapped SWCNTs by different diazonium salts. Quantum optical studies were performed on the diazonium doped-SWCNTs. We first investigated the photoluminescence (PL) spectra of doped-SWCNTs at room temperature and low temperature (4K). Individual, bright and stable emission peaks were observed from doped SWCNTs at room-T. The linewidth of the PL peaks decreased to 200 micro-eV at 4K. Next, Hanbury Brown-Twiss (HBT) experiment was performed to explore the single photon nature of the emissions. Strong photon anti-bunching behavior with purity above 99% were observed on the doping emission peaks in the range of 1100-1400 nm at RT(295K).

Development, Validation & Implementation of a Real Time Release Test for Dissolution

Sarah Nielsen, Yleana Colon, Stan Altan, Olav Lyngberg

Janssen Pharmaceuticals has designed and implemented a direct compression continuous manufacturing (CM) line for producing commercial product. One of the benefits of continuous manufacturing is the ability to make numerous "mini" batches of material which can show the correlation of process conditions and tablet critical quality attributes. A targeted design of experiments guided by a criticality analysis can specifically probe how process conditions and tablets physical properties correlate to dissolution. This information can then be used for building real time release tests. This talk will give an overview of the development, validation and practical implementation of a real time release test for dissolution.

Criticality of attribute measurement in the feed-frame for low dose compounds – a low dose formulation example
Benoit Igne

The development of continuous manufacturing for pharmaceutical oral solid dosage forms brings unique challenges associated with the monitoring and controlling of the line. In-line analytics provide means to measure product attributes in a relevant time frame to the process, potentially allowing for feed-back and feed-forward control to ensure overall product quality. However, traditional PAT techniques have limitations for low drug formulations and new solutions have to be found. In this presentation, the attribute monitoring of low dose final blend assay in a feed-frame will be presented. Using Light Induced Fluorescence Spectroscopy in a tablet press feed-frame, the homogeneity of a low dose formulation was monitored in-line and in real-time.

Building Robustness into Chemometric Models for Supporting Continuous Manufacturing
Caitlin Schram, Justin Pritchard, Kelly Swinney

Building robustness into chemometric models is important for implementation into an automated control strategy. Rigorous model development starts with careful development of the calibration set. An ideal calibration set is composed of data collected from materials prepared over a range of time and that covers the anticipated compositional, physical, chemical, and manufacturing process variability. Additionally, the calibration set should contain data from multiple like-for-like instruments spanning the manufacturer's specifications. Spectral pre-processing and range selection should be chosen to highlight relevant chemical differences while reducing the influence of physical differences. Vertex's approach to robust model development will be discussed along with examples.

Improving On-line Monitoring of Tablet Coating Process with Terahertz based Near-Infrared Coating Thickness Models

Shikhar Mohan, Noritaka Odani, Hanzhou Feng, James Drennen, III, Carl Anderson

Tablet coating is an important unit operation in the manufacturing of solid oral dosage forms. Coating thickness and uniformity are often considered critical quality attributes (CQA) for the coating process. For drugs sensitive to light, humidity, and oxygen, coating thickness dictates the chemical stability of the final product. Near-Infrared (NIR) spectroscopy has been demonstrated as a Process Analytical Technology (PAT) tool for monitoring coating thickness in a pan coating unit operation. A major disadvantage of NIR is that it is not a primary technique and requires a reference method. Typical reference methods for coating quality purposes include weight gain based values as a surrogate for coating thickness. This study investigates Terahertz hyperspectral imaging (THI) as a reference method to improve the accuracy of NIR coating thickness estimates in a non-destructive manner. Tablets comprised of a photolabile drug (Nifedipine) were manufacturing in-house and coated with 4% hypromellose, 2.4% titanium dioxide, 0.8% Talc, and 0.8% polyethylene glycol 6000 in a pan coater to improve the chemical stability. Two NIR models were generated with reference measurements of film amount calculated from tablet weight and loss-on-drying measurements (traditional) and coating thickness obtained from THI measurements. For the calibration set, tablets were randomly sampled during the spraying phase from coating times of 80 minutes, 100 minutes, 120 minutes, 140 minutes, and 160 minutes. The unique test set included samples from 100 minutes, 120 minutes, and 140 minutes. For the traditional reference measurements, 100 tablets were sampled from each calibration sampling point, while for the THI based reference measurement 10 tablets were sampled. Both methods used the same set of 5 tablets sampled from each test sampling point. The relative error of the NIR model using THI based reference values had significantly lower relative error, compared to the reference method based on tablet weight

1st principles, statistical, and chemometric modeling to understand, develop, control, and continuously verify drug product manufacturing

Brandye Smith-Goettler

Process analytical data and process data have been used in conjunction with off-line data (i.e. certificate of analysis, batch-records, in-process control testing and end-product testing) to develop a quality drug product manufacturing process. In a manner analogous to a symbiotic relationship; concurrent process analytical method and manufacturing development is an efficient means to yield real-time understanding -> leading to better designed development activities -> to transition from aggregating all data to modeling relevant data -> enabling process control -> improving process capability. As a drug product manufacturing process progresses from pilot scale development to process qualification; mathematical models are critical to inform, and even to inspire, decisions that lead to a highly capable drug product manufacturing process. Using an example drug product; the plethora of data acquired and diversity of models used will be detailed. For example, experimental design and statistical analyses are used to yield quality data. From this; residence time distribution, chemometric, and statistical/multivariate statistical process control models transition this quality data into process understanding. In late stage development; discriminant analysis, partial least squares, mechanistic, and statistical models are used to control the manufacturing process. Control as mentioned herein, refers to models for in-process control and for real-time release testing. Lastly, multivariate outlier diagnostics and batch-level multivariate statistical process control modeling can be used to maintain and continuously verify process performance in supply.

Applying Low Frequency Raman to QbD in Pharmaceutical Development

John Wasylyk, Robert Wethman, Ming Huang

Low frequency Raman spectroscopy has been used to study various polymorphs and can be applied to the design of crystallization control strategy. Extending the low frequency spectral region to include the fingerprint region, provides access to collective vibrations of molecules in the amorphous and crystalline states and yields valuable insight when differentiation of various forms is quintessential. Whether during process development or production, low frequency Raman bands provide greater sensitivity for detecting the onset of crystallization and has allowed differentiation of crystal types when multiple forms are possible. Applying this to Quality by Design (QbD) studies brings an increase in process understanding leading to developing optimal control strategy and avoid the many pitfalls that can occur when scaled-up to the production environment. A recent applications of in-line crystallization processes provided insight into establishing the ideal crystallization parameters. The parameters evaluated include temperature, mixing rate, seed levels and solvent variable. In-line and off-line QbD studies demonstrated both ideal and non-ideal conditions, ultimately yielding critical process knowledge. As a results of our studies, low frequency Raman has proven to be a valuable tool for at-line and on-line monitoring of active pharmaceutical ingredient crystallization and paves the way for robust production in a large scale facility.

Two-dimensional low frequency Raman correlation spectroscopy study of bioplastics
Isao Noda, Anjan Roy, James Carriere, Brian Sobieski, Bruce Chase, John Rabolt

The low-frequency Raman spectral region was used to study the evolution of crystalline state of poly[(R)3-hydroxybutyrate-co-(R)-3-hydroxyhexanoate] or PHBHx, which is a totally biodegradable plastic produced by microbial fermentation of vegetable oils. Raman bands associated with the lattice mode vibrations of crystalline lamellae were detected in the very low frequency/THz region. Their systematic intensity variations during the isothermal crystallization of PHBHx were contrasted to those of carbonyl vibrations using two-dimensional (2D) Raman correlation spectroscopy. Coordinated dynamic variations in the spectral features were observed with the crystallization process, and surprisingly detailed mechanisms for the development of crystalline structures were revealed.

Using low frequency Raman and THz absorption spectroscopy to understand order in metal-organic frameworks and solar cell polymers
Keith Gordon

THz absorption techniques have been used very successfully in the analysis of pharmaceutical materials, the utility of this spectral region has been demonstrated in both imaging and quantitative analysis. More recently the possibility of combining the relatively high sensitivity of CCD-based Raman spectroscopy with the low frequency regime has been realized with the introduction of volume Bragg gratings. These permit Rayleigh line rejection down to 9 cm^{-1} with minimal modification of normal Raman experiments. We have used two systems to investigate low-frequency Raman spectra for a number of systems, one of these allows for the simultaneous measurements of Raman scattering from 9 – 2400 cm^{-1}, the second through the use of a number of grating types can measure spectral windows as small as 200 cm^{-1} thus boosting spectral resolution. We have used THz Raman to evaluate the effect on order in metal-organic frameworks with incorporation of new linker units that facilitate frequency doubling. IN these studies we show that some added linker units disrupt the order int he framework and these appear to result in poorer frequency doubling characteristics. Those additives that maintain order show better performance. We have also examined order in solar cell donor-acceptor polymers. We have shown that H-bonding may be used to engineer rigid polymer domains that support phonon and show higher structural crystallinity and solar cell efficiency than less ordered and rigid polymers.

Discriminative and quantitative analysis of pharmaceutical polytypes using low-frequency Raman spectroscopy

Kentaro Iwata, Masatoshi Karashima, Yukihiro Ikeda, Motoki Inoue, Toshiro Fukami

Solid forms of active pharmaceutical ingredients (APIs) have attracted great interest in the pharmaceutical industry due to its impacts on process optimization, intellectual properties and regulatory requirements. Many analytical methods have been developed to characterize solid form of APIs such as powder X-ray diffraction and differential scanning calorimetry. Among them, vibrational spectroscopy is recognized as non-destructive analysis and has been applied to quality control and process analytical technology. Especially, Raman spectroscopy is preferably used for qualitative and quantitative characterization of APIs in formulations avoiding interference from water. Conventional Raman systems collect spectra in the fingerprint (FP) region, typically lying between 200-1800 cm$^{?1}$, where vibrations correspond to molecular skeleton and conformation. Low-frequency (LF) region is commonly limited down to 200 cm$^{?1}$ and Raman spectra in the region closely correlate with crystal structure of the molecule. Therefore, Raman spectra in the LF region can differentiate crystal structures more directly than in the FP region. The purpose of the present study was to apply LF Raman spectroscopy to discrimination and quantitative determination of polymorphs (polytypes) where common two dimensional superstructure piles up in different stacking mode. Due to similar API conformation and packing unit, Raman spectra in the FP region exhibited little difference between such polymorphs. LF-Raman spectroscopy was able to discriminate the polymorphs. Calibration curve was established using LF-Raman spectra of physical mixture containing two polymorphs with various ratios. Both peak shift and score of partial least squares (PLS) showed good correlation to the polymorphic ratio. The PLS approach had smaller error in the calibration curve than the peak shift approach because spectra resolution strongly impacted on the peak shift approach. Furthermore, quantitative analysis was also performed in the model formulation system. The API showed much stronger LF Raman vibration than the model excipients and the polymorphic ratio was also determined even at 10% (w/w) API content. These results demonstrated that LF-Raman spectra can be a novel tool promising for discrimination and quantitative determination of pharmaceutical polymorphic systems where another analytical method has not achieved.

Solid-state analysis of amorphous and crystalline forms using THz Raman spectrometry

Alison Nordon, Joanna Lothian, Pol Macfhionnghaile, Paul Dallin, John Andrews, James Carriere

Many pharmaceutical solids can exist in different forms, e.g. crystalline and amorphous. Hence, there is a need to be able to monitor physical form during the development of drug products. There are a number of techniques that are commonly used for solid-state analysis including x-ray powder diffraction (XRPD), solid-state NMR spectrometry and infrared (IR) spectrometry. However, some of these techniques are expensive (e.g. XRPD and NMR) and most cannot be used in situ. While Raman spectrometry can be used for in situ structural and chemical analysis, it is the low frequency (<200 cm^{-1}) region in the Raman spectrum that is most sensitive to solid form. This region is generally not observed in traditional Raman measurements owing to limitations with the notch filters used to block the Rayleigh scattered light. However, major advances in ultra narrowband filters and wavelength stabilised lasers have now made the so-called THz Raman region accessible. Therefore, in this work, a novel portable THz Raman instrument has been used for in situ analysis of solid form transformations in sulfamerazine. Sulfamerazine is a widely used sulfonamide antibacterial drug and has three known polymorphic forms. Form I can be transformed to form II by milling at room temperature and to an amorphous form by cryo-milling. THz Raman spectrometry was used to determine the optimum cryo-milling time for production of the amorphous form and to study its subsequent transformation to forms I and II over a period of approximately 100 minutes. In addition, it was shown that form I transforms to form II via an amorphous form when milled at room temperature. The concentrations of the different solid forms was determined from the THz Raman spectra using a classical least squares model. The advantages of using THz Raman spectrometry will be discussed.

Dynamic SERS nanosensor for neurotransmitters sensing near neurons

Jean-Francois Masson, Felix Lussier, Thibault Brulé, Marie-Josée Bourque, Louis-Éric Trudeau

Current electrophysiology and electrochemistry experiments have provided unprecedented understanding of neuronal activity. However, these techniques are suited for a small, albeit important, panel of neurotransmitters, such as serotonin and dopamine, but these constitute only a subset of the broader range of neurotransmitters involved in brain chemistry. Therefore, the need to develop a technique broadly applicable for the sensing of neurotransmitters near living neurons remains important. Surface-enhanced Raman scattering (SERS) provides a unique opportunity to detect neurotransmitters in close proximity of neurons. We have recently developed dynamic SERS (D-SERS) nanosensors to measure lactate and pyruvate near cells. These nanosensors are based on the decoration of patch-clamp-like nanopipettes with gold nanoparticles, offering a solid support that can be located accurately under a microscope. The nanosensors are thus highly compatible with current physiology experiments also relying on similar nanosensors based on electrochemistry and electrophysiology. In this paper, we will show that D-SERS can measure ATP, acetylcholine, GABA, serotonin, and dopamine, among other neurotransmitters, in a single experiment, with the potential of analyzing a greater number of neurotransmitters. The SERS spectra of these neurotransmitters were identified with a barcoding data processing method and time series of the neurotransmitter levels were constructed. The D-SERS nanosensor was then located near dopaminergic neurons. Detection of ATP and dopamine was realized in a series of K+ depolarization experiments showing elevated levels of both ATP and dopamine. Control experiments were also performed near glial cells, showing only a few basal detection events of ATP and dopamine. This paper demonstrates the potential of D-SERS to detect neurotransmitter excretion events near living neurons. In addition, the technique is also broadly applicable to different neuron or cell models for other pathologies.

Ultrasensitive detection of thyrotropin-releasing hormone based on azo coupling and surface-enhanced resonance Raman spectroscopy

Yukihiro Ozaki

Surface-enhanced Raman scattering (SERS) is highly sensitive enough down to the single-molecule level, and allows one to identify multiple components in a mixture. Ultrasensitive detection methods based on SERS have been successfully developed for some small molecules and biological samples. For SERS-based estrogen detection, a major problem is that the a?nity of these phenolic molecules to silver or gold surfaces is rather weak, leading to poor SERS signals when mixing the analytes with silver nanoparticles (AgNPs) directly. There is well known coupling reaction between phenol and diazonium ions, yielding products known as azo dyes. Thus, Pauly's reagent is often used for tyrosine detection, based on the optical absorption of the products, azo compounds, which have been proven to have strong Raman activity and a propensity toward binding to AgNPs. BY coupled with surface-enhanced resonance Raman scattering (SERRS), ultra sensitive phenolic estrogen detection with multiplexing capability would become possible by these azo dyes derived from phenolic estrogens because di?erent SERRS "?ngerprint" spectra of azo dyes correspond to di?erent phenolic estrogens. Studies have shown that many adverse health e?ects are associated with human exposure to dietary or environmental estrogens. Therefore, the development of rapid and highly sensitive detection methods for estrogens is very important and necessary to maintain hormonal concentration below the safety limit. Herein, we demonstrate a simple and rapid approach to detect trace amounts of phenolic estrogen based on surface-enhanced resonance Raman scattering (SERRS). Because of a coupling reaction between diazonium ions and the phenolic estrogens, azo compounds are formed with strong SERRS activity, which allows phenolic estrogen recognition at subnanomolar levels in solution. The proposed protocol is universal and highly selective, not only for phenolic estrogens but also for other phenolic molecules, even in complex systems.

Combat Forensics: Identification of Bad Actors with the aid of Microfluidic SERS

Augustus Fountain, Neal Kline, Ashish Tripathi, Rustin Mirsafavi, Martin Moskovits, Carl Meinhart, Jason Guicheteau, Jason Guicheteau

Combat operations over the past 15 years have uncovered the confluence of the worldwide drug trade and improvised explosive devices (IEDs). In many cases, drug trafficking is used to finance terrorist operations that include IEDs or drug traffickers have adopted IEDs to protect their illicit operations. Therefore the ability to detect trace quantities of illicit drugs or explosives and tie those materials biometrically to a "bad actor" has become increasingly important. Since combat operations often make it impossible to ship samples a quick, simple, and field forward, cost-effective method for detecting drugs of abuse, explosives, or their metabolites in biological fluids would be of great value to the United States military. In addition this same tool could be used in fighting domestic terrorism or in support of law enforcement agencies and heath care organizations in battling drug abuse. The United States Army continues to jointly develop a microfluidic SERS device with the University of California Santa Barbara. Surface enhanced Raman spectroscopy (SERS) is used to provide a sensitive, selective detection technique within the microfluidic platform employing metallic nanoparticles as the SERS medium. The work presented here describes updated efforts of the Edgewood Chemical Biological Center to optimize the microfluidic platform for detection of drugs and explosives.

Application of SERS in bioanalytical detection schemes

Dana Cialla-May, Karina Weber, Juergen Popp

Powerful detection schemes in bioanalytics are associated with the requirements for molecular specificity, high sensitivity and fast detection times. Surface enhanced Raman spectroscopy (SERS) is known to meet those requirements and the strong capability of this method in bioanalytical detection schemes is due to the enhancement of the molecular specific Raman fingerprint by several orders of magnitude using plasmonic active nanostructures. Within this presentation, the SERS-based analysis of food samples will be discussed. As an example, the ratio of lycopene and ?-carotene is estimated by using silver nanostructures for recording the SERS spectra. Based on a simplified model, the percentage of lycopene is determined in tomatoes [1]. Further on, the potential of SERS in food analytics is demonstrated by detection the vitamins B2 and B12 in cereals by combining extraction protocols with a SERS-based detection scheme [2]. In order to detect and quantify the illicit drug THC, the inside walls of capillaries are coated with SERS-active silver nanostructures following a microwave-based fabrication protocol. By simply dipping the capillary into the analyte-containing solution, the SERS-active area is incubated and the SERS-based detection is performed achieving a limit of detection in the range of 1 nM [3]. Finally, by combining SERS with microfluidic devices, high-throughput measurements are feasible, which will be introduced related to the topic of therapeutic drug monitoring. The broad spectrum antibiotic levofloxacin is detected in the complex matrix of human urine and the quantitative analysis is achieved down to a root means square error of prediction (RMSEP) between 0.058 and 0.16 mM for the different investigated urine samples [4]. [1] Radu et al., Analyst, 2016, 141, 4447-4455. [2] Radu et al., Talanta, 2016, 160, 289-297. [3] Yüksel et al., Anal Chim Acta, 2016, 939, 93-100. [4] Hidi et al., J. Phys. Chem. C, 2016, 120, 20613-20623. Acknowledgement: Funding of the research projects "JBCI 2.0" (03IPT513Y) within the framework "InnoProfile Transfer – Unternehmen Region", "EXASENS" (13N13856) and "InfectoGnostics" (13GW0096F) by the Federal Ministry of Education and Research, Germany is gratefully acknowledged.

A New Laboratory Instrument for Raman Detection of Thin Films on Planar Surfaces – Surface Plasmon Enhanced Surface Raman Spectroscopy (SPESRS)

Stephen Weibel, Emily Smith, Charles Kofi Adarkwa Nyamekye , Jonathon Bobbitt

Advancements in Raman spectroscopy and nanotechnology have enabled the construction and use of nano-structures and nano-modified surfaces, with Raman detection greatly enhanced by the generation of localized surface plasmons on the nanoscale. However there is need for applying Raman spectroscopy for characterization and detection on planar surfaces. SERS methods all in some way require a three-dimensional surface on the nano-scale. Many actual applications prefer the substrate surface of the system characterized to be two-dimensional or planar. We present instrument techniques utilizing propagating surface plasmons on planar metal surfaces producing the required electromagnetic enhancement allowing Raman spectroscopy measurements of thin films and monolayers. The surface assays presented show high reproducibility, quantitation, and sensitivity. This instrument allows spectroscopic detection of functional changes at a planar metal surface providing time-dependent binding information and simultaneously providing the chemical selectivity of Raman spectral detection. To distinguish from SERS using localized plasmons, we refer to the method as surface plasmon enhanced surface Raman spectroscopy (SPESRS). This measurement technique and product concept will utilize experimental and theoretical research developed more than 25 years ago, but long since over-looked. This work is combined with new technology along with an optical design for collection and detection of all the propagating plasmon-enhanced Raman scattered photons.

Neurological disease diagnosis with surface-enhanced Raman spectroscopy

Bhavya Sharma

Many health-related conditions are difficult to diagnose, including neurological changes due to stress, injury, or disease. New sensing technologies that are sensitive, selective, rapid, label-free, have multiplexing capabilities, and require little to no sample processing is an area of need. We believe surface-enhanced Raman spectroscopy (SERS), which is a highly sensitive, vibrational spectroscopy that can identify small molecules at low concentrations and provides fingerprint-like spectra can be used to aid in the early detection of disease. SERS requires little to no sample processing, is rapid, label-free, and SERS sensors can be designed to include multiplexing capabilities. We are investigating the detection of biomarkers of neurological activity in non-invasively collected biofluids. In particular, we are interested in detection of the effects of different neurological conditions on the concentrations of neurochemicals in the body using SERS.

The development of stable isotope probing with Raman spectroscopy for studying single cell bacterial metabolism

Roy Goodacre, Howbeer Muhamadali, Malama Chisanga

Bacteria and fungi very rarely live in isolation and are usually found in multi-species communities. Whilst, there is no doubt that microbes are an essential component of the maintenance of life?on earth, understanding the role of these microbes in situ?is currently a bottleneck as most microorganism can not be cultured in the laboratory. Stable isotope probing (SIP) is a potential tool for the characterization of bacteria at community and single cell level and linking organisms to their functional roles. This presentation will discuss our developments on the powerful combination of SIP with Raman spectroscopy as a way of quantifying the level of metabolism in bacterial cells at the single cell level by measuring the incorporation of combinations of the stable isotopes 13C, 15N and 2H into microbial biomass.

Raman spectroscopy as a tool for studying the role of lipids in prostate cancer
Lauren Jamieson, Mark Salji, Rachana Patel, Hing Leung, Karen Faulds, Duncan Graham

Raman spectroscopy is a powerful technique for biomedical application, allowing label-free imaging of biological samples at the molecular level. Biological lipid species are notoriously difficult to image in biology. Standard fluorescence spectroscopic techniques often use bulky fluorescent 'tags' that can interfere greatly with lipid distribution and metabolism, while techniques such as mass spectroscopy are destructive. Raman spectroscopy offers a label-free alternative, allowing lipid distribution and metabolism to be imaged intracellularly. In this work we characterise lipid species using the high wavenumber regions of the Raman spectrum, in contrast to the 'fingerprint' region typically focussed on in biological Raman experiments. We optimise high wavenumber Raman mapping to image and characterise biological lipid species in cell and tissue samples, with an aim towards gaining a new understanding of lipid metabolism in cancer as a new target for cancer therapy. Lipid synthesis is reported to be upregulated in cancer cells and, therefore, inhibiting lipid synthesis has displayed antitumoural activity.(1) We investigate the use of drugs that alter lipid synthesis, as a metabolic target for cancer therapy, using Raman spectroscopy. We compared the response of non-cancerous prostate, PNT2 cells, and cancerous prostate, PC3 cells to a variety of lipid altering drugs, with a view towards using this platform to screen for potential new drugs for metabolic cancer therapy that effectively target tumour over non-tumour cells. In addition, in collaboration with the Beatson Institute, we have employed our high wavenumber mapping technique to tissue samples from murine and human prostate, in order to better characterise and understand the effect of cholesterol level on response to castration. After establishing a robust Raman mapping protocol, the multivariate spectral data acquired was used to guide coherent anti-Stokes Raman imaging of single vibrational frequencies, in order to rapidly decrease acquisition time while obtaining the same analytical output, providing a platform more amenable to a clinical setting. (1) D.A. Tennant, R.V. Durán, E. Gottlieb, Targeting metabolic transformation for cancer therapy., Nat. Rev. Cancer. 10 (2010) 267-277.

Life, Death and Spectroscopy
Colin Campbell, Hannah Johnstone, Lauren Jamieson, David Harrison, Tim Morley

From a physical-chemistry perspective, life could be considered as a system that can maintain itself in a non-equilibrium state. Pretty simple really, but it's not a way that we typically think about biological systems. Similarly life as it applies to us (humans) is a 3D phenomenon, but most therapies are still tested using monolayers of cells growing on a plastic plate – that's not very physiological so its not surprising that therapies discovered using 2D cell culture don't always translate well in vivo.1 I will discuss recent advances in the use of spectroscopies to measure energetics in living systems (both 2D and 3D). Specifically I'll be discussing the use of Surface Enhanced Raman Spectroscopy (SERS) to measure redox potential and pH and the use of complementary techniques such as NMR and RNA sequencing to understand how redox and pH drive phenotypic changes at a molecular level. Our findings shed new light on the way that the physical microenvironment changes the fate of cancer cells.

Multi-modal imaging of erlotinib-nanoparticle conjugates in lung cancer cells.
Rachael Cameron

Small molecule inhibitors are frequently used to treat lung cancer, which causes the greatest mortality of all cancers worldwide. Clinicians have no short term means of assessing treatment response, other than to monitor symptom progression over time. During this period, non-responding patients' health can deteriorate before ineffective drugs have been identified. Nanomedicine is an emerging field demonstrating enhanced efficacies and could be one part of the solution, however penetration of cancerous cells and retention of the drug still requires investigation, especially in an era of personalised medicine. This research looks at the spatio-temporal distribution of anti-cancer drugs within lung cancer cells. Surface enhanced Raman spectroscopy (SERS) is used to investigate the drug-nanoparticle conjugates location. The Raman-silent region of a cell is from 1800-2800 cm-1. Localisation studies are possible by selecting the anti-cancer drug erlotinib. Erlotinib is an inhibitor containing an alkyne group that displays a molecular vibration unique from the cells fingerprint, it is active against the overexpressed epidermal growth factor receptor (EGFR) of cancerous cells. The distinctive Raman molecular fingerprint makes erlotinib inherently identifiable without modification with fluorescent markers or Raman reporters. This study aims to elucidate treatment efficacy by demonstrating drug localisation and retention in cancerous cells and highlight the benefits of nanomedicine drug delivery. SERS, TEM, fluorescence and other confocal techniques to correlate the conjugates' localisation and biomarker expression. The intention is to characterise the biological effect on downstream signalling pathways of the EGFR, preventing proliferation. In future, it is possible this technique could be used as a pre-clinical drug development platform to test nanomedicines.

High recognition specificity remote sensing of trace gases using IR/THz double resonance spectroscopy

Henry Everitt, Dane Phillips, Elizabeth Tanner, Frank De Lucia

IR/THz double resonance (DR) spectroscopy has been used as a laboratory tool to characterize collisional physics of gas phase molecular systems. We have recently proposed that this technique may be adapted for remote sensing of trace molecular gases in the atmosphere. Operation at atmospheric pressures and conditions presents several challenges including pressure broadening, atmospheric attenuation, and rapid molecular relaxation processes. We will show that atmospheric pressure broadening enhances the utility of this technique and generates a rich signature matrix that allows isotopic isomers to be easily distinguished at distances up to 1 km and concentrations as low as 1 ppm m. Attenuation by water vapor may be mitigated through the judicious selection of pump/probe frequency coincidences accessible at high pressures, and the ~100 ps relaxation timescales mitigate obfuscating atmospheric fluctuations. The frequency- and time-dependent effects of atmospheric pressures on DR signatures will be illustrated using prototypical methyl halide gases. The extension of this concept to remote sensing will be discussed with scenario-dependent performance predictions.

Open-Path Dual Comb Spectroscopy for Atmospheric Measurements

Nathan Newbury, Kevin Cossel, Eleanor Waxman, Gabriel Ycas, Fabrizio Giorgetta, Esther Baumann, Sean Coburn, Ian Coddington, Daniel HermanGregory Rieker

Dual-comb spectroscopy is a rapidly evolving new tool for broadband spectroscopy that exploits the advantages of frequency combs for precision spectroscopy. It relies on a laser frequency comb, whose output is composed of hundreds of thousands of perfectly spaced, discrete wavelength elements or comb teeth, which act as a massively parallel set of single frequency lasers with highly stable, well-known frequencies. In dual-comb spectroscopy, two such frequency combs are used with slightly offset repetition frequencies so that their interference on a photodiode leads to a measurable rf comb with a one-to-one correspondence with the optical comb teeth. If one, or both, of the frequency combs pass through a gas sample, then the samples absorption spectrum can be "read out" in the measured rf spectrum. The instrument resolution is quite narrow, set by the frequency comb teeth, while the spectral bandwidth can be quite broad, set by the frequency combs' bandwidth. In this way, dual-comb spectroscopy combines the advantages of tunable laser spectroscopy, e.g. of high resolution and long range, with the advantages of Fourier Transform Spectroscopy,
e.g. multi-species detection. We have developed a near-infrared dual-comb spectrometer that can be used over kilometer-scale open air paths to measure several atmospheric gases in the 1.6-micron spectral region with high precision and high accuracy. This system is being explored for regional carbon monitoring, methane leak detection, and spatial scanning (via a retroreflector mounted on a small quadcopter). In addition, we have extended the system operation to 2 microns with future coverage possible from 3 to 5 microns. This type of broadband open path dual-comb spectrometer will enable detection of an ever-wider range of molecules and isotopes. I will discuss the status of open-path dual-comb spectroscopy and its future potential.

Polarization-controlled white light LIDAR

Matthieu Baudelet, Shermineh Rostami, Martin Richardson

Filamentation is, simply put, a balance between nonlinear Kerr self-focusing and plasma defocusing creating a diffraction-free structure with high intensity core. One of the unique characteristics of these self-guiding light channels is the ability to transfer high intensities (?1013-1014 W/cm2) over extended distances, making them a promising tool for long range and/or stand-off applications such as and remote sensing. White light LIDAR has been the first sensing application of filamentation which takes advantage of the supercontinuum generated, and continuously broadened, along the propagation. This makes it difficult to obtain absolute absorbance measurements due to the lack of a spectral reference at the location of the scatterer. Creating the white light source at the location of the operator would take care of this problem. Such a broad source with the advantages of femtosecond LIDAR could be available using controlled filamentation in gas cells. We demonstrate a novel approach to creating this source by controlling the polarization of the filamenting pulse in molecular gases covering the spectrum from 500 to 900nm. The white light is generated using the Multi-Terawatt Femtosecond Laser (MTFL) facility at UCF. A molecular gas cell (nitrogen) is excited by a filament generated with elliptically polarized 50 fs, 3 mJ pulses with a spectrum centered at 800 nm. In order to cover an even broader range in the UV, we combined 400 nm beam with the 800 nm filamenting beam. The generated supercontinuum from both 800 and 400nm beams was then sent to a fully controlled 50-meter indoor propagation range. The LIDAR signal was collected by a 16" Ritchey-Chretien telescope coupled to a Czerny-Turner spectrometer equipped with an ICCD camera that could provide a 5ns (1.5 m eq.) temporal resolution. LIDAR signals using such white light will be shown for water as well as organic aerosols. This work is funded by the Army Research Office and the State of Florida.

Projecting High Power Density at Long Distance for Standoff Spectroscopy
Jean-Claude Diels

The challenges for performing standoff spectroscopy are reviewed. We will attempt to answer the following questions: • Can high resolution be achieved despite the limitation imposed by pressure broadening in the atmosphere? • Can the spectral lines be guided ef?ciently towards the detection? • Spectral detection: can the collection ef?ciency be maximized while preserving the resolution? • How to project the excitation source onto a remote sample? Remote focusing requires typically an extremely large aperture. Several approaches to create "diffractionless" beams, such as Bessel beams, require an equally large aperture. Exploiting the nonlinear properties of air to create self-guided beams of ?laments has been the anticipated solution for several decades. A number of unanticipated problems have delayed the practical implementation of this approach. A ?ne balance has to be found between the desired nonlinearity con?ning the pulse in space and time, and avoiding losses by dispersion, diffraction and undesirable nonlineari-ties. A solution is proposed that combines these requirements with affordability and compactness of the source.

Development and characterization of a short-wave infrared Conformal Filter hyperspectral imager for real-time standoff detection of materials
Shawna Tazik, Matthew Nelson, Patrick Treado, Srinivasa Narasimhan, Bernardo Pires, Martial Hebert

As autonomy proliferates into our daily lives, there is a growing demand for rapid, safe, and effective detection of materials in real-time with a high degree of autonomy. In an effort to address this need, ChemImage in collaboration with the Carnegie Mellon Robotics Institute has been developing a novel, adaptable, short-wave infrared (SWIR) hyperspectral imaging system for real-time standoff detection of materials, including hazardous materials (e.g., explosives, chemical warfare agents). At the heart of this system is the Conformal Filter (CF), which is a liquid crystal (LC)-based tunable filter that transmits multi-band waveforms. Building on concepts of multivariate optical computing, the CF is tuned electro-optically and dynamically to mimic the functionality of a discriminant vector for classification. The resulting integrated detector response approximates the detection response of conventional hyperspectral imaging with only two discrete measurements instead of hundreds to thousands. Real-time detection is achieved by operating two CFs in tandem within a dual polarization (DP) system, which exploits the polarization sensitivity of the LC filters and allows for simultaneous acquisition of the compressed hyperspectral imagery. This improved sampling rate coupled with advanced object recognition, semantic scene understanding, and image reconstruction algorithms enables on-the-move detection of targets. The development, characterization, and feasibility testing of the first generation SWIR DP-CF imaging sensor will be discussed, with a focus on its application to explosives and chemical threat detection.

Tip-enhanced Raman microscopy in deep UV

Atsushi Taguchi, Satoshi Kawata

Tip-enhanced Raman spectroscopy (TERS) is recently becoming indispensable as a nano-imaging technique for analyzing advanced nano-devices and nano-materials because it allows strong enhancement of weak Raman signals from the nanometric volume of a sample. The strong enhancement and photon localization is realized with plasmonic metallic tips. While gold and silver are commonly used as plasmonic metals, these metals behave absorptive dielectrics but not as metal in deep UV (DUV) wavelength region, which hampers the use of TERS microscopy in the DUV region where molecules such as nucleotides and proteins show characteristic resonance Raman scatterings. We have demonstrated TERS in the DUV using aluminum tip instead of gold and silver [1]. With the aluminum tip fabricated with vacuum evaporation and deposition, Raman scattering from adenine nanocrystal was enhanced with an enhancement factor of ~1,300 [1]. We have demonstrated the control of the localized plasmon resonance of aluminum and indium nanoparticles in the DUV wavelengths by changing the size [2,3]. Applications of DUV-TERS for imaging and analysis of nanomateirals such as nano-carbons and DNA will be shown. [1] A. Taguchi, et al, "Deep-UV tip-enhanced Raman scattering," J. Raman Spectrosc. 40, 1324–1330 (2009). [2] A. Taguchi, et al, "Tailoring plasmon resonances in the deep-ultraviolet by size-tunable fabrication of aluminum nanostructures," Appl. Phys. Lett. 101, 081110 (2012). [3] Y. Kumamoto, et al, "Indium for Deep-Ultraviolet Surface-Enhanced Resonance Raman Scattering," ACS Photonics 1, 598–603 (2014).

Monomeric Polyglutamine Structures that Evolve into Fibrils

Sanford Asher, David Punihaole, Ryan Jakubek, Riley Workman, Lauren Marbella, Patricia Campbell, Jeffry Madura

We investigate the solution and fibril conformations and structural transitions of the polyglutamine (polyQ) peptide, $D_2Q_{10}K_2$(Q10), by synergistically using UV Resonance Raman (UVRR) spectroscopy and Molecular Dynamics (MD) simulations. We show that Q10 adopts two distinct, monomeric solution conformational states, a collapsed ?-strand and a PPII-like structure that do not readily interconvert. This clearly indicates a high activation barrier in solution that prevents equilibration between these structures. Using metadynamics, we explore the conformational energy landscape of Q10 to investigate the physical origins of this high activation barrier. We develop new insights into the conformations and hydrogen bonding environments of the glutamine side chains in the PPII and ?-strand-like conformations in solution. We also use the secondary structure-inducing cosolvent, acetonitrile, to investigate the conformations present in low dielectric constant solutions with decreased solvent-peptide hydrogen bonding. As the mole fraction of acetonitrile increases, Q10 converts from PPII-like structures into ?-helix-like structures and ?-sheet aggregates. Electron microscopy indicates that the aggregates prepared from these acetonitrile-rich solutions show morphologies similar to our previously observed polyQ fibrils. These aggregates re-dissolve upon the addition of water! These are the first examples of reversible fibril formation. Our monomeric Q10 peptides clearly sample broad regions of their available conformational energy landscape. The work here develops molecular-level insight into monomeric Q10 conformations and investigates the activation barriers between different monomer states and their evolution into fibrils.

Hydrogen sulfide inhibits lysozyme fibrillation: DUV Raman spectroscopic study

Igor Lednev, Tatiana Quiñones-Ruiz, Manuel Rosario-Alomar, Juan López-Garriga

Proteins aggregation is of great biological and biochemical importance. Atomic-level structural characterization of protein aggregates is challenging because of the limitations of solution NMR and X-ray crystallography when applied to such insoluble, noncrystalline species. The problem is complicated further when protein aggregates occur in the dynamic equilibrium with the environment and cannot be separated and purified for the analysis without changing their structure and composition. We have demonstrated that Raman spectroscopy is uniquely suitable for structural characterization of proteins and protein aggregates. Specifically, deep ultraviolet (DUV) excitation allows for resonance enhancement of Raman scattering by amide chromophores, building blocks of the polypeptide backbone. The latter data reports on the distribution of the backbone dihedral angles and, consequently, the three-dimensional protein conformation according to Asher's semiempirical approach (J Phys Chem B 2006). Varying the excitation wavelength in the DUV region allows for probing the local environment of aromatic amino acid residues including phenylalanine, tyrosine and tryptophan. In combination with hydrogen-deuterium exchange and two-dimensional correlation analysis, DUV Raman spectroscopy became a powerful tool for studying protein aggregates' structure and dynamics. Specifically, we utilized DUV resonance Raman spectroscopy for probing the structure and dynamics of amyloid, insoluble aggregates of misfolded proteins, associated with many neurodegenerative diseases. Most recently, we investigated the effect of hydrogen sulfide (HS) on the formation of amyloid fibrils in vitro using a model fibrillogenic protein hen egg white lysozyme (HEWL). The addition of HS completely inhibits the formation of ?-sheet and amyloid fibrils, as revealed by DUV resonance Raman (DUVRR) spectroscopy. The inhibition mechanism will be discussed.

Sample photodegradation and protection in deep-UV resonance Raman spectroscopy

Yasuaki Kumamoto

Raman spectroscopy allows label-free, minimally-invasive molecular analysis of a biological sample, and has recently been used in biomedical research fields. Employing deep-UV light for the Raman excitation allows tremendous scattering enhancement of biological molecules up to 10^8 due to the electronic resonance, as well as fluorescence-free spectral acquisition due to the wavelength separation of Raman scattering and fluorescence of a sample. In deep-UV resonance Raman spectroscopy, however, sample photodegradation often occurs due to absorption of the deep-UV light by molecules in the sample, followed by ionization and reactions with surrounding molecules. This is the serious limitation of the technique in biomedical sciences. Here we present three approaches for sample protection from UV- and deep-UV-induced photodegradation. Trivalent ions of three lanthanide group elements, terbium, europium, and thulium, were added to a buffer solution immersing fixed cells. As a result, molecular photodegradation of the cells during deep-UV Raman measurement was significantly suppressed. Utilizing such protective effects of lanthanide ions, we successfully attained repetitive, two fold higher signal-to-noise ratio deep-UV resonance Raman imaging of cells. The underlying mechanisms of the protective effect can be explained by the energy relaxation of the excited molecules through the lanthanide ions before the excited molecules undergo ionization and/or reaction with surrounding molecules. Removing oxygen from the sample environment also suppressed molecular photodegradation of the cells during deep-UV measurement, by avoiding generation of reactive oxygen species around the excited molecules. Combination of these two techniques further suppressed the deep-UV-induced photodegradation of the cellular molecules. Additionally, we found that deep cooling of the sample was an approach for suppressing the molecular photodegradation under UV exposure, by decelerating the rates of photoreactions, which will in principal also work for a deep-UV-exposed sample. The three presented techniques can extend applications of deep-UV resonance Raman spectroscopy in biomedical sciences.

Deep Ultraviolet Resonance Raman (DUVRR) Spectroscopy of Protein Therapeutics

Sergey Arzhantsev, Chen Qiu

Quality control of formulated protein drugs presents unique analytical challenges compared to small molecule pharmaceuticals. Proteins are complex macromolecules with secondary and tertiary structures that may be easily disturbed, potentially resulting in loss of therapeutic efficacy. Deep ultraviolet resonance Raman (DUVRR) spectroscopy is capable of characterizing the higher order structure of proteins in the presence of excipients in aqueous media with minimal sample preparation. The resonance excitation provides an increase in Raman signal, specifically enhancing intensity of peptide vibration bands and completely removing fluorescence background. The formulated monoclonal antibodies used in the study underwent forced degradation with chemical and thermal stress and the resulting samples were analyzed using DUVRR spectroscopy. Chemical stresses included acidification and addition of surfactant to the formulated product. Changes were observed in the DUVRR spectra with each level of stress and multivariate statistical models were applied to identify trends within the data. A two-dimensional (2D) correlation analysis was also applied to the DUVRR spectra to visualize the detailed folding and unfolding processes of monoclonal antibodies.

Multi-channel detection for electrophoretic analysis of individual organelles

Edgar Arriaga, Erik Tyrrell, Katherine Muratore, Heather Brown

The electrophoretic analysis of isolated organelles provides a glimpse into the heterogeneity of the subcellular microcosmos. Such heterogeneity provides signatures with potential significance to understand cellular dysfunction in aging and disease. To describe such heterogeneities, we must conduct individual organelle measurements. Our laboratory introduced the use of capillary electrophoresis with laser-induced fluorescence detection (CE-LIF) for the electrophoretic separation and quantification of fluorescence reporters found in individual organelles. The approach has been successfully used for the analysis of mitochondria, lysosomes, autophagy organelles and nuclei in which a specific organelle type is detected/measured using one or two fluorescent reporters. One of the main limitations of the technology has been the inability to monitor more than one organelle type at the same time. To address these issues, we are currently developing technologies to monitor simultaneously three or more organelle types in a CE-LIF analysis. This presentation will give an update on the design and use of a modified 4-channel CE-LIF instrument for the analysis of up to four organelle types in a single CE-LIF run. The presentation will also mention technological issues that must be addressed for the successful implementation of this approach and emerging strategies for multi-parametric measurements of individual organelles.

Enhancing the Information Content of Single Cell Analysis on Microfluidic Devices using Optical Fiber Bridges

Christopher Culbertson, Damith Padabadige, Jalal Sadeghi, Jay Sibbitts

Microfluidic devices are changing the way cellular analysis is performed. They are capable of analyzing multiple analytes from single cell lysates at rates of up to 20 cells/min for >80 min. These rates are sufficient to generate statistically significant results in reasonable time periods. One potential application for these devices is the rapid profiling of kinase, protease, and cytokine activities in tumors. Tumor cells are heterogeneous due to the rapid mutation rates in the cells comprising and recruited to the tumor. Chemotherapy works best if the drugs used are optimized for the different cellular mutations contained in the cells in the tumor. Such information cannot be obtained from bulk cellular analyses. It must be obtained via the analysis of individual cells. We have developed a novel microfluidic device that integrates an optical fiber to provide 3 excitation and 3 detection spots using only 1 excitation source and 1 detector. This was made possible by taking advantage of a tunneling mode available in multimode fibers. With this detection system intact cell velocity, cellular lysis and lysate separation can all be tracked using a single detector. Being able to detect the lysis event and the completeness of the lysate injection is important as it allows one characterize how well a system is behaving, to monitor whether something begins to go wrong over the course of an analysis, and to identify analytes in a multicomponent separation based upon their absolute migration times. This device was used to Measure kinase, protease, and cytokine activity in cells.

Physical Properties of Bioparticles and High Resolution Separations with Dielectrophoresis

Mark Hayes, Shannon Hilton, Claire Crowther

Cells have a much more complex electric signature in the presence of an applied external field than is typically discussed in the introduction to most papers on electrokinetic manipulations of cells. This talk will focus on cataloging and characterizing theoretical and experimental physical features and the forces associated with them for cells undergoing electrokinetic movements. While much is known about the behavior of cells in electric fields, there also remains a good deal that is poorly characterized or completely unknown. The physical features are very important to understand as the resolution of cell separations improves, allowing unequivocal identification based on electrokinetic measurements alone. In addition to discussing the physical features of cells, new device designs which promise higher resolution will be presented.

Multi-spectroscopic NVU-on-a-chip system for developing efficient nanoparticle strategies for increased drug efficacy

Sagnik Basuray, Bhuvana Mohanlal, Victoria Harbour, Samuel Roy, Tarun Masimukku, Jocelyn Davis

Recently, the fields of microfluidics and biology have come together to produce the next generation of transformative devices collectively called Organ-on-chip. Organ-on-chip systems are at the frontier of preliminary drug screening, cell dynamics studies and toxicity testing, among others. There is increasing recognition that the full potential of organ-on-chip, especially for preclinical studies, lies in using the platform to represent 'whole body' physiology. This potential has compelled researchers to begin linking multiple organ-on-chip systems into a single entity. Successfully merging these systems requires understanding the role of electrokinetics, the intricate relationship between electrokinetics and cell seeding, the characteristics of cell growth and the formation of organ-mimicking cell layers using a multitude of cell types. These crucial aspects, however, have not been explored in detail. For example, application of well-known electrokinetic models, that explain phenomena arising from complex architecture of the chip, is needed to elucidate the cellular behavior of organ-on-a-chip systems. In terms of cell growth, it is not clear how shear stress distribution affects the growth of different cells types in an organ-on-chip system. Additionally, potentially useful electrokinetic and spectroscopic techniques are infrequently used or used with limited success in the devices. Dielectrophoresis, electrophoresis and others are not commonly used for building multicellular, organ-on-a-chip systems. Spectroscopy techniques, like electrochemical impedance spectroscopy and electrochemical testing, have been used in organ-on-chip with only limited success. These techniques are crucial to modeling cellular dynamics, for example, differentiating between the paracellular and transcellular processes in cell layers. In our lab, we are using well-established electrokinetic models and processes to look at an organ-on-chip approach to mimic the brain, specifically the neurovascular unit (NVU). To do so, we are combining shear studies with electrokinetic spectroscopic techniques, like electrochemical impedance spectroscopy, to develop protocols for building an NVU-on-chip. We have used an open microfluidic design with interdigitated electrodes to build an NVU-on-chip that has reproducible endothelial cell seeding. An electrokinetic technique, like dielectrophoresis, will be used to pattern the neuronal cells. This will allow us to seed neuronal cells that are physiologically relevant to the NVU. Further, the interdigitated electrodes facilitate a multiple spectroscopic technique that decouples the paracellular and transcellular pathways of the NVU. These analytical methods will enable us to develop nanoparticle design strategies based on the physical properties of nanoparticles like type, size, shape, charge and Zeta Potential using our NVU-on-chip. These design principles will increase the efficacy of drug-loaded nanoparticles to cross the blood-brain barrier leading researchers to target and cure neurodegenerative diseases like Alzheimer.

Improving the Separation Resolution of Insulator-based Dielectrophoresis

Claire Crowther, Mark Hayes

The field of separation science is driven by the desire to obtain concentrated, pure, and distinct targets from a diverse mixture of increasingly similar entities. Dielectrophoresis (DEP), a force resulting from non-uniform electric fields, exploits subtle differences in the electro-physical properties of analytes to achieve separations of micro and nano sized analytes. Insulator-based DEP (iDEP) utilizes various insulator shapes to achieve non-uniform electric fields. This technique has been used to manipulate several analytes including polystyrene spheres, cells, bacteria, proteins, DNA, and viruses. However, current insulator shapes do not induce similar forces across the lateral dimension of the microchannel, meaning that identical particles on different pathlines will experience different forces. The variations in the lateral forces results in partial trapping and low resolution separations. A novel design, that according to models and preliminary evidence, induces particle pathlines that improves separation efficiency and induces laterally-similar environments has been developed. Forty different insulator shapes were modelled and compared. To understand the forces relevant for separation, the electric field, gradient of the electric field squared, and the ratio of the two were examined. The resulting unique multi-length scale insulator system achieves separation forces comparable to currently utilized designs. The unique design is shown to compel all analytes to experience similar forces regardless of initial pathline, therefore improving separations. This results in higher resolution separations chiefly because partial trapping close to the insulating features will be minimized or eliminated. Predictive models will be shown and data presented which supports the behaviors noted in the models.

Glow Discharge Time of Flight Mass Spectrometry for fast multidimensional analysis of VOCs

Nerea Bordel, Jonatan Fandiño, Marcos Bouza, Jorge Pisonero, Alfredo Sanz-Medel

The capability to detect and identify VOCs in different samples is an issue of great interest in different research fields. In particular, it is known that tumour cells can generate unique Volatile Organic Compounds (VOCs) which can be localized in the surroundings of some altered tissues or, depending on the type of cancer, in the exhaled breath. Therefore, the detection of these compounds in exhaled breath can offer an excellent non-invasive diagnosis tool for some types of cancer. Although Glow Discharges (GDs) coupled to optical spectroscopy or mass spectrometry is an analytical technique typically used for direct solid sample analyses, the development of Pulsed Glow Discharges (PGDs) coupled to time of flight mass spectrometry (TOFMS) seems to offer interesting features for the analysis of VOCs. PGD-TOMS have demonstrated the capability of producing not only elemental but also fragmental and molecular ionization of different kind of samples introduced in the discharge. The PGD source is a dynamic plasma with different ionization mechanisms along the pulse period (e.g. pre-peak, plateau, and afterglow)2 and a proper use of these time regimes provides fundamental information to elucidate structures of different molecules. In this presentation, a new GD source specifically developed for the analysis of gas samples will be shown as well as the results obtained when different VOCs were analysed. This new design allows sample introduction close to the spatial afterglow where the plasma energy is low; therefore, the sample molecules are preserved of a complete fragmentation.

The volatile organic compounds used were aromatic hydrocarbons: benzene, toluene, ethylbenzene, xylene (BTEX). Mass spectra obtained during the plateau and the afterpeak provide information about the fragmentation patterns of the investigated molecules and their dependence with the time regime. Further, proton affinity nature of the molecules can also be investigated due to the enriched proton environment generated along the afterglow region. Finally, analytical characteristics were evaluated with solid phase microextraction-gas chromatography (SPME-GC) and with direct sample injection coupled to PGD-TOFMS for BTEX.

Recent Advances in Elemental Mapping through Glow Discharge Optical Emission Spectroscopy
Gerardo Gamez

Obtaining spatially resolved chemical information from solid surfaces is necessary to understand the underlying mechanisms of many systems, from novel materials to biological samples. On the other hand, current elemental mapping techniques suffer from throughput limitations, some requiring several hours or more for sufficient pixel density maps, and/or accessibility limitations, some requiring access to particle accelerator facilities for sufficient source brightness. Over the last few years, glow discharge optical emission spectroscopy has shown great potential for surface elemental mapping when the GD is operated in pulsed-power mode and at higher pressures. Typical GDOES advantages are maintained with the possibility of obtaining high-pixel density maps with at least three orders-of-magnitude higher throughput compared to typical techniques. Herein, the most recent developments in this area will be presented.

Comparison of VUV and NIR wavelengths for O and H in GD-OES compositional depth profiling
Arne Bengtson

In Compositional Depth Profiling (CDP) with GD-OES, a large number of applications require that the light elements C, O and H are measured. The most sensitive emission lines for O and H are found in the VUV spectral range at 130 and 121 nm respectively. For "low-end" instrumentation lacking VUV optics, alternative lines in the NIR spectral range are available; O 777 and 845 nm, H 656 nm. This spectral range has become more viable with the development of highly sensitive CCD detectors. In this work, the performance of the VUV and NIR lines for O and H are compared. The samples used are hot rolled steel and hot dip galvanized steel. It is shown that the CCD – NIR alternative is adequate for these CDP applications.

Compositional Depth Profile Analysis by Radio Frequency Glow Discharge Utilizing a Solid-State Spectrometer
Kim Marshall

Glow discharge spectrometry (GDS) has proven to be a very capable "bulk" analysis technique. However, it is the analysis of surface layers, also known as compositional depth profiling (CDP), where GDS truly excels. Historically, CDP capable glow discharge instruments have utilized multichannel photomultiplier (PMT) based detection. This is because solid-state CCD detectors have been considered either too slow or too insensitive for these types of analyses. This is particularly true when such analyses involve so-called "thin layers", as for example, in the case of measuring the magnetic layers of a computer hard disk. In 2013 we reported work in this forum [1] showing that a solid-state spectrometer is certainly capable of analyzing multilayer structures on computer hard drives. However, this work was carried out using DC-glow discharge. One of the factors that we noted at the time was that a radio frequency source (RF-GDS) might be better suited to this kind of analysis. In particular, the plasma startup should be significantly quicker. In the work presented here, a series of various layered materials will be analyzed by RF GDS on a newly released solid-state spectrometer, the GDS900. CDP analysis of thin layer materials such as the magnetic layers of hard drives will be presented and these results compared with those obtained by PMT based instruments as well as those obtained using DC glow discharge sources. 1) K. Marshall, G. Schilling and S. Chrispell - invited presentation, SCIX Milwaukee, WI (2013)

On the roles of metastable and quasi -metastable species in atmospheric noble gas dielectric barrier discharges

Felix David Klute, Sebastian Burhenn, Pascal Vogel, Antje Michels, Charlotte Lewis, Daniel Thurston, Beatrix Biskup, Paul Farnsworth, Joachim Franzke

The spatio-temporal development of capillary dielectric barrier discharges that are used for analytical applications was subject of detail studies in the recent years[1,2]. These studies have shown that different plasma phases develop that can be associated with different excitation mechanisms. The soft ionisation of molecules was associated with the early jet formation stage of the discharge while the dissociative excitation
of molecules could primarily be detected in the later coincident stage of the discharge. Although the characterization of these stages is already advanced and it is possible to either enhance or supress certain parts of the discharge[1] it is still not clear which reactive species are responsible for the different observables in the discharge stages. Time and space resolved measurement of the 1s2s ^{3}S He metastable state observed via laser
absorption of the 1083.03 nm transition reveal that this state plays a major role in the dissociative excitation of the coincident plasma stage. On the other hand these metastable states are most likely not involved in the ionisation of gas impurities that is a crucial factor in the formation of positive species in the atmospheric noble gas DBD as was stated in prior works[2]. A streamer like process driven by photoionisation of neutral N_2
seems to be more likely taking place instead of the penning collisions of N_2 with the 1s2s ^{3}S Hem. The source of the necessary high energetic UV radiation is in all probability the He* 1s2p 1P state that recombines resonantly to the ground state 1s2 ^{1}S He via the emission of 58.43 nm. This state is effectively trapped under
atmospheric pressure conditions due to constant reabsorption and reemission of the 58.43 nm line making it a quasi-metastable long living state until the photon is absorbed by a neutral N_2 that in turn becomes ionised

N_2^+. [1] A. Schütz, F. D. Klute, S. Brandt, S. Liedtke, G. Jestel and J. Franzke, Anal. Chem., 2016, 88, 5538-5541. [2] Klute, F. D.; Michels, A.; Schütz, A.; Vadla, C.; Horvatic, V.; Franzke, J. Anal. Chem. 2016, 88, 4701-4705. [3] Klute, F. D.; Schütz, A.; Michels, A.; Vadla, C.; Veza, D.; Horvatic, V.; Franzke, J. Analyst 2016, 141, 5842-5848.

Time-domain inelastic x-ray scattering from short-wavelength phonons

David A. Reis

X-ray free-electron lasers (FELs) are more than a billion times brighter than other synchrotron-based light-sources. Their peak brightness arises from a combination of a huge number of spatially coherent x rays packed in femtosecond duration pulses. This combination makes them particularly attractive for studying atomic-scale dynamics. Here we present a novel application of x-ray FELs to probe non-equilibrium lattice dynamics initiated by ultrafast optical excitation. This produces a spectrum of excited phonons spanning the Brillouin zone. Inelastic scattering from the nonequilibrium phonon distribution is probed directly in the time-domain with high spectral resolution. In addition to providing information about the transient state, it also gives parallel access to a wide range of wave vectors, all without the need for high-resolution crystal monochromator and analyzers. We give a brief overview of recent results and speculate on what may be possible in the near future.

Multidimensional Spectroscopic Probes of Vibrational and Vibronic Coherence in Photosynthetic Systems

Jennifer Ogilvie

Since the first reports of vibrational coherence in photosynthetic systems by Vos et al.[1]. there have been open questions about its possible functional significance. More recently, there has been renewed interest in coherent processes in photosynthetic systems following observations made by two-dimensional electronic spectroscopy (2DES)[2]. An outstanding challenge is to understand the physical origin and functional significance of the observed coherent dynamics. 2DES, which resolves the third order response of the system as a function of both excitation and detection frequency provides rich information about the physical origin of coherent processes[3-5]. I will discuss the use of multidimensional spectroscopies for studying coherence in photosynthetic systems, and summarize our current understanding of the physical nature of the observed coherence and its possible functional role in photosynthesis. 1. M. H. Vos, J. C. Lambry, S. J. Robles, D. C. Youvan, J. Breton and J. L. Martin, Proceedings Of The National Academy Of Sciences Of The United States Of America, 1991, 88, 8885-8889. 2. G. S. Engel, T. R. Calhoun, E. L. Read, T. K. Ahn, T. Mancal, Y. C. Cheng, R. E. Blankenship and G. R. Fleming, Nature, 2007, 446, 782-786. 3. V. Butkus, D. Zigmantas, L. Valkunas and D. Abramavicius, Chemical Physics Letters, 2012, 545, 40-43. 4. D. B. Turner, R. Dinshaw, K.
K. Lee, M. S. Belsley, K. E. Wilk, P. M. G. Curmi and G. D. Scholes, Physical Chemistry Chemical Physics, 2012, 14, 4857-4874. 5. F. D. Fuller, J. Pan, V. Butkus, A. Gelzinis, S. Senlik, D. E. Wilcox, C. F. Yocum, L. Valkunas, D. Abramavicius and J. P. Ogilvie, Nature Chemistry, 2014, 6, 706-711.

Monolayer Magnets
xiaodong xu

Since the discovery of graphene, the family of two-dimensional (2D) materials has grown to encompass a broad range of electronic properties. However, until recently 2D crystals with intrinsic magnetism were still lacking. Such crystals would enable new ways to study 2D magnetism by harnessing the unique features of atomically-thin materials, such as electrical control for magnetoelectronics and van der Waals engineering for novel interface phenomena. In this talk, I will describe our recent magneto-optical spectroscopy experiments on van der Waals magnets, chromium(III) iodide CrI3. I will first demonstrate the existence of isolated monolayer semiconductor with intrinsic Ising ferromagnetism. I will then show the layer number-dependent magnetic phases. The magnetic ground state evolves from being ferromagnetic in a monolayer, to antiferromagnetic in a bilayer, and back to ferromagnetic in a trilayer and thin bulk. Lastly, I will discuss the emerging spin phenomena in monolayer WSe2/CrI3 ferromagnetic semiconductor heterostructures, including ferromagnetic control of valley pseudospin in WSe2 via large magnetic exchange field, and optical analog of giant magnetoresistance effect.

Ultrafast reversal of the ferroelectric polarization
Roman Mankowsky, Alexander von Hoegen, Michael Först, Andrea Cavalleri

Mid-infrared optical pulses can resonantly drive selected vibrational modes in solids to deform their crystal structure. In complex oxides, this method has been used to melt electronic orders, drive insulator-to-metal transitions, or induce superconductivity. In a recent set of experiments, we found signatures of light-induced ferroelectric polarization reversal in LiNbO3 on femtosecond timescales. Instead of coupling to the polarization with an electric field, we used high field mid-infrared pulses to resonantly excite the highest frequency c-axis vibrational mode. Phase sensitive second-harmonic generation measurements revealed that following excitation the polarization transiently switches sign with a magnitude that is 40% of its equilibrium value before switching back. Density Functional Theory calculations predict a strong coupling of the excited mode to the atomic motions, which entail the ferroelectric polarization, providing a possible microscopic explanation of the observation.

Raman spectroscopy of collective excitations in correlated electron systems
Girsh Blumberg

Interaction between electrons in “strongly correlated materials” is associated with many scientifically important and technologically useful phenomena, including superconductivity and exotically ordered charge and spin phases. I will review the recent advances of Raman spectroscopy aimed to characterize these extraordinary self-organized electronic phases, determine the condition under which they are formed, identify the underlying microscopic mechanisms, and test the relevant theories.

Using novel deep Raman approaches to measure multiple characteristics of biomaterials at depth in scattering media / mammalian tissues
Nick Stone, Ben Gardner, Pavel Matousek

Deep Raman spectroscopic approaches are developing rapidly. Transmission and spatially offset approaches have been shown to have significant value for use in biomedical diagnostics. One of the benefits of using spatially offset Raman spectroscopy (SORS) has been the ability to identify the approximate depth of the inclusion. Here we introduce methods for depth identification using purely transmission Raman (TRS) measurements. Furthermore with SERS labels we can use SESORS approaches to provide information on both temperature and pH at depths of many cms. Combining all of these approaches allows us to provide probe depth, pH and temperature using SE-TRS.

Dynamic SERS imaging of intracellular environment
Kazuki Bando, Jun Ando, Nicholas Smith, Katsumasa Fujita, Satoshi Kawata

Surface-enhanced Raman scattering (SERS) on a metallic nanostructure enables us to detect molecular vibrational information with high sensitivity. The combination of SERS with a metalized atomic force microscopy (AFM) probe can be utilized for the surface analysis such as such as CNT, strained silicon with a spatial resolution on the metal nanometer scale. Here we report a technique to detect SERS from the inside of a sample such as a living cell by using a metal nanoparticle as a probe. Since metal nanoparticles can be taken into the cell body by endocytosis, SERS from locally existing biomolecules in the body can be detected. Following the endocytosis, metal nanoparticles are transported in the cell body associated with cellular functions. Continuous SERS measurement during the transportation of a particular metal nanoparticle provides the chemical and environmental information during the transportation pathway. In order to obtain the SERS signal continuously, we built a dynamic SERS imaging system, which detects the position of the metal nanoparticle in 3D, coupled with a spectrophotometer to measure SERS around the metal nanoparticle simultaneously. We employed a dark-field imaging system to obtain the position of the metal nanoparticle as a bright spot. The position information of the metal nanoparticle was recorded and a laser beam was kept focusing on the metal nanoparticle to obtain SERS spectra from the metal nanoparticle. We applied dynamic SERS imaging with using HeLa cell and achieved with a position accuracy of a few tens nanometer and a temporal resolution of 100 ms. The SERS spectra measured from the tracked metal nanoparticle changed with time, indicating the drastic change of the chemical environment during the intracellular transportation. We applied principal component analysis (PCA) and mean square displacement (MSD) analysis to the SERS dataset to extract the meaningful spectra that are associated with the motion of the metal nanoparticle. The combination of PCA and MSD analysis extracted the temporal behavior of SERS spectrum during the intracellular transportation.

Surface Enhanced Raman Spectroscopy (SERS) Optofluidics for Whole Cell Analysis
Marjorie Willner, Kay McMillan, Rachael Cameron, Duncan Graham, Peter Vikesland, Michele Zagnoni

Rapid single-cell surface enhanced Raman spectroscopy (SERS) optofluidics is an ideal method to measure spatial and temporal variation of whole cells in a non-destructive fashion. We have successfully prototyped a two-phase microfluidic based platform with droplet storage that allows for the screening of over one thousand individual droplets leading to the collection of high volumes of data (10,000 spectra/hour). For proof of concept we utilized an established biosensor assay with readily visible cancer cell targets enabling us to focus on the novel aspects of our SERS optofluidic platform including nanoprobe development, controlled sample handling, and development of in-house algorithms to automatically process and analyze SERS maps. Specifically, we studied the surface expression of N-acetyl neuraminic (sialic) acid residues on cancerous (PC3) and non-cancerous prostate cells (PNT2) using carbohydrate (lectin) functionalized gold nanoparticles. Droplet microfluidics, when combined with a confocal SERS setup, enables the 3D imaging of intact whole cells. From XY and XZ raster scans, high resolution maps are rendered and then mined using computer vision methods. Analyzing the SERS data from a whole map perspective enables us to shift the discussion of SERS as an imaging tool. Recently, we have applied our SERS optofluidic platform to study and image the uptake and potential metabolism of the anti-cancer drug Erlotinib. Using a label free assay, we exploit the inherent SERS signals of both Erlotinib and the cancer cells to simultaneously track drug delivery and release within the cell. Our work demonstrates the efficacy of using SERS optofluidics to investigate and image biologically active molecules both on the cell membrane and intra-cellularly.

Engineering Non-Cadmium I-III-VI Quantum Dots for Bioimaging and Sensing
Xiaoshan Zhu

Currently, a major portion of QDs are synthesized with cadmium, and coated with polyethylene glycol (PEG) for QD bioavailbility. Cadmium is toxic. PEG is sensitive to solution pH, salinity and oxygen levels, and tends to cause QDs to aggregate in physiological media, acidic, or salt-rich environments, degenerating the QD quality. I-III-VI QDs ($AgInS2/ZnS$ and $CuInS2/ZnS$ core/shell structures) possess low toxicity, and are a logical replacement for cadmium-based QDs. We have been working on the development of I-III-VI QDs for bioimaging and sensing. Our synthesis of I-III-VI QDs is based on thermal decomposition of less toxic precursors and can be easily scaled up for mass production for sustainable and reliable imaging and sensing experiments. Through nonstoichiometric composition adjustment and element doping into QDs, we synthesized I-III-VI QDs with reliable and controllable optical properties, including high QYs (> 40%) and long PL lifetime (sub-??s to ms). We also developed new zwitterionic amphiphiles and applied them to encapsulate I-III-VI QDs to achieve colloidal stability in proteinaceous solutions with wide pH/ionic ranges, low non-specific binding, and easily bio-conjugation. The surface modification also guarantees the QD super-stability against amphiphile-coating-layer degradation in cellular and tissue microenvironments. On the basis of these developments, we applied our I-III-VI QDs in cellular imaging to detect endogenous targets expressed on human brain tumor cells, and also demonstrated their applications in time-gated or time-resolved lateral flow immunoassay to diagnose melioidosis. Additionally, coupled with magnetic nanoparticles or magnetic element doping, we adopted our QDs to develop composite magnetofluorescent dual-imaging contrasts for both non-invasive magnetic resonance imaging (MRI) and fluorescence imaging.

Biosensors for Cellular Nucleic Acids Detection Using SERS
Pietro Strobbia, Bridget Crawford, Hsin-Neng Wang, Tuan Vo-Dinh

Single cell analysis allows for the study of molecular events responsible for cellular functions and can unveil their connections to the organism biology. Biosensors based on surface enhanced Raman scattering (SERS) can be used for these analysis and offer several advantages over other optical sensing platforms, including sensitivity and multiplexed capabilities. Nanosensors/nanoparticles have been used for SERS analysis in cells, however, the delivery of such biosensors relies on their uptake, which requires long incubation times and has different efficiencies among cell lines. Sensors based on tapered optical fibers can instead be inserted in single cells to detect target molecules. The combination of this technology with the sensing mechanism of nucleic acid based nanoprobes (i.e., inverse molecular sentinels) will permit the direct detection of nucleic acids within single cells. This paper presents the current development in tapered fiber-based biosensors for the detection of nucleic acid biomarkers in plant cells. The use of inverse molecular sentinels in plant cells was demonstrated. Sensors based on tapered fibers were fabricated and used to measure SERS from a single cell.

Development of a Comprehensive Flavonoid Analysis Computational Tool for Ultra High-Performance Liquid Chromatography-Diode Array Detection-High Resolution Accurate Mass-Mass Spectrometry Data

Mengliang Zhang, Jianghao Sun, Pei Chen

Liquid Chromatography and mass spectrometry methods, especially ultra-high performance liquid chromatography coupled with diode array detection and high resolution accurate-mass multi-stage mass spectrometry (UHPLC-DAD-HRAM/MSn), have become the tool-of-the-trade for profiling flavonoids in foods. However, manually processing acquired UHPLC-DAD-HRAM/MSn data for flavonoid analysis is very challenging and highly expertise-dependent due to the complexities of the chemical structures of the flavonoids and the food matrices. A computational expert data analysis program, FlavonQ, has been developed to facilitate this process. The program uses UV-Vis spectra for an initial step-wise division of flavonoids into classes and then identifies individual flavonoids in each class based on their mass spectra. Step-wise identification of flavonoid classes is based on a UV-Vis spectral library compiled from 146 flavonoid reference standards and a novel chemometric model that uses step-wise strategy and projected distance resolution (PDR) method. Further identification of the flavonoids in each class is based on an in-house database that contains 5686 flavonoids analyzed in-house or previously reported in the literature. Quantitation is based on the UV-Vis spectra. The step-wise classification strategy to identify classes significantly improved the performance of the program and resulted in more accurate and reliable classification results. The program was validated by analyzing data from a variety of samples, including mixed flavonoid standards, blueberry, mizuna, purple mustard, red cabbage, and red mustard green. Accuracies of identification for all samples were above 88%. FlavonQ greatly facilitates the identification and quantitation of flavonoids from UHPLCHRAM-MSn data. It saves time and resources and allows less experienced people to analyze the data.

Fingerprinting Food Ingredients For Authentication By Portable Vibrational Spectroscopy Technologies

Luis Rodriguez-Saona

Economic adulteration and counterfeiting of foods may cost the industry an estimated $10 to $15 billion per year resulting on reduced market size, increased costs (recall, liability, withdrawals), lost revenue or market share, damaged brand, and failed business or bankruptcy Current methods for testing adulterants and contaminants in foods are time-consuming, expensive, and labor-intensive, requiring complex procedures of sample pretreatment and well-trained technicians to operate expensive instrumentation. Portable and handheld vibrational spectroscopic techniques can provide unique fingerprinting capabilities to establish reliable authentication programs for high-value food ingredients vulnerable to economically motivated tampering or unintentional contaminants produced during food processing (i.e. acrylamide in snack products). This in turn would allow for untargeted detection of economic adulteration through monitoring any deviations in the unique profiles of pure raw ingredients using pattern recognition analysis. Testing done close to the original source and throughout the process would permit detecting adulteration before an ingredient has been diluted or combined with other ingredients. Our sensor technologies are directed at improving efficiency, throughput and reliability of critical processes, such as monitoring incoming raw materials and making "real-time" processing modification that could improve the quality of their products. The algorithms developed for screening adulteration and contaminants benefits food companies that will be able to implement rapid testing to streamline food quality assurance and will curtail the growing danger of food ingredient fraud due to the unconventional adulterants or substituted products used for these activities.

Rapid Prediction of Milk Powder Adulteration Based on Portable Vibrational Spectroscopic Devices

Betsy Jean Yakes, Sanjeewa R. Karunathilaka, Keqin He, Lea Brückner, Magdi Mossoba

Since the melamine crisis in 2007-2008 that caused illnesses in pets in the United States and renal complications in thousands of children in China, researchers have sought to develop rapid and field-usable methodologies to prevent adulterated milk products from reaching the public. Portable devices that use Raman and near infrared (NIR) spectroscopic readout are capable of on-site sample evaluation. While initial studies were for targeted detection of melamine, current state-of-the-art research is focused on creating nontargeted analytical methods that can not only detect the previously used adulterants but also those that are unknown. Challenges to developing these methods have primarily been due to the complex, natural variability of milk powder as well as the inability to spectrally identify an unknown contaminant. As such, chemometric data analysis is essential to the rapid screening of milk-powder, where diverse, authentic products are used to create a library of spectral signatures that can be compared to an unknown test sample for prediction of its authenticity. This presentation will highlight research based on spectra collected from three handheld devices and compare their performance to previously developed Raman and NIR chemometric models for benchtop instruments. Specifically, a portable Raman spectrometer that has fluorescence reduction was studied to determine its ability to overcome the natural background of milk powders, a handheld Raman spectrometer with on-board chemometrics was evaluated to determine if a fully field-deployable technique was achievable, and a handheld NIR analyzer was investigated to determine if its low resolution due to instrument miniaturization affected its ability to discriminate adulterated samples from authentic milk powders.

Greater Than The Sum of Its Parts—Exploiting Ambient Ionization Mass Spectrometry-derived Chemical Fingerprints for Complex Matrix Identification
Rabi Ann Musah, Justine E Giffen, Kristen L Fowble, Meghan G Fogerty

A continuing challenge in analytical chemistry is the identification of complex matrices, particularly those found in nature, such as animal-derived products and plant materials. In general, attempts to characterize complex matrices rely on identification of diagnostic singular components such as biomarkers, a process which requires extensive sample preparation. The steps involved, which can include solvent selection, extraction procedures and method optimization, cause bottlenecks in the analysis of composite materials. This negatively impacts several fields, including forensics, agriculture, and the food industry. However, the development of approaches to accomplish this task also has the potential to lead to new discoveries in other areas such as plant biochemistry, chemical ecology and environmental science. We posit that the ability to identify highly complex materials is greatly amplified by considering a broad range of detected molecules. In this regard, direct analysis in real time-high resolution mass spectrometry offers unique opportunities to routinely and rapidly detect molecules in complex matrices that span the dielectric constant spectrum. Furthermore, the application of chemometric processing tools to the mass spectral data enables identification of many types of materials. We demonstrate here how this approach can be applied to identification of illegally traded endangered woods, characterization of early life stages of necrophagous insects that are important in assessment of time-since-death in forensics investigations, studies of the chemical ecology of carrion flowers which mimic the smell of a decaying carcass, the identification of psychoactive plant drugs used as “legal highs” and mapping of small molecule spatial distributions in various materials.

Detection of Nitrogen-Containing Pollutants with a Chip-Based Sensor
Peng Zheng, Nianqiang Wu

Nitrogen-containing compounds are widely used in agriculture, pharmaceutical, manufacturing industries, etc. However, some compounds are either intrinsically hazardous or can be gradually transformed into hazards, leading to increasing health and safety concerns. This talk will show our recent efforts to develop a portable chip-based device for detection of nitrogen-containing hazard using surface-enhanced Raman scattering (SERS). The SERS sensor device is built upon hierarchical plasmonic nanostructures, which is optimized with assistance of finite-difference time-domain simulation.

Correlating Biophysical Properties of Amyloid Aggregates in vitro and in vivo by AFM-based Infrared Nanospectroscopy
Francesco Simone Ruggeri, Chris Dobson, Michele Vendruscolo, Tuomas Knowles

Today, more than 40 million people worldwide are affected by neurodegenerative disorders. Onset of these and more than other fifty human diseases is associated at the molecular level with the misfolding and aggregation of proteins into insoluble fibrillar protein aggregates, termed amyloids. The discovery of this link has caused a widespread interest in the biophysics and biochemistry of proteins aggregation both in vitro and in vivo. However, the molecular mechanisms of the disease onset remain elusive and there is contrasting evidence in favor of protective or by contrast toxic role of the large proteins intracellular inclusions or extracellular plaques in the brain. This knowledge gap is largely due to the characteristic heterogeneity and nanoscale dimensions of amyloid pathological aggregates, which preclude the routine use of biophysical assays for fundamental investigations of the correlation between the amyloidogenic nature of the aggregates in vitro and the molecular mechanism of toxicity in vivo. In the present work, we show that the combination of Atomic Force Microscopy and Infrared spectroscopy (AFM-IR) can be used to study at different biological scales the correlation between the chemical and structural properties of amyloid aggregates in vitro with their cytotoxicity in vivo. First, we will give example of the advance characterization at the single molecule scale of the structural properties of amyloid aggregates. We will focus on the systematic investigation of the heterogeneity and polymorphism of aggregated amyloid proteins when subjected to post-translational modification or when incubated with prescreened drugs able to inhibit or change their aggregation pathway. Then, as main step forward, we will expand the in vitro approach to characterize the structural properties of proteins aggregates inclusions in living organisms, investigating at the nanoscale the heterogeneity and the structural organization of intracellular and extracellular aggregates in whole organisms or in ex vivo tissues. The development of novel biophysical methodologies, able to bridge the investigation of morphological and structural properties of amyloid aggregates formation in vitro and ex vivo, is a fundamental avenue to address the challenge of developing pharmacological approaches able to delay or cure neurodegenerative disorders and effectively improve the lives of affected patients.

Advanced IR Nanospectroscopy to study lipids bodies in micro-organisms: toward a better understanding of metabolic pathways at stake.
Ariane Deniset-Besseau, Rolando Rebois, Alexandre Dazzi

Infrared spectroscopy is a powerful tool to specifically probe molecular vibrations of organic components without exogenous labelling and becomes even appealing to cellular and tissue biology. The coupling with microscopy and all the improvements done in the last decade have reinforced this trend. The main drawback is still remaining: the resolution limited by the diffraction. The solution to the problem was in the side of the near-field techniques. After several years of research in the field and considering all the previous methods limitations, we have developed at our lab an innovative infrared nanospectromicroscopy : AFMIR (Alexandre DAZZI, patent 2007) 1,2. AFMIR is a cutting-edge near-field technique coupling an atomic force microscope (AFM) with a tunable pulsed IR laser. This allows IR mapping of molecules of interest within a sample with a sensitivity of ten of nanometer. We will present the study of lipid bodies accumulated in the different microorganisms (bacteria, microalgae and yeast). Those lipid accumulations consist mostly in triacylglycerols (TAG) and steryl esters. It was possible thanks to our AFM-IR system to create sub-cellular chemical maps that allows label-free identification of TAGs inclusions in the cytoplasm. To carry out these studies new set-up configurations were required as well as the use of correlative imaging. We will also discuss the capability of AFM-IR technique to provide new insights into the constitution of the fatty inclusions and the role of TAGs in the morphological and metabolic differentiation that characterize those micro-organisms.

AFM-IR explores a chromatin role in formation of chromosomal aberrations

Ewelina Lipiec

The chromatin structure affects almost every process within the eukaryotic cellular nuclei [1] and it may play crucial role in the appearance of aberrations following damage after ionising radiation exposure. Previous studies confirmed, that critical DNA lesions, leading to chromosome breakage, are double strand breaks (DSBs) [2, 3], but the chromatin structure influence on their processing into the chromosome damage was not explored due to instrumental and methodological limitations. Hitherto, there are no reliable tools to decipher the molecular composition of chromosomes without additional staining. We have applied infrared nano-spectroscopy (AFM-IR) to generate infrared images of single chromosomes revealing detailed information on their molecular structure and composition. Principal Component Analysis (PCA) treatment of AFM-IR data has confirmed that chromosome areas containing euchromatin and heterochromatin are distinguishable based on differences in a degree of methylation. Heterochromatin has significantly more methylated sites compared to euchromatin. This approach allowed for a detection of DNA methylation in human chromosomes providing the location of the heterochromatin–euchromatin boundary with the spatial resolution reaching 12 nm (the order of AFM tip radius). An investigation into chromosomes extracted from cells treated by 3 Gy of 60 Mev protons confirmed that a distribution of chromosome aberrations in euchromatin and heterochromatin is not random. Infrared nano-scpectroscopy and Infrared nano-imaging were applied in order to localize places of chromosomal aberration formation such as chromosome breakage and areas of fusion in dicentric chromosomes. Full chemical information was obtain from areas of our interest. We detected an increased intensity of amide bands in areas of chromosome fusion, probably related with high protein expression involved in DNA damage repair. References: [1] T. Jenuwein, C. D. Allis, Science, 2001, 294, 2477. [2] G. Obe, P. Pfeiffer, J. R. K. Savage, C. Johannes, W. Goedecke, P. Jeppesen, A.T. Natarajan, W. Mart?nez-López, G.A. Folle, M.E. Drets, Mutation Research, 2002, 504, 17. [3] E. Lipiec, R. Sekine, J. Bielecki, W. M. Kwiatek, B. R. Wood, Angewandte Chemie, 2014, 53, 169.

In vivo AFM-IR nano-spectroscopic investigation of the bacterial cell wall

Philip Heraud, Kamila Kochan, David Perez-Guaita, Julia Pissang, Jhih-Hang Jiang, Anton Peleg, Bayden Wood

Atomic Force Microscopy Infrared (AFM-IR) spectroscopy is a rapidly evolving technique combining macromolecular compositional analysis, offered by IR spectroscopy, with nanoscale spatial resolution, enabled by Atomic Force Microscopy (AFM). AFM-IR has been successfully applied for the study of synthetic polymer composites, being often a mixture of various components and aiming at high quality performance in terms of strength or ductility. Here we extended this approach to bacterial cell walls, which are essentially natural polymeric composite structures. Six strains of bacteria were studied: *Staphylococcus aureus* G(+), *Bacillus subtilis* G(+), *Enterococcus faecalis* G(+), *Escherichia Coli* G(-) *Pseudomonas aeruginosa* G(-) and *Acinetobacter baumannii* G(-). Principal Component Analysis (PCA) was able to easily discriminate G(+) from G(-) bacteria based on marker bands from cell wall components. The most distinguishing feature of G(+) was a series of intense bands located at 1061, 1091 and 1116 cm^{-1}, assigned to complex sugar modes from the

peptidoglycan layer. In addition, contributions from vibrational modes from teichoic acid, a G(+) specific component of cell wall, were observed with bands at 1092 cm^{-1} ($?_s PO^{2-}$) and 1216 cm^{-1} ($?_{as} PO^{2-}$). We demonstrated, using *E. coli*, that even in the case of G(-) bacteria, which possess very thin cell walls
compared to G(+) bacteria, that it was still possible to study cell wall chemistry using AFM-IR spectroscopy. In addition, we report for the first time the use of AFM-IR spectroscopy to observe structural changes occurring in living cells. Specifically, we monitored the formation of the cell wall septum in living *S. aureus* cells prior to division, with the AFM-IR spectra of the newly forming structure clearly showing steadily increasing intensity of bands attributed to peptidoglycan and teichoic acid.

Nanoinfrared spectroscopy of cell membranes and extracellular vesicles

Leonetta Baldassarre, Valeria Giliberti, Alessandro Nucara, Paolo Calvani, Alessandro Rosa, Valeria De Turris, Mattia Musto, Loredana Casalis, Eglof RitterMichele Ortolani

Infrared absorption spectroscopy can provide label-free non-destructive identification of organic macromolecules "in vitro" but, even in its microspectroscopy version performed with synchrotron radiation, it lacks the necessary lateral resolution to unambiguously address biological nanostructures, such as the cell membrane patches or extracellular vesicles. Here we show that nano-infrared spectroscopy based on the photo-thermal expansion of samples illuminated by tunable quantum cascade lasers (11 to 5.5 micron wavelength range, 900-1900 cm^{-1}) under a scanning probe tip can provide true nanoscale resolution, if accomplished on ultra flat gold substrates and on isolated particles. Sub-micron sized single cell-membrane flakes are studied and the spectra of transmembrane proteins at different hydration levels are discussed. Single extracellular vesicles of 200 nm diameter are spectroscopically investigated to identify their nucleic acid cargo. Nano-infrared mapping of HeLa cells at the amide-I absorption (1660 cm^{-1}) will also be discussed.

Biological sex determination via elemental analysis by LIBS

Mauro Martinez, Abigail Woltering, Maria Andreoli, Lana Williams, Tosha Dupras, Matthieu Baudelet

Human hair has the ability to record changes or alterations in the human body and show the information as the rings in trees. Either organic or inorganic, these differences can highly contribute to knowing the origin of this hair. DNA is a particular material that is not easy to recover from hair (unless pulled from the source and with remaining follicular cells). Furthermore, with time and environmental alterations, this can be degraded. This prevents DNA to identify the sex in an individual using hair. The aim of this work is to use a human hair to predict the sex of the source. Using LIBS on the hair, we identified different minor elements of interest such as Ca, Mg, Si, Sr, Al and Fe. The absorption of these elements can reflect the metabolism change in the body that can be different between male and female individuals. Taking these spectroscopic information for this group of elements, the sex of the human source, and working with a population of 76 individuals between alive, cadaver and even mummy samples, we created a statistical model follow a binary logistic regression when the capability to predict with 75% of accuracy the sex of the human source in the studied population. This study substantiates the potential of the elemental analysis of hair as a proxy for human sex determination.

Elemental analysis of packaging tapes by LA-ICP-MS and LIBS

Claudia Martinez Lopez, Masataka Sakayanagi, Jose R. Almirall

Packaging tapes are often involved in criminal activities as they are used in drug packaging, kidnappings, ligatures, and other violent crimes. A calibration strategy for the semi-quantitative analysis of tapes was accomplished by the use of certified polymer standards. Thin films of these polyethylene standards were prepared utilizing the pellet form of the certified reference materials. The methods for the analysis of packaging tape backings by Laser Induced Breakdown Spectroscopy (LIBS) and Laser Ablation-Inductively Coupled Plasma-Mass Spectrometry (LA-ICP-MS) were developed and applied to eight samples of tapes manufactured in Asia. The use of LIBS for tape analysis is reported for the first time. The results of LIBS were compared to LA-ICP-MS to assess their complementarity for the analysis of this type of sample. The optimized methods can be applied to different types of tapes, and potentially to other polymeric materials. The use of these laser-methods allowed for the detection of up to 8 elements by LIBS, and 18 elements by LA-ICP-MS. LIBS proved especially useful in detecting elements that were problematic to LA-ICP-MS such as lithium and potassium. LIBS was able to differentiate two packaging tapes manufactured in Japan by two different manufacturers by their difference in lithium; these two tapes were undistinguished by LA-ICP-MS. The results show that LIBS is a feasible alternative to the already established LA-ICP-MS for forensic tape analyses.

Gunshot residues analysis using LIBS for the estimation of firing distance.

César Alvarez-Llamas, María López-López, Carmen García-Ruiz , Jorge Pisonero, Bordel Nerea

Gunshot residues (GSR) are particles produced during the discharge of a firearm. GSR are composed by a mixture of organic compounds coming mainly from the propellant and inorganic compounds from the primer. They are launched with the projectile resulting in a conical plume of particles that finally create a pattern on the target depending on different factors, including the type of ammunition, type of firearm and especially the firing distance [1]. The study of gunshot residue (GSR) patterns can assist in the reconstruction of shooting incidences. LIBS technique is proposed to perform this kind of studies, due to the excellent analytical capabilities of this technique to perfume fast and multielemental analysis with spatial resolution [2]. The present work is focused on the investigation of LIBS imaging for the visualization of the GSR patterns. In particular, it evaluates the simultaneous presence of Sb-Pb-Ba as a marker for GSR. Several clothing samples, which were shot from different shooting distances (10-150 cm), were mapped using LIBS technique, obtaining an elemental distribution of the GSR over the target. This distribution creates a pattern which could be correlated to the firing distance [3]. In addition, the recorded LIBS spectra were treated using univariate and multivariate methodology to estimate the firing distance. In the first case, the use of the averaged signals from Pb, Sb and Ba versus the shooting distances was considered. In the second case, a multivariate method (PCA) was applied to the hyperspectral data (using the entire recorded spectrum for each spatial coordinate). The multivariate image analysis permits to use the entire information from the recorded spectra with a reduced number of variables. It is demonstrated that PCA could allow a fast and easy way to represent the GSR spatial distribution. [1] O. Dalby, D. Butler, and J. W. Birkett, J. Forensic Sci., vol. 55, no. 4, pp. 924–43, Jul. 2010.
[2] D. W. Hahn and N. Omenetto, Appl. Spectrosc., vol. 64, no. 12, pp. 335–366, Dec. 2010. [3] M.López-López, C. Alvarez-Llamas, J. Pisonero, C. García-Ruiz, N.Bordel.. Forensic Sci In., 273, pp. 124-131, Apr 2017

Nanosecond-pulsed laser ablation for explosives characterization: emission and shock wave measurements

Jennifer Gottfried

The development of laboratory-scale methods for characterizing the properties of energetic materials, i.e., using only milligram quantities of material, is essential for the development of new types of explosives and propellants for use in military applications. Scale-up of these materials to multi-gram levels for conventional safety (e.g., impact sensitivity, electrostatic sensitivity, compatibility, etc.) and performance (e.g., detonation velocity, specific impulse, etc.) tests is costly and time-consuming. Laser-based excitation methods for initiating or exciting the energetic material offer several advantages for investigating the response of energetic materials to various stimuli: 1) very small quantities of material can be studied prior to scale-up synthesis, 2) no detonation of bulk energetic material is required, eliminating the need for expensive safety precautions, and 3) extensive diagnostics can be incorporated into the experimental setup to provide as much information as possible per shot. Direct excitation of a sample residue using a focused nanosecond laser pulse enables investigation of the high-temperature chemistry of the reacting material via emission spectroscopy [laserinduced breakdown spectroscopy (LIBS)],[1] evaluation of the material for propellant applications based on the laser induced deflagration,[2] and estimation of the performance of the energetic material based on the measured shock wave velocity with a technique called laser-induced air shock from energetic materials (LASEM).[3,4] The techniques presented offer exciting potential for the laboratory-scale characterization of the chemical reactions, detonation performance, or propellant potential of new energetic materials under development prior to synthesis scale-up. 1 J. L. Gottfried, "Laser-induced plasma chemistry of the explosive RDX with various metallic nanoparticles," Appl. Opt. 51, B13-B21 (2012). 2 E. S. Collins and J. L. Gottfried, "Laser-induced deflagration for the characterization of energetic materials," Propellants Explos. Pyrotech. DOI: 10.1002/prep.201700040 (2017). 3 J. L. Gottfried, "Influence of exothermic chemical reactions on laser-induced shock waves," Phys. Chem. Chem. Phys. 16, 21452-21466 (2014). 4 J. L. Gottfried, "Laboratory-scale method for estimating explosive performance from laser-induced shock waves," Propellants Explos. Pyrotech. 40, 674–681 (2015).

Nanomaterials in Energy Generation and Storage Devices

Meyya Meyyappan

Nanomaterials exhibit interesting physical, mechanical, electrical and other properties and as such, have been widely exploited for a range of applications. The energy sector is no exception and this talk provides an overview of the activities at NASA Ames Research Center in this domain. We have investigated triboelectric nanogenerators (TENG) to couple with wind energy for Mars applications as an alternative to solar energy and radioisotope thermal generators. We have studied the effect of Mars ambient (such as pressure, temperature, radiation etc.) on the output performance. In addition, the hysterisis behavior of TENG is analyzed since the output performance seems to be dependent on previous contact history. Storage of the energy from such intermittent source is critical in Mars as in terrestrial applications. Supercapacitors can provide the power density needed for the rovers and other applications that require quick start-stop options. MWCNTs directly grown on metal foils without any catalysts reduce the interfacial resistance and provide a very high 550 F/g for double layer capacitor operation. Finally, silicon nanowires have been fabricated using top-down approach and the asymmetric pillar-type NWs help in efficient light trapping and increasing the conversion efficiency in solar cells. These asymmetric SiNWs have also been highly valuable in constructing thermoelectric devices, and power conversion results will be presented. The author acknowledges Myeonlok Seol, Dong-il Moon, Jinwoo Han, Kihyun Kim, Chang-Ki Baek and Cattien Nguyen for their contribution to the results discussed in the presentation.

Hydrogen Evolution Reaction Catalyzed by Ruthenium Ion-Complexed Graphitic-like Carbon Nitride Nanosheets

Shaowei Chen

Development of effective electrocatalysts for hydrogen evolution reaction (HER) based on low-cost, earth-abundant materials is urgently needed. In the present study, graphitic-like carbon nitride (C3N4) nanosheets of 2.0 ± 0.4 nm in thickness were prepared by controlled heating of melamine in air, into which were then embedded ruthenium metal ions by refluxing C3N4 and RuCl3 in water by taking advantage of the strong affinity of ruthenium ions to pyridinic nitrogen. The formation of C3N4-Ru nanocomposites was confirmed in optical and X-ray photoelectron spectroscopic measurements, which suggested charge transfer from the C3N4 scaffold to the ruthenium centers. Significantly, the hybrid materials exhibited apparent electrocatalytic activity towards HER in acid and the activity increased with the loading of ruthenium metal centers in the C3N4 matrix. Within the present experimental context, the sample saturated with ruthenium ion complexation at the atomic ratio of ruthenium to pyridinic nitrogen of ca. 1:2 displayed the best performance, with an overpotential of only 140 mV to achieve the current density of 10 mA/cm2, a low Tafel slope of 57 mV/dec, and a large exchange current density of 72 ?A/cm2. This was superior/comparable to leading results reported in recent literatures with relevant non-platinum HER catalysts. In contrast, only minimal activity was observed when ruthenium was replaced with other transition metal ions (e.g., Fe3+, Co3+, Ni3+ and Cu2+). This was ascribed to the formation of Ru-N moieties where the synergistic interactions between the carbon nitride and ruthenium metal centers facilitated the adsorption of hydrogen, which was strongly supported by results from density functional theory calculations.

Plasmonic Enhanced Catalysis Based on Rh Nanostructures

Jie Liu, Xiao Zhang, Xueqian Li, Henry Everitt

Traditional heterogeneous catalysis that requires high temperatures to overcome reaction barriers often activates undesired side reactions that must be mitigated. Photocatalytic reactions on semiconductors can proceed near room temperature but are primarily used for reactions that require little selectivity. Here we introduce rhodium nanoparticles as a compelling new photocatalyst whose excellent catalytic and plasmonic properties combine to reduce the activation energy and selectively determine products for the important carbon dioxide hydrogenation reaction. The reduced activation energy and strong sensitivity to light cause the room temperature photocatalytic methane production rate to exceed the thermocatalytic rate at 350 C.

A graded nanoscale catalytic MoSx/TiO2 interface for hydrogen evolution
Jing Gu

The future of clean energy depends heavily on innovative breakthroughs concerning the design of efficient and stable systems for the conversion and storage of solar energy. Photoelectrochemical (PEC) water splitting is a promising solution in which energy collection and water electrolysis are combined into a single system. In the talk we will demonstrate that annealing a bilayer nano-interface of amorphous titanium dioxide (TiOx) and molybdenum sulfide (MoSx) deposited onto GaInP2 results in a photocathode with high catalytic activity (current density of 11mAcm?2 at 0 V versus the reversible hydrogen electrode under 1 sun illumination) and stability (retention of 80% of initial photocurrent density over a 20 h durability test) for the hydrogen evolution reaction. Our findings demonstrate the potential of utilizing a hybridized, nano-scale surface layer as a cost-effective catalytic and protective interface for solar hydrogen production.

Portable Standoff Detection of Materials Using Monolithic QCL Arrays
Mark Witinski, Romain Blanchard, Kalyani Krishnamurthy

This presentation introduces the spectroscopic concepts and results enabled by arrays of Distributed Feedback (DFB) QCLs, with each element at a slightly different wavelength than its neighbor. In portable optical systems, such as standoff threat detectors and in situ gas analyzers, this increases analyte sensitivity and selectivity by broadening spectral source coverage while also allowing for extremely fast all-electronic wavelength tuning with no moving parts. This talk will first present the QCL array and its packaging, then move into the description of an integrated prototype standoff detection system, and finally show condensed phase standoff threat detection results from a handheld system from over 1 meter. These data are each compared with legacy contact-based methods to ensure that the technique can be reliably deployed to handheld chemical analysis using suitable chemometric algorithms. The data show how monolithic and all-electronic tuning enables next-generation spectroscopes that are not only more robust and miniature than those that utilize external cavity-tuned lasers, but that are inherently more stable in terms the shot-to-shot amplitude and wavelength parameters. This enhanced stability increases signal to noise for a given configuration (pathlength, averaging time, concentration, etc…). Some discussion of how to maximize the benefits of high speed, highly reproducible tuning is presented, including mapping the tradespace available to the end user (for instance, the relationship between observation time, distance, certainty, concentration, etc...).

Handheld and Portable FTIR: After a Decade, What Have We Learned?

Norman Wright

Over the last 10 years, the use and acceptance of mobile analytical instrumentation has steadily increased. The need for analysis in non-traditional areas such as security, forensics and where the analysis is done better outside of the laboratory, has spurred this growth. In the case of Fourier Transform infrared (FTIR) spectroscopy, as an established technique, growth has been addressed by the development of both innovative instruments and methods. FT-IR spectroscopy has progressed to the point where instruments are now routinely taken into the field and measurements made away from controlled, laboratory environments. A key reason for the successful use of FT-IR in the field has been the development of lighter, more compact optical systems that provide portability, durability and stable operation in uncontrolled environments. This talk will discuss the state of out-of-lab FTIR instrumentation with a focus on proven applications that highlight those capabilities. These small, but very usable systems with advanced sample interface technology provide for easy use while expanding the range of potential applications.

The Use of Spatially-Offset Raman Spectroscopy (SORS) To Identify Unknown Threats Through Opaque Containers

Eric Roy

Handheld Raman-based chemical detectors have been used for more than a decade by Hazmat, Military, and Security personnel to identify unknown chemicals in the field. However, the scope of previous generation handheld Raman systems is limited to identifying materials that are either unpackaged or contained in thin, transparent containers (e.g. baggies, thin/clear bottles). When opaque containers (envelopes, opaque plastics) are encountered by these operators in the field, the package had to be opened and manually sampled prior to analysis, which decreases operational efficiency and exposes the operator to the potentially hazardous substance. Spatially Offset Raman Spectroscopy (SORS) is a new variant of Raman spectroscopy capable of determining the Raman spectrum of a material contained within opaque containers such as: colored and opaque plastics, paper, fabric and dark glass. The SORS technique collects two spectra (zero and offset points) and processes the information into representative Raman spectrum of the contained material. Cobalt Light Systems has productized the SORS technique into a handheld system (Resolve™), which is purpose-built for security personnel. In this presentation, we present various real-world scenarios where handheld SORS has been used to identify chemical targets (e.g. explosives, toxic materials, “white powders”) through opaque containers that cannot be addressed using traditional Raman-based detection systems. The results of these trials will be discussed in context of the operational relevance (capabilities & limitations) to missions encountered in the security applications.

Development and Implementation of a Pass/Fail Field-Friendly Method for Detecting Sildenafil in Suspect Pharmaceutical Tablets using a Handheld Raman Spectrometer and Silver Colloids

Adam Lanzarotta, Lisa Lorenz, JaCinta Batson, Cheryl Flurer

Handheld Raman spectrometers and silver colloids have been employed to develop a simple, fast, sensitive and effective pass/fail field-friendly method for detecting sildenafil in suspect Viagra and unapproved tablets. The method involves dissolving a portion of a tablet in water followed by filtration and addition of silver colloids, resulting in a solution that can be measured directly through a glass vial. Over one hundred counterfeit Viagra and unapproved tablets were examined on three different devices during the method development phase of the study. While the pass/fail approach was found to be 92.6% effective on average, the efficacy increased to 97.4% on average when coupled with the software's "Discover Mode" feature that allows the user to compare a suspect spectrum to that of stored sildenafil spectra. The lowest concentration of sildenafil in water that yielded a "Pass" in the lab was found to be 7.6 μg/mL or 7.6 parts per million (ppm). For the analysis of suspect tablets, this value was found to be as low as 10 μg/mL and as high as 625 μg/mL. This variability was likely related to the tablet formulation, e.g., concentration of sildenafil, presence and concentration of water-soluble and/or water-insoluble ingredients. However, since most counterfeit Viagra and unapproved tablets contain >50 mg sildenafil per tablet, such low concentrations will not be encountered often. Limited in-lab and in-field validation studies were conducted in which analysts/field agents followed the procedure outlined in this study for small sample sets. These individuals were provided with written instructions, a ~20 minute demonstration regarding how to perform the procedure and use the instrument, and a kit with field-friendly supplies (purified bottled water from a local grocery store, disposable plastic pipettes, eye-dropper with a silver colloid solution, etc.). The method proved to be 98.3% and 91.7% effective for the in-lab and in-field validation studies, respectively, which demonstrated the ruggedness, simplicity and practicality of the method.

MicroNIR PAT – Spectroscopic Sensor for Industrial Process Monitoring

Peng Zou

Interoperability, information accessibility, and decentralized decision-making are at the core of Industry 4.0. Spectroscopic technology is of great interest for process analytical technology (PAT) for industrial process monitoring and feed-forward or feed-back control because it can provide rich chemical and physical information in real-time while the measurement method is typically noninvasive. We will present the latest product developments and application case studies of our near infrared spectral sensors dedicated for PAT applications. The spectroscopic sensors are based on our proprietary LVF technology which provides necessary ruggedness and reliability for harsh process environments, a small form factor and high performance for pharmaceutical, chemical, food, feed and other industries to meet 4.0 design concept and requirement

Laboratory Investigation of Suspected Materials at Celgene

Ming Wang, Dong Xiang, Xiaoxuan (Jason) Shen

Celgene's anti-counterfeiting program is to ensure the security and quality of Celgene products, a commitment to always put patients first. Laboratory investigation of suspected materials is an essential component of Celgene's overall anti-counterfeiting program. At Celgene, laboratory investigation of suspected materials is authorized by global product security group and governed by security sample investigation procedures. Laboratory investigation not only needs to meet compliance requirements but also must have the capability to identify suspected materials from diverse and unusual sources in a timely manner. Presented here is our overall strategy for suspected material laboratory investigation to meet these objectives by utilizing Raman spectroscopy, FTIR, LC/MS/PDA, and other analytical technologies. A few case studies will be presented to illustrate this strategy.

The use of X-Ray Powder Diffraction (XRD) and Vibrational Spectroscopic Techniques in the Analysis of Pharmaceutical Forensic Samples

Mark Witkowski, Kelsey DeWitt

X-Ray powder diffraction (XRD) and vibrational spectroscopic techniques, such as Fourier transform infrared (FT-IR) and Raman spectroscopies, are well established analysis techniques with numerous literature references citing their application in the areas of pharmaceutical and forensic sciences. Individually, XRD, FT-IR and Raman provide useful and unique information about a sample. However, by combining the data from each individual technique, a comprehensive picture of the sample is obtained and can be very useful especially in cases of complex sample matrices. XRD and vibrational spectroscopic techniques, for the most part, are nondestructive and require a minimal amount of sample for analysis. Therefore, the remaining portion of sample can be retained for analysis by other destructive techniques (e.g. GC-MS or HPLC-UV). Depending on the order of analysis, the same portion of a sample can be analyzed by all three techniques (XRD, FT-IR and Raman) providing data on the same portion of sample. One example where this can be very useful is in the analysis of suspect counterfeit pharmaceutical dosage forms and in comparing the suspect counterfeit to an authentic product. The unique chemical and crystalline information obtained from XRD, FT-IR and Raman provides a profile of the suspect counterfeit dosage form which can be quickly compared to the same information derived from the authentic. This information from the suspect sample can also be used to establish a "fingerprint" that can be compared to future samples submitted for analysis. This presentation will discuss the application of XRD, FT-IR and Raman spectroscopies, the type of spectral information obtained and how the data can be combined when applied to variety of pharmaceutical forensic sample types (e.g. counterfeit pharmaceuticals, dietary supplements).

Multispectral approach for identifying counterfeit lifestyle and medicinal products of different formulations

Sulaf Assi

The last decade witnessed a foremost change in the use of medicines worldwide beyond life-saving purposes. The change involved the utilisation of medicinal products for lifestyle purposes. Hence, lifestyle medicinal products emerged among individuals and were intended mainly to improve physical appearance (e.g. weight-loss, sexual stimulants) and/or mental performance (e.g. smart drugs, antidepressants). Lifestyle and medicinal products can be from any origin (herbal or synthetic origin), and of any formulation type (tablets, powders, creams or solutions). Not only has the change in drug scenario played a role in the widespread of lifestyle products; but also, their ease of availability/access via the Internet. The Internet offered quick, convenient and anonymous ways for purchasing lifestyle medicines. Furthermore, many online pharmacies provide detailed information about lifestyle medicines to consumer thus saving time from visiting healthcare centres. Purchasing these medicines from the Internet is a major concern due to purity and quality issues. According to the World Health Organization, around 50% of medicines purchased from the Internet are counterfeit. One analytical technique is not sufficient for detecting counterfeit medicines. This urges the need for a multiple analytical approach in order to identify lifestyle medicinal products. Therefore, this work aims at developing a multispectral approach for identifying counterfeit lifestyle products. Three handheld and laboratory based spectroscopic techniques were used and included: Fourier transform infrared (FTIR), near infrared (NIR) and Raman spectroscopy. Spectral libraries of authentic lifestyle medicines and constituents were generated using the three techniques. Additionally, spectra of quantitative mixtures of constituents present in medicines were collected and stored. For identification purposes both in-built instrumental and offline classification algorithms were used. Quantification of different constituents in medicines was conducted using multivariate regression algorithms. The results showed that three techniques were complementary for identifying counterfeit lifestyle medicines. Raman was more selective to active pharmaceutical ingredients within a sample provided it was present within a detectable concentration range. On the other hand, additives were better identified using NIR and FTIR. Among the three techniques FTIR was the only destructive technique; yet, it required minimum sample amount. Accurate quantitative models were achieved with the three techniques.

Key indicators of falsified medicines
Susan Macha

Case studies of confirmed falsified medicines will be presented outlining data gathered from the forensic examination leading to the conclusion.

Screening Technologies for the Detection of Substandard and Falsified Medicines: Public Standards for Quality
Daniel Bempong, Lukas Roth

Medicine quality screening technologies are an invaluable tool in the fight against substandard and falsified medicines. They are used by almost all stakeholders in this field, including but not limited national medicine regulatory authorities, pharmaceutical manufacturers, global procurement organizations, distributors, wholesalers, pharmacies and even patients. Aligned with these organizations are an equal number of screening technology manufacturers working to capacitate these stakeholders to effectively assure the quality of medicines they produce, procure, distribute, regulate and take. There is however a dearth of information on what a potential screening technology user should be looking at when evaluating a technology for procurement and subsequent deployment. This presentation will introduce the collaborative work that the U.S. Pharmacopeia is undertaking to develop standards for the evaluation of these technologies and generate data on the capabilities of existing and emerging medicine quality screening technologies.

Coupling and Stacking Order of ReS2 Atomic Layers Revealed by Ultralow-Frequency Raman Spectroscopy
Chun Hung Lui

We investigate the ultralow-frequency Raman response of atomically thin ReS2, a special type of two-dimensional (2D) semiconductors with unique distorted 1T structure. Bilayer and few-layer ReS2 exhibit rich Raman spectra at frequencies below 50 wave number, where a panoply of interlayer shear and breathing modes are observed. The emergence of these interlayer phonon modes indicate that the ReS2 layers are coupled and orderly stacked. Whereas the interlayer breathing modes behave similarly to those in other 2D layered crystals, the shear modes exhibit distinctive behavior due to the in-plane lattice distortion. In particular, the two shear modes in bilayer ReS2 are non-degenerate and clearly resolved in the Raman spectrum, in contrast to the doubly degenerate shear modes in other 2D materials. By carrying out comprehensive first-principles calculations, we can account for the frequency and Raman intensity of the interlayer modes and determine the stacking order in bilayer ReS2.

Raman hyperspectroscopy of a biological stain for forensic phenotype profiling
Igor Lednev

Several novel methods to determine externally visible characteristics of body fluid donors have been developed in recent years based on DNA analysis. These tests can help forensic investigators make predictions about the appearance of a suspect or victim, such as their sex, race, hair color, or age. While their potential benefit is undeniable, these methods destroy the physical evidence in the process, need to be conducted in the laboratory and, consequently time consuming. We conducted a serious of proof-of-concept studies demonstrating that Raman spectroscopy has a great potential as a nondestructive technique to test for many of these characteristics. Our approach is based on multivariate data analysis of hyperspectral datacube acquired from heterogeneous samples of dry biological stains. We will present the results from our studies to determine the donor race of dry bloodstain and semen as well as donor sex from bloodstain ad saliva traces.

Transmission Raman Measurements using a Spatial Heterodyne Raman Spectrometer
K. Alicia Strange, Kelly C. Paul, S. Michael Angel

Transmission Raman spectroscopy (TRS) provides nondestructive, noninvasive, and unique molecular identification of the bulk contents of a heterogeneous sample. In TRS, Raman scattered photons exit the sample in a very large area, in a spot about 6 times the sample thickness. Typical dispersive spectrometers have small entrance slits and are not efficient at collecting light from a large area sample, which leads to low collection efficiency and reduced sensitivity in TRS in comparison to backscatter Raman. Recently, we demonstrated a new type of Fourier transform (FT) Raman spectrometer, the spatial heterodyne Raman spectrometer (SHRS), that has very high light throughput and a wide field-of-view. The SHRS is well suited for measuring large sample areas and could be useful for TRS measurements. The purpose of this paper is to demonstrate TRS measurements of highly scattering compounds using the SHRS and to make comparisons to a more conventional, high throughput dispersive spectrometer. Transmission Raman spectral intensities of ibuprofen measured using the SHRS were within a factor of 2.4 of the backscatter Raman intensity, while being more representative of the bulk material. The throughput was about eight times higher than an f/1.8 dispersive spectrometer, and the width of the area viewed was found to be seven to nine times higher. However, the signal-to-noise ratio was two times lower for the SHRS than the f/1.8 dispersive spectrometer, apparently due to high levels of stray light.

TERS with Angstrom Resolution
Richard Van Duyne

During the last few years, there has been an explosion of interest and activity in the field of nanoscale vibrational spectroscopy. The goal is to confine and manipulate light on the sub-nanometer length scale using the properties of the collective electronic excitations in noble metal nanostructures, known as surface plasmons. An improved understanding of the interactions between adsorbed molecules and local environment under various experimental conditions is having a significant impact in a number of research areas including electrochemistry, surface science, catalysis for energy conversion and storage, low dimensional materials, biomedical diagnostics, and art conservation science. In the first part of this lecture, I will provide some background materials on the basic physical concepts underlying plasmonic enhanced Raman spectroscopy with an emphasis on localized surface plasmon resonance (LSPR), tip-enhanced Raman spectroscopy (TERS), and the high spatial resolution of TERS. In the second part of this lecture, I will focus in on two recent advances in ultrahigh vacuum (UHV) TERS which illustrate the power of this emerging technique. First, new insights into the nature of a dynamic phase boundary involved in the room temperature (RT), UHV-TERS of the Ag tip/ N-N'-bis(2,6-diisopropylphenyl)-1,7-(4'-t-butylphenoxy)perylene-3,4:9,10-bis(dicarboximide) (PPDI)/Ag(100) system will be described. We have unravelled the orientation of PPDI molecules at the dynamic molecular domain boundary with unprecedented ~4?nm spatial resolution by UHV-TERS mapping. This topic illustrates that TERS provides access to molecular adsorption geometries in the region with little STM topographical information. Second, we have achieved unprecedented Angstrom-scale spatial resolution in TERS. At room temperature, the strong adsorbate-substrate interaction between the meso-tetrakis-(3,5-ditertiarybutylphenyl)-porphyrin (H2TBPP) and the underlying Cu(111) substrate leads to the formation of the bowl up/down conformations. Though simultaneous UHV-TERS and STM analysis on the neighboring conformational isomers, we have observed ~15 cm-1 spectral shift in one of the porphyrin-ring Raman modes. This last topic demonstrates that TERS reveals more detailed information on the adsorbate-substrate interaction which is not easily accessible by other techniques.

Accurate, High Speed Immunodiagnostics using SERS
Marc Porter, Jennifer Granger, China Lim, Aleksander Skuratovslky, Courtney Scaife, Jill Shea

This presentation details a methodology for the realization of rapid, highly portable immunoassays for low-level pathogen detection. The approach, which is termed Raman-Solid Phase Extraction (R-SPE), is based on a sandwich immunoassay construct that integrates nitrocellulose solid phase extraction membranes with modified gold nanoparticles (AuNPs) and surface-enhanced Raman scattering (SERS). The membranes are coated with capture antibodies; and the AuNPs are functionalized with a monolayer of a Raman reporter molecule (RRM) and a layer of secondary antibodies. The AuNPs act as indirect labels for pathogen quantification, and the intensity of the SERS response from the RRMs is used to indirectly measure pathogen levels in the sample. The analytical response is measured by means of a handheld Raman spectrometer. The merits of this approach will be demonstrated by an assay speed of 2-3 minutes and femtomolar levels of detection for alpha-fetoprotein (AFP), a marker for hepatocellular carcinoma, and the Category A pathogens botulinum neurotoxins A and B. The potential impact of this methodology as an addition to the point-of-need (PON) diagnostics toolbox will be discussed briefly.

SERS for Rapid Detection of Bacterial Infection
Steven Bell, Jessica Kelly, Virginia Blanque, Miguel Valvano, Sheila Patrick

Bacteria in the form of biofilms represent a major factor in impaired wound healing and the development of chronic wounds. Open wounds contain micro-organisms from either endogenous or exogenous sources. Generally the hosts' immune system can deal with this microbial threat, however problems arise when these microorganisms multiply, attach to the wound surface and invade tissues; forming a biofilm. A well-established biofilm will display resistance to destruction by the immune system causing inflammation, tissue necrosis, delayed healing and if left untreated, serious illness. Typically, if a clinician suspects infection of a wound, then a swab will be taken for culturing in a clinical microbial laboratory. This is a time-consuming technique as it relies on long incubation times for positive identification of micro-organisms. Therefore there is a clear need for the rapid identification of bacteria present in a wound, at an early stage. Herein, we demonstrate that Surface-enhanced Raman Spectroscopy (SERS) can be used for the point of care-identification of biofilm formation on a wound dressing. Initial studies centred on SERS of planktonic bacteria to establish reliable spectroscopic markers. Both gram-positive and gram-negative bacteria of aerobes and anaerobes were investigated. These included; Escherichia coli, Streptococcus milleri, Staphylococcus aureus, Enterococcus faecalis, Pseudomonas aeruginosa and Bacteroides fragilis. The literature in this area is contradictory with some early studies reporting features which were due to either the growth media or spectra which were dominated entirely by a single component at a single wavelength. We have established experimental conditions that allow us reproducibly obtain SERS spectra of the same bacterial strain. A second major issue which we have addressed is the molecular origin of the spectral bands. The number of potential sources is of course high, but we have found that the majority of the variation between bacteria can be accounted for as differences in the relative concentrations of a relatively small number of compounds. These ratios are also found to change with the growth cycle of the cell and contact time with the colloidal nanoparticles used for enhancement. These observations on model bacteria also carried through to bacteria grown on wound dressings where SERS spectra of biofilm sampled directly from the wound dressing where found to be similar the spectra of the planktonic bacteria used to infect the wound dressing. This opens up the possibility of using SERS as a rapid point-of-care diagnostic method for bacterial wound infections.

An innovative SERS approach for in situ and real time study of pesticide behaviors in live plants
Lili He, Tianxi Yang

Understanding the penetration, translocation, and persistence of pesticides is significant for effectively applying pesticides and reducing pesticide exposures from plant food. Herein, we developed an innovative approach for real-time and in situ monitoring of pesticide in live plants using surface enhanced Raman spectroscopy (SERS). Citrate coated gold nanoparticles that can penetrate into plant tissues were applied as probes for tracking of pesticides over time in live plants. For the penetration study, it was observed that systemic pesticide thiabendazole and the non-systemic pesticide ferbam can penetrate into the live basil leaves with depth of 225 ?m and 130 ?m, respectively. We found that intact leaves allowed for increased penetration when compared with damaged leaves. The larger older leaves (45 days) allowed deeper pesticide penetration than the smaller younger leaves (15 days). Internalized thiabendazole degraded apparently after 1 week, whereas obvious degradation of internalized ferbam became apparent after 2 weeks. For the translocation study that applied thiabendazole in a hydroponic system for growing tomato plants, we detected the thiabendazole signal after 1 day exposure with 200 ppm thiabendazole in the hydroponic system, and the signal appeared firstly along the midvein in the lowest leaves. Translocation of the pesticide into the trichome was also detected. When 10 ppm thiabendazole was applied in the system, we detected the first signal after 5 days. Signals from thiabendazole metabolites were observed after 6 days exposure of 200 ppm pesticide. In addition, we found a unique SERS peak at 737 cm-1 coming from nicotinamide adenine dinucleotide (NAD) and other adenine-containing materials might be related with plant response to the pesticide toxicity. The SERS method provides a rapid and effective way to study the behaviour and fate of pesticides in a plant system. The information obtained here could provide useful guidance for effective and safe applications of pesticides on plants.

Surface Enhanced Raman Spectroscopy for Rapid Selenium Monitoring in Flue Gas Desulfurization water

Merwan Benhabib, Mark Charles Peterman, Samuel Kleinman

Coal fired power plants produce large levels of toxic Selenium in the wastewater streams discharged from wet scrubber systems, such as Flue Gas Desulfurization (FGD) units. Federal regulation requires wastewater streams to contain less than 12-parts-per-billion (ppb) in total selenium before they are released in the environment. To comply, plants install Biological Selenium remediation units, which commonly digest only specific selenium species. Selenium occurs as selenide, selenite, selenate and selenocyanate all of which are environmental hazard and life threatening. Operators of FGD units are forced to send samples for expensive, time-consuming laboratory testing for selenium speciation and analysis. The cost, the distance and the delay limit the data available for accurate process control. To guarantee treatment performance the development of a method capable of selenium speciation and quantification in complex wastewater matrices is necessary, in particular the distinction between selenite and selenate. The standard laboratory technique is ion chromatography coupled to mass-spectrometry. It is an expensive and labor-intensive operation that is not adequate for a timely-response. The quantitative detection of selenate down to ppb-levels is shown for lab and FGD field samples. The apparatus, the procedure, its optimization, and its challenges are described. A pretreatment process necessary to remove the major interfering ions and the ultra-stable calcium carbonate nanoparticles present in FGD is discussed. Specificity of the analytical procedure to the +6 oxidation state is demonstrated with complete independence toward other selenium species. The technique employed is based on Surface Enhanced Raman Spectroscopy (SERS), it used active gold-nanoparticles coated with a cationic monolayer to enhance specificity to selenate and used a selenate isotopologue as internal standard to ensure high accuracy and repeatability. Quantitative analysis is demonstrated down to the 5-ppb regulatory level in complex-matrices in lab and in various wastewater samples. A side-by-side comparison with standard analysis equipment proves the high accuracy of this technique. The successful results obtained for both lab and field samples demonstrates the capabilities of this cost-effective analysis technology to generate quickly meaningful data that can be used to optimize the FGD treatment process.

Detection of CVD Biomarkers using Functionalised Nanoparticles and SERS

Kirsten Gracie, Samuel Mabbott, Steven Asiala, Jonathan Noonan, Neil MacRitchie, Gianluca Grassia, Pasquale Maffia, Karen Faulds, Duncan Graham

Cardiovascular disease (CVD) is one of the main causes of death in developed countries. Methods of prevention and multiple means of intervention exist, however there is a growing demand for more accurate and reliable methods for detecting CVD biomarkers during the early stages of the disease progression. In this work, surface-enhanced Raman scattering (SERS) is the chosen method for detecting several CVD biomarkers, both individually and simultaneously, *in vitro* and *in vivo*. Here, gold nanoparticles and gold nanostars are the main SERS active substrate used. Functionalisation involved both antibodies and aptamers as targeting moieties alongside small organic molecules as Raman reporters. Designing particles with different combinations of reporters and antibodies/aptamers allows for SERS to report on single or multiple biomarkers simultaneously. Results presented show the successful conjugation of gold nanoparticles with both antibodies and aptamers for the detection of three CVD biomarkers: ICAM, VCAM and P-Selectin. Additionally, ELISA-based detection methods have been developed for the detection of ApoA1 and ApoB using nanoparticles and nanostars.

Quantitative Molecular Orientation Analysis in Organic Semiconductor Thin films having a rough surface by pMAIRS

Takafumi Shimoaka, Nobutaka Shioya, Miyako Hada, Takeshi Hasegawa

Organic semiconductors used for thin-film electronic devices have a ?-conjugated functional group. Since the electrical conductivity and its anisotropy of the thin film depend on the molecular interaction and orientation of the functional group, the analysis of these parameters is important for developing a thin film device. Infrared (IR) spectroscopy is a very powerful tool to investigate the molecular interaction even in an ultrathin film. When IR spectroscopy is coupled with the p-polarized multiple-angle incidence resolution spectrometry (pMAIRS) technique[1], the molecular orientation is also investigated quite easily. pMAIRS provides both surface-parallel and -perpendicular components of each normal mode as the in-plane (IP) and out-of-plane (OP) spectra at a time from an identical sample, with which the molecular orientation is quantitatively discussed. On analysis of an organic semiconductor thin film using pMAIRS, the CH out-of-plane deformation vibration mode (?(CH)) of an aromatic ring in ca. 900-600 cm^{-1} has been revealed to be useful for quantitatively evaluating the ring orientation, since the ?(CH) mode is highly localized on the ring having the perpendicular direction to the surface.[2,3] In the present study, our recent studies on the analysis of representative organic semiconductor compounds, pentacene, a porphyrin derivative and a polythiophene derivative, are discussed. [1] T. Hasegawa, *Appl. Spectrosc. Rev.* **43**, 181–201 (2008). [2] N. Shioya et al., *Phys. Chem. Chem. Phys.* **17**, 13472-13479 (2015). [3] M. Hada et al., *Chem. Eur. J.* **22**, 16539–16546 (2016).

Analysis of Carbon Nanotube Films on Silicon by p-Polarized Multiple Angle Incidence Resolution Spectrometry (pMAIRS)

David Drapcho, Amir Mashal, Matthew Meyer, Nathaniel Safron

Semiconducting single-walled carbon nanotubes (s-SWCNTs) have been one of the most exciting materials over the past several decades in materials science research. A recently pioneered technique by Hasegawa introduces a new approach that can be used to characterize CNT films on double side polished silicon substrates via p-Polarized Multiple Angle Incidence Resolution Spectrometry (pMAIRS). This method utilizes a series of IR transmission spectra of CNT films on polished silicon (transparent substrate in the IR), collected at several oblique angles of incidence. Separate in-plane (IP) and out-of-plane (OP) infrared spectra can then be obtained by a decomposition of the series of spectra collected as a function of angle of incidence, allowing the direct computation of the surface orientation of the s-SWCNT functional chemistries. This technique can provide valuable information on s-SWCNT films ranging from the detection of impurities in s-SWCNT films to the molecular orientation of functional groups bound to the s-SWCNT sidewalls. While there are other techniques such as XPS, Raman, or Visible absorption spectroscopy that may provide similar information, none of these techniques deliver the combination of a high signal-to-noise, simple detection of surface functionalization on CNTs, nor the molecular orientation of functional attached to the CNT films, respectively. In this paper, we will present pMAIRS spectra of CNT films functionalized with various organic and inorganic chemistries and extract valuable data to provide solutions to material science researchers. *T. Hasegawa J. Phys. Chem. B, 106, 4112 (2002)

Recent progress of MAIRS

Takeshi Hasegawa

pMAIRS has recently been an established technique to reveal the molecular orientation in a thin film deposited on an IR transparent substrate, which is quite powerful to analyzed the molecular orientation for each chemical group using no optical parameters. Another benefit of using pMAIRS is that the orientation analysis can be performed in a thin film even with a surface roughness. Therefore, practical thin films prepared by the spin-coating and drop-cast techniques can be a good analytical target. In this talk, a recent progress of pMAIRS will be mentioned, which gets over some analytical limits of measurement time, water-peaks subtraction and fringes. pMAIRS is still being renovated.

Interfacial Molecular Structure and Proton Transport Characteristics of Polymer Thin Films for Fuel Cells
Yuki Nagao

Understanding the relationship between the interfacial transport property and polymer structure enables us to develop the performance of fuel cells. Many highly proton-conductive polymers are amorphous or amorphous-like. Recently the author found that the proton conductivity of the polymer thin films can be enhanced by the molecular orientation. IR p-polarized multiple angle incidence resolution spectrometry (pMAIRS) study is helpful to discuss the molecular orientation because it includes the structural information in the non-ordered parts. In this talk, the relationship between the interfacial structure in the polymer thin films and transport property will be discussed.

Impact of pMAIRS on Metal Oxide Nanowires
Takeshi Yanagida

Here I present our recent progress as to pMAIRS use for metal oxide nanowires with volatile organic molecules. Although pMAIRS is clearly powerful tool to measure vibrational properties of molecules on surfaces, the applications to metal oxide nanowires have not been performed. We found unique features of molecules on single crystalline metal oxide nanowires. In this talk, I will present some recent results as to these results.

Influence of polarization and ion crowding effects on nonlinear electrophoresis
Bruno Figliuzzi, Jeffrey Moran

When a charged particle is placed in an electrolyte, ions of opposite charges migrate and screen its surface, eventually forming a structure usually referred to as electric double layer (EDL). The ionic concentration in the EDL strongly influences electrokinetic phenomena including electro-osmosis and electrophoresis. When the ?-potential of the particle is of the same order of magnitude as the thermal voltage, it has been shown that the translation velocity of the particle is proportional to the applied electric field. With the development of microfluidics, large applied electric fields are however commonly encountered in lab-on-a-chip technologies. At large applied voltages, nonlinear ionic fluxes arise between the electric double layer and the bulk solution.
These ionic exchanges result in a polarization of the concentration field around the particle, which in turn drives a diffusio-osmotic flow at the surface. In addition, ion crowding and dielectric decrement phenomena occur at high ?-potential that modify in depth the charge repartition in the electric double layer surrounding the particle. Dielectric decrement refers to the decrease in dielectric permittivity caused by local ion polarization. In this situation, the velocity field, the electric potential and the ionic concentration in the immediate vicinity of the particle are described by a complicated set of coupled nonlinear partial differential equations. In the general case, these equations must be solved numerically. In this presentation, we introduce a numerical model for determining the electrophoretic velocity at high ?-potential and high applied electric field. We rely on this model to study two archetypal problems in electrokinetics, namely the electrophoresis of particles with fixed surface charges and the electrophoresis of ideally polarizable particles. The simulations demonstrate that crowding effects and dielectric decrement are of paramount importance to describe nonlinear electrophoresis of ideally polarizable particles, and that surface conduction significantly decreases the electrophoretic mobility at high zeta potential and at high applied electric field.

Electrokinetic Fingering: A Problem in Vector Laplacian Growth
Mohammad Mirzadeh, Martin Bazant

Pattern formation is ubiquitous in various physical and chemical processes and has fascinated scientists over the past couple of decades. Perhaps one of the most well-known examples is the striking patterns generated when a high-viscosity fluid is displaced by a low-viscosity fluid in a Hele-Shaw cell. This “viscous fingering” phenomenon was originally described by Saffman and Taylor and is also a contributing factor in limiting the extraction efficiency in secondary oil recovery. This instability is a manifestation of the Laplacian Growth phenomena where the interfacial velocity is related to the gradient of a harmonic function (here pressure) and the instability is attributed to the curvature-dependence of the gradient. More generally, however, the interface could be driven by multiple competing forces. In this talk, we will consider a class of problems which we term “Vector Laplacian Growth”. In these problems, the interface velocity is determined by the gradient of a vector-valued potential function through a generalized mobility tensor. Specifically, we will consider the linear electrokinetic response of two immiscible electrolytes in a Hele-Shaw cell where the interface motion is set by both the pressure-driven and electro-osmotic flows. By performing linear stability analysis and nonlinear numerical simulations, we find that the coupling between pressure and potential fields affects interfacial stability and, under certain conditions, can suppress viscous fingering. Remarkably, our theory suggests that for a given pair of electrolytes, the ratio of electric current to the flowrate controls the onset of instability. We believe that our findings could have implications in electrically enhanced oil recovery application and might help with understanding other similar multi-field driven instabilities.

Using Dielectrophoresis to Separate Protists Present in Termite Hindguts
Claire Crowther, Katalina Freeman, Mark Hayes, Gillian Gile

Termites can survive feeding on wood because of protists present in their hindgut, forming a symbiotic relationship between the insect and the microorganism. The specific makeup and dynamics of the protists community has not been fully characterized, some work has focused on the composition of the protists community in a particular termite or determining a protists affinity for specific species of termite. In order to further understand the morphological and molecular characteristics of the symbionts present in termite hindguts, the various species need to be isolated. One method is to culture a specific protist; however this is extremely difficult or currently impossible, depending on the species. Another common method is hand picking the cells but this is extremely time and labor intensive. A rapid and inexpensive separation system is desired, which has led to the development of a novel dielectrophoretic cell-separation device. Dielectrophoresis (DEP), a force achieved from non-uniform electric field, enables separations by exploiting subtle differences in the electrophysical properties of the analytes of interest. Insulators have been used to achieve the non-uniform electric field, known as insulator-based dielectrophoresis (iDEP). This method has been used to manipulate several nano and micro sized analytes; however this work seeks to separate the much larger protist cells that range in size from 10-250 μm. A larger iDEP device was developed, which has wider constrictions, a larger depth, and side channels to allow for easy removal of the analytes of interest. The design was modeled and confirmed to achieve the forces necessary for separation using a finite element modelling software (COMSOL Multiphysics) to interrogate the electric field, gradient of the electric field squared, and the ratio of the two. Channel molds have been etched from plastic using a CO2 laser and used for preliminary experimental work The novel design, models, and preliminary experimental results will be presented.

Mechanism of sequence-based separation of single-stranded DNA in capillary zone electrophoresis

Jia Zhao, Steve Cramer, Linda McGown

Separation of DNA has wide application in biology, medicine, human health and forensics. Genome sequencing, genotyping, phenotyping as well as metagenomic analysis of pathogenic organisms and microbial communities have all relied on methods for DNA separation, most commonly based on DNA fragment length. Fragments of same length can then be sequenced or separated by sequence. Even though researchers are working on new generation sequencing techniques, DNA sequencing is still tedious and sometimes not practical for certain samples. Separation based on sequence would be satisfying as a rapid way to acquire information about the sequence diversity of same-length strands. DNA separation based on length is readily achieved using sieving gel electrophoresis, while sequence-based separation is not as easy. Existing methods usually require sufficient differences in conformation or in resistance to thermal or chemical denaturation and sometimes even the knowledge of target sequences. A simple approach to sequence-based separation of DNA would be a powerful advance in DNA analysis, especially for genetically complex samples such as microbial communities. Previous work in our group showed that same-length fragments of single-stranded DNA differing by only one or two bases could be separated by using high salt concentration buffers in capillary zone electrophoresis (CZE). To investigate the mechanism of sequence-based separation, we first calculated the mobility of different strands, noticing only very tiny difference in mobility. Long migration time could be one of potential reasons for the separation. Thus, we explored the effect of migration time. Since CZE separates samples based on charge to size ratio, we also characterized the charge and size (conformation) of DNA using multiple biophysical techniques, such as fluorescence anisotropy, small angle X-ray scattering. At last, an empirical model of DNA mobility as a function of components was proposed, which could potentially to predict mobility of random ssDNA.

Initial Evaluation of a Commercially Available MicroFluidic Capillary Electrophoresis Mass Spectrometric Interface

Joseph Snodgrass, Alexander Langston, Eric Chan, Shuojia Bai, Eduard Luss, Ricardo Borjas

Initial experiences using a new commercially available microfluidic chip-based capillary electrophoresis (CE) interface for electrospray (ESI) mass spectrometry show encouraging results for intact mass characterization of proteins and peptide mapping applications. The interface is trademarked as the "ZipChip" interface for CE-ESI by

908 Devices, using technology developed by the J. Michael Ramsey group at the University of North Carolina. We have demonstrated ~4 minute electrophoretic separations of charge variant impurities of a ~26kDalton protein. The separations were comparable to those achieved by ion exchange (IEX) methods running 20-40 minutes. The ZipChip separation technique is fully compatible with mass spectrometric detection. In contrast, the high salt content mobile phases used for the ion exchange methods required fraction collection or other 2-dimensional techniques before informative mass spectral data could be obtained. Significant efficiency improvements have also been demonstrated for peptide mapping applications. In other work, we have seen efficient separation of various alpha, beta and gamma hemoglobin subunits. We discuss our experiences using a late-stage beta version of the ZipChip interface, utility of the data as part of full protein characterization and process development workflows, and improvements made by the vendor incorporated into the currently versions of the interface.

OPTIMIZATION AND CHARACTERIZATION OF THE SOLUTION ELECTRODE INTERFACE FOR IMPROVED PERFORMANCE OF THE SOLUTION CATHODE GLOW DISCHARGE

Stuart Schroeder

The solution cathode glow discharge (SCGD) is a spectrochemical plasma source that is particularly suited to the direct analysis of dissolved inorganic ions and may emerge as the preferred technique for on-line analysis relating to industrial process control applications. As in traditional glow discharge devices, the SCGD consists of a plasma sustained between and anode and cathode. However, unlike a traditional glow discharge, the SCGD consists of a liquid cathode that is the sample solution and operates at atmospheric pressure in an air environment. Many designs have been described showing how the acidified flowing waste sample solution becomes the electrical path to ground potential. A new design based on a porous ceramic wick as an interface between the base of the plasma and electrical ground will be described. This plasma engine has been engineered to be designed for manufacturability and also incorporates a simple structure to divert hydrogen gas from directly entering the plasma region. Stabilizing the dc circuit with a porous wick allows for reduced sample flow rates and higher power settings, resulting in lower detection limits for this technique. Detection limits for many elements compare well to the limits achieved with inductively coupled plasma atomic emission spectrometry (ICP-AES) and are near or below 1 ng/mL. For the alkali metals, the SCGD demonstrates superior sensitivity with detection limits approaching 1 pg/mL. Overall system design including the containment of acid vapors along with analytical figures of merit will be presented.

Industrial SCGD – a Technique for Inline Analysis of Metals in Liquids

David Malmström, Arne Bengtson

Solution Cathode Glow Discharge (SCGD) is a simple technique but has yet shown to be powerful with its compact size, cost efficiency and practical detection sensitivity. SCGD works by applying a high voltage between a sample solution and an electrode forming a plasma. The emitted light (atomic emission) is then utilized to identify and quantify the elemental composition of the solution. SCGD can show numerous benefits compared to Inductively Coupled Plasma (ICP) such as no need for a nebulizer, carrying gas or cooling system. Due to its simplicity, the system can easily be modified depending on application and analytical need and even made portable. The sensor also has the capability of controlling secondary process based on determined content and concentrations. A complete automatic instrument has been built with industrial applications in focus. Different hardware configurations and fundamental parameters, such as choice of pump, materials, plasma height, applied current, liquid flow etc. has been investigated and evaluated. Large efforts have been made to get a stable system with minimized stray light coming from the electrode. Large efforts have also been put into the software with the ability to control the hardware as well as analyse the outcome. The results showed an automated, inline cost efficient analysing sensor that has the capability of detecting low levels of metals in e.g. waste water or processing fluids coming from e.g. the steel industry to a much lower price than commonly used techniques.

Enhanced Performance of the Liquid Sampling-Atmospheric Pressure Glow Discharge Ion Source for Elemental Isotope Ratio Mass Spectrometry

Edward Hoegg, Garret Hart, David Koppenaal, George Hager, R. Kenneth Marcus

The liquid sample-atmospheric pressure glow discharge (LS-APGD) ion source has successfully been interfaced with Exactive Orbitrap mass spectrometers in order to develop a field deployable system capable of highly accurate and precise isotope ratio analysis. This system offers many positives over traditional elemental isotope mass spectrometers (i.e. thermal ionization and inductively coupled plasma multi collector instruments) including lower operational overhead, less required training and the ability to simultaneously measure samples with multiple elements present. Recently, the authors have reported that a natural U sample ($^{235}U/^{238}U = 0.007258$) can be measured as 0.007041 with a relative standard deviation of 0.158%. This level of precision is important because it preliminary meets the international target value set by the International Atomic Energy Agency. While these results are promising, the lack of research using either the LS-APGD or Orbitraps for isotope ratio mass spectrometry prevents the authors from making conclusions about the individual components thus limiting our understanding of the system. In order to further improve the accuracy and precision of isotope ratio measurements the LS-APGD will be scrutinized so that the sources of variability can be teased out. To this end the research presented here will focus on measuring the variation in the source voltage as a function of the operating parameters, including the flow rates, current, and electrode position. This data will be examined for any potential correlations with variations in the measured ion current which will be measured at the end of the Exactive's ion transfer capillary and prior to any filters or the Exactive's C-trap that would perturb the data. By isolating the sources of variability present, it is expected that the levels of precision will be further improved.

Analysis of Solid and Aqueous Samples for Atomic and Molecular Species with Solution-Cathode Glow Discharge Mass Spectrometry (SCGD-MS)

Jacob Shelley, Andrew Schwartz, Courtney Walton, Garett MacLean, Judy Wu, Gary Hieftje

Atmospheric-pressure glow discharges (APGDs) that utilize a flowing liquid as one or both electrodes have received increased attention lately as alternative sources for atomic emission spectroscopy. Though initially developed as alternatives to the inductively coupled plasma (ICP), these compact, low-power plasmas have been used more recently as tools for atomic and molecular mass spectrometry. Liquid-electrode discharges are unique in that they have the ability to create a wide-range of energetic species useful for atomization and excitation as well as desorption/ionization. These plasmas produce highly energetic species (e.g., ions, metastable neutrals, fast electrons, etc.), which can lead to high-energy physical or chemical processes to fragment and ionize molecules. Meanwhile, lower energy ionization reactions can also occur due to the abundance of collisional cooling that takes place at ambient pressures. This presentation will focus on the analytical capabilities of a liquid-electrode plasma called the solution-cathode glow discharge (SCGD). The SCGD is an APGD sustained on the surface of a flowing solution with ambient air serving as the discharge gas. The plasma volatilizes a portion of the liquid, and any chemical compounds contained therein, leading to desorption/ionization. Ions produced by the SCGD were sampled into a high-resolution Orbitrap mass analyzer. It was found that SCGD-MS could be used for the detection of atomic, molecular, and biological species directly from solutions. Furthermore, it was found that molecular species could be tunably fragmented through adjustment of SCGD operating parameters. For instance, peptides were tunably fragmented at atmospheric pressure, which led to 100% sequence coverage for many of the peptides examined. The power of this sequencing approach is further realized through coupling SCGD-MS with ultra-high performance liquid chromatography (UPLC). For the determination of elemental species, SCGD-MS was found to offer detection limits in the sub-ppb range for the analytes examined. The exceptional stability of the SCGD as an ionization source will be demonstrated through the measurement of isotope ratios for a number of elemental species. In the case of uranium, isotope ratio precisions with SCGD-MS were found to be on par with or better than the gold-standard approaches of thermal-ionization mass spectrometry (TIMS) or multicollector ICP-MS.

Pushing the limits: Exploration of pulsed Solution Cathode Glow Discharge.
Jaime Orejas Ibanez, Andrew J. Schwartz, Steven J. Ray

A simple but effective device based on a glow discharge in contact with a liquid has proof in the last decade to be a valuable analytical tool for elemental detection and quantification: the Solution Cathode Glow Discharge (SCGD). The simplicity of operation has not been a hindrance to offer interesting analytical figures: part-perbillion detection limits for many elements using Atomic Emission Spectrometry detection and even promising capabilities as a Mass Spectrometry ionization source. Moreover, the SCGD is typically powered by direct current continuous electrical supply of less than 1 kV potential and typically no more than 80 mA current intensity, limiting the power consumption below 80 W. Apart from maintaining low consumption, higher applied powers are normally non desirable to prevent malfunctioning due to high heat production. With the aim of pushing further its analytical capabilities, in our lab we have developed an experimental set-up capable of producing short time pulses (~ a hundred of ns) of increased applied potential and current, leading to a more energetic plasma with instantaneous power peaks above 500 W, without affecting the system due to high heat production. Additionally, the time scale of the pulse event can be short enough to provide different plasma characteristics, compared to the pure direct current SCGD version. The work to be presented will focus on the exploration of the changes in the discharge through the study of its optical emission spectrum in order to assess the analytical applicability of the pulsed SCGD.

Fluorescence microscopy for biomedical applications: Imaging nanoparticle-cell interactions
Christine Payne

Nanoparticles offer exciting new approaches for biomedicine ranging from drug delivery to cellular imaging. In the course of these applications, nanoparticles are exposed to a complex mixture of extracellular proteins that adsorb onto the surface of the nanoparticle. This "protein corona" dominates the interaction of nanoparticles with cells. We have investigated how proteins found in blood serum affect the cellular binding of protein-nanoparticle complexes. Using fluorescence microscopy, we find that the cellular binding of cationic nanoparticles is enhanced by the presence of serum proteins while the binding of anionic nanoparticles is inhibited. Competition assays confirm that these protein-nanoparticle complexes use distinct cellular receptors. Circular dichroism spectroscopy, fluorescence spectroscopy, and isothermal titration calorimetry show that the secondary structure of the adsorbed serum albumin is altered following adsorption on the surface of cationic nanoparticles. These structural changes redirect the albumin-nanoparticle complex to scavenger receptors. In comparison, the secondary structure of albumin is preserved following adsorption on anionic nanoparticles. The cellular binding trend is independent of nanoparticle composition: quantum dots, colloidal gold nanoparticles, and low-density lipoprotein particles all show the same behavior. This link between protein structure and cellular outcomes will provide a molecular basis for the design of nanoparticles for use in biomedical applications.

SERS Optophysiology of small organic metabolites in biosensing applications
Jean-Francois Masson, Felix Lussier, Benjamin Charron, Dimitris Missirlis, Joachim Spatz, Thibault Brulé

Monitoring small organic metabolites is of interest in a series of applications. For example, we are interested in detecting cellular metabolites near cells and small organisms such as algae. SERS is a prime technology for identifying and quantifying small organic molecules. We have recently designed a SERS nanoprobe capable of rapidly (ms range) acquiring the vibrational spectra of a series of small organic metabolites, such as ATP, urea, lactate, pyruvate, and neurotransmitters. In this presentation, we will expose our progress to locally probe the ATP concentration in the extra and intracellular environment of different cell types (cancerous, endothelial, fibroblast and nervous cells) in order to investigate the hypothesis regarding the low concentration of extracellular ATP. In order to analyze the complex and heterogeneous optical response by SERS-Optophysiology, a novel tool allowing a selective multiplex sensing of Raman active molecules was developed and will be presented. Potential applications for monitoring chemical cell perforation using saponin and for detecting small organic acids in algae cultures will also be presented.

Resonance Raman and Fluorescence Studies of Membrane Proteins
Judy Kim

Membrane proteins continue to be a focus of modern biochemical and biomedical research. Numerous complementary techniques reveal the structures and dynamics of these uniquitous biomolecules. Our group utilizes the resonance Raman spectroscopy and fluorescence quenching to probe the dynamics of membrane proteins. The focus of our experiments is a beta-barrel membrane protein and the mechanisms of folding into synthetic bilayers. Resonance Raman spectroscopy reveals changes in protein-lipid interactions, such as alterations in hydrogen bonding and local environment of anchoring aromatic residues, while fluorescence methods highlight tertiary structure and solvent accessibility. Collectively, these complementary methods elucidate the evolution of structure during insertion and folding of a protein into lipid bilayers.

Imaging lipid turnover rates in mouse brains with desorption electrospray ionization mass spectrometry
Paul Farnsworth, Richard Carson, Charlotte Lewis, Mercede Erickson, Anna Zagieboylo, Bradley Naylor, Kelvin Lee, John Price

Desorption electrospray ionization (DESI) mass spectrometry has proven to be an effective tool for chemical imaging on a sub-millimeter spatial scale. It is particularly effective for imaging lipids. We have exploited the imaging capability of DESI by combining it with stable isotope labeling and isotope ratio measurements to map spatially-dependent lipid turnover rates in mouse brains. Mice were initially injected with a bolus of saline D_2O and then were provided free access to D_2O enriched drinking water. The animals were sacrificed at
increasing intervals in a 40-day experiment, and brain sections were imaged by DESI mass spectrometry. From the changing neutromer distributions of four selected lipids we were able to determine the rates at which the lipids were synthesized and incorporated into the brain tissue on a spatially-selective basis. There were clear regional dependencies for three of the four test lipids, and those regional dependencies varied from lipid to lipid. In this presentation I will describe the results of this initial proof-of-principle experiment in detail, and discuss the potential of the technique for the study of diseases for which changes in lipid metabolism are implicated as causes or symptoms of the diseases.

Investigating Photophysical Properties Associated with Bumble Bee Venom Antimicrobial Peptide-Membrane Interactions
Matthew Roberson, Devin Smith, Simon White, Ian Wallace, Matthew Tucker

Recent interest in the use of naturally occurring antimicrobial peptides from insects as possible therapeutics and antifungals has emerged. Our research evaluates the structural and dynamic changes that arise from the antimicrobial peptide-membrane interactions. We have performed comparative spectroscopic studies on antimicrobial bombolitin peptides found in the venoms of different bumblebee species in the presence of membrane mimics as well as bacterial and fungal cells possessing various membrane compositions. The antimicrobial selectivity and action of both bombolitin peptides have been assessed via a variety of spectroscopic methods such as FT-IR, fluorescence, CD, and 2D-IR spectroscopy. Furthermore, we assessed the ability of these peptides to inhibit the growth of model bacterial (Escherichia coli) and fungal (Saccharomyces cerevisiae) organisms. Key differences in the helical content as well as the non-conserved residues of these peptides have led to variances in the selectivity for membranes of different lipid compositions.

Non-Destructive Recovery of Defaced Serial Numbers Using Infrared Thermal Imaging
John Kalivas, Ikwulono Unobe, Lisa Lau, Andrew Sorensen, Rene Rodriguez

Infrared thermography is useful in nondestructive evaluation of structural integrity. Subsurface defects produce non-uniform heat dissipation and this phenomenon is captured by an infrared camera. Pulsed thermography has emerged as the most widely used technique where the specimen surface is heated with a brief pulse of heat and surface temperatures are monitored with an IR camera. In this study, lock-in thermography (LIT) is used to recover defaced serial numbers based on the underlying deformation from the stamping or laser etching process. The collection of LIT phase images are unfolded (as is common with hyperspectral imaging) to form a sequence of principal component score images. In order to avoid selection of specific score images for further examination, all images are used in a data fusion approach for a consensus analysis. In order to identify a defaced number, multiple numerical library images are matched to each score image by a collection of similarity measures. Prior to computing similarity measures between a library number image and a defaced score image, all images are decomposed to Zernike moments by respective Zernike polynomials. Zernike polynomials form an orthogonal basis set allowing extraction of image features describing shape characteristics of an imaged object. Fusion is again the main tool for analysis of the multiple libraries and similarity measures providing consensus identification of a defaced number. Results are presented for a series of known defaced numbers on stainless steel as well as recovery of the VIN defaced on a stolen motorcycle.

The Use of LA-ICP-MS Databases to Estimate Likelihood Ratios for the Forensic Evaluation of Glass Evidence
Jose Almirall, Daniel Ramos, Ruthie Corzo

Laser Ablation-Inductively Coupled Plasma-Mass Spectrometry (LA-ICP-MS) has been shown to be an excellent technique for the discrimination of glass from different sources and for the association of glass originating from the same source. A match criterion comparing the known and questioned sample (for example, a t-test or a range overlap) is employed and if the glass samples are determined to "match" this is followed by the use of a verbal scale to report the forensic practitioner's conclusion. This approach has several disadvantages: a fixed match criterion suffers from the "fall off the cliff effect;" the rarity of an elemental profile is not taken into account; and the use of a verbal scale is subjective. An alternative approach includes the use of a continuous likelihood ratio that provides a quantitative measure of the significance of the evidence and accounts for the rarity of an elemental profile through the use of a glass database. In the present study two glass databases were used to evaluate the performance of the likelihood ratio; the first database comprised 420 vehicle windshield samples, while the second database comprised 398 known glass samples from casework. The two-level model proposed by Aitken, Zadora, and Lucy was used for the calculation of the likelihood ratio. However, this model led to unreasonable (too high or too low) likelihood ratios. A Pool Adjacent Violators (PAV) algorithm post-hoc calibration step was necessary in order to improve the accuracy of the likelihood ratio. The results of the calibrated likelihood ratio, and a comparison to the match criteria currently in use, will be presented as a viable alternative for the reporting of significance of glass evidence that is objective and quantitative.

Modeling, inference, and predictive calibrations for forensic spectroscopic data
Stephen Morgan, Michael Myrick, Edsel Pena

Statistical modeling of large spectroscopic data sets is complicated by the size if individual spectra (often several thousand or more wavelengths) and the need for samples to adequately characterize data rank. A variety of methods have been used to determine the rank of a spectral data matrix, but often the true rank is overestimated by a degree that is dependent on the method employed. Although we still use the common methods (e.g., PCA) as a check, determining data rank in this way is not how we plan to determine the minimum number of wavelength bands, nor does it offer insight into which wavelength bands should be sensed. Traditional statistical analysis has focused on hypothesis testing based on the Fisher and Neyman-Pearson paradigms. Stepwise modification of a model is one approach for assessing parameter weights; however, stepwise regression is only viable for a small number of parameters. Chemometric approaches to variable selection typically often involve calculating Student t-tests for two groups, or Fisher F-ratios for more than two groups, at each wavelength to judge the ability to discriminate groups of spectra. However, pursuing the goal of multiple testing in this way guarantees loss of control of the level of significance. Classical least squares fitting of regression models minimizes the sum-of-squares of residuals, which coincides with the maximum likelihood approach under normal error terms. Modern statistical problems involving large data sets have benefited from the development of large-scale methods for inference that outperform conventional frequentist-based methods by taking advantage of empirical Bayes strategies that combine variable selection, testing, and estimation to produce parameter estimates that reduce the variance and increase prediction accuracy. Alternatively, modern sparse modeling approaches do not experience ill-conditioning as a result of correlated data and near-singular matrices. General linear models that include multiple grouping related parameters, a large set of covariate effects, and perhaps even a Bayesian-related estimation approach is also commonplace in "big data problems."

Building Forensic Science on a ROC

Michael Sigman, Mary Williams

Many forensic science disciplines in the United States continue to provide reports and testimony using categorical statements, even in the face of pressures to move toward probabilistic statements which reflect evidentiary value. The 2009 report "Strengthening Forensic Science in America: A Path Forward" from the National Academies of Science pointed to the need to reform reporting strategies. The 2016 report on feature comparison methods by the President's Council of Advisors on Science and Technology made similar ascertains with regard to pattern evidence. Statements of evidentiary value, in the form of a likelihood ratio, are promoted by the European Network of Forensic Science Institutes' guideline on evaluative reporting, which was published in 2016. The Receiver Operating Characteristic (ROC) construct provides a natural bridge between the cultures of categorical and probabilistic reporting. An ROC curve is created by plotting the true positive rates and false positive rates for a series of binary decisions made at different decision thresholds. The likelihood ratio, as a measures of evidentiary value, can be related to slope of the ROC curve in several ways and provides a visual and conceptual relationship between categorical and probabilistic statements in forensic science. The role of the ROC curve as the foundation for a needed paradigm shift in forensic science will be reviewed in this presentation. ROC curves developed for the analysis of fire debris evidence will be presented as an example of this powerful technique.

Modeling and Performance Evaluation of Real-Time Standoff Hazardous Materials Detection
Arjun Bangalore, Mathew Nelson, Shawna Tazik, Robert Schweitzer, Patrick Treado

Detection of trace amounts of materials in complex backgrounds can be very challenging. However, recent advances in hyperspectral imaging (HSI) systems have improved automated detection and identification of a variety of targets in the presence of complex backgrounds. Operating near the physical limits of a HSI system capability is challenging, but is often enabled by rigorous modeling of detection performance. Specific performance envelopes we consistently strive to improve include: operating under low signal to background conditions; at higher and higher frame rates; and under less than ideal motion control scenarios. A standoff sensor architecture we have been maturing includes different types of imaging spectrometers such as a liquid crystal tunable filter, multi-conjugate filter, or conformal filter. This presentation will explore current generation standoff sensors, including recent modeling results and detection performance testing for a wide array of materials.

Sample Preparation Approaches for Laser Ablation Inductively Coupled Plasma Mass Spectrometry
Jhanis J. Gonzalez

Laser ablation (LA) as a direct solid sample introduction tool for inductively coupled plasma (ICP) mass spectrometry (MS) and optical emission spectrometry (OES) has become one of the most reliable alternatives to avoid acid dissolution and chemical separation steps conventionally associated with solid sample analysis. Certainly, phrases like 'no sample preparation' are often used in expressing the benefits of sample analysis by LA-ICP-MS, although, strictly speaking, these comments refer to no wet chemical sample preparation. Simple processing like milling, pulverizing, and pressing to a pellet (with or without binders or other additives) are common practice for laser ablation. However due to solid matrix variability and its influence on the ablation processes, matrix dependent elemental fractionation, lack of matrix matched standards for external calibration, and limitations to internal calibration because it is challenging to add and distribute spikes into solid samples. In this presentation, we will discuss some of the common approaches for simple solid sample preparation as well as we will introduce the concept of Matrix modification. The aim of this minimal sample preparation procedure is to transform the original matrix to a more consistent one for LA-based analysis, thus reducing the effects of matrix variability, while enabling the addition of tracers.

Initial Attempts Into Characterizing Surrogate Nuclear Fireballs
John Auxier II, Eric Francis, Lajos Majcos, Howard Hall

In the immediate seconds after the detonation of a nuclear weapon a large plasma cloud is formed. This plasma cloud has a unique environment due to the pressure and temperatures that are achieved. To simulate the high temperature efforts, collaborations between Oak Ridge National Laboratory have been leveraged to setup an industrial grade plasma torch. This work will leverage the existing collaboration and laboratory facilities to investigate the plasma chemistry in real time to determine what processes of formation (e.g. oxide, nitride, sulfide, etc.) occur in different regions of the plasma (e.g. edge, center, etc.) using chemical compositions for the fallout recipes derived using the methods developed previously. After particulate matter is formed in the plasma region, the experimental chamber is set up to collect samples for investigation by scanning electron microscopy (SEM) and other similar morphological and surface chemistry characterization techniques. This talk will detail the early measurements that have been made in the UV-Vis region with regard to detection of elemental species.

Uranium Isotope Ratios by Laser Absorption Spectroscopy
Alonso Castro, Joshua Bartlett, Vyacheslav Lebedev

We present our results on the development of a technique for the determination of isotope ratios in solid uranium samples. The method is based on performing laser absorption spectroscopy through a collimated atomic beam of uranium generated by a high-temperature micro-crucible in rough vacuum. Disposable micro-crucibles are fashioned from tantalum foil, which are resistively-heated to temperatures up to 2500 C. A laser beam from a compact, tunable, external-cavity diode laser (ECDL), with a center wavelength selected to match a specific atomic transition of uranium is intersected with the uranium atomic beam. By detecting the photon absorption through the atomic beam while rapidly tuning the laser frequency across a series of specific isotopes, a spectrum showing their relative abundance is obtained. The isotope shift, which depends on a specific atomic transition (typically a few GHz) is much larger than the sub-MHz ECDL linewidth. The method, being completely optical in nature, is immune to isobaric interferences, a substantial problem with mass spectrometry methods. We have demonstrated the determination of isotope ratios of metallic uranium and its oxides at various levels of enrichment.

Cat Litter, Nitric Acid, Plutonium, and $2.4 Billion Dollars: the microscopic investigation into the 2014 Contamination at WIPP

Jon Schwantes, David Atkinson, Edgar Buck, Carlos Fraga, Larry Greenwood, Bruce McNamara, Michael Minnette, Randal Scheele, Luke SweetJon Wahl

A recount of the vital role Pacific Northwest National Laboratory (PNNL) played into elucidating the cause of the radioactive contamination of the US's only transuranic waste repository, the Waste Isolation Pilot Plant (WIPP), that occurred on February 14, 2014, will be provided. The effort put forth by over 40 PNNL staff, employing numerous microscopic techniques, culminating in over 30,000 man-hours, resulted in revealing the underlying chemical mechanisms that caused the accident. On September 18th, 2016, 10 of those PNNL staff members were awarded the highest non-monetary award within Department of Energy, The Secretary's Achievement Award, for their role in the investigation.

Near-IR Quality Control of a Drug to Treat a Rare Tropical Disease

Robert Lodder, Mayte Hernandez-Murtillo

Ebola is characterized by a high fatality rate which has varied from 25% to 90% in past outbreaks. After an incubation period ranging from 2 to 21 days, nonspecific symptoms appear, including sudden onset of fever, weakness, vomiting, diarrhea, headache and a sore throat. A fraction of patients may later develop severe internal and external hemorrhagic manifestations and experience multiple organ failures. Laboratory findings include low white blood cell and platelet counts and elevated liver enzymes. Early supportive care with rehydration and symptomatic treatment improves survival. Until recently, Ebola outbreaks of occurred mostly in sub-Saharan Africa. However, there are indications that Ebola viruses may be more widespread. Sequencing of Ebola viral genomes from the recent outbreak in Africa demonstrated considerable genetic variation, not only when compared to the genomes of viruses from previous outbreaks, but also between virus samples collected during the latest outbreak, complicating vaccine development. BSN389 is an experimental therapy for active Ebola virus infections. BSN389 is not a vaccine, and has broad antiviral applicability to Ebola. It is stable, but poorly soluble, so formulation is very important in delivery. The wrong formulation can render the drug product useless. This paper describes how near infrared spectroscopy with integrated sensing and processing can be used as a process analytical technology to identify the proper formulation of BSN389 and to accelerate manufacturing.

Determination of the Volume of Interrogation of Powder for NIR Monitoring of a Continuous Powder Flow

Carl Anderson, Anik Alam, Zhenqi Shi, James Drennen

In-line monitoring of continuous powder flow is an integral part of the continuous manufacturing process of solid oral dosage forms in the pharmaceutical industry. Specifically, monitoring downstream from loss-inweight (LIW) feeders provides important data about the state of the process. Such measurements support control of the process and thereby enhance product quality. Near Infrared Spectroscopy (NIRS) is a potential PAT tool to monitor the homogeneity of a continuous powder flow stream in pharmaceutical manufacturing. Information regarding the volume of interrogation of NIR method is critical during stratified sampling to monitor a continuous process. Understanding the effect of process parameters on the volume of interrogation is also critical to design and optimization of a process to probe representative sample by the analytical method. In this study, a method was developed to determine the volume of interrogation of NIR method for continuous powder flow. Two types of materials, one possessing good flow properties and the other possessing poor flow properties, were used to in this study. Powder blends were set to flow from a loss-in-weight feeder at different flow rates and descendent tube angles. NIR transmission spectroscopy was used to determine the thickness of the powder stream. The thickness of the powder stream and volumetric flow rate was used to calculate the linear velocity of the powder stream. The volumes of interrogation for NIR methods were calculated from the linear velocity, penetration depth and volumetric flow rate. The effect of flow rate and tube angle on the volume of interrogation was studied. Powder flow rate was found to affect the volume of interrogation at a greater extent compared to the descendent tube angle. The difference between volumes of interrogation of the transmission and reflectance NIR method were dependent on the powder flow rate. At high flow rate, the difference was larger compared to the lower flow rate. This study provides an in-depth understanding about the volume of interrogation of NIR method for continuous powder flow and its interaction with process parameters. Such an understanding would be critical in demonstrating NIR method as a PAT tool for monitoring continuous powder flow.

Multivariate Curve Resolution (MCR) and PARAFAC Analysis of Infrared Specular Reflection Spectra of Quartz Particles: Strategies for Optimal Surface Sensing

Thomas Blake, Paul Gassman, Neal Gallagher

Infrared reflectance measurements have been used extensively to characterize soil samples, and the ability to acquire measurements at multiple angles promises to improve measurement selectivity, and separate physical (e.g., particle size, shape, packing density, etc.) and chemical information. It also provides the potential to enhance detection and characterization of anomalies and targets of interest. This study made infrared reflectance measurements of quartz samples with five narrow particle size ranges, three soil samples, and a piece of sandpaper using an FTIR and specular reflectance accessory. The measurements were made at twelve different angles of incidence. Light from the spectrometer was partially polarized. Although classification approaches can be based on a large number of available algorithms (e.g., PLS-DA), multivariate curve resolution (MCR) and parallel factor analysis (PARAFAC) have the unique ability to provide highly interpretable results. Interpretability is afforded because MCR is based on the classical least squares (CLS) model while PARAFAC can be considered a multi-linear CLS model and is a distinct advantage because it provides confidence in the underlying model and can be used to tune the model using a priori information (e.g., library measurements). This talk with show the results for a set of measurements on quartz and soil samples and compare MCR and PARAFAC results. The analysis shows optimal angle of incidence ranges for different particle sizes where contrast between volume scattering and Reststrahlen regions can be increased, thus improving detection of chemicals that may be on those materials.

Possible application of Near Infrared reflectance Spectroscopy to quantify a parasite (Perkinsus marinus) in oyster tissues (Crassostrea virginica)

Eric Guevelou, Jessica M. Small, Standish K. Allen Jr.

Disease resistance is the most important economic trait in agriculture and so too, in aquaculture. Since the 1930s, wild populations of the eastern oyster (*Crassostrea virginica*) have been affected predominantly by two major pathogens: *Perkinsus marinus* and *Haplosporidium nelsoni*. As resistance to infection of *H. nelsoni* was proved to be heritable, breeding selection through the last decades helped to develop resistant oysters. True resistance to the second pathogen *P. marinus* has been elusive and remains a critical breeding objective and a continued limitation for oyster farmers. This protozoan parasite proliferates when salinities are above 15ppt and when water temperatures are above 20 C, so is most notably an issue for eastern oysters during the summer and fall. Infection occurs through ingesting infective particles and in advanced infections, disruption of the host gut epithelia occurs as well as destruction of other tissues. NIRS technology provides a fast, safe, and eco-friendly alternative to traditional analytical methods which may need the use of chemicals that are toxic to humans and the environment. Since 2012, Aquaculture Genetics and Breeding Technology Center (ABC) has been developing NIRS technology for use in its breeding program and has developed several models suited for compositional analysis of moisture and glycogen in *C. virginica*. These models were developed using both homogenized tissues and freeze-dried tissues and showed a high correlation between the predicted and measured data. With all the advantages of NIRS, especially the speed of acquisition to quantify parameters of interest, ABC aims to determine the feasibility of creating quantitative models to measure the presence of *P. marinus* in *C. virginica* tissues. For this study we developed two models: one using homogenized tissues and the second using the same tissues after lyophilization. To the best of our knowledge, this is the first attempt to develop a NIRS model to quantify a parasite in oysters.

Using Near Infrared Spectroscopy Monitoring Heavy Crude Oil of Production
Toni Miao, Ajit Pradhan, Michael Moir, Eddy Lee

In this presentation, we will discuss applications of Near-IR spectroscopy and PLS analysis for monitoring production and transportation of heavy oil. Heavy crude oil resources are more than double the conventional light oil reserves worldwide. However, heavy crude oil exploitation is faced with technical challenges at all stages, from recovery/production to transportation and refining due to high viscosity alongside the presence of undesirable components such as asphaltenes, heavy metals and sulfur. Use of diluents for the recovery of heavy oil and bitumen began to evolve in seventies. Widely used diluents include condensates from the natural gas production, naphtha, kerosene, light crude oil, etc. Current methods for measuring the amount of diluent in produced fluids are slow and often inaccurate. Near-IR is sensitive to hydrocarbons and powerful for intermolecular identification when compared traditional Mid-Infrared (MIR) spectroscopy. This talk will focus on various applications of NIR spectroscopy and PLS analysis for monitoring production and transportation of heavy crude oil and bitumen in the presence of diluents.

Molecular laser-induced breakdown spectroscopy
Christian Parigger

Recombination molecular spectra are routinely measured subsequent to plasma generation with pulsed laser radiation. Typical laser-induced plasma histories are characterized by time-resolved spectroscopy methods. The recorded spectra show occurrence of atomic and molecular signatures including background contributions that depend on the laser pulse-width in standard ambient temperature and pressure environments. In this work, results from experiments are communicated with focus on molecular emission spectroscopy. The interests include carbon Swan spectra (C_2), cyanide (CN), aluminum monoxide (AlO), hydroxyl (OH), and nitrogen
monoxide (NO), to name a few. Some of these spectra are clearly discernible for time delays of 1 to 100 microseconds, however, recombination radiation of CN occurs within the first 100 nanoseconds for laser-induced breakdown in 1:1 CO_2:N_2 gas mixtures when using picosecond and nanosecond laser pulses. Aspects
of the analysis including Abel inversions and computational modeling of the plasma are discussed together with how to obtain inferences of molecular excitation temperatures. For nominal 6 nanosecond laser pulses and for pulse-energies in the range of 100 to 800 milli-Joules, expansion dynamics and turbulence due to shock phenomena are elucidated to address local equilibrium details that are frequently assumed for atomic and molecular species emission spectroscopy diagnostics.

Predicting molecules in laser induced plasma based on equilibrium plasma model
Igor Gornushkin, Sergei Shabanov, Ulrich Panne

A model of laser induced plasma includes the coupled Navier-Stokes, state, radiative transfer, and material transport equations and incorporates plasma chemistry through the equilibrium approach based on the use of atomic and molecular partition functions. Several chemical systems are modeled including ablation of metals and non-metals in nitrogen, argon and air atmospheres with the formation of molecules and their corresponding positive and negative ions. The model is used to obtain number densities of atomic and molecular species in the expanding plasma plume in both the single and double pulse operation modes. The results are compared to experimental observations obtained by optical imaging and tomography. The model and experiments elucidate mechanisms of formation of molecules in LIPs and provide recommendations for their detection.

Overcoming spectral interference in LAMIS with robust statistical spectral fitting
George Chan, Xianglei Mao, Vassilia Zorba, Richard Russo

Laser Ablation Molecular Isotopic Spectrometry (LAMIS) is a novel laser-based all-optical technique for isotopic analysis in ambient environment. This technique measures the molecular emission spectra of those radicals that are formed at a late time scale in the plasma formation process, when ablated atoms begin interacting with atmospheric species to create excited-state molecules (e.g., oxides). As molecular spectra typically exhibit two to three orders of magnitude increase in isotopic splitting compared to atomic transitions, the splitting can be readily measured with a relatively low-resolution optical spectrometer. One method to extract isotopic information from the measured LAMIS spectrum is multi-variable non-linear spectral fitting. This method fits the experimental spectrum to a dataset, which consists of all the rovibronic line positions of the molecule in interest. The fitting is typically performed through the common least-square approach. One well-known limitation of least-square fitting is that the fit is very sensitive to outlier (e.g., spectral interference). Experience showed that just a few relatively mild overlapping spectral interferences can notably degrade the analytical result. The solution to this outlier limitation is through the use of robust statistics – to identify the outliers, ignore them and perform least-square fitting on the remaining data set. In this presentation, the use of robust statistics for spectral fitting will be described, and their effectiveness to overcome spectral interference in LAMIS will be assessed.

Partial Least Squares Calibration Modeling towards the multivariate limit of detection for enriched isotopic mixtures via Laser ablation molecular isotopic spectroscopy
Candace Harris, Luisa T.M. Profeta, Codjo Akpovo, Ashley C. Stowe, Lewis E. Johnson

An IUPAC novel approach for determining the limit of detection via laser ablation molecular isotopic spectroscopy (LAMIS) for its viable sensitivity (down to ppm) and discrimination in isotopic analysis is presented here. The calibration set consists of experimental spectra from the laser ablation of single shot spectra of zirconium metallic particles ($^{94}Zr/^{90}Zr$) dispersed on the surface of silica substrates, boric acid mixtures with various isotopic ratios $^{10}B/^{11}B$ and uranium glass with $^{235}U/^{238}U$ isotopic ratios were used as the predictors for the model, respectively. For the analyte(s) of interest, the generated plasma using a Q-switched Nd:YAG excitation laser operating at 532/1064 nm has been examined for their molecular bands. Only bandhead structures with prominent intensities and wide energy separations were selected for this study. To ensure single particle spectra, the laser beam was focused to spot size of ~100 ?m on the sample and a CCD microscopy system was employed for imaging single particle laser induced plasma interaction. The calibration model was created using partial least squares regression (PLSR) and the parameters for such along with the instrumental signal variance from the detector were used in determining the limit of detection for the isotopic ratios. The multivariate limit of detection interval was compared to other methods for determining the LOD to demonstrate the effectiveness of the novel approach.

Chemical imaging of thin films with high spatial resolution by femtosecond laser-induced breakdown spectroscopy

Johannes D. Pedarnig, Christoph M. Ahamer, Kevin M. Riepl, Norbert Huber

We use femtosecond laser-induced breakdown spectroscopy (fs-LIBS) for spectrochemical imaging of thin films with high spatial resolution. Chemical images are obtained by recording LIBS spectra at each site of 2D raster-scans across the samples employing one fs-laser pulse (400 nm, 175 fs) per site. The diffraction images of the Echelle spectrometer used for LIBS are binned to reduce the read-out time of the intensified CCD detector and to increase the stability of the emission signals against peak drifts in the echellograms. The fs-LIBS spectra of Yttrium Barium Copper Oxide (YBCO) thin films on MgO reveal various atomic and molecular emission lines (film thickness 200-1000 nm). The chemical images obtained from element (Y, Ba, Cu, Mg) and molecular (Y-O) emission lines match very well the structure of the micro-patterned YBCO films. For thin layers of Copper (thickness 5-500 nm) on glass the intensity of Cu I emission lines and the size of ablation craters vary non-monotonously with film thickness. The emission efficiency, defined as the Cu I line intensity per ablated volume, strongly decreases for films thicker than the optical penetration depth. The Na I line intensity from glass increases exponentially with decreasing Cu film thickness. Possible mechanisms responsible for the observed Cu and Na signals are discussed. The achieved lateral resolution r = 6 ?m for YBCO films is among the best values reported for spectrochemical LIBS imaging. Acknowledgements: Financial support by the Austrian Research Promotion Agency FFG (imPACts Project 843546) is gratefully acknowledged.

High Photoelectrochemical Water Splitting Efficiencies: Materials Development and Measurement Challenges

John Turner, Todd Deutsch, James Young, Henning Döscher, Myles Steiner

In the 19 years since the first water splitting efficiency of >10% was reported utilizing a III-V tandem structure, very little has been done to improve upon that efficiency. Most of the work since that time has focused on a few metal oxides whose poor electronic properties eliminate them from a high efficiency system. Although the corresponding tandem cell bandgap configurations for high efficiency water splitting have been known for some time, it has not been possible to produce such optimized devices. However, recent advances in III-V growth driven by high efficiency photovoltaics has made it possible to construct tandem water splitting devices with these optimum bandgaps. Taking advantage of state-of-the-art inverted metamorphic multijunction (IMM) growth processes for the synthesis, III-V alloys can be grown that point the way towards meeting these high efficiency goals. The IMM devices are grown by metal organic chemical vapor deposition using a GaAs substrate and a transparent step-graded buffer layer to allow off-lattice growth. We used IMM growth to achieve a high-quality, lattice-mismatched lower bandgap (1.2 eV) InGaAs bottom cell that captures a larger fraction of the solar spectrum. This 1.8/1.2 eV bandgap combination of the GaInP2/InGaAs configuration is able to generate higher photocurrents, and thus efficiencies, than the 1.8/1.4 eV bandgap combination of the lattice-matched GaInP2/GaAs configuration. While the photovoltaic community has developed standardized methodology for efficiency measurements, the photoelectrochemical water splitting community has lagged behind. We have studied measurement challenges and systematic errors in assessing solar-to-hydrogen efficiencies for these tandem absorbers and have developed measurement protocols to address these. Using more rigorous protocols, we have confirmed over 16% STH on our most advanced IMM devices.

Nanoscale infrared, mechanical, and electrical characterization of surface defects in perovskite crystals via peak force tapping based characterization techniques

Xiaoji Xu

The photo-conversion efficiency of organometallic halide perovskite solar cells is limited by the presence of the surface defects that affect the generation, diffusion, separation and recombination of charge carriers. Nanoscale defects on the surface of perovskite crystal are not directly readily detectable by the current macroscopic characterization techniques, therefore remain unexplored in the characterization of perovskites.

Here, we implement a combination of scanning probe microscopy that delivers the infrared imaging, mechanical mapping, work function mapping, and conductivity mapping with nanometer-scale spatial resolution. The application of techniques allows correlative studies of the surface defects of $CH_3NH_3PbBr_3$ crystal. The findings shall help the interpretation of the photo-electric processing of perovskite in solar energy conversion.

Understanding Rechargeable Batteries by Synchrotron X-ray Spectroscopy at Multiple Length Scales
Feng Lin

The rapid development of energy storage technologies creates opportunities for various applications, including rechargeable batteries for transportation and "thinner/smaller" consumer electronics. The present work is targeted to understand the characteristics of phase transition and chemical evolution in lithium-ion battery materials. The study was conducted using in-depth electrochemical diagnostics, nanoscale-to-macroscale core-level spectroscopies as well as atomic resolution electron imaging. We reconstructed the 3D architecture of the lithiation/delithiation fronts for anode materials and found that heterogeneous phase nucleation prevails in the battery conversion reactions. Moreover, we investigated the structural evolution in prominent stoichiometric layered cathode materials under high-voltage operation conditions, and determined that the anisotropic phase transition from an R-3m structure to an Fm-3m structure at the surface cathode particles is partially responsible for performance degradation. This work sets a refined example for investigating energy materials using correlated core-level spectroscopies at multiple length scales and provides insights into materials synthesis and cell design.

Infrared Spectroscopy and Confocal Raman Microscopy Measurements in the Study of Materials for Energy Conversion
Carol Korzeniewski, Ying Liang

Infrared spectroscopy and confocal Raman microscopy techniques are being adapted for quantitation and structural characterization in the development of materials for electrochemical energy conversion. The presentation will discuss applications in the study of ion conductive membranes and membrane-catalyst assemblies. The aperture of a confocal Raman microscope defines a fixed volume within a sample, usually not more than a few femtoliters, from which the detected Raman scattering originates. Analyte within the well-defined probe volume can be quantified from confocal Raman spectra. Furthermore, scanning capabilities can enable quantitative spatial mapping of materials composition with resolution approaching the diffraction limit. In the study of catalytic membranes, confocal Raman microscopy is being adapted to characterize steps in the processing of biocatalyst electrodes. Membranes are prepared from dispersions of ionomers fully exchanged by tetra-n-butylammonium ions (TBA^+) to promote biocatalyst (enzyme) incorporation through enhanced size of the ionic domains and hydrophobic interactions. Confocal Raman microscope measurements confirm the presence of entrapped enzyme within dispersion cast membranes and the ability to quantify TBA^+ ions and protein. For the enzyme laccase, a common biofuel cell O_2 reduction
catalyst, features associated with ligands coordinating the enzymatically active copper center are present in Raman spectra between 400 cm^{-1} - 500 cm^{-1} and provide a possible route to interrogate mechanistic chemistry at the enzyme active site of biocathodes under fuel cell reaction conditions. Using a similar experimental approach based on confocal Raman microscopy, the ability to probe spatially across the junction connecting the anion exchange and cation exchange phases of a bipolar membrane is being investigated. Strong Raman scattering from perchlorate is enabling its use as a probe for mobile anions. In preliminary work, ?5 ?m depth resolution has been attained in mapping the transition region. In applications of infrared spectroscopy, measurements on thin (10 nm – 1,000 nm) film ionomer materials are being supported by Fresnel calculations to assist with spectral interpretation and identify perturbations originating from optical effects.

Developing Quantum Dot Solids for Next-Generation Photovoltaics
Matt Law

Colloidal semiconductor quantum dots (QDs) are attractive building blocks for next-generation solar photovoltaics (PV). In this talk, I will highlight recent progress in understanding, improving, and controlling charge transport within mesoscale assemblies of lead salt QDs, with the aim of achieving simultaneously efficient carrier multiplication (CM) and charge extraction to produce record-efficiency single-junction PV. I begin by discussing QD film fabrication, charge transport physics, and evidence that the carrier diffusion length is short and limited by electronic states in the QD band gap. Efforts to improve carrier mobility by enhancing inter-dot electronic coupling, passivating surface states, and implementing surface doping will be presented. Engineering the inter-QD matrix to produce QD/inorganic or QD/organic nanocomposites is introduced as a powerful way to optimize coupling, remove surface states, eliminate hysteretic charge trapping and ion motion, and achieve long-term environmental stability for high-performance, robust QD films that feature good CM efficiency. Prospects for overcoming the diffusion-limited recombination that seems to constrain diffusion length are discussed, as well as the possibility of band-like transport through extended electronic states and the importance of utilizing QD superlattices for this purpose.

Rapid identification of biotherapeutics with label-free Raman spectroscopy
Ishan Barman

Product identification is a critical and required analysis for biotherapeutics. In addition to regulatory requirements for identity testing on final drug products, in-process identity testing is implemented to reduce business risks associated with fill operations and can also be used as a tool against counterfeiting. Biotherapeutics, in particular monoclonal antibodies, represent a challenging cohort for identity determination because of their similarity in chemical structure. Traditional methods used for product identification can be time and labor intensive, creating a need for quick, inexpensive and reliable methods of drug identification. Here, driven by its molecular-specific and nonperturbative nature, we present Raman spectroscopy as an alternate analytical tool for identity testing. By exploiting subtle differences in vibrational modes of the biologics, we have developed partial least-squares-discriminant analysis derived decision algorithms that offer excellent differentiation capability using spontaneous Raman spectra as well as label-free plasmon-enhanced Raman spectra. Coupled with the robustness to spurious correlations due to its high information content, our results highlight the potential of Raman spectroscopy as a powerful method for rapid, on-site biotherapeutic product identification.

Screening technologies and the fight against substandard and falsified medicines
Lukas Roth

The fastidious and targeted screening of medicines is one of the crucial tools in a sustainable response to assure the quality of medicines worldwide. However, limited financial and human resources at National Medicine Regulatory Authorities in developing and the lack of availability of a standardized approach for the screening of medicines limit the effectiveness of this critical regulatory function. Consequently entities conducting surveillance often struggle to identify if and how screening technologies are appropriate for their particular applications. Confirmatory quality control testing is so effective because of the validated methods and standards that guide the testing and the well characterized qualified laboratory-based instruments that underpin these methods. Conversely, many of the myriad screening technologies are comparatively newer and less well understood than their laboratory-based counterparts and little work has been done to critically but objectively review their capabilities as they pertain to effectively and accurately screening medicines. With the increasing availability of these technologies, this knowledge gap needs to be addressed. This session will introduce an intervention being led by the U.S. Pharmacopeia to empower the potential users of screening technologies as they seek to fight substandard and falsified medicines.

CSI: GAT - Finding the "DNA" of Counterfeit Drugs Using Raman Spectroscopy

Jeremy Peters, Eugene Park, Ravi Kalyanaraman

Protecting pharmaceutical products from falsification and counterfeiting is not only of paramount importance to any pharmaceutical company but, more importantly, to the health and safety of our patients. Both portable and bench-top FT-Raman spectrometers have been extensively used by Global Analytical Technology (GAT) in Bristol-Myers Squibb (BMS) to rapidly screen and detect counterfeit drugs for the past ten years. However, determining the authenticity and chemical composition ("DNA") of suspect counterfeits samples can be a challenging endeavor when 23 different counterfeit investigations, comprising of 33 separate samples were to be handled in a short span of time. Using FT-Raman and LC-MS techniques, the initial screening of the first set of counterfeit samples resulted in the presence of sucrose, starch and a different active ingredient that is not supposed to be present in the product. This information was leveraged effectively in all other investigations by only measuring the FT-Raman spectra. Using the measured FT-Raman spectra and by using multicomponent search software, the chemical ingredients in all of the other counterfeit samples were identified accurately. The development of this technique has drastically reduced the labor and time involved for most cases of counterfeit investigations. We have learned from these investigations that there are always ways to improve the existing testing workflow and provide accurate information in faster time frames.

Raman technology for bio and pharmaceutical counterfeits
Anna Luczak, Ph.D., Ravi Kalyanaraman, Ph.D.

Any pharmaceutical drug manufactured for the purpose of curing a disease can be counterfeited. This is especially true for small molecule based medicine and has been a target for many years. With a rise of biologic based drugs available on the market, there is a new danger of fake medicine since biologics are also prone to be counterfeited and are reported as a new target in fight against fake medicine. Facing this endless battle, pharma and biopharma industries require a specific knowledge and tools to successfully differentiate between fake and legitimate product not only for small molecules but also for biologics. Efficient authentication of counterfeited small molecule medicine has become achievable for years now due to the current state of portable vibrational spectroscopic tools such as Raman and NIR spectrometers. On the contrary, the understanding of what is needed to successfully characterize biologics from their counterfeited counterparts still poses a major challenge and needs to be carefully explore for an efficient combating of this type of counterfeit medicine.

Going Digital: A Landscape View of Digital Technologies to Combat the Fake Medicines Trade
Tim Mackey

The globalization of the pharmaceutical supply chain has introduced new challenges, chief among them, fighting the international criminal trade in fake medicines. As the manufacture, supply, and distribution of drugs becomes more complex, so does the need for innovative technology-based solutions to protect patients globally. We conducted a multidisciplinary review of the science/health, information technology, computer science, and general academic literature with the aim of identifying cutting-edge existing and emerging "digital" solutions to combat fake medicines. Our review identified five distinct categories of technology including mobile, radio frequency identification, advanced computational methods, online verification, and blockchain technology. We also describe a newly launched UCSD project that will leverage the potential of computational mass spectrometry to detect and catalogue fake antibiotics. Digital fake medicine solutions are unifying platforms that integrate different types of anti-counterfeiting technologies (including chemical and lab-based techniques such as mass spectrometry, raman, liquid chromatography, etc.), improve information sharing and data collection, and are designed to overcome existing barriers of adoption and implementation. Investment in this next generation technology is essential to ensure the security and integrity of the global drug supply chain.

Development of SERS-Based Microfluidic Technology for Biomedical Diagnosis

Jaebum Choo

The development of surface-enhanced Raman scattering (SERS)-based microfluidic platforms has recently attracted significant attention in the wider scientific communities. SERS has highly sensitive detection capabilities, and microfluidic platforms provide many advantages such as automatic sampling and reduced sample volume. Therefore, the integration of SERS with microfluidic platforms offers wide application in chemical or biological analysis. Recently, we developed a fully integrated SERS-based microdroplet platform for the automatic immunoassay of specific target antigens. In this system, an efficient immunoreaction was achieved through sequential droplet generation, transport, and merging processes, while wash-free immunodetection was realized through the droplet-splitting process. Indeed, a novel multifunctional microfluidic platform capable of performing a complicated multi-step immunoassay process in nanoliter-scale droplets has been successfully developed in this work. This assay system also has many advantages including small sample consumption (less than 100 ?L), fast assay time (less than 10 min), and fully automated fluid controls. We anticipate that this fully integrated SERS-based microdroplet device will provide new insights in the development of a facile assay platform for various biomarkers.

SERS-based bioassay for detection of biomolecules

Young Mee Jung

Surface-enhanced Raman scattering (SERS) is a powerful analytical technique that enables ultra-sensitive assays for chemicals and biochemicals. SERS has been successfully used for a number of sensing applications, including the sensing of heavy metals, disease biomarkers and biological molecules. In particular, SERS-based biosensors have received extensive attention in the field of disease diagnosis. We have recently reported SERS-active substrates for detection of biomolecues and their versatle applications in highly sensitive SERS-based bioassay. SERS-based MTT assay for living cell viability in situ and the effect of the enzymatic reaction on the spectra of a GOx-conjugated SERS-active reporter molecule (4-mercaptobenzoic acid (MBA)) for glucose sending were also explored. In this presentation, results of SERS immunoassy and in situ SERS will be presented to demonstrate the potential, utility and versatility of ultrasensitve SERS-based bioassay.

Sensing Biomolecules using SERS platforms for Medical Point of Care Applications

Gerard Cote, Haley Marks, Monika Schechinger, Javier Garza, Brian Walton, Andrea Locke, Dandan Tu, Po-Jung Huang, Jun Kameoka

Timely detection of multiple biomarkers simultaneously is needed at the point of care to diagnose diseases such as acute myocardial infarction (heart attack), preeclampsia, and cancer as well as to detect toxic molecules in blood or biomarkers related to toxic exposures in order to implement the appropriate early treatment and significantly reduce the chance of mortality. Ideally, a point of care (POC) device needs to be fast, sensitive, reliable, and small enough to be used in the ambulance, clinic, or in the field. Surface enhanced Raman spectroscopy (SERS) is a sensitive optical technique that can potentially be used to quantify biomarkers of interest. The specific narrow peaks from the Raman reporter molecule SERS signal allow for potential multiplexing capabilities as different Raman reporter molecules can be used in functionalized nanoparticles with different biomarkers. In this work, we discuss development of functionalized nanoparticles, aggregation mechanism, and the SERS spectrum obtained from the Raman reporter and a biomarker functionalized nanoparticles to assess its potential to be used in a POC assay. Specifically, the optical properties of colloidal nanoparticles are manipulated, tuned, and monitored for plasmonic sensing applications by forcing the particles into a controlled, but reversible, aggregative state. This has been accomplished by conjugating nanoparticles to specific biological ligands with predictable binding affinities, such as aptameranalyte pairs, antibody-analyte pairs, or DNA-functionalized nanoparticles and competing molecules depending on the medical application. To improve and simplify the detection biomolecules, multiple plasmonic and magnetic nanoparticles were investigated and the SERS assay response in both a traditional liquid suspension and housed within a microfluidic chip platforms were analyzed and compared.

Quantitative SERS by "Hot Spot" Normalization

Haoran Wei, Weinan Leng, Junyeob Song, Marjorie Willner, Linsey Marr, Wei Zhou, Peter Vikesland

Surface-enhanced Raman spectroscopy (SERS) has been proposed as an ultrasensitive analytical technique for years. However, it is still challenging to make SERS fully quantitative due to "hot spot" heterogeneity across a SERS substrate. Although highly uniform SERS substrates have been reported, they are still difficult to be made at large scale and at reasonable cost. Recently, internal standards were incorporated into SERS substrates, which underwent the same variation as the target analytes and thus reducing the point-to-point SERS signal variation. However, the introduction of guest molecules as internal standards not only adds to the complexity of material synthesis, but also occupy "hot spot" volume and interfere with the target bands. Herein, we propose a universal approach that can fully address the problem for SERS quantitation by applying a pseudo Rayleigh band (?e) as internal standards. ?e originates from the spontaneous emission of laser and locates where the longpass filter cuts at. The intensity of ?e is extremely weak but can be substantially enhanced by SERS "hot spots" with the same magnitude as Raman scattering. The intensity of ?e linearly correlates with the "intensity" of the SERS "hot spots" and is independent of analytes. Applying ?e as internal standards ("hot spot" normalization) can reduce the signal variation resulted from the changing hot spot density, hot spot intensity, laser intensity, and focusing, thus leaving the concentration of analyte within "hot spots" the only dominant variant. The coefficients of variation of SERS signal across 100 μm × 100 μm maps were reduced from ~50% to ~10%, indicating "hot spot" normalization can turn non-uniform substrates into uniform ones. Owing to "hot spot" normalization, the linearity of the signal variation as a function of analyte concentration was significantly improved and the error bar of each point became much smaller. In conclusion, "hot spot" normalization is a universal approach for SERS quantitation. It significantly lowers the bar for making quantitative SERS substrates and pushes SERS from a semi-quantitative to a quantitative tool.

Surface Enhanced Raman Spectroscopy of Aerosol Particles

Vasanthi Sivaprakasam, Matthew Hart, Paul Lane, Gary Kushto, Jay Eversole

Raman spectroscopy is a widely used technique to measure vibrational and rotational molecular modes which can provide very specific spectral signatures for chemical identification. However, typically low Raman scattering cross-sections have been a barrier to using this technique for in-situ characterization of environmental aerosols or monitoring hazardous materials present in the air. By combining metallic nano-particles (MNPs) with aerosol particles, we have been able to observe Surface Enhanced Raman Spectra from single suspended aerosol particles. We have explored the comparison measurements of spontaneous Raman spectra with MNP-enhanced Raman spectra using identical materials under controlled conditions both as aerosols and as bulk liquid samples. The experimental repeatability of this approach has enabled investigation of spectral responses and enhancement factors, and their dependence on factors such as, MNP composition, MNP concentration and material composition. This presentation describes the design and construction of a Raman spectroscopy test-bed capable of interrogating single aerosol particles or bulk samples in a cuvette for comparison/validation studies. Aerosol particles are suspended in a linear Electrodynamic quadrupole (LEQ) trap, that provides ample time to study the weak spontaneous Raman. Aerosol particles are generated and charged using a customized nozzle-reservoir droplet generator, and particles ranging from 1 micron to 60 microns in diameter have been successfully generated and maintained. Droplets of materials such as glycerol, dibutly sebacate and ethyl cinnamate have been used for validation and calibration. Initial SERS data has been obtained for Pyridine and Rhodamine 6G dye. Plans to extend the study to other chemicals are under way. The latest spectral signatures and enhancement factors resulting from the addition of the MNP's to these materials will be presented.

Longitudinal monitoring of murine bone quality using transcutaneous Raman spectroscopy

Andrew Berger, Jason Maher, Marien Ochoa Mendoza, Guanping Feng, Hani Awad

Raman spectroscopy of bone provides information about the relative abundance of mineral (phosphate, carbonate) and collagen matrix. It has been shown that such spectra can be used to predict biomechanical properties of bone that are typically measured by fracture methods. In addition, our group and others have shown that it is possible to obtain meaningful Raman spectra of bone without removing the overlying soft tissue. This is accomplished by performing transcutaneous measurements with spatial-offset diversity; in our case, we model the variation in spectral shapes using libraries of bone and soft tissue spectra. With the ability to provide biomechanical property estimates both nondestructively and noninvasively, Raman spectroscopy is an attractive candidate for longitudinal monitoring of bone strength in individual animals. Here, we report our first results from anaesthetized mice. This work paves the way for studies of individual animals' responses to medications such as drugs intended to combat glucocorticoid-induced osteoporosis, a common problem in treatment of rheumatoid arthritis in humans.

Frequency Offset Raman Spectroscopy (FORS) for subsurface probing of scattering media

Antonio Pifferi, Sanathana Konugolu Venkata Sekar, Sara Mosca, Andrea Farina, Fabrizio Martelli, Paola Taroni, Gianluca Valentini, Rinaldo Cubeddu

The talk will present a novel technique – named Frequency Offset Raman Spectroscopy (FORS) – to investigate Raman signals in depth through highly scattering media [1]. While in Spatially Offset Raman Spectroscopy (SORS) higher depth sensitivity is obtained by increasing the source-detector distance [2], conversely in FORS this is achieved by reducing the absorption and scattering coefficients in the medium by properly tuning the excitation wavelength. The effectiveness of FORS in providing distinct depth sensitivities in the Raman signal will be shown on bi-layered tissue simulating phantoms exhibiting spectral changes both in the absorption and in the scattering coefficients. Measurements were achieved using a setup based on a tunable Ti:Sapphire laser and an axicon to produce a variable radius ring illumination permitting both FORS and SORS measurements in non-contact configuration. Although stem from different basic principles, similar results were obtained both in FORS and SORS in terms of increased depth sensitivity upon changing wavelength (FORS) or source-detection distance (SORS). An enhancement factor of 3 was achieved in both techniques for a practical case of a top 10 mm thick layer made of silicone added with ink over a bottom layer made of a marble (calcite $CaCO3$) slab. Also, a hybrid FORS+SORS approach was demonstrated combining the two techniques leading to an increased enhancement factor of 6 for the same phantom. Different advantages and limitations as well as complementarities of FORS and SORS will be discussed, with the former permitting in-depth discrimination with a fixed single-distance configuration yet requiring an excitation tunable system. Also, FORS leads to higher spatial resolution since large source-detector separations are not required. Conversely, the medium must exhibit spectrally variable optical properties to achieve significant depth discriminations. Potential future directions and applications on biological tissues will be discussed also based on the broad experience on in vivo tissue spectroscopy for clinical diagnostics exploiting time-domain diffuse optics. [1] S. Konugolu Venkata Sekar, et al., "Frequency offset Raman spectroscopy (FORS) for depth probing of diffusive media," Opt. Express (2017). [2] P. Matousek and N. Stone, "Recent advances in the development of Raman spectroscopy for deep non-invasive medical diagnosis.," J. Biophotonics, (2013).

Tracking of Reporter Functionalised Nanoparticles in Tissue Using Handheld Surface Enhanced Spatially Offset Raman Spectroscopy

Fay Nicolson, Neil Shand, Duncan Graham, Karen Faulds

Spatially offset Raman spectroscopy (SORS) provides chemical analysis at depth even when obscuring barriers such as plastic or tissue are present. As the collection probe is moved further away from the point of laser excitation, scattered photons from deeper layers begin to dominate the acquired spectra, thus giving rise to through barrier detection. Surface enhanced spatially offset Raman spectroscopy (SESORS) combines both SERS and SORS to yield enhanced Raman signals at much greater sub-surface levels. By using gold nanoparticles functionalised with a Raman reporter molecule, the "fingerprint spectrum" can be tracked at depths of several millimetres. Using a handheld SORS spectrometer, we have obtained Raman spectra from analytes through 21 mm of plastic, and SERS analytes through 28 mm of plastic. This was compared to a conventional handheld Raman spectrometer where signals were obtained at maximum thicknesses of 9 (Raman) and 19 (SERS) mm. Previous work reports the use of SESORS in the field of biomedicine, however the most commonly reported method uses extreme spatial offsets (i.e. transmission Raman). This approach is potentially unfeasible for some biomedical applications where it may not be possible to collect Raman scattering in a transmission mode. Here we show the promise of handheld SESORS using a back-scattering geometry, hence the collection optics are on the same trajectory as the incident laser. Our approach demonstrates that the SESORS technique can detect nanotags through at least 20 mm of tissue. Furthermore, multiplexing of SERS nanotags is possible at similar depths, also demonstrating the potential for multiplexed imaging using a back-scattering geometry.

Surface Enhanced Spatially Offset Raman Spectroscopy Detection of Neurochemicals Through the Skull
Bhavya Sharma

We are interested in developing Raman spectroscopy-based methods to non-invasively detect neurochemicals through the skull. We are using surface enhanced spatially offset Raman spectroscopy (SESORS), a powerful technique that combines the sensitivity of surface-enhanced Raman spectroscopy (SERS) with the ability of spatially-offset Raman spectroscopy (SORS), to probe subsurface layers through the skull. Here we present SERS measurements of neurotransmitters (melatonin, serotonin, and epinephrine) at varying concentrations followed by the SESORS measurements of the neurotransmitters at concentrations of 50 mM and 100 ?M in a brain tissue mimic through a cat skull. Principal components analysis was performed to distinguish between the surface bone layer and the subsurface layer, comprised of a brain tissue mimic modified with neurotransmitters, and to determine if each individual neurotransmitter could be accurately identified.

Deep Raman Spectroscopy: A novel platform to non-invasively measure physical properties deep within turbid samples
Ben Gardner, Nick Stone, Pavel Matousek

The development of the spatially offset Raman spectroscopy (SORS) has led to exciting new research and applications in a number of diverse areas including pharmaceutical quality control, security screening, healthcare and cultural heritage. To date, most applications of SORS have used it as a tool to recover chemical signals at depth to confirm or monitor a subsurface constituent's presence, e.g. a calcification in tissue.

However, the recovered Raman spectra encodes more than just the chemistry of a species at depth conveying also information related to the physical properties of a subsurface material. Here will be outlined applications of SORS to monitor physical properties of materials at depth including temperature (T-SORS) and pH (pH-SORS).

Microelectrode Collector-Generator Systems for Biomedical Applications: New Insights from Theory
Danny O'Hare , C.G. Bell, P. Seelanan

The advantages of electrochemical methods of analysis in many measurement situations are well established-low cost, real-time, ease of miniaturisation-additional advantages accrue when using electrochemistry to generate reagents, initiate reactions or control titrations. In typical systems, one electrode is used to electrolytically generate a reactive species e.g. hypobromite which reacts with amino acids or a proton flux to perform acid-base titrations. The extent of reaction is followed by a second electroanalytical device. Such methods were originally developed in hydrodynamic voltammetry to provide mechanistic insights but they have found application as sensors and as detectors in liquid chromatography. Microelectrode approaches can, if suitably engineered, reduce the impact of convection and offer the possibility of use in confined spaces and enable spatially resolved chemical distributions to be determined. We have described the construction of micro ring-disc electrodes and their application as flow sensors and more recently applied similar approaches to acid-base titrimetric collector-generator devices which have applications in determining the buffer capacity in biological samples and alkalinity in environmental water. Here we present a new theoretical treatment for a dual-disc device which consists of a constant current generator and potentiometric detection. Asymptotic mathematical approaches were used to generate a solution for the steady state concentration of hydrogen ions as a function of the proton flux, bulk concentration of base, pKa and the relevant diffusion coefficients. Key design rules emerge from this analysis that are not obvious from either experimental approaches or numerical solutions: most notably, there are stringent restrictions on the maximum reagent flux. Experimental verification is presented showing clear resolution of buffer capacity where the flux is restricted as indicated by the analysis. The results can be generalised for other devices and other chemistries to create titrimetric sensors capable of implantation.

Construction of DNA-Protein Hybrid Molecules for Bioanalyses
Eiry Kobatake

Proteins have various and high functions in living bodies, and they have been applied to various fields, such as diagnostic, therapeutic, clinical, biotechnological, bioengineering fields and so on. However, not only proteins but also DNAs have some unique properties except genetic information, such as hybridization, molecular recognition, amplification, and self-assembling functions. Therefore, wider varieties of applications of DNA-protein hybrid should be expected by utilization of combinatorial functions of protein and DNA. Recently, DNA-protein hybrid molecules have been used as versatile tools for a variety of applications in biotechnology and materials science. In particular, the utility of DNA-protein hybrid has emerged in bioanalyses to quantify target molecules. Despite the potential applicability of DNA-protein hybrid molecules, synthetic methods are still challenging. Chemical synthesis of DNA-protein hybrid has been performed successfully, although the incorporation of DNA into larger proteins is still poorly developed. Furthermore, these methods require several purification processes of the hybrid to remove excessive protein and DNA after each coupling step. To overcome these problems, we developed an easy-to-use and economical method for site-selective conjugation of a recombinant protein with single-stranded DNA (ssDNA). To make such DNA-protein hybrid molecules, we paid attention to Gene A* protein of coliphage ?X174. Gene A* protein can cleave the specific sequence of DNA, and a phosphate bond is formed between the 5'-phosphoryl group of the cleavage site and the tyrosine residue of the Gene A*. Therefore, Gene A* could be used to construct a DNA-protein hybrid site-specifically and covalently. Some DNA-protein hybrid molecules were constructed and they were applied to bioanalyses.

Producing New Sensing Molecules with Gene Technology – An Ascendant Topic
Xian-En Zhang

Producing new sensing molecules with gene technology" is the title of Dr. Cass' presentation at the ACS meeting in 1997 (Cass, AEG, Gilardi, G, Beaumont, NJ, Sadeghi, SJ. Abstracts of Papers of the American Chemical Society, 213:220-Anyl, Apr 13, 1997). Now, 20 years have passed, the topic is gaining widespread practice in analytical biotechnology. This talk briefs the collaboration between Professor Cass and my group and beyond. Early attempts including engineering enzyme proteins to improve the quality of enzyme electrodes. Compare to the conventional chemical modification methods, gene technology is more moderate, rational, efficient and reproducible in achieving quality biosensors, which had been demonstrated by many examples, such as the engineered maltose-binding protein for reagentless fluorescence sensing, fusion enzyme molecules for sequence enzyme electrode, improvement of homogeneity of analytical biodevices by gene manipulation, genetic modification of glucose oxidase for improving performance of an amperometric glucose biosensor, and so on. Methods involved include molecular evolution by random and site-specific mutagenesis, gene fusion, 3D protein structure based design, etc. In recent years, imaging molecular events in live cells become attractive because many scientists believe "seeing is believing". This has been realized by various molecular biosensors that are constructed by gene technology. A molecular biosensor may be regarded as a biomolecular system that produces a visible or quantifiable signal to indicate the existence of an occurrence of a biological substance or a biological event. The most common molecular biosensors include molecular beacon (MB), fluorescent resonance energy transfer (FRET) using fluorescent proteins, bioluminescent resonance energy transfer (BRET), and molecular fluorescence complementation systems, such as BiFC and TriFC. Most recently, gene editing and synthetic biology are introduced to create cell biosensors for more complicated sensing purposes. * This talk is to celebrate the Award of the Sir George Stokes 2017 Medal to Professor Tony Cass.

Metabolic enzymes as sensors for arsenic
Joanne Santini, Thomas Osborne, Cameron Watson, Thomas Warelow, Dimitri Niks, Russ Hille, Graham George, Tony Cass

Arsenic is a notorious poison and carcinogen of global importance. It is estimated by the World Health Organisation that greater than 140 million people across 70 countries drink water that contains arsenic above safe levels (i.e., 10 parts per billion). While arsenic remediation remains the answer to the problem, development of a low-cost, easy-to-use and sensitive method for detecting both toxic, soluble inorganic forms of arsenic (i.e., arsenite and arsenate) is essential for saving lives. Harnessing metabolic enzymes for their ability to use both inorganic forms of arsenic as energy sources is key to the development of arsenic biosensors. The respiratory enzyme arsenite oxidase catalyses the oxidation of arsenite to arsenate. The arsenite oxidase from *Rhizobium* sp. str. NT-26 is a heterotetramer consisting of a large catalytic subunit (AioA), which contains a molybdopterin guanine dinucleotide cofactor at its active site and a 3Fe-4S cluster, and a small subunit (AioB) containing a Rieske 2Fe-2S cluster. Two Aio X-ray crystal structures have been resolved and further characterisation of the enzyme's redox centres and its interaction with electron acceptors studied using a variety of techniques (e.g., X-ray aborption spectroscopy, EPR, stopped-flow spectroscopy). Less is known about the respiratory arsenate reductase which contains two subunits, ArrA (predicted to contain MGD at its active site and one 4Fe-4S cluster) and ArrB (predicted to contain four 4Fe-4S clusters). The redox properties of both Aio and Arr make them ideal for the development of electrochemical biosensors where arsenite oxidation or arsenate reduction can be measured as a current. This talk will explore the need for arsenic biosensors and the suitability of these arsenic-metabolising enzymes for this purpose.

Cytochrome P450 electrodes for drug metabolism screening and metabolite biocatalysis
Gianfranco Gilardi

In humans the majority of drugs and xenobiotics are cleared by the human hepatic cytochromes P450 (P450) that are the most important Phase I drug metabolizing enzymes. Inhibition of P450-mediated drug metabolism by a concomitantly administered second drug is one of the major causes of drug–drug interactions (DDI) and can cause serious adverse drug reactions (ADR) or an increase in toxic side effects. The availability of rapid and simple in vitro platforms able to measure the inhibitory effect of a drug is therefore very attractive. Interfacing P450 enzymes to electrode surfaces and electrochemically driving their catalytic cycle has proven to be very difficult. The breakthrough in electrocatalysis of these enzymes has been made when both the electrode and the enzyme have been engineered, in some cases mimicking the natural environment of the microsomal enzymes and the inclusion of their electron transfer partners. Examples of electrochemically-driven drug metabolism as well as drug–drug interaction tests involving human cytochrome P450 3A4 probed with erythromycin, ketoconazole, cimetidine, diclofenac and quinidine will be shown. Inhibition also plays an important role on P450 2A6 and the metabolism of nicotine with consequences on smoking habits. Inhibition by tranylcypromine on the activity of CYP2A6 towards the substrates coumarin nicotine, will be shown. References 1. S.J. Sadeghi et al. Bioelectrochemistry, 2012, 86, 87-91. 2. S. Castrignanò et al. Anal. Chem., 2014, 86 (5), 2760–2766. 3. G. Di Nardo, G. et al. Electrochem. Comm., 2015, 52, 25-28. S.J. Sadeghi et al. BBA - Proteins and Proteomics, 2011, 1814, 237-248

Conformer-selective spectroscopy of flexible molecules: when IR/UV double resonance techniques reveal peptide conformation
Michel Mons

The exquisite sensitivity of UV spectroscopy to the chromophore environment enables gas phase physicists to address selectively the several conformers of a flexible molecule, for example a small peptide molecule. After being vaporized by laser-desorption and cooled in a supersonic expansion, the UV-selected conformations can be characterized by their IR spectroscopy, especially in the NH stretch region, which bears much structural information, especially about the H-bonding network of these species. The single-conformation spectra obtained by IR/UV double resonance are well resolved due to the low rovibrational temperature achieved and eventually assigned by comparison with quantum chemistry calculations. The talk will provide examples of such spectroscopy, carried out on neutral peptides, in particular iconic secondary structures of protein chains, such as ?- ou ?-turns or helices, and will emphasize the interplay between experimental spectroscopic characterization and quantum chemistry molecular modelling. Review paper: E. Gloaguen & M. Mons, "*Isolated Neutral Peptides*" **Topics in Current Chemistry** 2015, 364, 225-270

Following electronic and structural dynamics with VUV photoelectron spectroscopy of liquids.
Christopher Arrell, Jose Ojeda, Luca Longetti, Majed Chergui

The use of high harmonic generation (HHG) and soft X-ray free electron lasers as a spectroscopic tool for following both electronic and structural dynamics of solvated species via photoelectron spectroscopy will be presented Specifically, the ultrafast VUV facility at the Lausanne Centre for Ultrafast Science (LACUS) at EPFL called Harmonium will be introduced. Harmonium is a new facility to conduct ultrafast photoelectron and photoion spectroscopy from liquids, clusters and solids (ARPES). The latest results from Harmonium from liquid studies will be presented, specifically the ultrafast charge transfer dynamics in solvated transition metal complexes. Recently within Switzerland the SwissFEL free electron laser has become operational. The cross-over between lab based PES techniques and the use of this exciting new tool will be discussed.

ATR FUV-DUV spectra of graphene polymer nanocomposites
Yukihiro Ozaki, Yusuke Morisawa, Krzysztof Be?, Justyna Grabska, Ichiro Tanabe, Harumi Sato

Recent introduction of attenuated total reflectance (ATR) technique to FUV-DUV spectroscopy by our group has opened a new area of electronic spectroscopy. It has enabled us to effectively study various kinds of samples in their native condensed phase. In the present study we investigated ATR FUV-DUV (140-300 nm) spectra of poly(3-hydroxybutyrate)(PHB), a natural biodegradable polymer and its graphene-based nanocomposite. Polymer-carbon nanocomposites are one of the most attractive modern materials in science and technology, offering significant possible advantages and prospects for future applications. Therefore, there is a high demand for new and highly capable sources of information about structural and electronic properties of these materials. ATR-FUV-DUV spectroscopy proves to be able to deliver unique data on high-energy electronic states of polymer-nanocomposites, while remaining non-invasive, universal and easily applicable. By employing quantum chemical calculations based on TD-DFT we demonstrate the capabilities of FUV-DUV spectroscopy in providing accurate and rich information about the electronic states and structure of both pure polymer and its graphene based nanocomposite. We studied numerous features including the differences between amorphous and crystalline domains of the polymer, the role of the interchain CH3•••O weak hydrogen-bonding of PHB on the observed absorption and its non-monotonic temperature dependence, the interaction with carbon-based nanostructures and new insights on the formation of the composite domains within the structure of the polymer. Several patterns observed in the experimental data could be explained in this study. (1) The FUV-DUV manifestations of crystalline and amorphous PHB; (2) high absorptivity of the polymer in the FUV region while low in the DUV region; and (3) the peak shifts and intensity changes upon the change of crystallinity. These findings subsequently helped to clarify (4) an unusual non-monotonic change in the peak FUV height, and relative FUV/DUV absorbance values observed in the temperature-dependent spectra of PHB below the melting point. These findings show the potential of spectroscopy of condensed phase in exploring electronic structure and molecular interactions in the area of polymer nanocomposites.

Changes in Electronic States of Organic Solids observed by Attenuated Total Reflectance Spectroscopy in the Far Ultraviolet region.
Yusuke Morisawa

Far- and Deep-UV spectrometer based on attenuated total reflection (ATR) enables us to measure spectra of liquid and solid samples easily and feasibly in the 140–280 nm regions. Changes in electronic states of organic molecules in the condensed phase are investigated by ATR-FUV. Intermolecular interactions between alkyl chains such as CH---HC should be reflected in the phase behavior of organic compounds. We measured the attenuated total reflectance spectra in the far- and deep-UV region of n-alkanes from room temperature to -50 C for studying changes in electronic states caused by phase transition. As to n-tetradecane, with decreasing temperature, the absorption band at 153 nm in the liquid phase becomes weaker and new bands appear at around 200 and 230 nm. Quantum chemical calculations were carried out to explain the spectral changes in far- and deep-UV region caused by phase transition.

Surface plasmon resonance sensors in far- and deep-ultraviolet regions using Al thin films
Ichiro Tanabe, Yoshito Tanaka, Koji Watari, Taras Hanulia, Takeyoshi Goto, Wataru Inami, Yoshimasa Kawata, Yukihiro Ozaki

The surface plasmon resonance (SPR) of Al thin films was investigated by varying the refractive index of the environment near the films in the far-ultraviolet (FUV, 120-200 nm) and deep-ultraviolet (DUV, 200-300 nm) regions. An original FUV-DUV spectrometer that adopts an attenuated total reflectance (ATR) system was used. The measurable wavelength range was down to the 180 nm, and the environment near the Al surface could be controlled. The resultant spectra enabled the dispersion relationship of Al-SPR in the FUV and DUV regions to be obtained. The FUV-DUV-SPR sensors are expected to have three advantages compared to visible-SPR sensors: higher sensitivity, material selectivity, and surface specificity. (1) Many materials have strong absorptions (i.e. large changes of the dielectric constant) in the FUV and DUV region even if these materials have no absorption in the visible region. Therefore, The FUV-DUV SPR sensor may have higher sensitivity than the visible SPR sensors. (2) Each material shows a unique spectral shape, even if these materials have no absorption in the visible region. Therefore, by using the FUV and DUV region, a material-selectivity can be achieved. (3) By using shorter wavelength (FUV and DUV) light than visible light, the measurement field (i.e. the evanescent wave range) is narrowly limited compared with the visible SPR sensors. Therefore, the FUV-DUV SPR sensor can achieve the surface-selective detection. In particular, the material-selectivity is a novel characteristic, which is not achieved by the visible SPR sensors.

Mann up, SERS Can be Useful!

Duncan Graham

SERS has been around for many years but has struggled at times to be used to it's full potential. This presentation will look at some examples of where SERS has been proven to be useful and in particular for biological analysis.

The technology behind coherent 2D IR spectroscopy and its application to amyloid diseases

Martin Zanni

The technology behind coherent 2D IR spectroscopy has dramatically improved over the past 10 years so that it can now be applied to a broad range of materials, biological systems, and most recently, tissues. In this talk, I will give an overview of some of the principles behind 2D IR spectroscopy, introduce new technologies enabling robust data collection and imaging capabilities, and then present studies examining the molecular mechanism of amyloid formation involved in type 2 diabetes and cataract formation.

Advances in 3D and High-Resolution LA-ICP-MS Bioimaging

Stijn J. M. Van Malderen, Brecht Laforce, Thibaut Van Acker, Olga Borovinskaya, Laszlo Vincze, Frank Vanhaecke

Low-dispersion ablation cells with >99% washout in <10 ms have recently been developed for laser ablationICP-MS. These ablation cells enhance the spatial resolution, sample throughput, and sensitivity, which enables new applications, e.g., tomography-supported 3D elemental reconstruction and sub-cellular imaging, to be explored. Similarly, zircon dating, a routine application for LA-ICP-MS, can also be further refined. Zircon crystals can capture geological information, e.g., late-stage growth, metamorphoses and weathering of the host rock introduce distinct crystal growth or erosion-induced alterations, which are identifiable by textural features in the mineral. These features include polyphase core–rim structures and inclusions, as well as evidence for Pb mobility. These hold relevance as the U-Pb decay systems upon which zircon dating relies, are affected by alterations in the concentration and distribution of radiogenic and common Pb in the zircon. In this work, a set of 5 zircons (~400 Myr old, Belgian origin, ~100 µm in size) were mapped using a 193 nm ArF* LA system (Analyte G2, Teledyne Photon Machines, Bozeman, MT, USA), equipped with the ARIS (Aerosol Rapid Introduction System) and a prototype of a next-generation low-dispersion ablation cell, coupled to an icpTOF (TOFWERK AG, Thun, Switzerland) ICP – time-of-flight (TOF) – MS unit. From the images, elemental and isotopic information was extracted from specific structural domains within the zircon by texturally-controlled sampling: the pixel size of 1.5x1.5 µm2 permits to excise inclusions from the data, as well as the isolation of individual crystal growth bands. Severe discrepancies in the age determination (using zircon 91500 as the primary reference material) were observed depending on the zone of the crystal sampled. By virtue of the full-spectrum acquisition, elemental correlations can be established for many elements within the individual zircon phases, and inclusions in the mineral phase can be identified. By overlaying 20 ablation grids (for line scanning), part of one zircon was visualized in 3D. Planar growth zones are observed to be slanted relative to the polishing plane; bulk sampling approaches which drill through these layers induce phase mixing, which has implications for age determination.

High repetition rate femtosecond laser ablation inductively coupled plasma Time-of-Flight mass spectrometry

Jhanis J. Gonzalez

Research into femtosecond laser ablation is based on improved performance expectations by controlling the amount of ablated material, particle size distribution, matrix dependence, and fractionation. Femtosecond laser ablation sampling into the ICP-MS has been shown to provide good chemical analysis precision and accuracy by reducing systematic errors related to the particle size distribution and resultant spikes in the ICP-MS response. A relaxation of matrix dependence also has been shown for the analysis of glass and metal alloy samples when using femtosecond laser pulses. In this presentation, the influence of repetition rate using low energy femtosecond laser pulses on sample throughput and LA-ICP-MS performance is discussed.

High-Resolution LA-ICP-TOFMS Imaging: Advances in Instrumentation and Data Analysis

Alexander Gundlach-Graham, Marcel Burger, Gunnar Schwarz, Jovana Teofilovic, Paulo Garafalo, Bodo Hattendorf, Detlef Günther

Elemental imaging by laser-ablation inductively coupled plasma mass spectrometry (LA-ICPMS) is a bench-top method that provides concentration-distribution maps of bulk-to-trace elements and isotopes across a sample surface. Applications of LA-ICPMS imaging range from determinations of native element distributions in tissues, to assays with isotope-labeled antibody tags, to imaging of hard materials such as geological specimens or advanced engineered materials. Despite the growth of LA-ICPMS elemental imaging applications, the ultimate scope of this method depends on advancing performance characteristics such as lateral resolution, depth resolution, elemental coverage, detection limits, time of analysis, and quantification capabilities. Recent developments in low-dispersion LA cell design and ICP-time-of-flight mass spectrometry (ICP-TOFMS) instrumentation enable the acquisition of large high-resolution multi-elemental quantitative elemental images within a day's work. In our lab, we routinely perform LA-ICP-TOFMS imaging with laser-spot sizes at or below 5 μm, pixel-generation rates up to 100 pixels/s, and multi-elemental image sizes in the megapixels. At a spot size of 5 μm, single-laser-shot detection limits are typically in the low μg g-1 range for the rare-earth elements and heavier; most lighter elements have detection limits in the 10s–100s of μg g-1. Importantly, the generation of large-format LA-ICPMS elemental images with fine-detail resolution offers new opportunity to accurately visualize and quantify the trace-to-major elemental and isotopic composition of micro- and macro-features across complex heterogeneous surface. In this presentation, we highlight the current state of our LA-ICP-TOFMS imaging method. We describe several LA-ICP-TOFMS imaging case studies to illustrate the importance of data handling and analysis. We explore the relationships between laser spot size, lateral resolution, and quality of elemental quantification within features of an elemental image. Ultimately, the relevance of high-resolution LA-ICP-TOFMS depends on the generation of accurate images and the development of methods to extract element or isotope abundance information to support image interpretation. Here, we aim to discuss current strategies, shortcomings, and future directions of high-resolution LA-ICPMS image generation and analysis.

ION DYNAMICS AND ABLATION MECHANISMS OF FEMTOSECOND AND NANOSECOND LASER PRODUCED PLASMAS

Prasoon Diwakar, Ahmed Elseid, Ahmed Hassanein

The dynamics of the ions emitted from ultrashort and long pulsed laser interaction with materials were studied. A series of successive experiments were conducted for six different elements (C, Al, Cu, Mo, Gd, and W) using 40 fs, 800 nm Ti: Sapphire laser over a wide range of laser fluences (0.8 J/cm2 to 24 J/cm2) corresponding to laser intensities (2.0 × 1013 W/cm2 to 6.0 × 1014 W/cm2). Foe nanosecond, a 1064 nm, Nd: YAG laser pulse was used to ablate pure Al, Fe, Co, Mo, and Sn samples using laser intensities varying from

2.5 × 1010 W/cm2 to 1.0 × 1011 W/cm2. Time-of-flight (TOF) ion profile was analyzed and charge emission and velocity dependence on material properties, laser pulse duration and incident laser energy were investigated. A theoretical model is proposed to predict the total ion flux emitted during femtosecond laser interaction that depends on laser parameters, material properties, and plume hydrodynamics. For further understanding of plume dynamics, laser optical emission spectroscopy was used to characterize the plasma by measuring the temporal and spatial evolution of plasma electron density ne and electron temperature Te. Time-of-flight measurements showed decreasing ion flux and ion velocity with increasing atomic weight, and multiple structure profile characterized by multiple peaks. Slow peak was found to follow shifted Maxwell Boltzmann distribution, while the fast peak was found to follow Gaussian distribution. In case of long laser pulse ablation heavy metal ion flux profile exhibited multiple peaks that was not observed in lighter metals. In contrast, for the case of short laser pulse ablation the multiple structure was observed regardless. Furthermore, the angular distribution of the fast ions was found to be different based on the pulse duration. This suggests that the physics of ion emission mechanism hence the ablation process are different. In addition to ion dynamics, ablation mechanisms of femtosecond and nanosecond laser produced plasmas are discussed in detail.

Atmospheric-pressure Plasma Assisted Reaction Chemical Ionization: A New Ion Sampling Interface for ICP-MS

Kaveh Jorabchi, Joseph Lesniewski, William McMahon, Hamid Badiei

Inductively coupled plasma mass spectrometry (ICP-MS) has become the method of choice for elemental MS. Detection of non-metals (particularly halogens) with ICP-MS, however, faces major challenges because of low ionization efficiencies, poor ion transmission efficiency at low mass and isobaric interferences. Recent advances in reaction cells and ICP-MS/MS have offered avenues to resolve some isobaric interferences. These ion-specific reactions, however, limit the simultaneous multi-elemental measurement capabilities. Moreover, inefficiency of the reactions significantly reduces ion currents, negatively impacting precision and transient signal measurements in chromatographic separations. In this presentation, we discuss development of a new ion sampling interface for ICP to resolve the fundamental challnges mentioned above for non-metal analysis. The new interface is based on plasma assisted reaction chemical ionization (PARCI) where plasma is used to convert atoms to predictable species and the ionization occurs in plasma afterglow. While the conventional ICP-MS interface relies on direct sampling of ions from the plasma with minimal ion-molecule reactions, the PARCI interface takes advantage of ion-molecule reactions to enhance the ionization of non-metals. In particular, we discuss detection of halogens as halides and polyatomic ions. We demonstrate that ICP-based PARCI provides sensitivities comparable to that of ICP-MS/MS for chlorine detection in compounds separated by liquid chromatography. We investigate factors impacting ionization and demonstrate that introduction of sodium along with methanol into the plasma leads to drastic enhancement of sensitivity for chlorine while acetonitrile causes signal suppression. The mechanistic explanation of these results along with performance of ICP-based LC-PARCI-MS for other halogens will also be discussed. It is of note that the new interface allows coupling of ICP to any atmospheric sampling mass spectrometer for high-sensitivity quantitative elemental and isotopic analysis, greatly enhancing the applications of elemental MS.

Mann in Shorts - The Future of Pocket sized Raman

Neil Shand, Clare Nixon, Terry Clark

Current developments and trends in spectroscopic equipment are yielding ever more compact and affordable instrumentation. New cutting-edge equipment exploiting consumer technologies in conjunction with techniques such as Surface Enhanced Raman Scattering (SERS) techniques can provide very powerful in-field methods of analysis. This paper will highlight how these developments are creating new in-field capabilities for detection and identification of materials that pose a threat to Homeland Security such as explosives, chemical weapons and precursor chemicals.

3D SERS Imaging
Yukihiro Ozaki, Sanpon Vantasin

Raman imaging is a powerful tool for collecting molecular information from an area of interest. The variations in the spectra obtained in the area can provide information such as the distribution of targeted molecules, homogeneity of the sample, lattice consistency, etc. This allows Raman imaging to be feasible in the studies of polymers, nanomaterials, biological samples, and various other kinds of samples. The vibrational information of the molecules in the area can be correlated with structural information from microscopy techniques to explore the visible and invisible parts in much deeper aspect. Incorporating the technique of surface-enhanced Raman spectroscopy (SERS) into Raman imaging results in SERS imaging. This technique allows low-concentration molecules to be visualized. SERS imaging is also useful for SERS substrates with nonuniform hotspot: by averaging SERS signals form a large area instead of a single point, the signal intensity can be obtained with considerably less selection bias. Three-dimensional (3D) imaging is a technique that combines 3D Raman imaging with SERS. To proceed 3D SERS imaging, we developed highly symmetric nanoporous silver microparticles for 3D SERS imaging. Our substrate is purely chemically synthesized and dose not bond to solid substrates. The microparticles can be embedded in any kind of solution/polymer that does not react with silver. The microparticles have intrinsic nanopores and SERS activity. Therefore, they do not need any nanoparticle decoration. This is useful because it reduces the risk of nanoparticle detachment when embedded into solution/polymer systems. Since the particles are not bonded to solid substrate, they can be easily dispersed throughout the sample to allow 3D SERS imaging anywhere in the sample, and not just the bottom part. The utility of this 3D SERS substrate is demonstrated by embedding the particles in a polystyrene/polyvinylpyrrolidone (PS/PVP) 2-layer polymeric system.

What is that Mann wearing?
Karen Faulds, Kirsten Gracie, Hayleigh Kearns, Sian Sloan-Dennison, Duncan Graham, Roy Goodacre

Surface enhanced Raman scattering (SERS) is an analytical technique with several advantages over competitive techniques in terms of improved sensitivity and multiplexing. We have made great progress in the development of SERS as a quantitative analytical method, in particular for the detection of DNA. However, the lack of quantitative data relating to real examples has prevented more widespread adoption of the technique. Detection of specific DNA sequences is central to modern molecular biology and also to molecular diagnostics where identification of a particular disease is based on nucleic acid identification. Many methods exist and fluorescence spectroscopy dominates the detection technologies employed with different assay formats. Another advantage of SERS over existing detection techniques is that of the ability to multiplex which is limited when using techniques such as fluorescence. We have clearly demonstrated the ability to identify the presence of a mixture of 6 analytes in solution using data analysis techniques. Here we demonstrate the development of new molecular diagnostic assays based upon SERS which have been used successfully for the detection of DNA sequences from bacterial pathogens associated with meningitis using modified SERS active probes. The probes have been designed to give a specific SERS response resulting in discernible differences in the SERS which can be correlated to a specific DNA hybridisation event. The simultaneous detection and quantitation of 3 pathogens within a multiplex sample will be demonstrated for the first time. We have also recently demonstrated the detection of pathogenic bacteria in food using functionalised nanoparticles combined with magnetic separation and SERS (Figure 1). Metallic nanoparticles have been functionalised with bio-recognition molecules (antibodies and lectins) which are specific for a bacterial strain and a Raman reporter to enable the SERS detection. A different Raman reporter was used for each strain of bacteria; therefore a SERS signal was only obtained when the SERS active nanoparticle binds specifically to its bacterial target. The aim is to develop a multiplexing assay where three bacterial strains can be simultaneously identified in a sample matrix.

The Taming of The SERS

Katsumasa Fujita

One of the major issues in the application of surface-enhanced Raman scattering (SERS) spectroscopy is the difficulty in controlling and stabilizing the enhancement effect. In addition to preparing the metal nanostructure providing a high enhancement, the affinity of analytes to the metal surface is also an important factor. To tackle this issue, we have examined the combination of alkyne Raman tag and SERS spectroscopy. Alkyne has a unique Raman band at the silent region and can be easily separated from Raman signal from other molecules. It also has an affinity to the metal surface such as gold and silver enough to provide stable Raman spectrum measurement. We have utilized the alkyne tag technique to detect drugs incorporated into living cells. Gold nanoparticles are introduced as agents to report drug arrival by SERS of the alkyne. Time-lapse SERS imaging of alkyne revealed the dynamics of the drug in the cells. We also applied the alkyne tag technique to determine a small molecule biding site in a protein. We tagged a small molecule by alkyne and applied to a protein to make a covalent bond between the small molecule and the protein. After digesting the protein, peptides that covalently conjugated with the small molecule were extracted by detecting the alkyne SERS signal. The extracted peptide fragment was analyzed by mass spectroscopy, and the binding site of the small molecule was successfully determined. We also confirmed that the stable alkyne SERS signal allows the detection of low concentration molecules at nano-molar level. These experimental demonstrations show that the combination of alkyne tag and SERS is a promising direction to the effective utilization of the high sensitivity of SERS measurement.

The Great Rock and Roll Swindle

Roy Goodacre, Howbeer Muhamadali, Malama Chisanga

Although not to be confused with the lessons learnt by Malcolm McLaren, the manger of the British punk rock band Sex Pistols, this talk will 'transflect' McLaren's 10 lessons and how these have analogues with Duncan Graham transition from a lone organic chemist into a highly worthy prize winner whose team continue to have an exceptional research presence in surface enhanced Raman scattering (SERS) for bioanalysis. Oh, and I shall cover some our own bioanalytical work in SERS as well…

3D Patterned Paper Immunoassay Coupled with SERS for the Sensitive Detection of Malaria Biomarkers

Laura Frame, Karen Faulds, Duncan Graham

Malaria is a mosquito borne parasitic infection caused by Plasmodium parasites. It is a widespread problem with reports that in 2015 there were over 200 million cases of malaria globally, with 429,000 deaths reported. 90 % of these deaths occurred in sub-Saharan Africa. Therefore, as with many other diseases a simple diagnostic test for fast and accurate point-of-care detection is desirable. Immunochromatographic lateral flow tests are commonly employed to detect malaria specific antigens due to their low cost and portability. However, many are used predominantly as singleplex assays, with multiplex tests requiring large sample volumes. 3D paper based devices, which use a combination of lateral and vertical flow, are an attractive alternative to the traditional lateral flow model and present the capability of allowing multiplex detection of different biomarkers within a single, small sample volume. The sensitivity of these paper devices can also be enhanced by utilising surface enhanced Raman scattering within the detection step. Here we present a 3D patterned paper immunoassay coupled with SERS for the sensitive detection of the malaria specific HRP-II antigen. Gold nanoparticles were synthesised and conjugated to antibodies which specifically recognised the corresponding target antigenic molecule. The detection layers were then analysed using both large benchtop and hand-held Raman spectrometers, showing the potential for portable detection. Low detection limits of the studied antigen have been observed, which highlights the capability of these combined methods being employed in malaria detection.

MECHANICAL RECYCLING OF WASTE PLASTIC STREAMS IN ZHEJIANG PROVINCE, CHINA

Peter Summers, Zheng Wang, Fu Gu, Bin Wang, Michael George, Philip Hall

Plastic is one of the largest and the fastest growing waste streams in China, the management of which has received significant attention worldwide. China has become the world centre for recycled plastics, with the total market value of the industry in China growing steadily over recent years.(1) As the world's largest plastic manufacturer and consumer, China plays a key role in developing sustainable solutions to globalised waste plastic streams. Generally, there are four routes for plastic disposal treatment: mechanical recycling (material recovery), chemical recycling (cracking, gasification, hydrogenation and pyrolysis), energy recovery (incineration) and landfill. We are developing spectroscopic identification, and processing techniques to improve the mechanical recycling of waste plastic for high end applications in Zhejiang province, an eastern coastal province which produces and consumes the largest proportion of plastics in China. During the first three months in 2017, 2.13 million tonnes of plastic raw materials were produced in Zhejiang province,(2) and (in 2015) an average of 42.1 million US dollar worth of waste plastics are imported into Zhejiang monthly.(3) The growth of the plastic recycling industry not only brings economic profits, but also causes serious environmental and health problems and in many cases it is currently still more economical to use manual sorting rather than automated procedures. As a result, employees can be exposed to toxic chemicals during recycling, and high concentrations of persistent organic compounds are often found in residential areas around plastic recycling centres. (1) National Bureau of Statistics of the People's Republic of China, www.data.stats.gov.cn/index.htm (accessed Apr 20, 2015). (2) National Bureau of Statistics of the People's Republic of China, www.data.stats.gov.cn/index.htm (accessed Apr 29, 2017) (3) Statistics of domestic trade provided by Department of Commerce of Zhejiang Province, www.zcom.gov.cn/col/col1393/index.html (accessed Oct, 20 2015).

New Strategies for Low-Cost Energy Storage for the Grid: a Size - Exclusion Approach Using Polymer Colloids

Joaquin Rodriguez-Lopez

Modern society and its technological progress depend on the reliable supply of electrical energy. Ensuring this supply in a clean, cost-effective and safe manner to a growing population is a challenge that puts new scientific ideas at work. The recent years have seen an aggressive growth in renewable inputs such as solar and wind. However, these sources are intermittent, which is undesirable for the reliable operation of the electric grid.[1] It is here were inexpensive, yet high performance grid energy storage becomes indispensable. Our collaborators at the Joint Center for Energy Storage Research[1,2] have estimated that it is possible to reach an installed cost $100 per kWh, in line with the projections of the Department of Energy for the widespread deployment of grid energy storage. Inexpensive energy is attainable by exploring new battery design principles. Nonaqueous Redox Flow Batteries (NRFBs) are emerging technologies that supply high energy density and versatile power delivery in compact formats, and enable the leveling of the irregular input of alternative energy installations into the grid. However, a viable adoption of NRFBs requires timely solutions to challenges in understanding and controlling the complex nano-scale interactions that emerge between charge-storing units such as ions, electrons and redox species, and battery structures such as electrodes and membranes. My group's research directly addresses energy storage challenges through the creation of a novel size-exclusion flow battery design (Scheme 1) that uses versatile redox active polymer particles.[3] This scheme simultaneously solves several challenges: it replaces expensive ion-exchange membranes for affordable nanoporous separators, it enables the use of a variety of electrolytes and solvents, it suppresses deleterious redox component crossover, and is conceived to work with inexpensive redox materials. Our discussion on redox flow batteries will highlight some of the challenges facing the creation of superior technologies for grid energy storage. [1] Crabtree, G. AIP Conference Proceedings 2015, 112, 1652. [2] Darling, R. M.; Gallagher, K. G.; Kowalski, J. A.; Ha, S.; Brushett, F. R. Energy Environ. Sci. 2014, 7, 3459.

[3] Burgess, M.; Moore, J. S.; Rodríguez-López, J. Acc. Chem. Res. 2016, 49, 2649.

InSciEd India: Using K-12 Education to Renew Connectedness to Nature,
Seth Thompson, Christopher Pierret

Improving K-12 STEM education is key to nurturing a new generation of scientists and science-literate citizens across the globe. Integrated Science Education Outreach (InSciEd Out) is an educational initiative originating from work in public schools in Rochester, Minnesota, focusing on the relevance of science education to both individuals and community, improved understanding of science, and environmental and cultural perspective. We seek to extend student science beyond rote knowledge, empowering students to apply science solutions to critical and topical community-based problems, and to advance research and improve health within communities we work with. Using the prescription education framework of InSciEd Out, we have implemented cutting-edge environmental science experiences within several partner schools in India. Here we present the prescription education framework used in the InSciEd Out program and how its application has renewed the conversation about the fundamental connections between humans and their environment in India.

A compelling homecoming-zebrafish as a tool for high-quality, low-cost science and education in India
Chris Pierret

The health of our local and global community is directly correlated with education. Integrated Science Education Outreach (InSciEd Out) partners researchers, educators, and their local communities. As part of a scientific relationship between Mayo Clinic and India's Council for Scientific and Industrial Research, InSciEd Out was selected as a method to 1) improve opportunities for scientific inquiry by Indian K-12 students; and 2) re-introduce the zebrafish to scientists in the country of its origin. The transparent development of zebrafish has been particularly effective in challenging misconceptions about life sciences, cells, and heredity. Specifically, experiencing life at this new resolution enhances the understanding of cells, organs, systems, and whole organisms as the products of a genetic process. Our current work with zebrafish in India surrounds the clinical themes of infectious disease (specifically Dengue Fever), environmental toxicity (heavy metals and plasticizers), and pediatric obesity. We are also exploring work in mental health and addiction (a core goal of the Mayo Clinic/CSIR relationship). Transgenic zebrafish are a valuable resource in this relationship. A host of immunologically relevant transgenics support this work. Moira Whyte's Tg(MPO:GFP) line, for one, allows live viewing of neutrophils during inflammatory response. Aparna Chaudhari's Tg(MT-Ia1:DsRed2) line gives a view into heavy metal contamination of a watershed. Even obesity can be richly explored with Tg(Betaactin:zAgR) lines from Youngsup Song and Roger Cone. Each of these fish show a response or condition that allows students access to health and disease like never before. In the InSciEd Out partnerships in India, the necessity of bringing the zebrafish to the classroom has also bloomed into relationships in higher education. At each site, InSciEd Out facilitators act as a liaison to the zebrafish model for undergraduate and graduate school host facilities. These relationships allow the transgenics to be used along with the rich genetic tools like CRISPRs, TALENs, and Morpholinos. The genome is wide open to exploration with these new tools. InSciEd Out provides a scaffold for meaningful work between the scientific and clinical zebrafish researchers and the communities they serve, and the potential for rapid scaling of this approach around the globe.

Open-path, quantum cascade laser spectroscopy for atmospheric measurements in power-constrained platforms

Mark Zondlo, Levi Golston, Dana Caulton, Da Pan, Kang Sun, David Miller, Lei Tao

Advances in quantifying greenhouse gas and air pollutant emissions require measurements that can capture high spatiotemporal variabilities and sample a representative fraction of emission sources. Top-down approaches using aircraft or tall towers with atmospheric inversion models have made great progress in constraining regional to global scale emissions, but inherent in these estimates are accurate knowledge of aggregate sources within the scale of interest. However, emission inventories from the bottom-up are often limited to small sample sizes that may not be representative, and the measurements themselves are often challenging to make accurately at local scales (10-100s m). Therefore, emissions measurements require both sampling local sources accurately over time and being able to sample representative numbers of local emissions. To these ends, we have developed open-path, mid-infrared laser-based sensors for key greenhouse gases (CH4, N2O, CO2), air pollutants (NH3, CO), and atmospheric tracers (ethane, water vapor). The sensors use quantum cascade lasers (CO, N2O, NH3), interband cascade lasers (CH4, C2H6), and Sb lasers (CO2, CH4). Multi-harmonic wavelength modulation spectroscopy is used for sensitive detection and minimizing interferences from atmospheric pressure-broadened line shapes. An in-line reference cell is also used to help distinguish between artifacts from instrument noise versus real atmospheric variability. The sensors have been deployed onboard UAVs and towers for spatiotemporal measurements at individual sources. We have deployed these sensors in natural ecosystems such as forests, wetlands and grasslands as well as near anthropogenic settings such as corn fields, cattle feedlots, cities, and the natural gas supply chain. Furthermore, because of the small sizes and low power requirements of each of the sensors, the sensors can be combined onto a common passenger vehicle to chemically "fingerprint" sources in complex settings and identify the shape of the distribution from many individual sources. Examples to be shown include the prevalence of "superemitting" well pads in the Marcellus Shale to vehicle emissions of ammonia in different US and Chinese cities. In both cases, current emission inventories greatly underestimate the "fat-tails" of these distributions, showing the need for both accurate and representative measurements of local emission sources.

Mid-Infrared Process and Emission Monitoring

Peter Geiser

In this presentation, we will report on new challenging applications for mid-infrared tunable laser absorption spectroscopy (TLAS). In the last decades, TLAS has proven to be an excellent choice for many applications – even under very harsh conditions – where measurements directly in the process (in-situ) are required. The deNOx-process is a method to reduce NOx emissions from combustions. Either gaseous ammonia or urea is used to convert NOx and ammonia into water vapor and nitrogen. The amount of ammonia needed depends on the NOx content of the process gas. Especially in a dynamic process where the fuel composition changes rapidly the variation of the NO concentration can be quite high. Typically, the amount of not-converted ammonia is measured to control a deNOx-process ("ammonia slip") since NIR NO measurements usually suffer from severe water vapor interference at gas temperature above 300 C. Industrial users, however, would like to measure the nitric oxide concentration directly in the raw gas. The mid-infrared absorption band of NO around 5.2 μm allows interference-free measurements of NO within a deNOx-process and results from a measurement campaign will be shown demonstrating a detection limit of below 1 ppm m in the raw gas. Another challenging task is the continuous emission monitoring of tetrafluoromethane (CF4), a very strong greenhouse gas. The aluminum smelting process is one of the largest anthropogenic contributors to atmospheric CF4. CF4 does not absorb in the near- but only in the mid-infrared with a strong band located at around 7.79 μm. A quantum cascade laser sensor was developed using a unique Wavelength Modulation Spectroscopy approach. During laboratory assessment, a detection limit of well below 10 ppb m was demonstrated. Afterward, the CF4 sensor was installed at an Alcoa aluminum smelter. A co-located Fourier Transform Infrared Spectrometer allowed direct comparison measurements and general agreement between the two methods was observed. This led to the conclusion that the developed in-situ quantum cascade laser based sensor has the potential to continuously measure CF4 at aluminum smelters.

Novel Improvements and Applications using Tunable Infrared Laser Direct Absorption Spectroscopy

Scott Herndon, Mark Zahniser, David Nelson, Barry McManus, Joanne Shorter, Rob Roscioli, Tara Yacovitch, Christoph Dyroff, Conner Daube

Tunable Infrared Laser Direct Absorption Spectroscopy (TILDAS) has benefitted from improvements in quantum cascade and interband-cavity lasers in recent years. Following a discussion of recent advances, several applications of the Aerodyne TILDAS spectrometers in field applications will be discussed. Highlights from emissions characterization, photochemistry and biosphere interaction studies will be presented.

Broadband multi-heterodyne spectroscopic chemical sensing in the mid-IR and THz with quantum and interband cascade laser frequency combs

Gerard Wysocki, Jonas Westberg, Lukasz Sterczewski, Link Patrick

Laser spectroscopic detection of large molecular compounds with unresolved rotational-vibrational spectroscopic signatures in the mid-infrared requires spectroscopic techniques that can provide broadband coverage. Broadband multi-heterodyne spectroscopic (MHS) chemical sensing with quantum (QCL) and interband cascade laser (ICL) frequency combs opens up potential application space by enabling spectroscopic technology that can address detection of broadband absorbers (with >50cm^{-1} features) while providing high spectral resolution (<0.001cm^{-1}) suitable for detection of small molecules that exhibit well-resolved spectral lines. The same MHS detection can be realized with novel THz QCL combs to provide access to rotational bands of many molecules. Often broadband mid-IR absorbers exhibit well-resolved rotational components in the THz spectral region, which can result in significant improvements in chemical specificity if both mid-IR and THz signatures can be probed. Thus, development of spectroscopic sensing technologies that can address both spectral regions is of great importance. Here we demonstrate recent progress towards broadband high-resolution spectroscopic sensing applications with QCL and ICL frequency combs using MHS techniques. Examples of mid-IR MHS spectra of isobutene and ammonia in the mid-IR region accessible with QCLs operating at ~8.5?m as well as methane and ethylene measured with ICLs operating at ~3.2?m will be discussed. A balanced detection based MHS system capable of molecular absorption spectroscopy as well as molecular dispersion spectroscopy achieves noise-equivalent absorption down to ~$5x10^{-5}$/Hz$^{1/2}$ and fast spectral acquisition down to 10 μs/spectrum. THz MHS system was realized with dispersion compensated THz-QCLs developed by Burghoff et al. (Nature Photonics 8, 2014). The MHS system is realized by combining radiation from two QCL combs on a hot electron bolometer (Scontel). Multiheterodyne spectroscopy of NO_2, NH_3 and H_2O is performed and the system achieves an instantaneous

spectral coverage of ~100 GHz, 10 μs/spectrum time resolution and a short-term noise-equivalent absorption of 7×10^{-4}/Hz$^{1/2}$. Most recent spectroscopic results as well as system performance will be discussed in details.

Infrared Spectroscopic Method for Uranium Isotopic Analysis in UF6 Gas

K. Alicia Strange , Patrick O'Rourke, William Spencer, Nicholas Deroller, Steven Serkiz, Leigh Martin, Darrell Simmons

The International Atomic Energy Agency (IAEA) administers monitoring of declared facilities for uranium hexafluoride (UF6) isotopes which requires sufficient sensitivity to verify the UF6 enrichment. Near real-time analysis is currently accomplished through a radiometric technique using gamma ray detection which allows for a field deployable system. However, the radiometric technique lacks sensitivity, is sample size dependent, and has background field limitations. Confirmatory analysis using mass spectrometry offers high sensitivity yet requires extensive sample preparation and is time consuming. Infrared spectroscopy can be used to probe elemental isotopes, because the difference in the atomic weight of the isotopes results in a shift in the absorption wavelength. At SRNL, we are developing a new infrared (IR) spectroscopic technique to measure the isotopic ratio of gaseous UF6 using a quantum cascade laser (QCL). The recent advancement and commercialization of QCL with ultra-fine spectral resolution (~0.002 cm-1) provides lasers with adequate resolution to excite UF6 molecules with enough accuracy to resolve the 0.65 cm-1 235UF6/238UF6 isotopic shift. Therefore, a spectrometer for wavelength discrimination is not required, and the proposed IR system can be made compact and robust to allow for a field deployable system capable of near real-time measurements. The talk will discuss the instrument development, demonstrate the current progress with spectra of gaseous samples, and discuss the operating parameters required to achieve the necessary sensitivity.

Direct Coupling of SPME to Mass Spectrometry
Janusz Pawliszyn, German Augusto Gomez Rios

The attempts have been made to scrutinize samples without sample pre-treatment, or with minimum sample preparation resulting in direct sample introduction to mass spectrometry techniques. Variety of applications has been developed, from monitoring chemical transformations to discerning between healthy and cancerous tissues at the surgery room. However, no technique is perfect, and predictable limitations of direct sample to MS for analysis of complex matrixes, such as ionization suppression, poor sensitivity at trace levels, and narrow linear dynamic range, have been the driving force toward the development of methods that efficiently integrate sample cleanup, analyte collection, and ionization. This new era of mass spectrometry (MS), where sample preparation devices are directly and efficiently coupled to MS instrumentation, has given rise to a growing branch of innovative research where micro/nano-extraction approaches excel. Solid-phase microextraction (SPME), a worldwide recognized green sample preparation technique, was certainly not the exception. In talk I will emphasize diverse SPME-based devices recently developed in my laboratory for the extraction/enrichment of analytes of interest from complex sample matrices, which can be directly coupled with mass spectrometry instruments for rapid analysis. These technologies include Coated Blade Spray (CBS), SPME-transmission mode-direct-analysis-in-real-time (SPME-TM-DART-MS), SPME-nano-electrosprayionization (SPME-nano-ESI-MS) and SPME Open-Port Probe (SPME-OPP-MS). Total analysis time did not exceed 5 minutes and sample volumes ranging between 1-1500 ?L were used. Sampling/sample-preparation is performed either by inserting SPME device directly into the investigated object, spotting the sample onto the SPME-device or by immersing the SPME-device in a vessel containing the sample. Despite short extraction times, limits of detection in the pg/mL to sub-ng/mL range were obtained, while good accuracy, and linearity were attained for all the investigated analytes (i.e. therapeutic-drugs, pharmaceuticals, and pesticides) in PBS, urine, plasma, serum, blood, saliva, grape-juice, orange-juice, milk, tissues and ground-water. In addition, the development of these technologies and their evolution in terms of geometry, coating characteristics, and thermal/spray stability, as well as its initial application to the study of tissue and homogenized cells will be presented as well.

Single-use Cartridges for Protein and Drug Detection
Nicholas Manicke, Chengsen Zhang, Brandon Bills, Greta Ren, Trevor Glaros

I will describe some recent efforts in our group to develop prototype sample cartridges which combine mass spec sample preparation and ionization steps in a single simple device. This will include analyzing designer drugs such as fentanyl analogs and spice cannabinoids on a paper spray cartridge with built-in solid phase extraction. Another paper spray cartridge to be presented is a blood fractionation cartridge in which a dried plasma spot is obtained after applying a drop of blood to a membrane built-in to the device. I'll also describe the development of a mass spec sampling cartridge designed for targeted detection of proteins from complex matrices such as plasma. Selective enrichment of protein targets is performed on cartridge passively by capillary action and gravity with no robotics or fluid handling devices. Ionization occurs with nanoESI like characteristics from a carbon nanotube coated porous polymer. Finally, I'll show some work to create a device to combine aerosol capture and direct ionization from the capture substrate to detect airborne chemical warfare simulants.

Improvement in Ambient Mass Spectrometry Sensitivity via Computational Fluid Dynamics
Jin Young Song, Allen R. White , Brian T. Molnar, Jacob T. Shelley, Gary M. Hieftje

Analysis and optimization of the fluid mechanics of atmospheric-sampling mass spectrometers and their ionization sources can yield increased sensitivity and efficiency. The movement of air and helium between the ionization nozzle and mass spectrometer entrance orifice influences transmission and sensitivity. The present study utilizes computational fluid dynamics models, validated with experimental data, to inform improvements of the efficiency and sensitivity of a mass spectrometer with a FAPA ambient mass spectrometry source.

Direct Extraction and Detection of Analytes from Biofluids Using Functionalized Paper Substrates
Abraham Badu-Tawiah, Deidre Damon, Tatiana Velez

Ambient ionization techniques were introduced in 2014 as a means for direct analyte detection from complex mixtures without prior sample treatment. For substrate-based ambient ionization methods such as paper spray – having reduced signal lifetime due to rapid solvent evaporation – the ability to sensitively detect multiple analytes in a single experiment will be an added advantage. In this presentation, two paper-based sample processing/extraction approaches are described for direct detection of proteins (e.g., histidine-rich protein 2 (HRP 2) malaria antigen) and small molecules (e.g., pyruvate for diagnosis of liver injury) from biofluids such as blood and serum. First, we will discuss an integrated 3D microfluidic paper device for multiplexed immunoassay-based detection of protein biomarkers from untreated biofluids. This device is capable of reagent storage, biofluid splitting, antibody reaction, control of assay time, and on-chip MS detection. This mass spectrometry-based immunoassay approach detects tagged small molecules instead of direct analysis of the intact protein. In so doing, MS signal can be amplified while also reducing instrumentation requirements. Two such tags will be presented: ionic probes for sensitive detection and isobaric peptides for tandem MS analysis of multiple analytes in a single experiment. Second, a new paper spray methodology is described, based of 2D hydrophobic paper substrate that enables on-line liquid/liquid extraction of small organic compounds from biological fluids placed on the dry hydrophobic paper triangle. The use of ethyl acetate spray solvent, which is immiscible with the biofluid, for the extraction process allows solvent pooling on a dry hydrophobic paper and enables in-situ electrostatic-spray ionization, after the application of 3 kV DC voltage. Compared with traditional paper spray performed on hydrophilic paper, over 100x increase in sensitivity is achieved for illicit drug analysis from fresh biofluids when using the new hydrophobic PS method. Factors influencing this performance are discussed. The method was applied to diagnose liver injury by monitoring alanine transaminase enzyme activity in the presence L-alanine.

Photo-Oxidation Studies of Polyester Coatings
Allison Pymer, Rebekah Scott, Pam White

Polyester-based coatings are traditionally considered to exhibit poor weathering performance compared to alternative technologies and are not a typical choice for outdoor exposure applications. Recent weathering data from polyester coatings that incorporate Eastman Tetrashield(TM) protective resin systems demonstrate that this assumption is not always true. This presentation will discuss how the rate of polyester coating degradation may be evaluated by monitoring the photo-oxidation value of polyester coatings, which is calculated from infrared spectroscopic measurements conducted in parallel with accelerated weathering studies. This innovative method provides coatings formulators an alternate technique to help predict and characterize the failure methods of coatings.

Development of Process Analytical Solutions for Real-time Monitoring of Continuous Flow Reactors
Brian Marquardt, Natasha Hippler, Michael Roberto, Patrick Witham, Kendra Cochran

Continuous Flow Reactors (CFRs) are flow cells that are optimized for the continuous production of a target chemical compound. Currently, production of pharmaceuticals relies on large-scale batch processes that are intensified from the laboratory scale. Compared to batch reactors, CFRs have greater energy efficiency due to their superior mixing schemes and heat transfer properties. As a result of this efficient heat transfer, reactions that require cryogenic temperatures in batch, such as the Swern oxidation, can frequently be performed at much milder temperatures in CFRs, significantly reducing temperature control costs. Though these features make CFRs attractive to process-intensive industries, they are currently limited by a lack of on-line analytical technologies suitable for small-volume, high-throughput measurements. A program in our company has been ongoing to evaluate the use of process analytics (PAT) and custom sampling systems for the validation of CFRs. An example reaction includes a four reagent sampling system interfaced to a continuous low-flow reactor to enable reaction monitoring using a process Raman spectrometer. Real-time process control of the Swern reaction was achieved by monitoring a series of process control parameters that included flow rate, stoichiometry, reactor temperature and pressure. Establishing control of the process allowed for optimization, minimizing the production of "off spec" product and removing the need for off-line quality control analytics. Real-time chemical modeling was performed to determine product quality using multivariate statistical methods. Results from these experiments demonstrated that the oxidation, which requires cryogenic temperatures in batch, can be performed at cost effective temperatures in continuous flow. This presentation will effectively demonstrate the use of process analytical tools for the optimization and control of CFR production systems across a number of reactions.

Online Analysis of the De-hydration of Glycol using a novel solid state FTIR spectrometer
Dan Wood, Jonathon Speed

Glycol dehydration of raw materials is an important part of the petrochemicals production process. However the glycol used must be thoroughly dried itself before recirculating back in to the front end of the drying process. Common techniques for measuring the water content of glycols are either slow (such as Karl Fischer) or are not sensitive enough to ensure that enough water has been removed. Here we present the use of a vibration-resistant FTIR spectrometer to monitor the water content in ethylene glycols. By combining the spectra with simple chemometrics the limt of detection was found to be 1000ppm, meaning that 99.9%triethylene glycol was reliably measured

Molecular Modeling in In Situ Monitoring Spectra Analysis
Xianghuai Wang, Ronen Weingarten, Xiaoyun Chen

In situ reaction monitoring is widely used in industrial applications to elucidate reaction mechanisms. One challenge for this analysis is to correctly assign new vibration frequencies in the spectra to several possible intermediates. In some cases, the intermediates are difficult to isolate and purify, and thus the control spectrum of each species cannot be obtained. This brings about additional challenges when trying to understand the reaction kinetics. Over recent years, molecular modeling has been proven to be a powerful tool to predict physical properties of molecules, especially for those which are unstable or hard to purify. In this talk, we show that molecular modeling can be employed to help predict the vibration frequencies for intermediates in a reaction between phenyltrimethoxysilane (PTMS) and octamethylcyclotetrasiloxane (D4).

Chemometric modeling and classification of IgG monoclonal antibodies utilizing Drop Coat Deposition Raman
Eric Reichard, Cara Fowler, Jeff Denault

The pharmaceutical utilization of monoclonal antibodies has increased in the past three decades due to a better understanding of various disease states and advancements in technology. Each year an increasing number of monoclonal drug products are approved with many more in development. As a result, rapid techniques are needed in order to accurately detect and distinguish originator monoclonal products from competitor and/or counterfeit products. It is also critical to distinguish across the portfolio of monoclonal products for a given manufacturer. Recently, Drop Coat Deposition Raman (DCDR) has been demonstrated as a viable technique to characterize monoclonal antibodies, whereby the protein is concentrated and segregated from other interfering compounds like buffer components and impurities. A resolved Raman spectral fingerprint of the native protein can be acquired in order to identify the antibody as well as to obtain structural information. An evaluation of the DCDR technique with regards to the characterization of similar IgG monoclonal antibodies will be presented. Chemometric techniques, including Principal Component Analysis (PCA) and Partial Least Squares Discriminant Analysis (PLSDA), used to analyze the Raman spectrum and classify the monoclonal proteins for identification purposes will be discussed.

Monitoring Protein Structural Changes using Raman Spectroscopy
Marinella Sandros

About a quarter of protein-based therapeutics are introduced subcutaneously and are formulated in liquid solutions with high protein concentrations in order to minimize the injection volume (< 2 mL). This concentration level can easily elevate protein-protein interactions and eventually lead to aggregation. Protein agglomeration can produce unwanted side-effects and minimize the drug efficacy. Thus, it is vital to have analytical tools that can monitor change in size, shape and structure of proteins in order to improve decision-making skills during the drug formulation process. In the lower concentration range (<10 mg/mL), circular dichroism (CD) and fluorescence are well established techniques for monitoring conformational changes. However, there are limited techniques to assess structural changes in high concentration antibody formulations when in their natural physical state (without dilution). The work presented here, investigates the potential of Raman spectroscopy as a tool to monitor the propensity for protein aggregation and found that specific signature Raman bands can be used as biomarkers. The exact position and intensity of some Raman bands are influenced by inter and intra molecular effects therefore permitting the unique ability to monitor dynamic fluctuations as a result of external factors.

Probing the conformation and orientation of disulfide groups in proteins and protein aggregates by means of normal and polarized Raman spectroscopy
Igor Lednev

Disulfide bonds play a significant role in preserving the physiologically functional three-dimensional form of proteins. Raman spectroscopy is uniquely suitable for characterizing the integrity and conformation of disulfide groups. Nonresonance Raman spectra of disulfide-containing proteins exhibit bands at 510±5, 525±5, and 540±5 cm$^{?1}$, which could be typically assigned to S?S stretching vibrations, ?(SS), in gauche?gauche?gauche (ggg), gauche?gauche?trans (ggt) and trans?gauche?trans (tgt) conformations of the C?C?S?S?C?C segment, respectively. For example, Gaussian deconvolution of the Raman spectra of naive insulin in the 470?580 cm$^{?1}$ range reveals that the ratio between the peak areas assigned to the ggg and ggt conformations was approximately 2:1 that is in good agreement with the protein crystal structure. We utilized near-IR Raman spectroscopy to probe disulfide bonds during the formation of amyloid fibrils, protein aggregates associated with various neurodegenerative diseases. For example, the secondary and tertiary structure of insulin changes dramatically as a result of its fibrillation in vitro. An ?-helical protein is converted into mainly ?-sheet form and yet all three disulfide bonds of native insulin remained intact during the aggregation process, withstanding scrambling. Polarized Raman spectroscopy showed strong orientation of disulfide bonds in insulin fibrils, indicating their association with the fibril core. We have recently found that Lysozyme disulfide bonds undergo significant rearrangements in the presence of hydrogen sulfide. Raman bands corresponding to disulfide (RSSR) vibrational modes in the 550?500 cm$^{?1}$ spectral range decrease in intensity and are accompanied by the appearance of a new 490 cm?1 band assigned to the trisulfide group (RSSSR) based on the comparison with model compounds. The presented evidence indicates that hydrogen sulfide causes the formation of trisulfide bridges, which destabilizes lysozyme structure, preventing protein fibrillation. As a result, small spherical aggregates of unordered protein form, which exhibit no cytotoxicity by contrast with lysozyme fibrils.

Drop Coat Deposition (DCD) Confocal Raman Spectroscopy: Waking Sleeping Beauty with Coffee Ring for Protein Characterization
Ravi Kalyanaraman, Jeremy Peters, Anna Luczak, Varsha Ganesh, Eugene Park

Drop Coat Deposition (DCD) is a pre-concentration sampling technique that is known to increase the signal-to-noise (S/N) ratio in Raman spectroscopy. In spite of improved S/N ratio, this technique has been underutilized in the past for protein structural characterization. This pre-concentration technique has been extensively used in this study to produce protein deposits in the "coffee ring". The protein in these deposits were found to be well hydrated, thus maintaining the protein in its native structural form. The confocal Raman microscope provided highly resolved visual image of the "coffee ring" and thus offered the spatial resolution to collect the Raman spectrum of the protein from the ring in its native state. High quality Raman spectra from model proteins and biologics drug products were collected and the Amide I vibrational band, which is typically in the region of 1620 to 1720 cm-1 was used to determine the secondary structure distribution of proteins. This technique has also been evaluated to monitor protein degradation, such as acid, heat and disulfide reduction. This talk will feature all the above applications of DCD technique coupled with confocal Raman spectroscopy.

Vibrational sensing of albumin glycation: a route to facile detection of proteins and biologics
Ishan Barman

In the past decade, considerable attention has been focused on the measurement of glycemic markers, such as glycated hemoglobin and glycated albumin, that provide retrospective indices of average glucose levels in the bloodstream. While these biomarkers have been regularly used to monitor long-term glucose control in established diabetics, they have also gained traction in diabetic screening. Detection of such glycemic markers is challenging, especially in a point-of-care setting, due to the stringent requirements for sensitivity and robustness. A number of non-separation based measurement strategies were recently proposed, including photonic tools that are well suited to reagent-free marker quantitation. Here, we will focus on Raman spectroscopy based characterization of these glycemic markers. We examine the underlying principles and the utility of these approaches as reagentless assays capable of multiplexed detection of glycemic markers and also the challenges in their eventual use in the clinic.

Rapid In Vivo Raman Spectroscopy for Endoscopic Lung Cancer Detection

Haishan Zeng

Lung cancer is the leading cause of cancer-related deaths worldwide, with a five year survival rate of less than 20%. If lung cancer can be diagnosed earlier, the survival rate can be improved significantly (>70% for patients with stage (0) or 1A disease). However, this is rarely the case at current clinical practice, as most patients are diagnosed at an advanced stage where there remains a lack of curative treatments. Currently the most sensitive method for localizing lung cancers in central airways is autofluorescence bronchoscopy (AFB) in combination with white light bronchoscopy (WLB). The diagnostic accuracy of WLB+AFB for high grade dysplasia (HGD) and carcinoma in situ is variable depending on physician's experience. When WLB+AFB are operated at high diagnostic sensitivity, the associated diagnostic specificity is low. Raman spectroscopy probes molecular vibrations and gives highly specific, fingerprint-like spectral features and has high accuracy for tissue pathology classification. In this study we present the use of a real-time endoscopy Raman spectroscopy system to improve the specificity. A spectrum is acquired within 1 second and clinical data are obtained from 280 tissue sites (72 HGDs/malignant lesions, 208 benign lesions/normal sites) in 80 patients. Using multivariate analyses and waveband selection methods on the Raman spectra, we have demonstrated that HGD and malignant lung lesions can be detected with high sensitivity (90%) and good specificity (65%). This is a significant improvement compared to WLB+AFB, which has a much lower specificity of 18% -32% when the sensitivity is set at 90%. We envision that an algorithm derived from a database of lung Raman spectra would be capable of classifying a lesion in less than half a second, making this method feasible for real-time lung cancer diagnosis. Different from subjective interpretation of the

WLB/AFB images, the Raman algorithm represents an automatic, objective diagnosis based on quantitative Raman spectral analysis. Point Raman measurement on lesions identified by WLB+AFB imaging could become a new clinical tool for real-time detection of lung cancer/precancerous lesions.

Forward-adjoint Monte Carlo modeling of SORS sampling depth in bone and soft tissue

Andrew Berger, Guanping Feng

Spatially offset Raman spectroscopy (SORS) enables chemical assessment of the lower layer of a two-layer system. The spectrum of bone can, for instance, be determined through overlying soft tissue. In this approach, however, it can be unclear how deeply the approach samples into the bone layer. We have developed a forward-adjoint Monte Carlo computer model to calculate the relative contributions from different depths in a Raman measurement with a specified source-detector separation. Comparisons to a published "direct" model demonstrate consistency and increased speed. In combination with experiments, the new method was able to determine the absorption and scattering properties of a two-layer system. The method then able to make a prediction of how the ratio of lower layer/upper layer signal changes with different source-detector separations, and the prediction agreed closely with experiment.

Raman Microspectroscopic Analysis of Pancreatic Endocrine Precursor Cells Derived from Human Embryonic Stem Cells

Michael Blades, H. Georg Schulze, Stanislav O. Konorov, Ali Asadi, Timothy J. Kieffer, James M. Piret, Robin F. B. Turner

The successful differentiation of human embryonic stem cells (hESCs) to insulin-positive cells is now placing their therapeutic use within reach. To meet the stringent regulatory quality and safety requirements imposed on the use of therapeutic cell products, the final product attributes of biologics must be verified along with validation of the entire process to ensure a predictable and well-characterized production process. hESC-derived pancreatic endocrine precursor cells constitute both a critical stage in the production sequence and the likely therapeutic product; thus, a careful characterization of their attributes is essential. In this presentation we discuss the use of Raman microspectroscopy to characterize single cells and small cell clusters generated by a directed differentiation protocol applied to hESCs.1 Because insulin contains disulfide bonds, in contrast to the other major pancreatic hormone, glucagon, our major focus was on the 480 to 600 cm-1 region. Our objectives were to examine the potential of RM as a process analytical technology in determining the uniformity of cells, assessing the presence of undesirable cells, evaluating target cell insulin content, and investigating the possibility of yield assessment. The 480 to 600 cm-1 spectral region contains information about different types of disulfide bonds and also has glycogen (485 cm-1) and DNA (495 cm-1) marker bands. Therefore, it may serve as a reporter spectral region for unwanted cell types characterized by inherently or transitory high glycogen concentrations. The results show that the disulfide region can play a valuable role in determining a Raman fingerprint characteristic of insulin-positive cells that could be important for assessing both successful differentiation during production and the final therapeutic cell product quality. The ability to detect insulin in cells provides a way to assess the quality as well as the yield of a batch of cells obtained by directed differentiation. Thus we conclude that Raman spectroscopy has considerable promise as a PAT to monitor process performance as well as product attributes.

In vivo interstitial Raman and diffuse optical spectroscopy to improve the safety and accuracy of brain biopsy procedures

Frederic Leblond, Joannie Desroches, Fabien Picot, Michael Jermyn, Michael Pinto, Sami Obaid, Marie-Christine Guiot, Kevin Petrecca, Brian Wilson

In situations where suspected cancer is located deep into the brain or when imaging results are inconclusive, biopsies are performed to determine if surgical excision is required and to help optimize chemotherapy and radiotherapy planning. There is an inherent limitation with current core needle biopsy approaches: surgeons are unable to insure samples are representative of the cancer because of potential geographical misses leading to non-diagnostic samples. This is important because of health risks associated with repeat biopsies namely hemorrhages leading to permanent losses of motor or cognitive functions. To address this unmet clinical need, we designed an in situ intraoperative optical needle using high wavenumber Raman spectroscopy for cancer detection and diffuse optical tomography for blood vessels detection. To insure practical integration into the surgical workflow, the optical fibers were embedded with a commercial core biopsy needle allowing optical measurements to be made prior to sample collection without disrupting the regular clinical process. Using this instrument we demonstrate deep brain in vivo interstitial high wavenumber Raman spectroscopy and diffuse optical tomography measurements in swine showing the instrument can reliably detect brain tissue spectra and blood vessels distributions. In a separate proof-of-principle study involving 19 glioma patients, we found brain cancer can be detected in situ during surgery using high wavenumber Raman spectroscopy with sensitivity and specificity of 90% and 80%, respectively. These studies set the stage for clinical translation of this highly specific optical molecular imaging approach for targeted biopsy collection in brain to improve safety and diagnostic yield.

A SERS-based strategy for multiplexed detection of cardiovascular disease biomarkers in vitro, ex vivo, and in vivo

Steven Asiala, Jonathan Noonan, Kirsten Gracie, Gianluca Grassia, Neil MacRitchie, Pasquale Maffia, Paul Garside, Iain Mcinnes, Karen FauldsDuncan Graham

Cardiovascular diseases (CVD) are the leading cause of death globally. Surface-enhanced Raman spectroscopy (SERS) has been demonstrated as a useful tool for the detection, and in some cases quantification, of biomarkers indicating a diseased state. This is accomplished through the use of biofunctional nanoprobes, consisting of a gold nanoparticle core, one of a number of SERS reporter molecules, an encapsulation layer, and a unique biomarker-targeting moiety at the particle surface. In this work, we demonstrate multiplexed CVD biomarker imaging and detection, utilizing four varieties of antibody/reporter nanoprobes—anti-ICAM/BPE, anti-VCAM/PYOT, anti-P-Selectin/PPY, and IgG Isotype/DP—with the first three targeting CVD biomarkers, and the fourth serving as an internal standard for non-specific binding. The particles are deployed to bind biomarkers in three biologically relevantly environments— *in vitro,* for imaging in human endothelial cell lines, *ex vivo*, for imaging in diseased human tissue samples, and *in vivo*, for detection in mice implanted with human tissue grafts. We also report progress on a relative quantification protocol, based on a direct classical least squares (DCLS) model, to be used to delineate between healthy and diseased states based on biomarker expression. Taken together, these experiments serve as a step toward the identification of CVD *in vivo* via SERS.

Stress Measurement Through Tip Enhanced Raman Spectroscopy

Razvigor Ossikovski, Marc Chaigneau

Characterization of the strain present in semiconductor structures such as strained Si (sSi) on insulator (sSOI) or epitaxially grown SixGe1-x on Si (SiGe/Si) is of special interest for the microelectronics industry. The stress state in crystal materials can be determined by using techniques such as X-ray diffraction, transmission electron microscopy and Raman spectroscopy. The main advantages of the Raman technique over the first two are its non-destructive nature and no need for sample preparation. Anyway Raman spectroscopy only gives an average value of the strain, and the optical resolution is not suited to current problems of the microelectronics industry. Thus, by combining oblique incidence Raman scattering geometry with a polarization control and a Scanning Probe Microscope equipped with a nano-antenna, it is possible to increase the Raman spatial resolution and determine the stress value in strained structures. After a brief review of the pros/cons and the recent advances of the TERS technique for this application, we will present an experimental study of misfit strain down to the nanoscale by Tip Enhanced Raman Scattering (TERS) of lithographically defined SiGe nano-stripes on Si(001) substrate. During the TERS experiment, we monitored the intensity and the frequency of the locally enhanced Si-Ge and Ge-Ge Raman modes across a single nano-stripe, giving the perpendicular strain profile (tensile and maximum at the centre of the nano-stripe, decreasing close to zero at the edges) with a lateral resolution of <20 nm. 3D Finite Element Modeling calculations are successfully compared to the experimental results.

Tip-Enhanced Raman Scattering Beyond Chemical Nanoscopy
Patrick El-Khoury

Trapping light waves as a result of their resonant interaction with free electrons in the conduction band of metals is the concept behind several emerging nanophotonic technologies. The localization of surface plasmons using engineered metal nanostructures and their interaction with molecular polarizability tensors have afforded single molecule detection sensitivity in surface-enhanced Raman scattering (SERS), and more recently, chemical imaging within one molecule through tip-enhanced Raman scattering (TERS). Single molecule detection sensitivity can be achieved by taking advantage of electromagnetic enhancement factors exceeding 10^{10}, amenable at plasmonic nanojunctions where molecules are sited and interrogated. This talk will highlight some of the operative mechanisms which govern optical spectroscopy at plasmonic nanojunctions, and their impact on what is experimentally observed in SERS and TERS. The concepts will be illustrated through model studies which take advantage of a wealth of microscopic, spectroscopic, and computational tools available at PNNL. Results of multimodal multiscale chemical imaging measurements, as well as on on-going efforts to push the limits of sensitivity and resolution in TERS will be discussed.

TERS as an Analytical Probe of Local Physicochemical Properties
Erin Wood, Maruda Shanmugasundaram, Angela Hight Walker, Katherine Tyner

The pursuit of sub-diffraction physicochemical information such as few and single molecule detection and nanoscale strain has spanned disciplines and industries. These efforts have spurred the development of techniques that promise superior resolution such as scanning probe based vibrational spectroscopy. Raman spectroscopy, when co-located with atomic force microscopy (AFM), or tip-enhanced Raman Spectroscopy (TERS), is a particularly encouraging technique for applications within semi-conductor, strain-sensing, and pharmaceutical industries due to both AFM and Raman being established within these communities. This talk presents overviews of applications of TERS spanning these three industries, beginning with efforts to use a lithographic test structure composed of SiGe to evaluate comparability between scanning probe based instruments of different designs as well as promote our understanding of appropriate methodology to ensure accurate interlaboratory comparisons. We follow with efforts to accurately probe applied mechanical strain to as-grown epitaxial graphene on SiC. Both projects were efforts led by the National Institute of Standards and Technology (NIST), in Maryland. We finish with efforts led by the Food and Drug Administration (FDA) to use TERS as a method to help foster understanding of complex drug formulations by using an example of an ophthalmic emulsion drug product and compare the TERS results to other efforts to characterize the emulsion drug product, such as CytoViva enhanced darkfield-hyperspectral imaging. This cross-disciplinary work, including the novel application of TERS to emulsions shows potential for the technique to be used in a variety of classes of materials. Furthermore, critical comparisons of TERS to other analytical techniques will help to foster an understanding of the place it holds within the scientific toolbox .

TMDCs on metals: Importance of nanoscale heterogeneity for thin film optoelectronic devices.
Deep Jariwala, Michelle Sherrott, Andrey Krayev, Harry Atwater

Over last few years, transition metal dichalcogenides (TMDCs) attracted significant and still growing attention of the research community based on the wealth of physical phenomena observed in these materials and potential applicability of TMDCs in future electronic, opto-electronic and other types of devices. One of many potential applications of TMDCs is the thin film photovoltaics. It has been demonstrated that only 13nm thick layers of these materials deposited on reflective metallic substrate can absorb almost 100% the light in visible range. Vertical WSe2-MoS2 heterosctructures deposited on gold demonstrated record values of quantum efficiency for such thin films. In order to understand and improve photovoltaic performance of such systems, it's absolutely important to be able to characterize their properties at nanoscale. We report application of tip enhanced Raman scattering (TERS) cross correlated with other channels provided by scanning probe microscopy, such as the surface potential, conductivity and photocurrent for characterization of WSe2 exfoliated to gold. In the course of TERS mapping of WSe2 flakes we observed presence of unusual nanoscale domains. While WSe2 demonstrates resonant Raman response at 638nm excitation, the TERS spectra of mentioned domains featured a single peak at around 250 cm-1, typical for non-resonant conditions. Distribution of these domains correlated extremely well with the surface potential maps, non-resonant areas being negatively charged compared to adjacent areas of WSe2 with resonant response. We excluded the possibility of the non-resonant domains to be 1-T (metallic) phase based on the fact that these domains conduct much worse compared to the adjacent areas. Correlating TERS maps with concurrently recorded photocurrent, we observed that the domains of both types generated significant photocurrent, but of different polarities. Based on this we can conclude that in exfoliated layers of WSe2 there exist intrinsic nanoscale semiconducting domains with the charge carriers of opposite sign. This heterogeneity must be taken into consideration at least in experiments with devices based on exfoliated few layer WSe2.

Tip-enhanced nano-imaging of 2D materials with improved resolution

Dmitri Voronine

Nano-optical imaging of two-dimensional transition metal dichalcogenide (2D TMD) materials may provide detailed information about optoelectronic properties of edge states, grain boundaries and defects with a high spatial resolution approaching the scale of single atoms. We perform nano-optical characterization of several monolayer and few-layer 2D materials such as MoS2, WS2, MoSe2, WSe2, and their alloys and heterostructures using tip-enhanced Raman scattering (TERS) and tip-enhanced photoluminescence (TEPL) spectroscopies. We investigate the limits of spatial resolution and demonstrate few-nanometer TERS and TEPL imaging on the scale of few atoms. We also use TEPL to investigate nanoscale band gap tunability due to the chemical reactivity of the edges of 2D materials. Quantum plasmonics is shown to provide unprecedented picometer-scale indentation control of excitons and trions via tip-sample interaction and to improve the imaging contrast. These results may provide new insights into the local structural and dynamical properties of 2D materials at the atomic scale.

Non-invasive Spectroscopy for the Analysis of Cultural Materials in the Field

Christian Fischer

Cultural materials represent an important corpus of evidence that has survived the effects of time and its study provides the means for a better understanding of ancient cultures as well as avenues for the preservation of this unique patrimony. More specifically, scientific research on these materials is key to the identification and sourcing of raw materials, the understanding of technological processes, and is often a prerequisite for conservation. Owing to the nature of the materials, their complexity, diversity and origin, this field of research is by essence multidisciplinary. The large array of analytical techniques employed for the characterization of these materials, from basic to high-end laboratory instrumentation, illustrates perfectly the close and interwoven connections between archaeology and the physical and natural sciences. However, in many cases, location and accessibility constraints as well as sampling restrictions impose serious limitations on their analysis. Therefore, field-deployable technologies are often the only way to collect scientific data on cultural materials. In this paper, we focus on the combined use of two non-invasive and portable techniques: X-ray fluorescence (pXRF) and Fiber Optics Reflectance Spectroscopy (FORS) in the visible and near infrared spectral range, for elemental and molecular analysis. Advantages and limitations are discussed with examples on stone, ceramics and pigments from various archaeological and cultural contexts around the world.

The Kenya Red Ochre Chemistry (KROC) Database: Integrating geochemistry and ethnography for a new approach to archaeometric provenience studies

Andrew Zipkin, Craig Lundstrom, Stanley Ambrose, Gideon Bartov, Alyssa Dwyer, Alex Taylor, Mercy Gakii

Radiogenic strontium isotope ratios for skeletal remains are widely used to investigate prehistoric mobility and residential biographies. Characterization of elemental "fingerprints" is the most common tactic for provenience studies of archaeological artifacts. We have united these two technological approaches with ethnographic research to investigate prehistoric and current ochre use in Kenya. Geochemical provenience studies are a valuable tool for investigating past acquisition, transport, modification, and use of ochre pigments. Ethnographic insights into modern ochre use allow us to refine our interpretations of past uses. We are constructing a radiogenic strontium isoscape to complement geological source elemental composition analyses for provenience studies of ochre in the Kenya Rift Valley (KRV). Our ultimate objective is to assess whether ochre sourcing can contribute to reconstructing mobility, social and symbolic exchange, and interaction networks during recent human cognitive evolution. Guided by Kenyan Maasai, Dorobo, and Samburu informants who have used ochre sources, we sampled 56 geologic ochre deposits between the Tanzanian border and the south end of Lake Turkana in 2012-2016. These sources are relevant to provenience studies of ochre artifacts from Middle Stone Age through Neolithic sites and recent rock art pigments in the KRV. Four main modes of ochre geological deposit formation were identified: (1) Late Quaternary hydrogeothermal vents, (2) later Neogene ferruginous hydromorphic paleosols thermally altered by overlying lava flows, (3) ancient Proterozoic metamorphic ferruginous ores, and 4) nodules from ferruginous cements precipitated in faults in Proterozoic metamorphic rocks. Samples were characterized for their 87/86Sr ratios using MC-ICP-MS and for their elemental composition using LA-ICP-MS and EPMA. 87/86Sr results were used to construct isotope prediction surfaces for the KRV using simple and ordinary kriging analysis. Geostatistical analysis indicates that ochres associated with older rock formations differ from those associated with younger volcanics based on higher and more spatially homogeneous 87/86Sr ratios, facilitating regional scale provenience. Multivariate statistical analysis of elemental fingerprint data for ochre source and artifact samples facilitates fine scale (<10 km between sources) provenience. Our most significant archaeological result is identification of two unusually distant sources of ochre from a Neolithic cremation burial site in the Kenya Rift Valley.

Identification of Historic Carbonaceous Media on Drawings using Raman Microscopy and Macroscopic X-Ray Fluorescence Scanning

Nathan Daly, Michelle Sullivan, Lynn Lee, Karen Trentelman

Historically, works of art on paper have proven challenging objects for scientific study. Due to the lower concentration of materials used in these works, sampling is not possible as removal of a sufficiently large sample would be visibly and irreversibly damaging to the work. As a result, conservators and curators rely upon close observation as their primary means of differentiating drawing media. However, because this is a subjective approach, this is a challenging, if not impossible, task. With the industrial revolution of the 19th century came a diversification of drawing materials and methods available to artists of the time. The works of these artists, deceptively simple in appearance, represent complex mixtures of black media including charcoals of varying hardness, assorted crayons, chalks, inks, and pastels. To address the problem of accurate characterization of black drawing media curators, conservators, and scientists at the J. Paul Getty Museum and Getty Conservation Institute have initiated a collaborative study of 19th century French drawings. To develop an effective, noninvasive, analytical framework for the study of 19th century drawings, studies were first conducted on mock-ups prepared with historic and modern drawing materials using a suite of techniques to characterize the materials present and understand the complex media-substrate interaction. Micro-Raman spectroscopy was used to identify carbonaceous materials by examining two characteristic spectral features, the graphitic G band and defect-associated D band. The vibrational frequencies and intensity ratio of these two bands can be matched to reference pigments for identification. Macro X-ray fluorescence (XRF) scanning was used to identify inorganic elements such as silicon in chalks and to distinguish media by spatial localization. Additional *in situ* characterization using Fourier Transform Infrared spectroscopy and X-Ray Diffraction supported media classification made by Raman and XRF. Multivariate statistical analysis of the data provided a means to further distinguish classes of drawing media. Collectively, these analytical techniques successfully characterized the materials used by 19th century French artists, providing insight into the development of modern drawing. Additionally, this work provides a framework to address the broader need within cultural heritage studies for noninvasive analytical methods to study low concentration materials *in situ*.

Effects of Sample Preparation on XRF Measurements of Ancient Mortars

Mary Kate Donais , Mina Alrais, David B. George, Eric Smith

Mortar samples from the Coriglia, Castel Viscardo excavation site (Umbria, Italy) were analyzed via portable energy dispersive X-ray Fluorescence spectrometry (XRF) in order to determine whether the form of the samples effected the archaeological conclusions drawn from the data. Samples were examined intact, ground, and as fused disks. Data were analyzed using multivariate statistics through principal component analysis (PCA). The conclusions drawn from the PCA agreed among the three sample forms, informing the research team that future studies can rely on *in situ* data collection and not require sample collection and transport to the lab. This is ideal for both the preservation of the site's cultural heritage and the elimination of the need for permits to remove and transport samples for laboratory-based measurements.

Tuning chiral and mechanical properties in assemblies of plasmonic nanocrystals

Vivian Ferry

Coupled assemblies of plasmonic nanostructures exhibit unique optical properties that are highly sensitive to the surrounding electromagnetic environment. This talk will discuss assemblies comprised of heterogeneous materials, and particularly of different metals, where applied stimuli change the properties of one component, and ultimately tune the properties of the entire system. The first part of this talk will discuss switchable optical properties of plasmonic assemblies that combine different metals. We have recently shown that chiral optical properties can be observed from chiral assemblies of plasmonic metals with different individual resonances, and that off-resonant scattering induces circular dichroism. For example, in assemblies comprised of two Au perpendicular bars and a disk placed above one bar, where the disk is a different material with a different resonance from the bars, circular dichroism is observed at the resonant frequency of the Au. By applying stimuli that selectively change the properties of the disk, the circular dichroism of the whole assembly can be tuned and switched. The second part of this talk will discuss the dynamic properties of assemblies of plasmonic nanocrystals upon ultrafast excitation. After irradiation with an ultrafast pulse, the absorbed power is eventually dissipated to the environment through acoustic oscillations of the nanocrystal. We have recently used ultrafast electron microscopy to study the excitation of these mechanical modes at a single assembly level, showing that unexpected behavior is seen at the interface between two metallic structures

Optical Devices based on Plasmonic Nanoparticles

Mahmoud Mahmoud

Plasmonic nanoparticles-polymers hybrid materials are used in different optical (polarizers, filters, and light wavelength modulators) and sensing (optoelectrical and optomechanical) applications. Plasmonic nanomaterials of different designs and compositions are prepared by new chemical techniques at a large scale. The Langmuir-Blodgett technique is used to assemble colloidally prepared plasmonic nanoparticles of tailored designs and structures with controlled organization on different substrates.

Bimetallic nanostructures: decoration and alloying effects on plasmonic properties
Emilie Ringe, Josee Daniel, Dayne Swearer, Lauren McCarthy, Anjli Kumar, Sadegh Yazdi, Denis Boudreau

The ability to combine multiple metals in a single nanostructure opens possibilities for multifunctional materials; combining plasmonic cores with catalytic surfaces can enhance catalytic processes, combining the same cores with an oxide surface can lead to enhanced biocompatibility. However, the heavily damped plasmon resonances of many catalytically active metals (e.g., Pt, Pd) and the high refractive index of oxide coatings may prevent efficient coupling of the enhanced electric fields of plasmon with nearby molecules. The understanding and design of efficient multimetallic structures is thus critical to harness their plasmonic properties. Here, correlated optical and electron microscopy approaches are used to characterize bimetallic plasmonic systems such as transition metal-decorated Al and Au nanocrystals as well as semi-hollow, liquid-filled Ag-M (M = Pd, Pt, Au) nanorods. We will show that plasmonic nanomaterials incorporating catalytically active but heavily damped metals sustain multiple size-dependent LSPRs that are strongly localized at the particle tips or edges. We will also discuss how shape and composition can be used to manipulate resonances, and how they affect plasmon sensing and plasmon-enhanced catalysis.

Plasmonic Nanoparticles: Synthesis and Catalytic Application

Supriya Atta, Laura Fabris

The size- and shape-dependent physical properties of gold nanostars exhibit tunable plasmon resonance, and the anisotropic growth of TiO_2 onto gold nanostars can further enhance their applications in the field of solar cell and catalysis. We will discuss here the facile synthesis, and optical properties of seed-mediated 6-7 branched stars and surfactant free stars over a range of controlled spike length from 50 to 110 nm diameter with a controlled morphology which can be achieved by changing the concentration of chemicals required in this synthesis. The seed-mediated gold nanostars synthesis route produces a number different spike architectures, from rod-shaped to conical shape, and from pointy tips to spherical tips. We observed that the morphology of the gold nanostars depends on the concentrations of the seeds, $AgNO_3$, Ascorbic acid, hydrochloric acid (HCl), and Triton-X (for 6-7 branched stars). The effect of these chemicals was found to be interdependent. Thus, they produced the interesting combinations of various shapes. We have also seen that $AgNO_3$ is one of the key parameters, which determines the final size and shape of the gold nanostars. We compare the experimental results to the scattering and absorption spectra calculated with finite element modeling. We will also show our most recent results in which anisotropic growth of TiO_2 on gold nanostars surface, gives a better contact between Au and TiO_2 which could reduce the Schottky barrier between the Fermi level of the metal and the conduction band of the semiconductor TiO_2, instead of depositing TiO_2 onto gold nanostars. In this study, we have first synthesized amorphous TiO_2-coated gold nanostars and then crystallized the TiO_2 by heat treatments with preserving the sharp morphology of the protruding spikes. We also investigate the effect of ions (Cl-, and F-) on crystallization of TiO_2 onto gold nanostars. Compared to other nanoparticle-based catalyst systems, gold nanostars promise to become a much more versatile and tunable Vis-NIR photocatalyst.

Synthesis of Bimetallic Hollow AgM Nanoparticles, Structure and Composition Analysis
Josee R. Daniel, Sadegh Yazdi, Lauren McCarthy, Emilie Ringe, Denis Boudreau

Plasmonic nanomaterials have found applications in multiple fields such as catalysis, molecular sensing, imaging and single molecule spectroscopy, among others. The performance of a given nanomaterial in terms of absorption, scattering or local electric field enhancement can be optimized for a specific application by controlling physical properties such as size, shape and composition. In this context, hollow bimetallic nanostructures provide additional variables, including shell thickness and metallic component ratio, which can be manipulated to finely tune the plasmonic and optical properties of the structures. Galvanic replacement is an efficient and versatile synthesis strategy for the preparation of hollow and semi-hollow nanostructures with enviable plasmonic properties. This presentation will discuss our recent progress on the synthesis of hollow bimetallic nanostructures using silver nanoparticles as sacrificial templates and partial galvanic replacement by more noble metals (Au, Pd, Pt). These structures could find applications in various fields such as reprogrammable sensors, responsive materials, optical memory units, molecular sensing, cellular imaging and in situ microscopy of chemical reactions.

Capillary Electrophoresis-Based Enzyme Assay of Acetyl Coenzyme A Carboxylase

Thu Nguyen, Alexandra Evans, Grover Waldrop, Samuel Douglass Gilman

A simple, direct, off-column capillary electrophoretic (CE) assay was developed and optimized to study enzyme activity and inhibition of acetyl-coenzyme A carboxylase (ACC). This enzyme catalyzes the first committed step and rate-limiting step in fatty acid synthesis in bacteria, plants, and animals. Malonyl-CoA, a product of ACC plays a critical role in fatty acid synthesis and regulation of fatty acid oxidation. Therefore, ACC is a potential target for treatment of obesity, diabetes, and other metabolic diseases as well as for antibiotic drug development. In CE assay, spectral interference is minimized by separation of substrates, products, inhibitors, and assay matrix. Compared to commonly used coupled assays for this enzyme, this assay is less susceptible to inhibitor interference because coupling enzymatic reactions are not needed. The developed CE assay was applied to study inhibition of ACC. Several synthetic and natural compounds that inhibit ACC-catalyzed reaction were identified.

Sequence Based Separation of DNA using Microfluidic Chip Electrophoresis

Wyatt Stevens, Jia Zhao, Linda McGown

Sequence Based Separation of DNA using Microfluidic Chip Electrophoresis Wyatt Stevens, Jia Zhao, Linda McGown DNA analysis is an important part of biology, medicine, and forensics. Using sieving gels, strands of DNA can easily be separated by size. In some cases, especially in mutation analysis or complex samples such as microbial communities and biofilms, same-length strands may include several different sequences that could be investigated to provide further information about the genetic and biological diversity in the sample. Unfortunately, separating DNA by sequence is much more difficult than separation by length, and so the same-length strands generally must instead be sequenced. This can be an expensive and time consuming process, and may not be possible for all samples. Previous work in our group demonstrated the separation of single-stranded DNA 76-mers and 15-mers that differed by only one or two bases simply by using capillary zone electrophoresis with running buffers containing high salt concentrations [1,2]. In our current work we are taking what we have learned from capillary zone electrophoresis and applying those conditions to a microfluidic chip platform. We are using microfluidic chip gel electrophoresis to separate 76-mers that only differ by one or two bases. For our work in capillary zone electrophoresis it takes on the order of an hour to achieve separation. The microfluidic chip on its own cannot achieve this due to the much shorter length of the separation channel. We overcome this by introducing a gel phase to slow the sample and give it the time it needs to separate under the electrophoretic conditions. We use either an agarose gel with high salt concentration or a guanosine-based gel for the gel phase. These gels have been formulated so that they do not function as sieving gels. Our long term goal is to develop two-dimensional separation of DNA by length and by sequence using the microfluidic chip platform. [1] Zhang, X., McGown, L.B.*, Sequence-Based Separation of Single-Stranded DNA at High Salt Concentrations in Capillary Zone Electrophoresis, Electrophoresis, 37, 2017-24 (2016). [2] Zhang, X., McGown, L.B.*, Sequence-based Separation of Single-Stranded DNA by using Nucleotides in Capillary Electrophoresis: Focus on Phosphate, Electrophoresis, 34, 1778-1786 (2013).

Capillary Electrophoresis of Intact Erythrocytes for Anti-Doping Analysis
Christopher Harrison, Jessica Torres, Thirada Kingphua, Shane Alexis Apostol, Sangho Yun

Among the various doping methods that have been employed by athletes the use of autologous blood transfusions, storing and re-infusing one's own cells, remains among the most difficult forms of doping to directly detect. Unlike traditional, homologous, blood transfusions, the re-infusion of cells from the same athlete leaves very little biochemical, or physical evidence of the manipulation of the blood. However, capillary electrophoresis can provide the means for performing separations that of analytes with minimal physicochemical differences, when properly applied. We will present the steps taken in our lab to improve the capillary electrophoretic separation of intact erythrocytes with the goal of identifying doped athlete blood samples. These separations exploit the minor changes that occur to erythrocytes while in the storage media used to preserve blood; namely changes in cell size, the outer phospholipid membrane, and the glycation of proteins. To exploit these differences, and to yield a separation of the different erythrocytes, various capillary coatings and buffer additives have been employed. The range of results and our increased understanding of the electrophoresis of these intact cells will be presented.

Assessing hydrophobicity of antibody-drug conjugate payload molecules utilizing lipid-surfactant mixtures
Suvi-Katriina Ruokonen, Marina Redon, Emilio Gonzalez-Jalonen, Filip Ekholm, Susanne Kristina Wiedmer

Antibody-drug conjugates (ADCs) are promising state-of-the-art biopharmaceutical drugs for selective drug-delivery applications and treatment of diseases such as cancer. ADCs consist of drug delivering monoclonal antibodies linked to cytotoxic payload molecules. Problems associated with the removal of hydrophobic cytotoxic compounds from the intracellular environment have driven the development of novel hydrophilic payload molecules, without losing the essential cytotoxic ability of the drugs. In order to assess the hydrophobicity of novel ADC payload drugs capillary electrophoresis, and particularly electrokinetic chromatography, were utilized. In electrokinetic chromatography the hydrophobicity is estimated based on analyte partitioning between an aqueous phase and a pseudostationary phase. ADC payload drugs are known to affect the cell mortality intracellularly but they could not interact with standard biomimetic phospholipid vesicles (liposomes) used as a hydrophobic pseudostationary phase in electrokinetic chromatography. This data indicates that the compounds were not able to penetrate the liposome membrane. Therefore, the main goal of this study was to determine how the interactions between the payload molecules and the pseudostationary phase are influenced by the shape, size, and composition of the lipid and surfactant aggregates utilized as a pseudostationary phase. The pseudostationary phases composed of neat phosphatidylcholine lipids with different acyl chain lengths, neat negatively charged surfactants, and mixtures of these. The lipid-surfactant mixture ratios were selected based on their different coalescent behavior: a high concentration of phospholipids yield negatively charged vesicles, while high concentrations of surfactants yield negatively charged micelles. From the data we were able to extract valuable information on the hydrophobicity of the studied ADC payload molecules.

Capillary electrophoresis for quantitative determination of ionic liquids and their degradation products
Susanne Wiedmer, Joanna Witos, Antti Rantamäki, Jesper Långbacka

Amidine and guanidine–based superbases have become an essential area of research in organic synthesis and in various industrial processes. The superbases themselves are hydrolytically unstable in contrast to their conjugate acids that are much more stable. Due to their physicochemical properties and reactivities, much attention has been paid towards the different combinations of superbase-derived acid-base conjugate ionic liquids (ILs). A novel IL, [DBNH][OAc] (1,5 diazabicyclo[4.3.0]non 5 ene acetate), has shown to be an excellent cellulose solvent that affords suitable spinning stability and outstanding fibre properties in the IONCELL-F air-gap spinning process. [DBNH][OAc] can dissolve high concentrations of cellulose at moderate temperatures (70-80 C) with only minor cellulose degradation. Nevertheless, the potential degradation products of [DBNH][OAc], which are formed by initial hydrolysis, should be studied in detail to develop the industrial IONCELL-F recycling concept. Recent research has shown that a similar IL called [MTBD][OAc] (7-methyl-1,5,7-triazabicyclo[4.4.0]dec-5-ene acetate) with excellent biomass-dissolving properties might be even superior to [DBNH][OAc] in the spinning process. Capillary electrophoresis (CE) is a common analytical separation technique where charged analytes, under the influence of an electric field, migrate in solution in a narrow capillary according to their charge, size, and molecular structure. Among the notable advantages of CE, the most important are high separation sensitivity and efficiency, high-speed analysis, versatility, robustness, excellent mass sensitivity, and low reagent consumption. The aim of this research was to develop a CE methodology suitable for quantitative determination of novel ILs and their degradation products.

A method for fabricating a through-microhole by using a near ultra violet femtosecond laser
Shoichi Kubodera, Masahiko Shiraishi, Kazuhiro Watanabe

The fabrication of a microhole inside an optical fiber by using a femtosecond laser drilling was profitable for optical sensing. The microhole worked as a spectrometer cell with a sensing volume of several picoliters. The spectrometer cell has capability of measuring refractive index of liquid samples using broad band propagating light in the core layer of the optical fiber. It's preferable that the microhole penetrates through the optical fiber because of introducing the liquid samples smoothly. The glass optical fiber which had 62.5 ?m of the core diameter and 125 ?m of cladding diameter was used for fabrication of a through-microhole. A deep microhole penetrating through the glass optical fiber was obtained after femtosecond laser processing. A pulse train of near ultra violet 400 nm femtosecond laser launched into the glass optical fiber. A repetition rate and a pulse duration were 500 Hz and 350 fs, respectively. A numerical aperture of objective lens was 0.65. After 180 shots of femtosecond laser with energy of 70 ?J, the through-microhole was created. The glass optical fiber with processed part was fragile because the damage of material was high. In order to minimize the damage of the material, pulse energy was reduced. The 130 shots of femtosecond pulse train with energy of 30 ?J irradiated into the glass fiber optic. Although the deep microhole was created, the microhole didn't penetrate the glass optical fiber. After irradiation of 130 shots, the glass fiber optic was rotated 180 degrees. The second pulse train with 130 shots to be opposite to the first laser direction irradiated into the glass fiber optic. The two direction irradiation fabricated the through-microhole with mechanical strength than the one direction irradiation.

A deterministic ratchet for sub-micrometer particle separation
Dai Hyun Kim, Jinghui Luo, Edgar Arriaga, Alexandra Ros

Resolving the heterogeneity of organelle populations by size is important when the size is an indication of abnormal biological properties leading to disease. Accessing size heterogeneity in the sub-micrometer regime is particularly important to resolve populations of sub-cellular species or diagnostically relevant bioparticles. Here, we present a dielectrophoresis-based ratchet migration mechanism in a non-linear, yet symmetrical post array applying the separation capabilities of particles in the sub-μm size range and demonstrate the application for organelles in a microfluidic device. First, a numerical model simulating the migration behavior of six different particle sizes (0.15, 0.28, 0.40, 0.52, 0.87 and 1.5 ?m) is developed in the DEP-based ratchet device combining a non-linear microfluidic post array and asymmetric periodic driving electrical potential. It demonstrated that the large particle group (>0.87 ?m) showed ratchet migration and the small group (<0.52 ?m) normal migration, resulting in two opposing migration directions for the two groups. The simulation is performed with more than thousand particles allowing to extract statistical parameters characterizing separation. With only ten seconds of migration time, the resolution over 1.7~2.7 is achieved. The resolution could be enhanced just by extending device size. Based on the simulation parameters, we demonstrate the deterministic ratchet migration experimentally for 0.87 μm polystyrene beads whereas normal migration is observed for 0.28 μm beads. Further, we apply the same conditions to polydisperse-sized C57BL/6 mouse hepatic mitochondria as a model for fractionation of organelle subpopulations into relevant size fractions. The deterministic ratchet mechanism is demonstrated with the large mitochondria (>0.6 ?m), whereas normal migration is observed for the small organelles(0.20.7 ?m) in contrast to the normal migration of small liposomes (<0.2?m), indicating the technique can be used to effectively separate each organelle into subpopulations by size. This is the first experimental realization of directing differently sized sub-micron particles into opposite directions by DEP-based ratchets with substantially improved average migration velocities as compared to our previous absolute negative mobility approach. Consequently, the versatile DEP-based ratchet is a powerful separation method and can be applied to a wide range of separation problems in which critical size selectivity is required, such as in the fractionation of organelles and their subpopulations, in nanotechnology-based applications or as fractionation method for protein nanocrystals.

Neural Stem and Progenitor Cells Separation Based on Insulator-based Direct Current Dielectrophoresis
Yameng Liu, Mark Hayes

Neural stem and progenitor cells (NSPCs) are an extremely important group of cells that form the basis of the entire neural system and they have shown potential through transplantation for repairing damage caused by diseases such as cerebral palsy, spinal injury and Parkinson's disease. Developing an understanding of their molecular level decision making machinery is important for basic biology knowledge and allowing improved transplant outcomes. The purification and separation of NSPCs is a necessary procedure before transplantation and enables certain fundamental studies. Current schemes for separation are inadequate to the complexity and diversity of cells and cannot produce pure populations for study or transplant. Some of these limitations stem from lack of markers to distinguish cells. This research aims to separate NSPCs according to their ultimate mature cell fate (neurons or astrocytes) using insulator gradient dielectrophoresis (g-iDEP). To develop the technique and demonstrate capabilities, an important class of cells was chosen. Some of the most common NSPCs are cortex cells from mouse embryos. These have similar biology and biochemistry to human NSPCs. A common source was taken on day 12 (E12 cells) which are rich in neurogenic progenitors. These can be chemically manipulated with N-acetylglucosamine (GlcNAc) to cause them to be enriched with astrogenic progenitors. These cells were exposed to the microdevice for separation. Several parameters were investigated to allow for isolation of the cells in the device including the selection of the DEP buffer solution and voltages applied. The numbers of cells captured at different positions in the channel of DEP were also recorded. Preliminary data is consistent with an increased number of astrogenic progenitor cells from the GlcNAc treated E12 cells, at the expense of neurogenic progenitor cells. In this way, progenitor cells with different ultimate fates can be recognized and separated. The purified progenitor cells have potential in improving transplant and studying the mechanism of differentiation. This offers benefits to basic research in which cells with specific characteristics are necessary.

Spectroscopic Analysis of the Maya Blue Pigment
Jeremiah Lopez, Jorge Lopez, Carlos Diaz-Moreno

The Maya Blue pigment has been known for its distinct color and ability to resist degradation over centuries throughout the harsh climate of Mesoamerica, however, these properties have yet to be fully explained due to the unique bond between the indigo and palygorskite structure. Using the surface techniques of X-Ray Fluorescence (XRF) and X-Ray Photoelectron Spectroscopy (XPS), we find information regarding the electronic structure that will allow this phenomenon to be understood to a further extent.

FTIR Microscopic Analysis of Plant and Animal Tissue Residues on Stone Tools
Bing Luo, Gilliane Monnier, Ellery Frahm, Kele Missal

The analysis of residues on stone tools can yield important insights into the tool-using behaviors of Paleolithic hominins. The ambiguity of residue identifications using visible-light microscopy has led to the development of additional techniques for their characterization. Reflectance-based Fourier transform infrared (FTIR) microspectroscopy is a technique with great potential to aid in nondestructive residue identifications. However, reflectance spectra of such systems sometimes exhibit complex features. These can result from combinations of specular and diffuse reflection, and from contributions from the substrate underneath. In the process of developing this method for archaeological applications, we systematically analyzed a variety of plant and animal tissue samples on a metallic mirror substrate as well as on stone substrates (flint and obsidian). The plant samples included wood bark, wood pith, grass leaves, starch and resin. The animal tissue samples included skin, meat, fat, hair, blood, feather barbules, fish scales and bone. As expected, for the measurements on the metal mirror where the samples were deposited as tiny fragments, the spectra were consistent with those obtained using the transmission method. For measurements on stones, however, the quality of spectra varied depending on the thicknesses and particle sizes of the samples. In general, when specular reflection dominated, the spectra exhibited derivative-like features and the Kramers-Kronig transformation was needed to make them comparable with absorbance spectra. When diffuse reflection dominated and the samples were sufficiently thick, the spectra were closer to the standards collected on the metal mirror, but some samples gave very low reflectance. In either case, when the plant or animal tissue layers were thin, the spectra were distorted by the reststrahlen band from the stone. We conducted thorough spectral assignment and discussion in each case and conclude that upon careful examination, reflectance FTIR microspectroscopy can produce detailed chemical information for complicated materials in archaeological study. All of these spectra are available in our online database, hosted by the University of Minnesota, along with the VLM images we gathered of these residues (http://z.umn.edu/ftir).

Potential G-quadruplex Forming Aptamers using a Genome-Inspired Reverse Selection Approach

Kathleen Morrissey, Linda McGown

Aptamers have become one of the leading alternatives to antibodies as affinity reagents to a wide variety of targets including proteins. They were first discovered in 1990 using the process known as Systematic Evolution of Ligands by Exponential Enrichment (SELEX). In this process, oligonucleotides are selected from a combinatorial library for high affinity, highly selective binding to a specific target. SELEX has discovered many aptamers but it does come with limitations that can hinder the process of aptamer discovery, particularly in the case of potential aptamers that form multi-tier G-quadruplexes (G4). G-quadruplex-forming sequences are found throughout the human genome and merit investigation as potential aptamers. In our lab, we have developed a new genome-inspired, reverse selection approach that allows us to explore these and other challenging secondary structures. In the reverse selection approach, a specific DNA sequence from the human genome is used for affinity capture of proteins from natural pools such as nuclear extracts. The nuclear extracts are incubated with streptavidin magnetic beads coated with biotin-modified G4 oligonucleotides or control oligonucleotides. A rapid screening for captured proteins is performed using MALDI-TOF MS to determine if there is any affinity capture of proteins specific for the G4 DNA. If affinity capture is established, the capture experiments are repeated on a larger scale and captured proteins are separated by gel electrophoresis. Bands specific for G4 DNA are analyzed by LC-MS/MS to produce a list of possible protein matches. This list is used to guide our selection of proteins for further interrogation using Western blots. Here we will present our results for genomic-G4 forming sequences from the promoter regions of the human oncogenes c-myc, Rb, and VEGF as potential aptamers toward proteins in nuclear extracts from the basal breast cancer cell line MDA-MB-468.

Biophysical Characterization of Antibiotic Resistance
Shannon Huey Hilton, Mark A. Hayes

Without proper diagnosis, it is difficult to give a patient the correct treatment that will allow them to become healthy again. Low-resource settings provide unique challenges for the development of diagnostics. To give guidance in addressing these challenges, the World Health Organization has developed the ASSURED criteria, stating that diagnostics for such settings must be affordable, sensitive and specific, user-friendly, rapid, robust, equipment-free, and deliverable to end users. One challenge faced in low-resource areas and the developed world alike is the testing of susceptibility of bacterial infections. Lack of access to such testing leads to inappropriate use of antibiotics, which in turn leads to more resistance. Microfluidics and the use of dielectrophoretic (DEP) forces has the potential to effectively meet these requirements and provide such an answer. The high-resolution separation of cells by DEP enables the concept of direct pathogen identification. Theory suggests that DEP can resolve very small differences in both electrokinetic and dielectrophoretic mobilities, which can arise from small physical differences between populations of cells. Here, data will be presented showing the differentiation of Staphylococcus aureus based on antibiotic resistance, a very subtle physical difference between two strains. This dielectrophoretic test takes minutes, while current methods to determine antibiotic resistance take 24 hours or more. Other explorations will include the location of the biophysical manifestation of resistance. Sensitivity to small physical differences enables the use of dielectrophoresis in specific and sensitive diagnostics. This ability to distinguish and separate based on such subtle differences allows for the possibility for this technology to be used in this capacity in the future.

OPTIMIZATION OF COLD ATMOSPHERIC PLASMA AND ELECTROPORATION FOR CANCER CELLS

Prasoon Diwakar, Arianna Avellan, Liesl Krause, Rohil Jain, Cagri Savran, Tatyana Sizyuk, Ahmed Hassanein

An emerging technology for medical applications is cold atmospheric plasma (CAP) which is generated using various gasses in a "pen" to create room temperature plasma and then carry the effluents and species. Success has been shown when CAP is applied to oncology treatments, accelerated wound healing, pathogen disinfection, and various material-changing effects. However, the mechanisms behind these effects are still speculative. This study uses multiple diagnostic techniques including fast photography, and optical emission spectroscopy (OES) to characterize the plasma properties to understand the chemical species generated in CAP which interact with the cells leading to their apoptosis. Plasma rotational and vibrational temperatures were measured using the atomic and molecular lines as measured using OES. In addition to CAP, electroporation is another technique which has been used for cancer treatment through a minimally invasive procedure and much reduced side effects. This study focuses on the optimization of parameters of Electroporation for efficient apoptosis of cancer cells followed by application of both CAP and electroporation in sequence and simultaneously with the goal of understanding the synergy of both techniques. Three cancer cell lines were used: MCF-7 (human breast cancer cell line), KB (human mouth/cervical cancer cell line), and LNCaP (human prostate cancer cell line). For electroporation, each of these cancer cell lines were exposed in vitro to different pulse lengths (50 ?s- 10ms), number of pulses (10-50), and electric field strengths (1 kV/cm and 2 kV/cm). Viability of these cancer cell lines was analyzed as a function of time using microscopy and fluorescence. Further studies are currently in progress to understand the viability of the cell lines at different times after application of CAP and electroporation as well as to measure the effectiveness of the synergy of both techniques on cancer cell lines. Detailed discussion of the interaction mechanisms of both techniques with cancer cell lines are presented.

31P Magnetic Resonance Spectroscopy for In Vivo Illness Analysis

Liesl Krause, Frederick Damen, Craig Gorgeon, Joseph Rispoli

Magnetic resonance imaging (MRI) and spectroscopy (MRS) techniques typically exploit the high-proton (1H) content of organic substances to generate a nuclear magnetic resonance signal. Conversely, molecules lacking hydrogen contribute no signal in standard MRI or MRS. While specialized methods can interrogate non-proton nuclei, 31P has proven especially useful in biological imaging due to the presence of high-energy metabolites. High-energy phosphates such as phosphocreatine (PCr), adenine triphosphate (ATP), phophorylethanolamine (PHOS), and phosphorylcholine (ChoP) can be probed by exciting phosphoros-31 (31P) instead of 1H. Still, proton coils are required to acquire "scout images" to define the region of interest. Since the two nuclei have different Larmor frequencies, the coils must be electromagnetically decoupled to prevent detuning and noise coupling. In this work we developed a 4-cm diameter 31P surface coil with passive decoupling, for use with a commercial Bruker 1H volume coil. Coil quality factor (Q) was measured using S11 measurements on a network analyzer (E5071C, Keysight Technologies). A capacitive pi network circuit was used to tune the coil to the resonant frequency for 31P at 7 tesla (7T), 121 MHz, and to impedance match the coil to 50 ?. With no load on the coil, the Q of the matched and tuned coil was 166 Loaded with a 113.4-grams 15 mM phosphoric acid phantom, Q dropped to 19.9, meaning the coil is properly designed for a sample noise dominant coil. Validation was performed with a Bruker 7T superconducting magnet (Bruker Biospec 70/30 USR System) using a 15 mM phosphoric acid phantom and a single pulse acquisition in order to obtain a peak at 0 ppm. A phantom with 1% ATP, 49% phosphoric acid, and 50% distilled water is being optimized to evaluate this coil with more biologically-relevant sample composition. ATP is expected to have three distinct peaks upfield to the phosphoric acid peak, and will be present in all functioning tissue. Ultimately, using dual-tuned coils for MRS will facilitate acquisition of spectra in vivo from murine hearts. Future work will use measurements of the relative levels of ATP and PCr within myocardial tissue to characterize cardiac bioenergetics.

Correlation between ionic liquid cytotoxicity and ionic liquid-liposome interactions

Suvi-Katriina Ruokonen, Alexandra Robciuc, Joanna Witos, Alistair King, Sami Hietala, Susanne Wiedmer

Ionic liquids have received much attention during the past few decades due to their unique chemical and physical properties. They are often quoted as having low vapor pressures, low flammabilities, and they are thermally and chemically stable. Therefore, they are widely used in various pharmaceutical and industrial applications. Despite their "green reputation", the environmentally unfriendly features of ionic liquids have been under a lot of investigation recently. The underlying mechanism of the toxicity and the physicochemical properties of ionic liquids affecting biomembrane interactions are still largely unknown. In this work, cytotoxicity data was compared with results obtained from physico-chemical and spectroscopy techniques with phospholipid vesicles (liposomes) as model biomembranes. The aim was to gain novel information on the underlying mechanism behind the cytotoxicity of ionic liquids. First, the cytotoxicities of eight ionic liquids and a reference compound were assessed using human corneal epithelial cells. To investigate the influence of ionic liquids on cell rupturing, hemolysis was performed using human red blood cells, and the results were compared with liposome integrity data. The effect of ionic liquids on the size and the zeta potential of negatively charged liposomes, composed of phosphatidylcholine and phosphatidylglycerol, were assessed in order to obtain information on the changes in the size and surface charge of the liposomes. In addition, pulsed field gradient nuclear magnetic resonance spectroscopy was utilized for studying the movement of the ILs with and without the presence of liposomes. The result show that the investigated ionic liquids can be divided into two groups based on the cytotoxicity mechanism: cell rupturing ionic liquids and cell wall penetrating ionic liquids.

A Temporal Study of Cell Death Signaling Responses to Cold Atmospheric Plasma and Electroporation in Human Cancer Cells

Danielle Krug, Prasoon Diwakar, Ahmed Hassanein

As cancer continues to be one of the world's leading killers, efforts to replace expensive and invasive therapies that cause harmful side effects led to the exploration of cold atmospheric plasma (CAP) as an efficient and selective treatment for the annihilation of cancerous tissues. CAP acts by producing reactive oxygen and nitrogen species within irradiated cancer cells, initiating the intrinsic apoptosis pathway through oxidative stress. Another treatment that is also emerging is electroporation (EP), which compromises plasma membrane stability through application of strong electrical fields, resulting in transmembrane mass transfer or, in severe exposure, cancer cell death. The combination of these two methods has not been thoroughly investigated. Although the apoptotic pathway and its components responses to cold plasma have been previously explored, there is still a lack of knowledge in the mode of signal propagation and the timeframe in which affected cells begin to feel the plasma effects. It has been previously observed that the addition of electroporation to treatments augments the plasma effects, increasing cancer cell death events. For this study, three cancer cell lines are used—MCF-7 (human breast cancer cell line), KB (human mouth/cervical cancer cell line), and MDB-231 (triple negative human breast cancer cell line). CAP and EP are used in sequence on these three cell lines to achieve synergistic effect. Plasma diagnostics using optical emission spectroscopy (OES) and fast photography reveal the species produced in CAP plume which the human cancer cells are exposed to during the plasma-cell interaction. Plasma rotational and vibrational temperatures are calculated using the atomic and molecular lines as measured using OES. Through an extended time, flow cytometry is used to assess the incidence and stages of apoptosis occurring in cells after CAP and EP treatments. Inverted fluorescent microscopy tracks global cell death through activation and inactivation of Caspase-3/7 complex, the penultimate signaling event in cellular apoptosis. Detailed discussion of the apoptosis pathways and mechanisms of cell death signaling due to CAP and EP combined treatment are discussed and presented.

Investigating cytotoxicity of ionic liquids using mammalian and bacterial cells

Antti Rantamäki, Suvi-Katriina Ruokonen, Alexandra Robciuc, Corinna Sanwald, Susanne Wiedmer

Ionic liquids (ILs) are organic salts utilized in several industrial applications for instance as replacements for organic solvents. They have been considered as more environmentally friendly alternatives for organic solvents, since there is a lower risk of atmospheric release due to their low vapour pressures. In contrast, due to their water solubility, ILs may be potent toxicological hazard for a wide variety of organisms from microscopic aquatic organisms to humans. Since ILs are structurally a diverse group of compounds, their toxicity varies from very toxic to harmless. The toxicological risk abides throughout the whole IL life cycle from synthesis to disposal. Therefore it is also essential to acknowledge the potential hazards related to the possible degradation products of the ILs. Overall, in research and development of novel ILs for differing academic and industrial applications, their toxicological effect is of importance due to the environmental responsibilities defined by ethical reasons and legislation. Here we have utilized three differing organisms for testing the toxicity of novel ILs: Chinese hamster ovary (CHO) cells, human corneal epithelial (HCE) cells, and Vibrio fischeri (VF) marine bacteria. CHO cells are an ideal model for mammalian epithelial cells and are utilized in a myriad of biological and medical applications, whereas the HCE cells represent a more specialized epithelial cell line. With CHO and CHE cells the cytotoxicity is determined using alamarBlue cell viability assay, whereas the toxicity determination with VF is based on the decay of bioluminescent light emitted by the bacteria. With all organisms, the toxicities were determined as median effective concentrations EC50. The differing characteristics of the CHO, CHE, and VF cells may affect the determined EC50 values greatly. Therefore, in order to give a more realistic insight into the overall IL toxicity, it is of importance to study the toxicities using different cell lines.

Drug Monitoring and Toxicology: Quantification of Antifungal Drug Voriconazole in Human Plasma and Serum by High-Performance Liquid Chromatography with Fluorescence Detection

Peter Tang

A simple and rapid high-performance liquid chromatographic (HPLC) method with fluorescence detection (FL) for the determination of voriconazole has been developed. Organic solvent acetonitrile is used to precipitate protein and extract voriconazole and internal standard ketoconazole in human plasma and serum using a single dilution step. Following protein precipitation, solvent extraction and centrifugation, voriconazole and ketoconazole in the extract are quantitated by direct injection onto the HPLC/FL system. Limit of quantitation and linearity (0.1-10 μg/mL) of the method adequately cover the therapeutic range for appropriate antifungal drug monitoring. This method has displayed essential improvements such as allowing a small portion of the extract to be analyzed (10 μL) and completing a chromatography in <7 min per injection when compared to most published HPLC/FL and HPLC/UV methods. This method would be of interest to analytical and clinical laboratories equipped with the HPLC/FL systems because it avoids time-consuming procedures such as solvent evaporation to dryness and residual reconstitution prior to HPLC analysis. Also, the sensitive method is set to inject a small volume of extract, which allows loading more injections onto the HPLC system without replacing the in-line filters or guard cartridges. Over all the method is simple, rapid, and cost-effective.

A Tale of Two Local Sites: Simultaneous Probing of Two Locations in MLCK Upon Binding
Natalie Fetto, Ian Wallace, Matthew Tucker

Prior studies of 4-Cyanophenylalanine (4-PheCN) have shown that this non-natural amino acid chromophore can be incorporated into peptides to provide a fluorescent probe of the local environment. Herein, we extend the family of cyanophenylalanine spectroscopic reporters by showing that 2-PheCN and 3-PheCN have intrinsic fluorescence properties that provide information about peptide dynamics. We demonstrate through solvent dependence studies that these reporters can monitor the degree of hydrogen bonding and/or hydration in the surrounding environment. Therefore, multiple nitrile derivatives of phenylalanine can be incorporated along a single peptide chain to observe the site-specific changes by selectively excitation. The binding of myosin light chain kinase (MLCK) to calmodulin is monitored at two locations, simultaneously, utilizing these fluorophores.

Confocal Raman Microscopy of Silica- and Cyano-Supported Phospholipid Bilayers within Porous Chromatographic Supports
David Bryce, Jay Kitt, Joel Harris

A wide variety of biological processes are dictated by molecular interactions which occur at phospholipid membranes. These interactions make stable supported phospholipid bilayers useful for assessment of biologically relevant molecular interactions, and as potential stationary phases for advanced separations applications. In planar supported phospholipid bilayers the surface-chemistry of the support has been shown to have a dramatic impact on the physical characteristics of the bilayer including changes in phospholipid diffusion and flip-flop rates, as well as broadening and separation of the melting transitions. In an effort to develop high surface area, biomimetic substrates for use in vibrational spectroscopy and chromatographic investigations we investigate the impact of substrate surface chemistry on phospholipid bilayers deposited on the pore walls within chromatographic silica particles. To accomplish this, in this work we employ confocal Raman microscopy for the physical characterization of model phospholipid bilayers in both bare-and cyanoderivatized high surface area chromatographic silica supports. In both cases structural indicators in the Raman spectra of the bilayers above and below the phase transition are in close agreement with those collected from free solution vesicle bilayers. Despite this, significant impacts on the melting transition of the supported bilayers were observed in bare silica supports. Here, a broadened and separate melting transition is observed, consistent with the separate melting of the proximal and distal leaflets. However, the cyano-supported bilayers exhibit a single, sharp melting transition, consistent with the melting transition measured in vesicles of the same phospholipid. Preliminary application of these materials as platforms for measurement of analyte-bilayer interactions will be presented.

Investigating the Competence Stimulating Peptide-Induced Quorum Sensing Circuit in Streptococcus mutans
Chowdhury Raihan Bikash, Sally R. Hamry, Yftah Tal-Gan

Streptococcus mutans is a gram-positive human pathogen that is responsible for dental caries with an economic cost over $100 billion per year in the U.S. *S. mutans* uses a 21-amino-acid peptide termed the competence stimulating peptide (CSP) to assess its population density in a process termed quorum sensing (QS) and initiate pathogenic phenotypes such as biofilm formation and bacteriocin production. Our goal is to develop CSP-based QS modulators capable of attenuating the infectivity of *S. mutans*. To this end, we plan to conduct a systematic replacement of the amino acids in the CSP sequence and identify structural features that are required to drive QS activation. This poster will present our recent results that include alanine and D-amino acid scans of the CSP sequence.

Genome Inspired Approach to Aptamer Discovery Using Genome-Wide-High-Resolution ChIP Techniques

Casey Fong, Linda McGown

Aptamers have amazing potential as pharmaceuticals and affinity reagents, but are often limited by current selection methods that rely on combinatorial libraries and repeated rounds of amplification that can create sequence bias, consequently, under represent certain secondary structures. This is especially true for G-quadruplexes. Sequences capable of forming secondary structures that can serve as aptamers are abundant in the human genome, and their affinity and selectivity for specific proteins has been honed through eons of evolution. Using high resolution Chromatin Immunoprecipitation (ChIP-exo) techniques, we can view and quantify protein binding sites at single nucleotide resolution across the genome and screen their potential to form secondary structures using predictive algorithms, and then analyze their protein binding affinities in vitro. The protein Nucleolin (NCL) has been shown to bind DNA G-quadruplex forming sequences both in vitro and in situ in live cells, and has been studied as a potential therapeutic as demonstrated by the antibody 4LB5 and G-quadruplex forming aptamer AS1411. NCL plays a pivotal role in many cancers as both a transcription factor and cell surface receptor. Through our approach, novel aptamers can be designed from genomic sequences for which NCL has high affinity, and NCL's role on gene expression can be better understood. Here we performed ChIP-exo on the ERBB2(+), P53(-), and WRN(-) estrogen driven metastatic breast cancer cell line BT-474, and have identified a variety of potential G-quadruplex forming sequences that could serve as aptamers to NCL. In addition, we have identified new genes that NCL interacts with, many of which are involved in the progression of cancer.

Analyzing the Effect of Simulated Heart Attacks on the Protein Profile and Structure of a Mouse Heart's Left Ventricle Through the Use of Matrix Assisted Laser Desorption/Ionization Mass Spectrometry Imaging

Emma Chowdhury

In this experiment, MALDI MS Imaging of proteins was used to compare the protein profile of six different samples of mouse heart, three from control hearts and three from hearts that had experienced simulated heart attacks. A tissue section was taken from each type of heart (one control, one test) one week, two weeks, and four weeks after the simulated heart attacks resulting in a pair of tissues from Week 1, Week 2, and Week 4. The control tissues are wild type controls without any simulated heart attack and the test tissues are left-anterior descending (LAD) ligation affected. The purpose of this experiment is to attain the molecular difference and distribution of the left ventricle of the mouse heart sections between the control tissue and the test tissue. The imaging slide with all the six tissue sections was prepared with a six-step rinse and sprayed with 20 mg/ml sinapinic acid matrix, followed by rehydration. Rapiflex MALDI TOF MS was utilized to obtain the molecular images and overall average mass spectra of the left ventricles. The mass spectra of the ventricular tissues showed a variation of the protein profile in each pair of tissues from Week 1, Week 2, and Week 4. In particular, the comparison of the spectra from the Week 2 tissues showed abundant peaks in the 4-15 kDa mass range of the test heart section, while there were few peaks in that range of the control heart section. The spectrum difference from the Week 4 tissues was also significant, in that the peak at ~15,500 Da in the test heart section is notably higher than its corresponding peak region in the spectrum from the control heart section. It is concluded that the LAD mouse heart model showed changes on both the structure and protein profile of the left ventricular walls compared to wild type mouse heart controls.

Developing a Non-Invasive Spectroscopic Technique for Detecting Liver Damage

Katherine Ember, Gabriel Onsicu, Fiona Hunt, Hannah Johnston, George Georges, John Hallett, Karen Faulds, Stuart Forbes, Colin CampbellLauren Jamieson

As our largest internal organ, the liver is responsible for the breakdown of toxic compounds, synthesis of multiple hormones and also plays a vital role in metabolism. However, liver disease is the only major cause of death in the UK that is increasing annually and it remains a worldwide problem. It is clear that novel methods of diagnosis are required; symptoms are fairly generic, whilst blood tests may show liver function to be normal even after significant damage has occurred. Moreover, in 2013, approximately 25,000 liver transplants were carried out worldwide and yet surgeons lack a quantitative means of determining whether a liver is suitable for transplantation or not. Here, we present our work towards a non-invasive, label-free method for detecting liver damage and assessing liver composition with molecular specificity. To establish whether spectroscopy can be used to detect liver damage, we have been using a porcine model of donation after circulatory death. After the heart ceases contracting, the liver is subject to a period of warm ischemia, when there is no circulation and oxygen delivery to the organs. We obtained in vivo spectral data from nine pigs using a 785nm handheld Raman spectrometer prior to death and after 45mins of warm ischemia, and observed significant specific spectral changes. Furthermore, we took liver and bile duct biopsies before and after ischemia for analysis using Raman microspectroscopy, coherent anti-Stokes Raman spectroscopy (CARS), nuclear magnetic resonance spectroscopy (NMR) and immunostaining. By applying these techniques to serial tissue sections, blood and bile, we have been able to detect morphological and biomolecular variations in samples. Moreover, we have demonstrated the ability of Raman spectroscopy to detect features such as bile duct cells (cholangiocytes), erythrocytes and lipid droplets in a label-free manner.

Magnetic Resonance Spectroscopy: A Tool for Assessing Concussion Injuries

Nicole Vike , Jonathan Tang, Thomas Talavage, Riyi Shi, Joseph Rispoli

With the increasing awareness of sports-related concussion injury and the need for better diagnostics and prevention, it is necessary to investigate the biological mechanisms of these injuries. Magnetic resonance imaging (MRI) is a powerful clinical tool used to diagnose a range of medical issues non-invasively. This method is particularly useful for exploring brain functionality, connectivity, and biochemistry. Because of the noninvasive nature of MRI, it can be employed to look more closely at how sports-related concussions and traumatic brain injuries (TBI) affect high contact-sports players. Specifically, magnetic resonance spectroscopy (MRS) can be used to investigate metabolic disturbances in the brain caused by disease and injury. The fundamental method underlying MRS is nuclear magnetic resonance (NMR) which uses the electromagnetic properties of atomic nuclei to obtain a spectral map signature to a specific molecule. One molecule, acrolein, which has not yet been investigated with MRS, has been shown to increase in rat brains following TBI. Acrolein is a volatile ?? unsaturated aldehyde shown to have damaging effects on cellular functions following injury to the central nervous system. Currently, only invasive and sacrificial procedures have been used to successfully quantify acrolein in the brains of animals. By instead using MRS, a novel, noninvasive method could be utilized to more quantitatively assess the severity and location of traumatic insult to the brain. We have already established that acrolein is detectable with MRS in a 7 Tesla pre-clinical animal MRI system (Bruker Biospec 70/30 USR System) with a Bruker volumetric rat head coil. These measurements were determined by injecting 39 μL of 1 M acrolein into a 39 mL spectroscopic whole-brain phantom. Signature acrolein peaks, previously quantified with 1H NMR and available in the Human Metabolome Database, were observable at 6.5 ppm and 9.4 ppm. Once further measurements are established in phantoms and animal models and proper quantification methods are developed, the protocol could be translated to assess humans suffering from TBI and prevent further, more serious, injuries from occurring.

Investigating the Potential of Silver Hydroxyapatite as an Implant Coating for Preventing Prosthetic Joint Infection

Luke DeHart, Austin Dinkins, Mary Tecklenburg

Substantial progress in medicine has dramatically increased life expectancy in developed countries. Alongside this increase in life expectancy, there has also been a drastic increase in the number of people who suffer from degenerative joint disease; the ultimate treatment for this disease is a total joint arthroplasty. It is estimated that one to two percent of these patients will develop a prosthetic joint infection at some point after their treatment. Prosthetic joint infections are associated with a fivefold increase in mortality rate and a significant economic impact. Utilizing silver integrated hydroxyapatite as a biocompatible and biodegradable surface coating for prosthetic implants could prevent biofilm formation and the development of a prosthetic joint infection without the use of antibiotics. Three forms of hydroxyapatite were synthesized, pure hydroxyapatite, silver-doped hydroxyapatite, and silver nanoparticle infused hydroxyapatite. Hydroxyapatite was synthesized using the wet chemical method with a 1.67 ratio of calcium to phosphate. The silver-doped hydroxyapatite was synthesized in the same way as the hydroxyapatite, but a stoichiometric amount of silver nitrate was added so that the ratio of silver plus calcium over phosphate equaled 1.67. Silver nanoparticles embedded in polyethylene were mixed with hydroxyapatite powder and the resulting mixture was heated to 350 C to burn off all the polymer leaving pure hydroxyapatite with embedded silver nanoparticles. The silver content of each sample was characterized by x-ray fluorescence. The size and morphology of the silver nanoparticles, prior to and after heating, was characterized by transmission electron microscopy. The antimicrobial properties and prevention of biofilm formation for each of the different types of hydroxyapatite pellets will be tested. The results will give insight into novel methods to prevent the development of biofilms on prosthetic implants.

Structure-Activity Relationship Studies of the Competence Stimulating Peptide (CSP) Signal Utilized by Streptococci Species

Sally Hamry , Chowdhury Raihan Bikash, Yifang Yang, Bimal Koirala, Lucia Sanchez, Naiya Phillips, Yftah Tal-Gan

Quorum sensing (QS) is a type of signal-mediated communication used by bacterial species such as Streptococcus pneumonia and Streptococcus mutans to determine their population density. The signaling molecule used in the QS circuit for S. mutans and S. pneumoniae is the competence signaling peptide (CSP). In QS, the signal concentration is directly proportionate to the bacterial population density. Once the signal molecules reach a certain concentration, representing high cell density, they can effectively activate a signaling cascade resulting in a population-wide alteration of gene expression. Processes which are under the control of QS include virulence factor secretion, development of competence, sporulation and biofilm formation. Consequences of these microbial communities communicating through QS include antibiotic resistance and the development of pathogenicity. In this study we aim to develop CSP-based QS modulators against the S. pneumoniae and S. mutans circuitries as a means to mitigate infections caused by these two pathogens. This was achieved by conducting full alanine and D-amino acid scans to identify key residues and chiral centers of peptide analogs synthesized using Fmoc Solid Phase Peptide Synthesis (SPPS). Additionally Beta-Galactosidase cell-based reporter assays were performed with CSP analogs to identify QS inhibitors and activators. Through this, potent inhibitors were found for both CSP1 and CSP2 in S. pneumonia. Additionally, residues, which play a key role in binding to the ComD receptor, were identified in the CSP chain of S. mutans.

Bacterial identification in ex vivo human fetal membrane biofilms using Raman microspectroscopy

Oscar Ayala, Ryan Doster, David Aronoff, Jennifer Gaddy, Anita Mahadevan-Jansen

Streptococcus agalactiae, also known as Group B *Streptococcus* (GBS), is a leading infectious cause of morbidity in newborns and mortality in immunocompromised adults. GBS is a Gram-positive opportunistic pathogen that colonizes the gastrointestinal and genitourinary tract of up to 50% of pregnant women.

Currently, rectovaginal cultures are used to screen pregnant women at 35-37 weeks of gestation for GBS colonization, which is a known risk factor for invasive perinatal infections. Although the sensitivity and specificity of these cultures are high (>89%), this strategy does not capture women with premature birth, which have a higher risk of perinatal GBS infection. An intrapartum point-of-care test able to detect the presence of GBS is strongly needed. To determine the potential of using Raman spectroscopy to detect the presence of GBS, three models were evaluated. First, Raman microspectroscopy was used to characterize the spectral signature of GBS bacterial colonies. Second, GBS biofilm was evaluated on a quartz slide to determine if spectral biomarkers of GBS could be identified within biofilm material. Lastly, GBS cells were co-cultured with human fetal membranes *ex vivo* and then evaluated using Raman microspectroscopy to determine if GBS could be detected within human tissues. Our results show strong Raman peaks at 1120 cm^{-1} (C-C stretch) and 1504 cm^{-1} (Uracil) that were consistent across multiple GBS colonies. In addition, these peaks were visible in both the *in vitro* biofilm model and *ex vivo* infected fetal membrane tissues, while absent in controls. These findings motivate the use of RS to detect GBS *in vivo* throughout pregnancy, providing physicians with vital point-of-care information.

Spectroscopic investigation of early tissue pathology in combat-wounded patients diagnosed with heterotopic ossification

Krystine Hill, Katherine E. Cilwa

Heterotopic ossification (HO) is a disease characterized by the abnormal formation of mature bone in areas of soft tissue. This can result in chronic pain, tissue entrapment, and limited function at affected joints. Incidence of HO is especially high in wounded service members inflicted with blast-related combat injuries involving amputation. As a result of HO, service members can experience difficulty using prostheses and reduced mobility. As of now, radiography is the primary diagnosis method; however, once mineralized HO is visible on an X-ray or MRI, early intervention is no longer feasible and the treatment available, namely surgical excision, may cause complications such as recurring HO. Previous studies demonstrated the ability of Raman spectroscopy to identify molecular markers of HO in muscle tissue that are associated with the start of chondrogenesis within one week after injury in a physiological animal model of combat-related HO. Such molecular markers include increased collagen levels, indicating error in the wound healing process. Early identification of altered tissue pathology using Raman spectroscopy may guide clinical diagnostics and decision making. In this study, human tissue biopsies collected during serial wound debridements within 1-3 weeks following combat-related injury were analyzed using a fiberoptic probe coupled Raman spectrometer with 3 mm spot size and 830 nm excitation. Spectra were collected using 2 second acquisitions, 25 accumulations, and 3 collection locations per biopsy. Spectra were processed using custom MATLAB® scripts; processing included spatial averaging, fluorescence and adipose signal subtraction, normalization to the methyl/methylene scissoring band at 1445 cm-1 as a marker of total protein content, and peak fitting. Biopsies were categorized according to HO related outcome and severity and Raman metrics analyzed. Raman metrics measuring changes in protein disorder/order (disordered and ?–sheet vs helical proteins) were altered following injury and may serve as markers of early HO formation. Raman spectroscopy has the potential to serve as a non-invasive method in clinically identifying patients at risk of developing HO by detecting molecular markers characteristic to wound healing error, such as increased collagen levels. Non-invasive methods are crucial in avoiding subsequent trauma to patients.

Dried Blood Spot and Microwave-Induced Combustion: The Perfect Combination for Iodine Determination in Human Blood

Fabio Duarte, Samuel Waechter, Paula Dalla Vecchia, Karine Reinke, Erico Flores

Venous blood is one of the most commonly biological material used in laboratorial tests. Since venous puncture is an invasive procedure, it requires special care and a qualified professional. In this sense, looking for less invasive procedures for blood collection from newborns and children, dried blood spot (DBS) sampling protocol have been used. DBS consists of the collection of a drop from body peripheral regions (e.g., fingertip), and then transfer to a filter paper. Regarding to the blood analysis, iodine is an essential micronutrient for the synthesis of T4 and T3 hormones, which are important for many body functions. In this way, the aim of this study was to evaluate the microwave-induced combustion (MIC) combined to DBS for subsequent iodine determination in human blood by

ICP-MS. DBS sampling was developed using a specific filter paper (M-TFN) and several conditions before and after sampling were careful evaluated. For the digestion of DBS by MIC, blood sample mass was weighted on the filter paper, 50 ?L of 6 mol L-1 NH4NO3 (combustion aid) were added to the paper and 6 mL of the absorbing solution (water, 10, 25, 50 or 100 mmol L-1 of NH4OH) were added into the digestion vessel. After MIC, iodine stability in the digests was also evaluated. The application of DBS protocol was performed with aliquots of capillary blood (fingertip) of 10 volunteers. For results comparison, microwave-assisted ultraviolet digestion (MAD-UV, 150 mg of blood + 100 µL 14.4 mol L-1 HNO3 + 2 mL H2O2) was used. Regarding to MIC, better agreement with reference method (MAD-UV) of iodine was achieved using 50 mmol L-1 of NH4OH as absorbing solution. The accuracy of overall procedure was also evaluated by spike of L-thyroxine before MIC (recovery of 99 ± 7%). No statistical difference (Student t-test, $p < 0.05$) was observed comparing DBS/MIC with MAD-UV. The limit of quantification by DBS/MIC was 8 ng g-1 and iodine concentration in the samples ranged from 30 to 70 ng g-1. The proposed protocol (MIC and DBS) was fit for porpoise, providing accurate results for iodine in blood using a simple sampling method.

Using Attenuated Total Reflectance Infrared Spectroscopy to Monitor Changes in Protein-Substrate Binding of Human Liver Pyruvate Kinase

Reid Brenner, Charles Wurrey, Aron Fenton

There has been recent interest in human liver pyruvate kinase (hL-PYK) because of its potential to limit hepatic glucose production (via gluconeogenesis) and therefore prevent fasting hyperglycemia in individuals with diabetes. When activated, hL-PYK converts phosphoenolpyruvate (PEP) and MgADP into pyruvate (Pyr) and MgATP. The enzyme itself is controlled allosterically by different allosteric effectors which activate or inactivate the enzyme. Fructose 1,6-bisphosphate (FBP) activates hL-PYK whereas it is inhibited by alanine (Ala) and phosphorylation. For rational design of allosteric drugs, specifically for hL-PYK activation, there is further need to understand which protein-substrate interactions are involved in allosteric regulation. This presentation is the next phase of a long-term study of how interactions between the enzyme (hL-PYK) and substrate (PEP) contribute to allosteric regulation. To understand how hL-PYK affects PEP, difference attenuated total reflectance infrared (ATR-IR) spectroscopy was used to study the binding of PEP to hL-PYK in real time. ATR-IR spectra of PEP were recorded in solutions of hL-PYK in the presence and absence of the allosteric effector FBP. Using the assignments of PEP in aqueous solution from the first phase of our project, vibrational frequency shifts of PEP were quantified and used to determine which functional groups in PEP provide the most significant contributions to allostery of hL-PYK.

Raman spectroscopy for monitoring cervical remodeling in pregnant women in vivo

Christine M. O'Brien, Elizabeth Vargis, Amy Rudin, James C. Slaughter, Kelly A. Bennett, Jeff Reese, Anita Mahadevan-Jansen

During pregnancy, the cervix must undergo significant biochemical remodeling to allow for successful labor and delivery. This process has been divided into four distinct phases: softening, ripening, dilation, and repair. *In vivo* quantitative assessment of the cervix would be useful in identifying patients at risk of preterm labor and providing increased time for clinical intervention to delay or prevent preterm birth. We hypothesize that Raman spectral signatures from each phase of cervical remodeling are distinct and reflect physiology of the changing cervix. To test this hypothesis, 62 patients receiving prenatal care were recruited. Raman spectra were measured using an *in vivo* Raman system with a fiber optic probe throughout pregnancy and post-partum. To flexibly model the spectral changes over time, a restricted cubic spline with 70 degrees of freedom was developed. This approach highlighted spectral regions that varied throughout pregnancy and was used for selecting peaks for further analysis. These regions were analyzed using a generalized linear model to provide quantitative analysis of a longitudinal data set. Furthermore, Raman spectra were fit to a non-negative least squares model based on pure Raman spectra from actin, blood, collagen type I, cholesterol, glycogen, and water. Raman peaks related to extracellular matrix proteins (1003, 1055, 1248, and 1657 cm^{-1}) significantly decreased ($p<0.05$) throughout pregnancy, indicative of extracellular matrix breakdown. In addition, peaks corresponding to blood (1233 and 1563 cm^{-1}) significantly increased, agreeing with reports revealing increased vascularity as the cervix prepares for delivery. In the post-partum cervix, significant increases in peaks corresponding to actin (1003, 1339, 1657 cm^{-1}), glycogen (1125 and 1335 cm^{-1}), and extracellular matrix proteins were observed. These signatures may be due to increased immune cell infiltration as part of the repair process, fueled by glycogen and other anabolic molecules. This foundational study demonstrates sensitivity of Raman spectroscopy to biochemical dynamics of cervical remodeling that occur with pregnancy. Raman spectroscopy has the potential to aid global efforts to close the gaps in our understanding of cervical maturation, and ultimately reduce the incidence, morbidity, and mortality caused by preterm birth.

Selectively Trapping Residual Vancomycin with Peptide Analogs to Prevent Colonic Bacterial Stress

Brittany Russ, Ryan Mull, Yftah Tal-Gan

Vancomycin is an important last-resort antibiotic utilized to fight off serious clinical pathogens like methicillinresistant Staphylococcus aureus (MRSA). Vancomycin works by binding to the terminal amino acid sequence acetyl-L-Lysine-D-Alanine-D-Alanine found in bacterial cell walls through a combination of intermolecular hydrogen bonds and hydrophobic interactions.1 This action prevents the bacterial cell wall from cross-linking, inhibits further cell wall synthesis, and can lead to cell death.1 During treatment vancomycin is usually administered intravenously at high doses, due to the antibiotic's low absorption at infection sites. As a direct result, vancomycin not only travels through the body via plasma in blood, rather it is accumulated in high concentration in the colon, attacking off-target bacterial populations, prior to being eliminated through the excretory system.3 Consequently, strains of vancomycin-resistant Enterococci (VRE) inhabiting the human colon have emerged due to selective pressure and became associated with significant morbidity and mortality.2 In this study, the goal is to eliminate vancomycin activity in the colon by orally co-administering a peptide analog that will accumulate in the colon and inhibit vancomycin as it reaches the colon. A library of peptides was created by incorporating specific functional group modifications on the side chains, to increase the binding affinity to vancomycin. These peptides were synthesized to determine structure-activity trends and create a competitive inhibitor to vancomycin in cultures of Enterococci species such as E. faecalis and E. faecium. The aim of this research is to preserve the effectiveness of vancomycin as an antibiotic by selectively minimizing unnecessary exposure to natural human microflora, while also allowing the drug to treat its intended infection. References 1. Yao, N. H.; He, W. Y.; Lam, K. S.; Gang, L. Conformational Studies of Resin-Bound Vancomycin and the Complex of Vancomycin and Ac2-L-Lys-D-Ala-D-Ala. J. Comb. Chem. 2005, 7, 123-129. 2. Horwitz, E.; Tal-Gan, Y.; Temper, V.; Shapiro, M.; Chiam, G.; Amnon, H. Chemical Trapping of Vancomycin: A Potential Strategy for Preventing Selection of Vancomycin-Resistant Enterococci. Microb. Drug. Resist. 2012, 18, 109-115. 3. Currie, B. P.; Lemos-Filho, L. Antimicrob. Agent. Chemother. 2004, 48 (11), 4427–4429.

Chemical Trapping of Vancomycin: A Method to Attenuate the Selection of Vancomycin Resistance in Enterococcus

Ryan Mull, Brittany Russ, Yftah Tal-Gan

The rapid spread of multi-drug resistant bacterial pathogens is one of the greatest obstacles faced by modern medicine. Several different antibiotics, known as last resort antibiotics, with the ability to treat these multi-drug resistant pathogens have been identified, and their use is tightly regulated to prevent the development of resistance. One of these antibiotics of last resort, the glycopeptide antibiotic vancomycin, is critical for the treatment of methicillin-resistant Staphylococcus aureus (MRSA). While transferable resistance to vancomycin is rare, the commensal colonic bacteria enterococcus possesses vancomycin resistance genes that have been found to be transferrable to multi-drug resistant pathogens such as MRSA. In order to preserve the effectiveness of vancomycin it is important to develop methodology that will attenuate antibiotic resistance while preserving its activity against multi-drug resistant pathogens. Following treatment, vancomycin accumulates in the colon and persists for weeks or months where it acts upon the natural colonic microflora, including enterococci, allowing for the selection of antibiotic resistance genes. Thus, it is vital to find ways to inhibit vancomycin's activity within the colon to reduce the selective pressure for these genes. Vancomycin naturally binds the C-terminal K-a-a motif of bacterial pentapeptide cell wall precursors. Modification of the native vancomycin tripeptide target could produce a molecule with increased binding affinity to vancomycin, competitively binding vancomycin and reducing its activity against the colonic microflora. Novel peptide-based vancomycin inhibitors containing carboxylic acid derivatives or aromatic ring containing compounds, as well as several truncated dipeptides, have been synthesized using Fmoc-based solid phase peptide synthesis (SPPS). These peptides' affinity to vancomycin has been compared against the native tripeptide in several cell-based assays, including minimum inhibitory concentration (MIC) cell-based assays, to determine which functional groups best enhance binding affinity. Additionally, these modifications have given insight into what structural components are required for vancomycin binding, and have produced several compounds that are able to completely eliminate the activity of vancomycin in a 20:1 molar ratio.

Investigate photo-induced skin damages and skin protection provided by different UV filters using FTIR Spectroscopy analysis

Samuel Gourion-Arsiquaud, Arianna Cozzi

Background: The skin is composed of two major layers: the epidermis, unvascularized epithelial layer, and the dermis, vascularized layer with a rich supply of capillaries, nerves, hair follicles etc. The epidermis plays a key role as it is our first protective barrier from external stresses like UV exposure or exogenous contamination. Its principal cells are the keratinocytes, which differentiate as they migrate toward the stratum corneum (SC). This superficial layer of the epidermis, typically consists of ~ 10 to 20 layers of terminally differentiated anucleated corneocytes embedded in an intercellular lamellar lipids. The skin barrier function is related to both the unique lipid composition of the SC and their complex structural organization. Changes in the SC may result in significant modification of the skin barrier functions and therefore modify the actives penetration into the skin. Aim: The objectives of the present study are 1) to assess the effects of UV exposure (10h-100h) on isolated SC form pig and human skin and 2) to investigate the penetration of different UV filters Avobenzone and Octocrylene (free and encapsulate in a gel sol silica capsule) inside skin samples. Methods: Both FTIR Spectroscopy and ATR-FTIR Spectroscopy Imaging were used to address these questions. FTIR spectrometer (Nicolet 6700 Thermo Scientific) equipped with temperature controlled transmission cell was used for studying SC lipids conformation and packing transition by analyzing the CH2 peaks position between 2800 cm-1 and 3000 cm-1. The accessory is able to create a gradual heating of the SC samples. The whole instrument was constantly purged with Nitrogen in order to remove water vapor. In all the experiments the FTIR spectra were acquired as a function of temperature from 5 C to 95 C and recorded every 2 to 3C . All spectra were collected with a spectral resolution of 4 cm-1 and 64 scans accumulation. During the data collection the sample temperature remained constant within +/-1 C . Each experiments were triplicated. By studying the frequency variation as a function of the temperature, the phase transitions in which the system goes from an ordered to a disordered conformation were observed. FTRI images were acquired with a Spectrum Spotlight 400 imaging system (Perkin Elmer Instruments, Shelton, Conn., USA) using a MCT (mercury-cadmium-telluride) focal plane array detector placed at the image focal plane of an IR microscope.

FTIR images were collected in the reflective mode at a spectral resolution of 4 cm-1 in the mid-infrared (MIR) region from 4000 cm-1 to 850 cm-1 with a spatial resolution of 6.25 x 6.25 ?m. Results: Our data show that progressive UV exposure and UV filter application may cause changes in the SC. these changes could significantly alter the skin barrier proprieties. The FTIR images have highlighted significant UV filter penetrations inside the SC after sunscreen applications. The encapsulation technology provides for a reduction in these unwanted penetrations.

Raman spectroscopy and neural network to detect the glucose level
jorge castro-ramos, Naara González-Viveros, Pilar Gómez-Gil, Fabián Villa-Manríquez, Francisco Gutiérrez-Delgado

In recent work, we developed a method to detect the high and low levels of glycated hemoglobin. The goal is to detect the levels of it. We start by developing a method to detect the glucose levels. We prepare a set of different concentration of glucose and by using their Raman spectra as input in a neural network algorithm we recover the concentration as output.

A Systematic Investigation of the Structure-Activity Relationships between GBAP and the fsr Quorum Sensing Circuit in Enterococcus faecalis
Brooke Gantman, Dominic McBrayer, Crissey Cameron, Yftah Tal-Gan

Enterococcus faecalis is found naturally in the human gut and is intrinsically resistant to vancomycin, an antibiotic of last resort. It causes more than 170,000 cases of antibiotic-resistant infections in the United States annually. *E. faecalis* is also able to spread its resistance genes to other bacteria, such as *methicillin-resistant Staphylococcus aureus* (MRSA). It is vital to find new therapeutic routes to prevent growth and infection of these bacteria, while minimizing the development of resistance. By altering a non-vital process associated with pathogenicity, the pressure to develop resistance is lessened. A new strategy targets a bacterial group-beneficial trait, quorum sensing (QS), to attenuate pathogenicity. Gelatinase-Biosynthesis Activating Pheromone (GBAP) is the macrocyclic signaling peptide that activates the *fsr* QS circuit in *E. faecalis*, inducing the formation of biofilms and production of virulence factors that initiate infection of the host. By systematically altering the structure of GBAP, new information about the structure-activity relationships between GBAP and its QS receptor were discovered. Solid phase peptide synthesis was used to prepare three libraries of GBAP analogs in order to methodically study the importance of the stereochemistry and side chains of each residue, as well as the importance of the exocyclic tail. Biological assays were performed to determine each analog's potency for activating QS. Results revealed residues with side chains and/or stereochemistry that were vital to activating QS. Additionally, the exocyclic tail was shown to be completely expendable. Further experimentation will be performed to further explore the structure-activity relationships of GBAP and its receptor.

Handheld Laser Induced Breakdown Spectroscopy (LIBS) for Field Archaeology
Luke DOuglass, Mary Kate Donais, David George

Handheld Laser Induced Breakdown Spectroscopy (LIBS) was used for the elemental characterization of archaeological materials from multiple excavation sites within the Orvieto region of Umbria, Italy. Use of a portable instrument for this work was essential considering the challenges and regulations regarding transport of cultural heritage objects from their country of origin. The first application to be presented is the analysis of bucchero, a style of Etruscan pottery shown in the picture above left, from two different archaeological excavation sites. Over twenty-five different sherds from each site were analyzed at three locations per sherd. Elements in the samples were identified followed by multivariate statistical analysis to explore chemical differences among the samples. As a second application, handheld LIBS also was used for the in situ characterization of Roman mortars on walls within a Roman bath complex (picture above right). These data also were evaluated using the multivariate statistics to better understand the construction sequence and structural relationships within this area of the site.

Detection of anti-toxin heavy-metal tagged antibodies using laser induced breakdown spectroscopy (LIBS)

Carmen Gondhalekar, Eva Biela, Bartek Rajwa, Euiwon Bae, Valery Patsekin, Jennifer Sturgis, Huisung Kim, Iyll-Joon Doh, Larry StankerPaul Robinson

Introduction: Food resources are both a biosecurity issue and a concern to the population due to the adverse health effects of contaminants and the time-intensive detection process. Our goal is to reduce the time and cost of contaminant detection by replacing fluorescent molecular tags (typically used for detection) with heavy-metal (HM) molecular tags. The advantage of this approach is that many HMs are not naturally occurring in biological samples, reducing background effects, and have a narrower signal bandwidth than fluorescent tags, facilitating mutiplexing. HM can be conjugated to antibodies against chemical or biological targets such as food contaminants, permitting detection of the contaminant typically with mass cytometry (CyTOF) and ICP-MS. Laser-induced breakdown spectroscopy (LIBS) offers a low-cost and modular alternative for HM-labeled antibody detection. This study is supported by the USDA and aims to detect HM-labeled antibodies using a LIBS system for the goal of developing a low-cost bioassay and analytical tool used by inspectors for contaminant detection. Methods/Results: A bioassay was developed to tag toxins with HM-labeled antibodies. First, toxins are immobilized on a 1µm bead functionalized with anti-toxin antibodies. A HM-labeled anti-toxin antibody is then introduced to the sample to bind to the toxin. CyTOF was used to show the HM-labeled antibody successfully bound to the toxin. Future work involves optimizing the bioassay and detecting the HM antibodies using LIBS. The LIBS system features a 1064nm Q-switched laser beam operating at 70mJ with a single 20ns pulse width focused onto a sample in a helium gas environment by a 400mm focal-length converging lens. Data were collected using two spectrometers (covering 220nm-630nm range). Current work involves testing AAS standards for each of the metals conjugated to the antibodies (Lu, Pr, Au, Dy, Yb, Nd and Gd) to optimize instrumentation parameters. Data analysis uses a custom R program to identify the sample peaks after baseline correction, data normalization, background subtraction, noise reduction, threshold selection, data standardization, peak location and comparison to HM standards. Conclusion: Combined, the bioassay, LIBS instrument and data analysis program will provide a low-cost and multiplexed alternative for conventional chemical and biohazard detection.

Bacterial Mounting and Concentration Techniques to Translate Laser-Induced Breakdown Spectroscopy into a Clinical Setting

Alexandra Paulick, Naila Rahman, Steven Rehse

Laser-induced breakdown spectroscopy (LIBS) is a rapid elemental analysis technique that has been used for the detection and identification of bacterial pathogens. There is a high demand for real-time identification of bacteria and the capabilities of LIBS for this are promising. Optimizing the bacterial mounting techniques prior to LIBS analysis in a clinical setting is currently underway. This includes the sample preparation steps of separating bacterial cells from the other unwanted cells that would most likely be present in a biological specimen. It also includes lowering the minimum number of bacterial cells required for detection and accurate identification by LIBS. This poster will present our efforts to accomplish these goals. We have investigated the effectiveness of the detergent Tween 20 as a possible anti-clumping agent for bacterial cells deposited on a nitrocellulose filter paper by observing the LIBS signal from *E. coli* specimens deposited both with and without Tween. The ability to quickly separate a contaminant from a bacterial suspension through the use of centrifugation and filter media with different pore sizes will also be presented. In addition, the capability of a metal cone designed to concentrate bacterial depositions during centrifugation onto a smaller area of the filter paper will be demonstrated.

Laser induced breakdown spectroscopy for elemental analysis of the thyroid
Condon Lau, Irfan Ahmed

The thyroid is an important hormone regulation organ. It secretes hormones, such as triiodothyronine and thyroxine, which are crucial for regulating metabolism and protein synthesis. Important elements for thyroid function include iodine, sodium, calcium, and potassium. Improper levels of these elements can lead to a range of thyroid disorders, including hypo and hyperthyroidism. Laser induced breakdown spectroscopy (LIBS) is developed to assess iodine in the thyroid in situ, along with other important elements such as sodium, calcium, and potassium. Adult male Sprague Dawley rats (N=30) were employed. Subjects were fed with an iodine diet through the drinking water consisting of 0.05% iodine for 0, 6, and 12 days (N=5 each). The thyroids were harvested after the treatment period. The thyroids were directly extracted from the remaining subjects and immersed in iodine solutions of 0, 500, and 1000ppm (N=5 each) for 24hrs at room temperature. The iodine solutions were prepared by dissolving crystal iodine (Sigma Aldrich) in de-ionized water to the required concentration. All extracted thyroids were immersed in saline, snap frozen, and stored at -80?C for 24hrs. The LIBS setup consisted of a 1064nm pulsed laser (CFR200, Quantel, 8ns duration, 200mJ energy) focused to a 10μm spot. The emitted light after a pulse was collected by a six channel fiber (2000μm diameter) bundle positioned 35mm from the focus and 45? from the laser beam. The fibers delivered the light to a six channel spectrometer spanning 200–900nm with 0.1nm resolution (MX2500+, Ocean Optics). Thyroid samples were placed at the focus and six spectra were acquired and averaged. Pronounced iodine, sodium, calcium, and potassium emissions are observed at approximately 746, 589, 395/422, and 766/770nm, respectively. The iodine emission is higher in 500 and 1000ppm immersed thyroids than in 0ppm immersed thyroids, as expected. The iodine emission is highest after 0 days iodine diet, lowest after 6 days, and shows recovery towards normal after 12 days. Emissions from other elements also vary with iodine emission. This surprising observation follows the Wolff–Chaikoff effect as ingestion of excess iodine reduces thyroid iodine. This development provides a novel method for in situ elemental analysis of the thyroid.

Metal Ablation by Laser Irradiation Under Different Ambient Conditions
Ahmed Elsied, Paysoon Dieffenbach, Prasoon Diwakar, Ahmed Hassanein

Several experiments were designed to study the effect of material properties, ambient conditions, as well as laser fluence on metal ablation. Ablation was induced using 1064 nm, 6 ns Nd:YAG laser with laser energy ranging from 2 mJ up to 300 mJ. The first experiment was conducted at vacuum to study the effect of laser fluence and material properties on metal ablation, using wide range of laser fluence (2 J/cm^2 up to 300 J/cm^2) for Al and W targets. The second experiment was conducted at 1 atm pressure using two different ambient gases (air and argon), to study the effect of ambient gas in laser-metal ablation process. The third experiment was conducted at two different pressure 10 Torr ant 760 Torr using same ambient gas, Ar, to investigate how ambient gas pressure impact the laser-metal ablation process. The amount of mass ablation, crater shape, melt formation were measured using White Light Profilometer (WLP). These measurements were used for comparing the ablation process for experiments described above. The experimental results were supported by an analytical model which shows good agreement. The analytical model provides a trend for the ablation yield. This trend was found to be independent on the ambient conditions and ablated material. At low laser fluence, the ablation yield was found to be a logarithmic function of incident laser fluence while, at high laser fluence the ablation yield was found to be a linear function of incident laser fluence. The significance of ambient pressure was more pronounced than the effect of ambient gas type. Detailed mechanisms of these effects will be presented in this study.

Characterization and Classification of Pharmaceutical Tablet Coating by Laser Induced Breakdown Spectroscopy (LIBS)
Lanfang Zou, Xiaodong Bu, Brittany Kassim, Yun Mao

Laser Induced Breakdown Spectroscopy (LIBS) has emerged as an innovative tool for quantitative and qualitative elemental analysis in pharmaceutical research. The sampling process, laser ablation (LA), involves a high-power pulsed laser beam directed and focused onto a solid sample to convert a finite volume of the sample into excited vapor and aerosol constituents. As the plasma cools, the excited constituents relax to their ground state and emit light at characteristic wavelengths of the elements present. LIBS has several advantages including little to no sample preparation, fast measurement time, broad elemental coverage, and versatile sampling protocols. Samples for LIBS experiments can be any form: solid, liquid or gas. Importantly, LIBS can detect both organic and inorganic elements of the sample simultaneously, which is impossible in common techniques such as ICP-OES or ICP-MS. With these attributes combined LIBS has become of significant interest in pharmaceutical research. Herein, the potential use of LIBS for rapid characterization of tablet coatings will be illustrated, including the investigation of the both intra-and inter-tablet variability. Firstly, the thickness and uniformity of the tablet coating, which has a significant effect on the coating performance, was assessed by LIBS. Additionally, the technique was used to locate contaminants on the coating, as well as to identify the specific elemental contaminants present. Further, principle component analysis (PCA), which can effectively reduce the high-dimensional LIBS data onto a lower dimensional coordinate system and thus discriminate among different samples was utilized to visualize differences in tablet coatings. To date, there are only a few examples using LIBS in the pharmaceutical industry, and no literature on thorough investigations of both intra- and inter-tablet coating variability by LIBS.

Latest advances in Tandem LA – LIBS Instrumentation: Leading the innovation in laser ablation chemical analysis with simultaneous LIBS and LA-ICP-MS
Chunyi Liu

The J200 Tandem LA – LIBS Instrument from Applied Spectra is a true innovation in laser ablation solid sampling that provides simultaneous LA-ICP-MS and LIBS analysis to expand elemental coverage and increase concentration dynamic range. In this presentation, the latest advances in LIBS detector, ultrashort laser source, and LA and LIBS sampling cell are presented for advanced analytical capability. These hardware advances, combined with integrated data analysis software, enable ground-breaking analytical capability to measure every element in the periodic table in one laser sampling including H, N, O, halogens and others that are difficult for ICP-MS alone.

Ultrafast LIBS for 3D chemical imaging of Li-Ion batteries
Huaming Hou, RIck Russo, Vassilia Zorba

Li-ion batteries are the most widely used energy storage devices in consumer electronics and the main storage technology in electric vehicle applications. Direct chemical analysis of Li-ion batteries provides critical information on factors that determine their overall performance, including energy density, specific energy and cycle lifetime. Despite progress in the chemical analysis of Li-ion battery materials, significant challenges remain, including the ability to analyze samples in their inherent environments and to provide spatially resolved chemical information. In this work we use femtosecond LIBS coupled with advanced visualization capabilities for the elemental imaging of Li-ion battery components including anodes, cathodes and solid electrolytes. Two-dimensional (2D) layer-by-layer mapping, 2D cross-sectional imaging and three-dimensional (3D) volume rendering of major and minor elements are presented. The elemental distributions are correlated to the electrochemical performance and serve to identify strategies to improve next-generation high energy density Li-ion technologies.

Microfluidic electroporation for gene delivery and cellular analysis
Chang Lu

Electroporation is a simple physical technique that permits release of intracellular proteins, nucleic acids and other metabolites for analysis and delivery of exogenous reagents such as genes for cellular manipulation. In this talk, I will discuss our development of microfluidic platforms for conducting electroporation for various applications. We developed a simple microfluidic flow-through technique that conducts electroporation under constant voltage for gene delivery. We showed that hydrodynamics such as Dean flows could be introduced to dramatically influence gene uptake and its distribution on the cell surface. We also demonstrated the release of genomic DNA and RNA from cells for analysis. Finally I will discuss the examination of protein translocation by combining electroporation and flow cytometry.

High Throughput Microfluidic Genetic Transformation
Paulo Garcia, Rameech McCormack, Cullen Buie

Genetic engineering and synthetic biology design microorganisms to create high value chemicals such as pharmaceuticals, biofuels, and other compounds. One of the biggest challenges to advancing the field is the difficulty of genetic transformation (insertion of DNA or extracellular material into bacteria). The difficulties are due to low efficiency, low rate of cell survivability, significantly high electric fields (15 kV/cm or higher), and the small volumes (~50 μL) that can be reasonably transformed using the current state of the art. Electroporation, transiently disrupting the cell envelope with an electric field, is frequently used in industry and academia for cell transformation. This pore formation allows for the uptake of DNA into cells. We have meticulously tried to improve the electroporation process and have successfully developed a high throughput technology that has the potential to increase the speed of the electroporation process 10,000X. In this talk we will discuss our unique, scalable device design that integrates microfluidic principles in a potentially low cost and scalable platform.

Tunable Insulator-Based Dielectrophoresis for Biomolecule and Bioparticle Pre-concentration and Separation
Alexandra Ros, Daihyun Kim

The vast complexity of biological particles requires efficient strategies for separation, fractionation and pre-concentration in order to study variations of biomolecular signatures qualitatively and quantitatively. Insulator-based dielectrophoresis (iDEP) is a relatively new method to capture bioparticles and biomolecules for further analysis or separate analytes based on their polarization and resulting translational motion in electric field gradients. iDEP can be integrated in microfluidic devices with the advantage of facilitated microfabrication strategies avoiding the need of embedded electrodes and ill reactions in the capture device. However, the obstacles or constrictions needed for generating high electric field gradients for iDEP are pre-determined with the microfluidic design and often limited by the resolution of standard photolithographic approaches. Dielectrophoretic forces can thus not be adjusted for varying media (i.e. body fluids or tissue extracts) or varying analyte properties, especially if sub-μm bioparticles or biomolecules are probed requiring large dielectrophoretic forces for trapping or separation. To overcome this drawback, we developed a tunable iDEP method allowing to adjust the necessary dielectrophoretic forces with a flexible constriction, where the constriction width varies the generated electric field gradient. The constriction width is tuned via membrane deflection in elastomer microfluidic devices achieving dielectrophoretic forces within a range of five orders of magnitude. We demonstrate that this approach can be used for trapping μm-sized and sub-μm sized beads as well as biological particles and biomolecules. We achieved dielectrophoretic trapping of liposomes smaller than 1 μm in size and demonstrate selective DNA DEP trapping behavior in tunable iDEP devices. This novel strategy is thus applicable to a wide range of biological analytes including biomolecules, can be achieved with standard soft lithography, and allows in situ adjustment of dielectrophoretic forces through the tunability of each independent elastomeric membrane.

Condensed Phase Separation Science and Microfluidics
Mark Hayes

Many common research and medical care processes involve or could be helped by isolation and concentration of pure populations of molecules or cells. The unique capabilities of microfluidics, including ultrafine spatial control of fluid movement and precise application of external fields, allows for significantly improved performance compared to traditional isolation and concentration systems such as filtration, chromatographic and electrophoretic, among others. An additional advantage of microfluidics is provided by the high relative field intensity available because of the reduced length scales. The precision of flow and external fields, combined with high intensity, allows for strong gradients in chemical potential (molecular scale) and ultrafine control and differentiation of particles (mesoscale). These features allow a large array of options in terms of isolating, purifying, and concentrating populations of like-molecules and/or particles. These species can be molecules, peptides, proteins, aggregates, viruses, exosomes, nucleotides, microbes, bacteria, and cells, along with any other inorganic or bioparticle within the same size range. In this presentation, I will show the underlying theoretical basis suggesting greatly improved isolation and concentration of targets and data from small molecules (fluorescent dyes), proteins, viruses, bacteria and cells will be discussed.

Microfluidic Analysis of 3D-Culture Derived Extracellular Vesicles
Mei He, John Sibbitt, Zhao Zheng

The 3D-cultured cell models have been recognized as the more realistic for translating into in vivo study, particularly for understanding tumor pathogenesis. The living tumor cells continuously produce extracellular vesicles (~1 ?m) and exosomes (30-150 nm) which have been recognized as the presentation of cell origins and exhibit intrinsic payload for mediating immune-responses and tumor microenvironment. However, current conventional 3D culture and exosome production protocols (e.g, ultracentrifugation and filtration) are un-scalable, often labor-intensive and time-intensive, and in low-yield and purity (< 25%). In this work, we report a novel microfluidic 3D cell culture approach for studying derived extracellular vesicles. The microfluidic integration of 3D cell-culture with downstream extracellular vesicle harvesting and molecular analysis provides a rapid and automated platform for investigating tumor behavior and microenvironment. Current microfluidic exosome isolation approaches either handle limited quality of exosomes in microliter scale, or the processed exosomes are bound to solid surface/particles and unable to stay intact. In this work, the imunomagnetically-captured exosomes are harvested on-demand at downstream, with a precise microfluidic control of continuous-flow over milliliter scale of volumes. We anticipate this technology could establish advanced tumor models for developing cancer therapeutics and diagnostics.

Spatio-temporal detection of arsenic in a dielectric barrier discharge by optical emission spectroscopy
Sebastian Burhenn, Jan Kratzer, Joachim Franzke

The detection of arsenic, primarily the quantification of its highly toxic inorganic compounds is of great interest due to its occurrence in many groceries such as rice and fish. By means of inductively coupled plasma mass spectrometry (ICP-MS) and atomic fluorescence spectrometry (AFS) low detection limits of arsenic have already been achieved. Compared to ICP-MS, applying optical emission spectroscopy (OES) to a capillary shaped dielectric barrier discharge (DBD) is a simpler and more cost effective method. However, the mechanism for the atomization and excitation of arsenic compounds in the DBD is not fully understood. Tracking the spatial and temporal development of the arsenic signal along the discharge axis will give insight into ongoing processes inside the plasma. This knowledge enables to adjust the discharge parameters and its design in order to improve the detection limit. In the presented setup arsenic in liquid samples is converted via hydride generation into volatile arsine which is introduced in the DBDI consisting of a glass tube with soldered electrodes. The advantage of hydride generation is the minimization of matrix effects and the isolation of arsenic(III) from the sample. The high temporal stability of the discharge enables to use a fast gated iCCD-Camera with nanosecond resolution to obtain spatial and temporal emission signals. The spectral resolution is provided by an interference filter at the arsenic 234 nm line mounted in front of the camera.

Comparison of LTP-, FAPA-, and LA-hFAPA-MS for Direct Surface Analysis
Christopher Kuhlmann, Jacob T. Shelley, Carsten Engelhard

Ambient desorption/ionization mass spectrometry (ADI-MS) has emerged as an important analytical field for direct sample analysis since 2004, when desorption electrospray ionization (DESI) was introduced [1]. The main goal of ADI-MS is to minimize sample preparation procedures (e.g., solvent extraction) and to overcome the lengthy analysis times, often encountered with chromatographic separations, by directly desorbing the target analytes from the sample surface. To meet this task, high-resolution mass spectrometry is used to both separate analytes and detect them, with the aim of being able to differentiate a diverse mixture of analyte and matrix molecules in a sample. While more than 50 ADI sources have been descried over the years little work has been done to compare the different desorption and ionization processes to designated applications. In the current study, three plasma-based ambient desorption/ionization sources, the low-temperature plasma (LTP) probe [2], flowing atmospheric-pressure afterglow (FAPA) [3], and laser ablation halo-flowing atmospheric-pressure afterglow (LA-hFAPA-MS) [4] will be compared for their applicability to selected analytical problems. Consequently, desorption and ionization properties for selected analytes will be discussed according to the specific sample matrices. In addition, the capabilities of the tested ADI-sources for molecular imaging will be also explored. First, LTP- and FAPA-MS will be compared for direct detection of active agents in pharmaceutical tablets on thin-layer chromatography (TLC) plates. In this study, different stationary phases for TLC, e.g., normal-phase and reversed-phase, will be compared for their capability to be coupled with these ADI sources. Additionally, molecular mapping of separated mixtures via TLC-ADI-MS will be used to enable faster analyte screening by avoiding organic solvents and time consuming extraction processes. Consequently, this analytical tool will be used for quantitative analysis of selected analytes via isotope dilution analysis. Second, the performance of different plasma-based ADI sources for molecular mapping of liquid crystals and additive materials inside a liquid crystal display (LCD) will be compared. For this purpose, LTP- and FAPA-MS were used to perform fast molecular screening with minimal or no sample preparation. Additionally, LA-hFAPA-MS is used to investigate small sample spots with a spatial resolution in the micron-range. Therefore, laser ablation enables high spatial resolution and reproducible sample introduction, while halo-FAPA opens the door to a wider range of detectable analytes due to a lager variety of ionization mechanisms like Penning ionization, proton transfer, charge transfer, and ion or electron attachment, in contrast to conventional atmospheric-pressure ionization sources (e.g., electrospray ionization, or atmospheric-pressure chemical ionization). [1] Takats, Z.; Wiseman, J.M.; Gologan, B; Cooks, R.G. *Science* **2004**, 306 (5695), 471–473. [2] Harper, J.D.; Charipar, N.A.; Mulligan, C.C.; Zhang, X.; Cooks, R.G.; Ouyang, Z. *Anal. Chem.* **2008**, 80 (23), 9097–9104. [3] Andrade, F.J.; Shelley, J.T.; Wetzel, W.C.; Webb, M.R.; Gamez, G.; Ray, S.J.; Hieftje, G.M. *Anal. Chem.* **2008**, 80 (8), 2646–2653. [4] Shelley, J.T.; Ray, S.J.; Hieftje, G.M. *Anal. Chem.* **2008**, 80 (21), 8308–8313.

Recent Advances in Solution-Cathode Glow Discharge Mass Spectrometry
Andrew Schwartz, Jacob Shelley, Courtney Walton, Kelsey Williams, Gary Hieftje

Rising to prominence in the last two decades, atmospheric-pressure glow discharges where one or both of the electrodes is replaced with a flowing, conductive liquid have emerged as one of the most promising alternative plasmas for optical emission spectrometry (OES). In contrast to more conventional sources, such as the inductively coupled plasma (ICP), liquid-electrode glow discharges are compact, utilize low, direct-current power, require no compressed or flowing gases, and enable the direct analysis of liquids, obviating the need for sample-solution nebulizers. Though many geometries of solution-electrode glow discharges exist, one design, known as solution-cathode glow discharge (SCGD), has engendered significant attention due, in part, to its analytical performance. In addition to the earlier-described benefits, the SCGD exhibits analytical figures of merit that rival radially viewed ICP-OES, where sub or single part-per-billion detection limits are achieved for several elements Though the SCGD has been extensively investigated as a source for OES, its application as an ion source for mass spectrometry (MS) is in its formative stage of study. Despite this, early investigations have revealed that the SCGD shows great promise as an ion source for atomic, molecular, and biological mass spectrometry. Analytes, introduced in the flowing liquid, as an ambient gas, or as a solid held near the plasma, are readily vaporized and ionized by the discharge, resulting in mass spectra that contain species that range from elemental ions and their adducts to both intact ions and fragments of small organic molecules and labile biomolecules. Most excitingly, the "hardness" of ionization is tunable through simple adjustment of SCGD current. This enables either intact molecular ions or fragments (useful for structural elucidation) to be preferentially formed within the discharge. In this presentation, we will provide an overview of the early SCGD-MS studies and also present the most recent work towards its expanding role as a versatile, inexpensive ion source.

Filamentary Discharges in DBDI Lead to Soft Ionization for Mass Spectrometry

Luzia Gyr, Felix David Klute, Joachim Franzke, Renato Zenobi

Atmospheric pressure plasma ionization coupled with mass spectrometry is a widely-used technique due to its capability for fast, direct and sensitive detection of organic molecules. Many different API methods have been developed, but there is still a lack of understanding their ionization mechanisms. Fundamental studies of the plasma ionization source would push its development forward, allowing, for example, to overcome matrix effects and difficulties for quantification. In this work, the active capillary plasma ionization source based on a dielectric barrier discharge was characterized using optical emission spectroscopy. It contains a stainless steel capillary inserted into a glass capillary, working as an electrode, and a copper ring surrounding the glass capillary, functioning as the counter-electrode. A cold plasma is generated by applying a sine modulated high voltage to the outer electrode and using nitrogen or air as discharge gas. All optical measurements were performed line of sight facing the plasma source using either a ns-gated iCCD camera or UV-VIS spectrometer (commercial Ocean Optics USB 2000 or Echelle) in the range of 200-850 nm. The active capillary plasma ionization source was characterized by optical emission spectroscopy and imaging. The filamentary nature of the plasma was confirmed by taking an iCCD image of the plasma on time scales shorter than a single discharge period. Filamentary discharges are produced by direct ionization of nitrogen, since no Penning ionization is possible. The filaments lead to a high carrier density, although due the geometry of the plasma source, it is still applicable for soft ionization. Further, the properties of the filaments were study by monitoring the transitions of NO (A2? --> 2?) and N2 (2nd positive system and Meinel band) using optical emission spectroscopy. In conclusion, the characterization of the plasma not only helped to improve the understanding of the ionization mechanism but also allowed to optimize the ionization performance of the plasma source when coupled with the mass spectrometer.

Development of an Ambient Air Micro Discharge for soft Ionisation – The Flexible Tube Micro Plasma Ionisation (FTμPI)

Sebastian Brandt, Felix David Klute, Alexander Schütz, Carolin Drees, Wolfgang Vautz, Joachim Franzke

Plasma based ambient air ionisation techniques for mass spectrometric applications became a popular and wide spread method over the last decade. A challenge that often occurs for these applications is the coupling between the ionisation source, sample introduction like GC/HPLC and the detection device due to e.g. incompatible gases or gas flows, distancing or dimensioning. These problems might lead to increased analyte losses or insufficient ionisation efficiencies reducing the overall signal yield. The flexible tube micro plasma ionisation (FTμPI) is an attempt to create a highly efficient discharge that can be coupled easily to different detection devices such as mass spectrometers (MS) or ion mobility spectrometers (IMS) among a very small footprint on the experimental arrangement. It will be shown that the FTμPI can be built easily from commercial available material creating a very small, low cost, robust and low consumption discharge. The voltage and plasma carrying parts are completely insulated by a tubing to reduce the operation risk for the user and the discharge. The small dimensions and its flexible nature make the coupling to chromatography and detection devices a simple task as well. The presentation will concentrate on the geometrical features and on a comparison to different established ionisation sources such as DBD or LTP. Although the discharge diameter is reduced drastically, the ionisation efficiency of the new FTμPI is increasing due to the new discharge design and better coupling properties. A coupling with a GC will be shown. In respect to the special geometry of the discharge the question of reactive species provoking soft ionisation will be discussed as they can be transported over distances of several centimetres keeping the properties for soft ionisation without fragmentation. Furthermore a short introduction of future applications will be given.

1D- and 2D-Time-resolved Spectroscopic Studies in Conventional and Supercritical Fluids
Mike George

The combination of photochemistry and fast IR spectroscopy in conventional and supercritical fluids is used to shed new light on chemical processes. In particular we will use both ultrafast 1D- and 2D-IR spectroscopy as a powerful tool for examining the relationship between structure and function of biomolecules and the excited states and reactive intermediates formed in catalytic systems.

Probing the instantaneous ion binding configurations K+ ion channel selectivity filter using 2D IR spectroscopy
Huong Kratochvil, Martin Zanni, Alvin Annen, Kim Matulef, Jared Ostmeyer, Eduardo Perozo, Benoit Roux, Francis Valiyaveetil, Martin Zanni

Potassium ion channels facilitate the flow of potassium across cellular membranes, which have many important implications in physiological processes such as signal transduction and osmotic regulation. Recent advances in two-dimensional infrared (2D IR) spectroscopy have enabled us to probe these membrane ion channels and address questions about their mechanisms of conduction with residue specificity and at ultrafast timescales. We use a combination of protein semisynthesis, 2D IR spectroscopy, and MD simulations to address the longstanding question of ion permeation through the filter of a potassium ion channel. By incorporating heavy 13C=18O isotope labels into specific sites in the selectivity filter of a semisynthetic KcsA channel, we probed the ion binding configurations with 2D IR spectroscopy. With the help of MD simulations, we were able to test against existing models of ion permeation and show that ions and water alternate through the filter, effectively ruling out mechanisms in which ions occupy adjacent binding sites. Additionally, we show that the changes in the intracellular gate of the channel underlie changes in ion affinity of the filter. Our work reveals that the ion occupancy in the S1 binding site is sensitive to the opening and closing of the intracellular gate, pointing to complexities in the conformational dynamics of the selectivity filter and its dependence on the state of the intracellular gate.

2D IR Vibrational Probe Pairs for Determining Structure and Dynamics in Biomolecules
Matthew Tucker

2D IR spectroscopy has been utilized to directly observe of conformational dynamics of biological systems for both equilibrium and non-equilibrium pathways. Observing specific structural changes associated of active biomolecules provides insight into the molecular movements leading to their functionality. Well-positioned probe pairs provide global structural rearrangement while simultaneously detecting the dynamics at key contact points where biological function takes place. A variety of 2D IR probes, including cyano-modes, ring modes, and azido- modes [1] will be discussed, focusing on the coupling between different pairings of these vibrational reporters as well as their sensitivity to frequency fluctuations resulting from the local environment changes. The associated mechanisms of coupling will be discussed for each coupled pair. [1] Andrew J. Schmitz, David G. Hogle, Xin Sonia Gai, Edward E. Fenlon, Scott H. Brewer, and Matthew J. Tucker, J. Phys. Chem. B 120, 9387-9394 (2016).

Probing systems at the interface of biology and nanotechnology with 2D IR spectroscopy
Lauren Buchanan

A fundamental question in biochemistry is how protein structure changes and affects function. Two-dimensional infrared (2D IR) spectroscopy is a powerful tool for studying molecular structure and motion. When implemented with mid-IR pulse shaping in conjunction with isotope labeling, 2D IR spectroscopy is capable of providing residue-specific structural resolution of protein dynamics in real time. In this talk, I will describe initial efforts in my research group to use this experimental approach to systematically examine the relationship between sequence and structure for short peptides. These studies will provide insight into the mechanisms of folding and self-assembly of synthetic peptide sequences and facilitate the rational design of peptide nanomaterials. Additionally, I will discuss work in the group to precisely determine the structural changes that occur when proteins adsorb onto nanoparticles. We aim initially to correlate these structural rearrangements to adverse biological responses observed in studies of nanomaterials being explored for biomedical applications, then ultimately to predict responses for novel materials.

Direct Charge Transfer across Organic Semiconductor and Metal Interfaces
Wei Xiong, Bo Xiang, Yingmin Li, Huy Pham, Francesco Paesani

Direct charge transfer across molecular interfaces promotes charges from donors to acceptors in one step and can enable new mechanisms for molecular electronics, light harvesting and photocatalysis. Despite its potential, its experimental demonstration and understanding are rare, which has constrained applications of direct interfacial charge transfer in energy materials. To probe charge transfer at hidden interfaces, we implement transient vibrational sum frequency generation (VSFG) spectroscopy to gain conformational specific insights on interfacial charge dynamics. We observe transient electric-field-induced VSFG signal, which indicates a charge transfer process at spin-coated prepared organic semiconductor-metal interfaces. The photon-energy and polarization dependence of pump pulses further indicates that electrons are directly transferred across interfaces. Frequency-resolved transient spectra reveal that, even though the interfaces are prepared without deliberate orientation control, a sub-ensemble of surface molecules can adopt orientations for direct charge transfer. Experimental observations are supported by density functional theory calculations, which show that although many orientations can exist at interfaces, at parallel orientation, electronic orbitals are delocalized across interfaces. The observation of direct charge transfer at complex interfaces paves the way for a new foundation of implementing direct charge transfer in energy materials.

The Chemistry Diversity Initiative: A Graduate Student Program for Success at Purdue University
Jean Chmielewski, Colby Adolph, Stella Betancourt, Reena Blade, Chris Pulliam

The Department of Chemistry at Purdue University has a rich tradition of diversity in our graduate student population. Over 100 graduate students from groups that are underrepresented in science have obtained Ph.D. degrees in our program. The Chemistry Department at Purdue has been ranked 2nd in the nation for the percentage of underrepresented minorities (URMs) in our graduate student body and was ranked #1 in the country for the growth in the number of URM students in our program from 1995-2009.[1] More recently, however, we have observed a significant reduction in the number of URM students entering our graduate program. To address this alarming problem, we have instituted a multi-prong approach for the recruitment and retention of graduate students from URM groups in the Chemistry Department. Our Chemistry Diversity Initiative (CDI) plan has 3 phases focused on attracting applicants from targeted universities, mentoring students through the acceptance process and a mentoring plan designed for the successful matriculation and professional development of these students through our graduate program. We are closely monitoring the success of these initiatives, as we believe the CDI will act as a model for success for departments across Purdue and other universities. [1] Laursen, S.L.; Weston, T.J., J. Chem. Ed., 2014, 91, 1762-1776.

Retaining underrepresented groups at both the undergraduate and graduate level
Susan Olesik, Terry Gustafson, Thomas Magliery

Our department and other college programs have launched extensive efforts to recruit and retain underrepresented students at the undergraduate and graduate levels. These efforts include increasing mentoring and tutoring of students, modifying courses so that they include primarily active learning methods as well as providing additional support systems to retain at risk students. Data on the progress on many of these initiatives will be discussed in this talk. In addition illustrations of the resultant student successes will be provided.

Global initiatives for conducting medicine anti-counterfeiting research: Working with multidisciplinary backgrounds worldwide
Sulaf Assi

Medicine anti-counterfeiting is a global expanding problem affecting all countries worldwide. The problems associated with the use of counterfeit medicines range from ineffectiveness at their easiest and lethal effects at their worst. Despite advances in technology worldwide, a single approach could not address counterfeiting problem. This is particularly an issue as medicine counterfeiting can occur with any class of medicine, any formulation type and contain different impurities. Tackling counterfeiting problems needs a multidisciplinary global effort with diverse field of expertise among academia, industry and manufacturers. This work presents global strategies and approaches to anti-counterfeiting among diverse researchers. The strategies operate at different levels including: identification/monitoring, tackling and dissemination. Identification of the spread of counterfeit medicines was conducted by monitoring the web on regular basis and by carrying out prospective studies in different countries (questionnaires, interviews). Tackling the problem of counterfeit medicines was undertaken at two levels. The first level was by developing rapid techniques for identifying fake medicines, and the second by evaluating awareness campaigns for the public regarding the dangers arising from the use of counterfeit medicines. Hence, the second level contributed to dissemination of the work not only among the scientific audience but also to the public (including patients and healthcare professionals). This was accomplished via scientific articles, online magazines, leaflets and social media platforms. Overall, the approach was successful especially with the diverse expertise and culture of the researchers that contributed to disseminating the information to larger audience.

Diversity, Equity, and Inclusion in Analytical Chemistry
Megan Schmale

How can graduate students promote diversity, equality, and inclusion in chemistry? How can you influence students to become involved, and maximize the impact you have on advancing diversity and inclusion? From the perspective of a graduate student, I will discuss my personal experiences in promoting diversity, the importance of strategy, and how to encourage and maintain student involvement while growing this community.

Label-free tissue classification by QCL based IR-Imaging
Klaus Gerwert

Infrared imaging is an emerging tool for label-free, non-invasive characterization of tissue, cells and body fluids.1 The approach is applied to fresh frozen and formalin fixed paraffin embedded tissue samples. For the entities colon, bladder and lung, spectral data bases are established to characterize tissue in an automated workflow with sensitivity and specificity of over 95%.2 For the therapeutic decision of the clinicians a differential diagnosis is essential. Therefore, in an important next step a predictive differential diagnostics is provided. We were able to differentiate between the predictive subtypes of the adenocarcinoma in lung cancer.3 Furthermore the grading of colon cancer could also be classified.4 In the next step the label-free annotated tissue is dissected by laser microdissection and the homogenous samples are analyzed by proteome with the MPC to provide beside spatial also molecular resolution. This is successful applied to mesothelioma subtypes.5 All currently used biomarkers in immunohistochemistry in pathology today are identified. This validates the power of this approach. In the next step a QCL-based microscope is used. It shows the same quality of spectra, but measuring-time is drastically reduced by a factor of 90 below 30 min. This improvement opens avenues for clinical applications. 1 Gerwert K., Großerüschkamp F., Ollesch J. (2014), SPIE Newsroom, DOI: 10.1117/2.1201-312.005297. 2 Kallenbach-Thieltges A., Großerüschkamp F., Mosig A., Diem M., Tannapfel A., Gerwert K. (2013), J. Biophotonics, 6(1), 88-100. 3 Großerüschkamp F., Kallenbach-Thieltges A., Behrens T., Brüning T., Altmayer M., Stamatis G., Theegarten D., Gerwert, K. (2015), Analyst, 140(7), 2114-20. 4 Kuepper C., Großerueschkamp F., Kallenbach?Thieltges A., Mosig A., Tannapfel A. and Gerwert
K. (2016), Faraday Discussion, DOI 10.1039/C5FD00157A. 5 Großerueschkamp, F., Bracht, T., Diehl, H. C., Kuepper, C., Ahrens, M., Kallenbach-Thieltges, A., Mosig, A., Eisenacher, M., Marcus, K., Behrens, Th., Brüning, Th.,

Theegarten, D., Sitek, B., Gerwert, K. (2017), Scientific Reports, 7, 44829, DOI:10.1038/srep44829

Quantum Cascade Laser (QCL) hyperspectral imaging applied to live cell analysis
Bayden Wood, Dale Christensen, Ellen Lowry, Philip Heraud, David Perez-Guaita

Quantum Cascade Laser (QCL) microscopy combines the power of a laser source with an ultra fast infrared (IR) detector to perform real time live imaging at discrete wavenumber values. While a number of studies in the realm of biology have focused on tissue sections, few studies have investigated the potential of the technique for live cell imaging. The power of the laser, which is approximately 6 times brighter than a conventional infrared beamline on a fourth generation synchrotron, can easily punch through two calcium fluoride windows, media and a large cell. In this study we demonstrate the power of the technology to image the giant *Micrasterias sp.*. from the order Desmidiales. We compare fixed and live cells and expose the cells to different environmental stresses including UV light, nutrient starvation and dehydration. Hitherto almost all IR QCL images have been generated by either discrete wavenumber values or alternatively a ratio of two wavenumber values. Here we demonstrate for the first time hyperspectral QCL IR imaging where we combine the the discrete wavenumber values into a spectrum in multivariate space and use this approach to generate multi-composite IR images that can resolve subcellular features in the giant algal cell. The combination of live cell imaging using QCL microscopy and hyperspectral image reconstruction opens the door for wide ranging applications of the technology in the biological sciences.

On-Chip Quantum Cascade Laser/Detector System for Remote Gas Sensing

Rolf Szedlak, Andreas Harrer, Benedikt Schwarz, Martin Holzbauer, Johannes Paul Waclawek, Hermann Detz, Aaron Maxwell Andrews, Werner Schrenk, Bernhard LendlGottfried Strasser

Infrared spectroscopy is a reliable tool for sensing applications in various fields from industry to medicine. Most molecules exhibit distinct absorption spectra in the molecular fingerprint region which increases the demand for powerful light sources in this spectral range. Quantum cascade lasers (QCLs) have proven to be adequate candidates providing tailorable emission characteristics in the mid-infrared and terahertz spectral region. Therefore, these lasers have been successfully implemented in several spectroscopic techniques. The typical sensing setup consists of a light source, an analyte interaction region and a separate external detector. We present a novel sensing concept which combines laser and detector on a single chip. Bi-functional quantum cascade heterostructures are utilized in order to compensate the intrinsic blue-shift of a QCL in detector mode. The resulting spectral overlap provides efficient coupling between laser and detector. Our concept consists of two concentric monolithically integrated ring structures with a second order distributed feedback grating. The latter facilitates vertical light emission as well as detection of vertically incident light. Hence, these rings can be operated as lasers when biased and as detectors when unbiased. In the sensing configuration one of the rings is used as the laser and emits light which is guided through a gas cell before it is back-reflected by a flat gold mirror. The back-reflected light traverses the gas cell once again and is incident on the sensor chip where it is detected by the other ring. We present proof-of-concept gas measurements of isobutane and isobutene demonstrating the working principle of our compact sensor concept. Combination of surface-active lasers and detectors on a single chip enables a crucial miniaturization of sensing systems and could pave the way for compact hand-held remote sensors.

Time-resolved spectroscopy of biological samples using QCL dual-comb technique
Markus Geiser, Markus Mangold, Pitt Allmendinger, Andreas Hugi, Filippos Kapsalidis, Pierre Jouy, Jérôme Faist

Quantum Cascade Lasers (QCLs) can be locked into frequency combs, which are very powerful tools for spectroscopy, combining high spectral bandwidth with fast acquisition by employing multi-heterodyne dual-comb spectroscopy. We demonstrate an integrated mid-infrared spectrometer combining a time resolution of 1 µs with an optical bandwidth of 85 cm-1 centered at a wavelength of approx. 8.3 µm. The spacing between optical elements is on the order of 0.5 cm-1. This development is well suited for time-resolved spectroscopy on biological samples in the mid-infrared range, where comb operation has been challenging to achieve with non-QCL frequency combs. As a benchmark, protein structural changes, specifically the formation of the light-activated L, M and N-states in bacteriorhodopsin – which only have µs to ms lifetimes – have been recorded in a single shot. The spectra as well as the kinetics show good agreement with those from step-scan FT-IR measurements. This approach greatly reduces the complexity of time-resolved bio-spectroscopy measurements in the mid-infrared currently requiring many repetitions.

Field-resolved spectroscopy in the mid-infrared region: A highly sensitive tool for molecular fingerprinting

Marinus Huber, Wolfgang Schweinberger, Liudmila Voronina, Syed Ali Hussain, Christina Hofer, Michael Trubetskov, Mihaela Zigman, Ferenc Krausz, Ioachim Pupeza

Mid-infrared (MIR) vibrational spectroscopy is a versatile tool for identifying molecular fingerprints of organic compounds (OC) in gases or liquids. With frequency-domain methods, measuring a small concentration of an OC requires the ability to detect a correspondingly small change of the source intensity. Consequently, relative intensity noise (RIN) of the source limits the detection sensitivity. In this contribution, we demonstrate that field-resolved spectroscopy (FRS), based on phase-stable MIR femtosecond pulses and coherent detection can overcome this limitation. As an inherent time-domain method, FRS allows for the detection of the free-induction decay of the OC under test in a background-free manner. To test the performance of FRS for sensing applications, a glucose dilution series is measured and a more than 10 times lower limit of detection is obtained compared to a state-of-the-art FTIR spectrometer (Bruker Vertex 70). An outlook on measurements with even higher sensitivity and a discussion of the specificity of FRS will be given at the conference.

Digitalization of Drill Core Samples for Mining Exploration using LIBS

Francois Doucet, Lutfu Ozcan, Altan Muftuoglu, Dominique Doucet

Detailed geological models and mineralogy knowledge greatly assist the mining industry in locating mineral resources, and can reduce the need for expensive drilling, saving time and money. LIBS logging technology provides a highly selective and sensitive means of gaining this information from materials and drill cores derived from drilling. Until recently, geologists, metallurgists, and geotechnical engineers had to manually interpret and record minerals found in these materials. This manual method – the visual interpretation of drill cores – is labor intensive and subject to human error. LIBS technology enables a systematic and selective way
– based on the physics of atomic emission spectroscopy (AES) – to minimize or eliminate human error via automation. Small gold nuggets, which cannot be seen by the human eye (i.e. less than 150 µm), are revealed by LIBS imaging. An intelligent, automated instrument that can flag gold nuggets and map different geochemical deposits in drill cores has been developed for the mining industry. When benchmarking this drill core analyzer against existing technology, it is possible to see the significant advantage that can be achieved with this technology in the mining industry. First, LIBS imaging provides the highest selectivity for elemental imaging through atomic emission spectroscopy. Second, the surgical resolution of laser ablation, the ease of scanning a large area and adapting to any sample shape and size, makes the instrument the best solution for permanent core logging. Hence, logged data can be used to discover new exploitable mines in the future using data mining, when other element demand becomes economically viable to extract. LIBS logging technology has the potential to eliminate the need for keeping drill core samples in core shacks as prescribed by the authorities. The results obtained using this drill core analyzer enable the logging of 200 to 1000 meters of drill core samples per day. The results presented in this paper demonstrate that miniscule gold nuggets can be easily outlined on drill core samples coming from different types of mineralogical assemblages (e.g. pyrite, marcasite, arsenopyrite). In addition, the results show that different typical geochemical signatures can be identified using an intelligent algorithm. The high

throughput analysis demonstrated here showcases LIBS logging technology as a potential game changer in the field of mining exploration.

Mapping Analyte Distributions in Surrogate Nuclear Melt Glass Using Laser-Induced Breakdown Spectroscopy and Micro X-Ray Fluorescence

Michael Shattan, Ashley Stowe, Kathryn McIntosh, John Auxier II, Christian Parigger, Howard Hall

Multiple organizations are producing surrogate nuclear melt glass beads to support the needs of the nuclear forensics community. Mapping of analytes in surrogate nuclear melt glass can provide insight into chemical migration during the melting and cooling process, as well as provide information on the process's consistency and repeatability. Melt glass samples were produced using recipes developed by the University of Tennessee that simulate debris from hypothetical nuclear detonations in New York City, NY and in Houston, TX. The samples were made into thin sections and the distribution of analytes mapped using both Laser-Induced Breakdown Spectroscopy (LIBS) and Micro X-Ray Fluorescence (micro-XRF). The melt glass is visually heterogeneous and the results of the both mapping techniques are used to quantify the level of heterogeneity of each analyte. The paper compares the utility of these two techniques for melt glass production monitoring.

Laser-induced breakdown spectroscopy in analysis of elemental distributions

Saara Kaski, Sari Romppanen, Heikki Häkkänen

The expertness of laser-induced breakdown spectroscopy (LIBS) in the fast analysis of untreated samples is well demonstrated in the construction of 2D and 3D maps of elemental distributions. Information based on the atomic emission, which can be observed from most of the elements in periodic table, can be used to deduct the sample composition. Material distributions in paper and paper coatings are important factors which determine the characteristics and quality of paper. For the construction of a 3D map, high resolution especially in the depth is needed to obtain representative sampling from the thin layers of different coatings. Thus, the microLIBS approach using low irradiance of 0.1 GW/cm^2 has been developed in University of Jyväskylä already in the 1990s. It has been demonstrated that the spatial distributions of coating pigments effectively show e.g. the changes in the spreading of the coating mixtures in varied industrial conditions. LIBS spectra can be used extract the mineralogical information of rock samples. The detection range has been extended to the vacuum ultraviolet (VUV) and near infrared (NIR) spectral regions. A 2D map representing the mineralogy and geological texture can be used to discriminate and visualize rock in the macroscopic scale. The grain size and shape of the ore minerals in addition to the location of the gangue minerals can complicate the further beneficiation of the ore (grinding, enrichment, flotation etc). Especially in the fine-grained areas the macroscopic identification of minerals may be very challenging, even impossible and the complementary methods may be time-consuming or the instruments are expensive. The chemical variations in minerals are very common and LIBS also offers possibility to detect the chemical alterations.

Precision Nanopipetting for Single Cell MALDI

Kermit Murray, Fan Cao, Fabrizio Donnarumma, Randy Duran, Jean-Baptiste Decombe

Matrix-assisted laser desorption ionization (MALDI) mass spectrometry is a powerful tool for the rapid analysis of biomolecules from complex biological samples with minimal sample preparation. We are developing methods for precise sampling of the contents of isolated single cells and deposition of the material onto small matrix spots for MALDI analysis. The MALDI matrix is deposited on a conducting microscope slide using matrix sublimation under high vacuum to create a deposit of precisely controlled thickness. Matrix spots 20 to 50 µm in diameter are created either by sublimation and laser ablation of excess matrix material or by sublimating matrix with a thin foil pinhole mask. Single cells are isolated using an automated single cell sorting microscope (MMI CellEctor) and their aspirated contents are deposited on a matrix spot on the microscope slide. Cellular material is drawn into a 20 μm inner diameter pulled capillary and volumes between 10 and 50 nL are typically sampled. The liquid is deposited onto the slide matrix spot array and allowed to dry. The spots are analyzed on a Bruker Ultraflextreme MALDI time-of-flight mass spectrometer with a 1 kHz repetition rate 355 nm Nd:YAG laser that is focused to match the matrix spot size. Initial experiments have demonstrated a limit of detection a peptide standard of 20 zmol range with negligible spreading of the deposited liquid on the matrix spot. The sensitivity is being tested on isolated single cells to demonstrate the ability to detect peptides, proteins, and other biomolecules. The overall goal is to develop a method for sampling of isolated cells onto an array of MALDI matrix pedestals for rapid single cell profiling.

Single-cell Mass Spectrometry: From Subpopulations to Subtypes
Thanh Do, Troy Comi, Stanislav Rubakhin, Sage Dunham, Jonathan Sweedler

Even when derived from identical genetic blueprints, single cell heterogeneity exists in morphologically homogeneous cell populations. The classification of cell subtypes and the search for rare cells, which for decades were akin to finding a needle in a haystack, have become tractable with the emergence of high-throughput and sensitive measurement techniques. The Sweedler Laboratory has developed an optically-guided mass spectrometry method suitable for volume-limited samples such as single cells. In this talk, I will discuss our recent success to characterize hundreds to thousands of individual cells using two mass spectrometry platforms: secondary ion mass spectrometry (SIMS) and matrix assisted laser desorption / ionization. We have shown that a range of cell sizes, from *Aplysia californica* neurons larger than 75-?m, to 20-?m dorsal root ganglion (DRG) cell bodies, to 7-?m rat cerebellar neurons and single islet cells is suitable for our techniques. I will focus on single cells dissociated from DRG, an important component of the peripheral sensory-motor system. Using SIMS, we illustrate the efficacy of our approach to classify single cell populations and subpopulations based on lipid and metabolite contents. With MALDI, we can gain additional information on lipids, peptides and even small proteins. The ability to interrogate individual cells and to perform high throughput analysis would provide invaluable information on how networks of cells and associated supporting cells can work together to confer emergent properties.

Cell-by-cell Metabolic Analysis of the Xenopus laevis embryo
Erika Portero, Rosemary Onjiko, Sally Moody, Peter Nemes

Discovery-based chemical characterization of the metabolome in single-cells holds the potential to advance our understanding of molecular processes taking place during normal and impaired development of the vertebrate embryo. The development of next-generation mass spectrometry (MS) technology capable of exceptional sensitivity allows for the interrogation of metabolic activity among single-cells. Recently, we developed and validated an in situ capillary microsampling technique which is minimally intrusive to the normal development of the live embryo. Here, we demonstrate that microsampling with capillary electrophoresis (CE) electrospray ionization (ESI) MS is able to detect metabolic activity differences between cells in the same embryo. Our system consists of a stereomicroscope used to identify dorsal and ventral cells located on the left side of the 16-cell *Xenopus laevis* embryo. Using fabricated borosilicate microcapillaries, a 10–15 nL portion of the cell content were aspirated from each cell and immediately transferred to a metabolite extraction solvent. Metabolites in the cell extract were separated and ionized using a custom-built microanalytical CE-ESI system and detected using a time-of-flight high-resolution mass spectrometer. This strategy allowed us to measure metabolites with a 60-amol lower limit of detection and a quantitative reproducibility of <20% relative standard deviation. About 250 molecular features (with unique m/z values and migration time) were detected between the cells, 55 of which were identified as metabolites based on their accurate m/z, migration time, and tandem MS comparison to chemical standards and databases. Multivariate analysis by PCA and HCA supplemented with statistical analysis (ANOVA) of the resulting metadata revealed metabolic differences between the cells. For example, dorsal cells (precursors to the brain) were found to contain higher amounts serine than ventral cells. In agreement, serine is a known neurotrophic factor during neuronal development. Diverse metabolite gradients were found among the probed cell types, which now can be used to design functional studies to help understand the implications that metabolism has during early embryogenesis.

Quantification of Ant-Cancer Compounds in Single Bladder Cancer Cells using the Single-probe MS Technique
Shawna Standke, Ning Pan, Naga Rama Kothapalli, Anh Le, Anthony Burgett, Zhibo Yang

Quantification of drug compounds in cells is traditionally done using a cell lysate prepared from large cell populations, and only averaged results are obtained. Single cell mass spectrometry (SCMS) is a powerful tool that has the potential for creating personalized treatments for cancer patients since it can be used to quantify the amount of anti-cancer compounds in individual cells. This study uses the Single-probe, a microscale sampling and ionization device, coupled with a mass spectrometer to estimate concentrations of various anti-cancer compounds inside of live, single bladder cancer cells. Human bladder cancer cell lines (T24 and SW-780) are used as models. The cell lines are grown in microwells on glass chips, and they are treated with anti-cancer compounds under various conditions. Cells are rinsed with fresh media to avoid the detection of drug molecules in the treatment media, and they are analyzed using the Single-probe SCMS technique. Both cell selection and penetration are monitored using microscopes. An internal standard, such as an isotopically-labeled form of the anti-cancer compound, is added into the sampling solvent at a known concentration. Drug compound and internal standard are simultaneously detected in experiments. The amount of anti-cancer compound inside individual cancer cells is estimated by considering the signal time, concentration of internal standard, flow rate, cell volume, and the intensities of both the analyte and internal standard. The accumulation of the anti-cancer compound can be seen by monitoring different time points. Since metabolites can also be monitored, this technique can be used to determine the effectiveness of patients' dosages in the future.

Cellular and Sub-cellular Level Localization of Lipids and Metabolites Using Two- and Three-Dimensional High-Spatial Resolution MALDI Mass Spectrometry Imaging

Maria Emilia Dueñas, Adam Klein, Liza Alexander, Marna Yandeau-Nelson, Basil Nikolau, Jeffrey Essner, Young-Jin Lee

Mass spectrometry imaging (MSI) has recently advanced to allow the visualization of metabolites at cellular and sub-cellular resolution, unveiling unprecedented details of the metabolic biology. Recently, we have achieved high-spatial resolution MALDI-MSI down to 5 µm by modifying the laser optics in our linear ion trap-Orbitrap mass spectrometer. Here, we demonstrate two applications of this ultrahigh-spatial resolution MALDI-MSI system for the mapping of small metabolites at the cellular and sub-cellular level. In the application to maize leaves of two different inbred lines, B73 and Mo17, distinct heterogeneous cellular localizations were observed for phosphatidylglycerols (PG) between two photosynthetic cell types. For example, PG-32:0 is preferentially localized to the bundle sheath cells in B73, while a more uniform distribution between both the bundle sheath and mesophyll cells is present in Mo17. This is in contrast to PG-32:1, which is present almost homogeneously between the two cell types for both B73 and Mo17. Interestingly, maternal inheritance is exhibited within the reciprocal hybrid lines, B73xMo17 and Mo17xB73, such that the localization of B73xMo17 is similar to the distribution in the B73 parental inbred, whereas that of Mo17xB73 resembles the Mo17 parent. Also discussed will be the quantitative distributions of fatty acyl chains in PG and sulfoquinovosyldiacylglycerols (SQDG) along the developmental gradient from the four genotypes, studied by MALDI-MSI. This platform is also applied for the 3D MSI of newly fertilized zebrafish embryos. This is the first time MALDI-MSI has been applied for three-dimensional chemical imaging of a single cell. In order to visualize diverse classes of compounds, positive mode is performed for odd numbered sections using 2,5-dihydroxybenzoic acid as a matrix, while negative mode is used for even numbered sections using 1,5-diaminonaphthalene as a matrix. We will also discuss how to achieve reliable relative quantification in 2D- and 3D-MALDI MSI data, as well as localization changes of some lipids at various cell developmental stages. Overall, this presentation will demonstrate that high-spatial resolution MALDI-MSI analysis can be directly applied to multicellular plant or animal tissues to uncover cell-specific or sub-cellular metabolic biology that has not been possible using traditional metabolomics methodologies.

Comparison of lab, portable and process gas chromatographs for on-line analysis of R&D reactions
Eric Schmidt, Anna Sandlin, Linda Heinicke

Understanding the feed and product compositions of gas phase R&D processes can be challenging, especially for compounds with low boiling points or reactive species. Application of on-line analytical measurements to observe the chemistry near the point of reaction and in near real time can significantly enhance the ability to determine optimum yield, ensure process control, understand process upsets, and characterize unknown streams. This presentation will compare and contrast the application of three different on-line GC systems in an R&D environment including a portable or "micro" GC; Laboratory GC; and a process GC. All three systems are useful in providing the analytical information needed to understand, optimize, and probe chemical problems, but each has advantages and disadvantages. The results will illustrate the utility of on-line process GC in reducing the time necessary for process development and in many cases enable the discovery of unanticipated results that would have otherwise been dismissed.

Conventional gas detection technology and the challenges in industrial applications

John Wilson, Ulf Ostermann

A gas detector is a device that detects the presence of gases and vapors in an area, often as part of a safety system. The equipment can interface with a control system so a process can be automatically shut down or it can operate stand-alone often with audible and visual annunciation where it can sound an alarm to operators in the area where the hazard is present. Gas detectors can be used to detect combustible and toxic gases as well as oxygen depletion or enrichment. This type of device is used widely in industry and can be found in locations where processes involve such gases and vapors and the risk of leak or accumulation is possible. The detector uses alarm levels which are set to provide early warning of a potentially hazardous environment either from toxic or flammable conditions. For combustible gas detection these levels are established based on the Lower Explosive Limit (LEL) – the minimum concentration required to sustain combustion assuming the presence of an ignition source and sufficient oxygen. For toxic gas/vapor detection the levels are based on the occupational exposure limits established by NIOSH (National Institute for Occupational Safety and Health). Although many technologies are available, the most common and conventional are catalytic combustion, infrared and electrochemical. Understanding the limitations and challenges for each sensor technology and its applicability to the situation is critical to implementing a safety strategy. Each technology based on the mode of operation needs careful consideration particularly with respect to environmental conditions, cross sensitivities and potential for unrevealed failures.

Open Path UVDOAS Ambient Air Monitoring for Petrochemical Applications
William (Bill) Pearman, J.D. Tate

Real time measurements of critical process parameters has been and remains an important focus for Process Analytical Technology (PAT) across the petrochemical industry. However, ambient environmental responsibly challenges the PAT community to also look outside the process to detect and quantify contributions of those processes to the environment. Historically, Gas Chromatography using a distributed point collection systems has been then primary source for detection. This technology remains critical to the detection of contributions to the environment, but it lacks the ability to truly quantitate contributions outside the immediate process area. Open Path (OP) systems, coupled with the data from the GC data from localized points in the process area, does provide a much clearer picture of overall environmental effects. The data provided by GC-OP systems provides both a forensic – location and magnitude of an event – and predicative data – direction and magnitude of the event leaving the process area – to all concerned parties. This presentation focuses on the recent deployment of an Open Path UV Differential Optical Absorption Spectroscopic (UVDOAS) system as a perimeter monitor for a petrochemical production facility in the Gulf Coast region. Technology selection, installation, calibration, validation and overall performance to date will be discussed.

FT-MRR for trace chemical detection and chiral characterization in reaction solutions
Justin Neill, Brent Harris, Robin Pulliam, Matt Muckle, Shelby Fields

FT-MRR millimeter wave and microwave instrumentation is a recent addition to the spectroscopy toolbox for process analysis. It is a long wavelength, high resolution technique that generates spectral fingerprints from the quantized angular momentum states of freely rotating gas phase molecules. The fundamental advantage for chemical analysis is the specificity of FT-MRR spectra to molecular structure and the high selectivity in complex mixtures. Isomers, diastereomers, and enantiomers can be distinctively analyzed. Common interferences or weak absorbers in laser-based spectroscopy techniques do not pose the same challenge for FT-MRR spectroscopy. We will present applications of FT-MRR spectroscopy to the trace detection of important manufacturing impurities in the chemical and pharmaceutical industries. The applications we present include: 1. gas phase contaminants like carbonyl sulfide, hydrogen sulfide, and ammonia with sub ppm detection limits 2. thermal evolution of residual solvents and genotoxic impurities from drug powders by automated library matching, and 3. chiral purity characterization in a continuous manufacturing process.

Comprehensive GC x GC Analyses Using a Process GC
Anna Sandlin, Ademola Idowu, Bill Winniford, Eric Schmidt, JD Tate, Linda Heinicke

Multi-component hydrocarbon analysis is a prevalent gas measurement made within the petrochemical industry. Typically, these analyses are accomplished using process gas chromatographs (GCs). Process GCs generally operate in a production environment and have higher electrical certification requirements (e.g. Class I Div II certification) vs. general laboratory purpose instruments such as benchtop GCs. In addition, many common lab GC characteristics such as high signal detection frequencies compatible with narrower capillary column peak widths as well as temperature and pressure programming during an analysis are not designed into a typical process GCs. A comprehensive, two dimensional (2D) approach could be leveraged to simplify complex chemical analyses with benefits in both reduced analysis time and detection limits. Some of the identified shortcomings with current process technology (1D) include: speed of analysis (could be up to 1 hour for complex sample) and limitation on the number of identifiable/measurable components. Potential benefits of a process 2D GC include faster analysis, improved separation power, improved detection limit, and the ability to perform chemical group mapping. The goal of this technology development effort was to demonstrate concept feasibility of converting a standard process GC to be capable of comprehensive GC x GC analyses.

Protein Dynamics via a Non-perturbing Site-specific Infrared Probe
Farzaneh Chalyavi, Andrew Schmitz, David Hogle, Matthew Tucker

The ability to detect changes in the local electrostatic environment of proteins is pivotal to determining the dynamic nature involved in many biological processes. The utility of tyrosine ring breathing mode as a vibrational probe for protein dynamics is investigated by measuring its sensitivity to the changes in local electric field. The ring mode is influenced by direct and indirect interactions associated with the charge distribution of the surrounding solvent molecules. Molecular dynamic simulations were implemented to obtain a correlation between the electric field induced by the solvent on the ring mode and the observed vibrational bandwidth. Trp-cage, a model peptide system containing Tyr, was then synthesized to assess the efficacy of this correlation to detect changes in the electric field by comparing the native fold and denatured states of this peptide. The 2D IR spectrum of tyrosine in dimethylsulfoxide (DMSO) and water (D2O) further indicates the vibrational dynamics are significantly different due to the dephasing mechanisms of the vibration.

Label-free quantification of IgG1 and IgG4 in mixtures using Raman Spectroscopy
Mekhala Spencer, Yun Xu , John Welsh , Peter Levison

Multiplex detection of proteins with Raman spectroscopy has been reached with great success to very low concentrations when using Raman reporters. Conventional methods for higher concentrations found in formulations include Bradford/Lowry assays, enzyme-linked immunosorbant assays (ELISA), bio-layer interferometry and surface plasmon resonance. These methods are highly selective and accurate but require protein to be immobilised or captured. We propose that Raman spectroscopy could be a fast, accurate, and label-free method for quantifying multiple proteins in a single formulation, without the need of immobilisation. The aim was to accurately and simultaneously predict different amounts of two IgG subclasses (Human IgG1 and IgG4) in a mixture using Raman spectroscopy with multivariate data analysis.

Confocal Raman microscopy for investigation of bilayer-analyte interactions at nanopore supported phospholipid bilayers
David Bryce, Jay Kitt, Joel Harris

Molecular interactions with phospholipid bilayers play a major role in a wide range of biological processes. These processes range from the regulation of small molecule uptake (i.e. drugs and toxins) to cell signaling, enzyme expression, and immune response. Typically, investigation of molecular interaction with phospholipid bilayers rely on fluorescently labeled analytes for detection or on-column chromatographic measurements, in which the nature of interactions are inferred from elution trends. These methods fail to provide insight into the impact of interaction on the structure of the phospholipid bilayer. Here we report on the coupling of the high spatial selectivity of confocal Raman microscopy with phospholipid bilayers deposited on high surface area silica supports for detecting phospholipid bilayer analyte interactions. Utilizing these materials provides phospholipid concentrations in the confocal volume as high as 0.5 M, generating a Raman spectrum which is easily detected. The well-characterized surface coverage of phospholipid within the particle pores provides an internal standard, allowing quantification of analyte, and spectral information provides insight into structural changes occurring in the bilayer with analyte interaction. Recently, we have employed these materials to investigate a variety of biologically-relevant processes. Several of these applications will be presented, including measurement of bilayer-partitioning of drug molecules, detection of protein-ligand binding, and measurement of bilayer association of a signaling peptide, glucagon-like peptide 1.

Metal induced folding patterns of α-synuclein **assemblies**
Heather Lucas

The correlation between conformational states of ?-synuclein (?S) and their relationship to Parkinson's Disease (PD) etiology have evolved in recent years to adopt monomeric, oligomeric, and tetrameric forms. Traditionally, the conformation of ?S was described as an intrinsically disordered monomer that aggregates to form ?-sheet fibrils, potentially through oligomeric intermediates. Tetrameric ?S has been characterized as both ?-helical and aggregation resistant, likely functioning as a passive storage form of the protein. The pathological conformation of ?S remains controversial but each functional fold and dysfunctional misfold must originate from the monomeric protein. Membrane-protein interactions are thought to facilitate the transition of monomeric ?S to the ?-helical tetramer due to the similarity in folding patterns. Physiological levels of cerebral biometals, such as copper, iron, and manganese, change in various regions of the brain with aging and with advanced disease states. Metal dyshomeostasis has long been linked to PD but most of the work predated the discovery of the tetramer and the acceptance of ?S as being post-translationally acetylated at the N-terminus. The presented research will highlight metal-induced changes in the folding pattern of ?S and the distinguishable differences observed between various metal oxidation states. Unique changes are observed in the protein structure and size upon the introduction of stoichiometric quantities of individual metal ions based on biochemical and biophysical analyses. Infrared spectroscopy studies, circular dichroism spectral trends, immunoblotting techniques, and light scattering measurements have revealed that metal-bound ?S can follow distinct pathways towards toxic oligomers or ?-sheet fibrils depending on the metal. Furthermore, a clear involvement of ambient oxygen in the production of post-translational modifications has been observed. Such metal-induced conformational changes result in altered membrane interactions, which could have physiological implications. Taken together, these results contribute to the ongoing conversation within the field about the role of metals and ?S in PD progression.

Towards a Rapid and Selective Nanoparticle-Based Assay for the Assessment of Biopharmaceutical Glycosylation

Craig Ward, Karen Faulds, Daniel Bracewell, Duncan Graham

The biopharmaceutical industry is highly competitive and growing, with the majority of biopharmaceuticals being glycoproteins. Glycosylation is the most common and structurally diverse post-translational modification in secreted proteins. These carbohydrate structures can have an effect on the physicochemical properties, thermal stability and reactivity of the therapeutic proteins, with some oligosaccharide structures reported to cause serious adverse effects. Therefore there is a need to maintain consistency of carbohydrate moieties throughout scale up processes and between batches, and consequently fast and accurate glycosylation analysis is vitally important. Lectins are proteins with a highly specific carbohydrate recognition domain in their structure making them particularly interesting. We have demonstrated that lectin-coated nanoparticles are capable of detecting glycoproteins both in solution and in lateral flow assays. When these nanoparticles are functionalised with a Raman active molecule, surface enhanced Raman scattering (SERS) can be used for a novel, rapid, and selective assay for identifying, and differentiating, between glycans on the surfaces of therapeutic glycoproteins. SERS is attractive because it demonstrates the ability for multiplexing which may be useful for proteins with complex glycans. Enzyme-linked immunosorbent assays (ELISAs) have been used in the detection of many different biomarkers with high sensitivity and selectivity. We have been developing an ELISA which, upon detection of a glycoprotein, induces growth of gold nanoparticles. Nanoparticle growth can be tracked using ultraviolet-visible spectrophotometry and provides low limits of detection. The use of highly scattering gold nanostars in SERS experiments in combination with these ELISAs is also explored and has the potential to push these limits of detection even lower.

Time-gated Raman spectroscopy and 3D data analysis for quantifying fluorescent pharmaceuticals

Tiina Lipiäinen, Jenni Pessi, Parisa Movahedi, Tapio Pahikkala, Jukka Heikkonen, Mari Tenhunen, Lauri Kurki, Jouko Yliruusi, Anne M. JuppoClare Strachan

Time-gated Raman spectroscopy enables the separation of Raman-scattered photons from photoluminescence (including fluorescence) photons, and therefore allows analysis of samples that are too fluorescent for conventional Raman spectroscopy. The technique is based on a pulsed laser and time-dependent rejection of fluorescent photons that arrive at the detector slightly later compared to Raman scattered photons. The aim of this study was to investigate the potential of time-gated Raman spectroscopy for quantitative analysis of fluorescent pharmaceutical solids. This involved the application of established chemometric data analysis methods, and more recently adopted methods that optimize the use of the data in both the spectral and time domains. Mixtures of different solid-state forms of the drug piroxicam were measured using a time-gated Raman spectrometer (TimeGate Instruments, Finland). Three different forms of piroxicam – ?, ?2 and monohydrate – were prepared and combined in ternary mixtures according to mixture design. The three forms fluoresce to varying degrees. The Raman instrument was equipped with a picosecond pulsed laser (532nm), a time-gated single photon counting (CMOS-SPAD) array detector and a sampling probe, and allowed measurements in ambient lighting. The system enabled acquisition of Raman spectra with fluorescence rejection as well as simultaneous measurement of time-resolved fluorescence spectra. Traditional partial least squares (PLS) models based on the standard normal variate (SNV) transformed Raman shift (2D) data resulted in root-mean-squared error of cross validation (RMSECV) values of 4.1%, 4.5% and 3.8% (for forms ?, ?2 and monohydrate, respectively). Meanwhile, kernel-based regularized least squares (RLS) regression methods with greedy feature selection, which optimized the use of data in both the Raman shift and time dimensions (3D data), could improve the performance of the quantitative analysis (RMSECV values of 1.6%, 1.2% and 1.4% for forms ?, ?2 and monohydrate, respectively). Overall, time-gated Raman spectroscopy is a useful tool for analyzing fluorescent materials and quantifying their mixtures. Standard multivariate analysis techniques allowed quantitative analysis, while more recently adopted, kernel-based methods provided superior quantitative results. Overall, the technique shows potential for relatively routine (solid-state) analysis of photoluminescent pharmaceuticals during drug development and manufacturing.

Evaluation of transmission Raman spectroscopy for detection and quantification of low level drug content in pharmaceutical tablets

Rajesh Morampudi

Content uniformity test for drug substance is necessary to ensure consistency of pharmaceutical dosage units where tablets and capsules have <25 mg or <25% of active pharmaceutical ingredient (API). Conventional analytical technique such as High performance liquid chromatography (HPLC) is used for chromatographic separation and detection of analyte of interest. These destructive methods may include extended periods of sample preparation to extract the analyte which may lead to errors in final results. Alternatively non-destructive methods like NIR and Raman spectroscopy combined with multivariate analysis can be used to replace these traditional methods. The sensitivity of transmission Raman spectroscopy for detection and quantification of API in a pharmaceutical tablet is evaluated in this work. A tablet formulation with very low concentration of API with a label claim of 1mg which is equivalent to 0.54%w/w was chosen. A central composite experimental design is used to prepare calibration tablets and analyzed by transmission Raman spectroscopy. Predictions models were built and optimized using partial least squares (PLS) regression and data processing. Cross-validation of calibration samples is performed to optimize the PLS model. Further the analytical method for assay and content uniformity of API using Raman spectroscopy is validated for specificity, accuracy and precision. An external validation set that represents commercial scale batch of tablets is used for validation studies of prediction model. The prediction error for API content is found to be 0.009%. The content uniformity of API in tablets manufactured at commercial scale are compared with results from HPLC and found to be within the acceptable range. Transmission Raman is sensitive enough to detect and quantify API at very low concentration of 0.54% w/w with prediction errors of less than 2% of target concentration.

Crystalline Content Determination in Solid Dispersions by Raman Spectroscopy –Effect of the Analyzed Sample Volume on Qualitative and Quantitative Results

Fran Adar, Mathieu Boiret, L Netchacovitch, E. Ziemons

While it is well known that Raman spectroscopy can provide detailed information on chemical and crystalline composition, the particle sizes of commercial pharmaceutical products are considerably larger than usual sampling volumes, making predictions of composition in these mixtures fraught with error. There are several ways to overcome this situation. On a Raman microscope, the use of the DuoScan™, a mechanism that enlarges the size of the excitation spot and the volume from which Raman light is collected, improves the situation immensely. Measurements of the content of a hot-melt extrusion pharmaceutical made with a standard Raman microscope, a microscope equipped with the DuoScan™, and a bulk Transmission Raman apparatus will be compared. Multiple samples were prepared and have been examined with varying amounts of amorphous vs. crystalline content. The data were analyzed using PCA and PLS modelling, and the improvement of the predictions with increasing analyzed volume was documented. The improvement in the predictions using a transmission geometry rather than back-scattering will be quantified.

0 , 180 - Why the scattering geometry makes a difference when applying Raman Spectroscopy

Sean J. Gilliam, Francis Esmonde-White, David Strachan, Ian R. Lewis

Many excitation/collection geometries have been used since the advent of Raman spectroscopy. Classical dispersive Raman used a 90 collection approach in order to reduce of the amount of excitation laser power entering the spectrometer. In the 1970's, a (180 collection) backscattering geometry was developed, using prefilter spectrograph stages to improve filtering. Backscattering geometries simplify sample positioning and allow for Raman microscopy, as well as in-situ probes and stand-off designs combining both excitation and collection optical pathways. In the 1990's and 2000's, new optical filter hardware enabled a surge of wide area (large volumetric) PhAT™ backscatter as well as forward scattered transmission Raman designs (0 collection) technology. Applications were developed for which enhanced volume sampling in scattering samples was needed – primarily for the characterization and quantification of heterogeneous solid samples – both in a moving and non-moving format. When shifting from a sampling friendly backscattering geometry to a more constrained transmission Raman setup limits sampling options. In this presentation, pros and cons of different scattering geometries are described via experimental and computational comparisons – answering the question: When should these various Raman sampling techniques be used?

Determination of Propensity to Crystallize of Amorphous Solid Dispersions by Transmission Raman Spectroscopy

Holger van Lishaut, Frank Theil, Sankaran Anantharaman, Johanna Milsmann

The preparation of an amorphous solid dispersion (ASD) by dissolving a poorly water-soluble active pharmaceutical ingredient (API) in a polymer matrix can improve the bioavailability by orders of magnitude. Re-crystallization of the API in the ASD though is an inherent threat for bioavailability and therefore patients. In this study, we demonstrate the high potential of transmission Raman spectroscopy (TRS) as non-destructive method to assess API crystallinity in ASD formulations and an approach to determine the relative effectiveness of the stabilizing formulation. To this end, hot-melt extrudates of itraconazole and fenofibrate were spiked with defined amounts of crystalline drug. TRS was then applied to build a chemometric model for the quantification of the crystalline content. Furthermore, the crystal growth of the amorphous API was followed under elevated temperature and humidity by TRS in individual dosage forms. Not only does this approach allow for the determination of crystal growth kinetics, but also makes it possible for the first time to manufacture samples with targeted amount of actually grown crystalline content. In this talk, the application of TRS to quantify crystalline content in solid dosage forms is demonstrated. Furthermore, this technique is applied to follow the crystal growth in individual tablets of an itraconazole and a fenofibrate formulation under stress stability conditions. Based on these findings, the targeted growth of crystalline API in amorphous ASD-formulations is demonstrated. The work presented not only demonstrates the feasibility of TRS to gain quantitative information on the crystallinity of API in solid dosage form, but also allows deriving a measure for the propensity to recrystallize under given stability conditions. The targeted crystallization of amorphous tablets allows for the first time the simulation of re-crystallized ASDs which is a valuable model system for biopharmaceutical studies on the impact of crystallinity. All authors are employees of AbbVie. The design, study conduct, and financial support for this research were provided by AbbVie. AbbVie participated in the interpretation of data, review, and approval of the publication.

Hyperspectral AFM-IR imaging of malaria infected cells

David Perez-Guiata, Bayden Wood, Philip Heraud, Shirly Espinoza Herrera, Christian Doerig, Jose Garcia-Bustos, Kamila Kochan, Don McNaughton

Atomic Force Microscopy-Infrared (AFM-IR) spectroscopy is fast becoming a powerful technique to study the molecular phenotype of cells and tissues at the nano-scale. One limitation of the technology is that the maps generated are univariate in nature and only show either the absorbance plotted against a single wavenumber value or a ratio of two absorbance values. Here we apply for the first time hyperspectral multivariate image analysis to generate multicomponent AFM-IR maps to resolve subcellular structural information from *Plasmodium falciparum* infected red blood cells at different phases of the erythocytic life cycle. This was achieved by converting the discrete spectral points into a hyperspectral line spectrum prior to multivariate image reconstruction. The approach was used to generate multi-compositional maps of subcellular structures in infected red blood cells including the food vacuole, lipid inclusions and the nucleus by their chemical localisation according to the intensity of haemozoin, haemoglobin, lipid and DNA IR marker bands, respectively. The high spatial resolution of AFM-IR combined with hyperspectral modelling enables the direct detection of subcellular components without the need for cell sectioning or the addition of any immunological or biochemical staining agents. Hyperspectral AFM-IR has the capacity to be used as a phenotypic probe of the malaria parasite to study its life cycle and ultimately to investigate the phenotypic response to antimalarial drug resistance.

Tip-Enhanced Raman Spectroscopy: An Emergent Tool for Probing Biology and Electrochemistry at the Nanoscale

Dmitry Kurouski, Igor Lednev, Tanja Deckert-Gaudig, Volker Deckert, Michael Mattei, Richard Van Duyne

Tip-enhanced Raman spectroscopy (TERS) is a unique label-free spectroscopic technique that has single-molecule sensitivity simultaneously providing sub-nanometer spatial resolution of the analyzed specimen. This is due to 1) extremely high confinement of an electromagnetic field at the apex of the scanning probe and 2) precise positioning of the scanning probe at the surface of the analyzed specimen. Most of chemical groups have unique vibrational bands in Raman spectra. Therefore, in the obtained TERS image a particular chemical structure, as well as a co-localization of such structures, can be clearly visualized. This makes TERS highly desirable for biological applications. For instance, TERS can be used to probe protein secondary structure and amino acid composition on the surface of amyloid fibrils. At the same time, TERS is capable of monitoring electrochemical and catalytic reactions at the nanoscale. Specifically, TERS allows to measure rates of electrochemical reactions at various surface defects, such as adatoms, step-edges and vacancies. These facts highlight the growth of TERS as a powerful analytical technique with far reaching practical implications.

Synchrotron Infrared Nano Spectroscopy and applications of SINS in Biology, Physics, Catalysis, and more

Michael C. Martin

Recent developments in near-field imaging have enabled infrared spectroscopic measurements to move well beyond the diffraction limit. In particular, synchrotron infrared nano-spectroscopy (SINS) combines the broad bandwidth, high brightness, and spatial coherence of synchrotron infrared light with scattering type, scanning near-field optical microscopy (s-SNOM) to enable sensitive vibrational spectroscopy spanning the entire mid-and far-infrared regions with nanometer spatial resolution [1]. The Advanced Light Source (ALS) at Lawrence Berkeley National Laboratory operates two infrared beamlines with SINS instruments that are available to general users, who in turn have used the technique to investigate a variety of soft and hard matter systems, including proteins, organic semi-conductors, bio-minerals, 2D materials, and phase-change materials [1-6]. We apply the technique to several biofilm and geological samples that demonstrate the full power of hyperspectral nano-imaging and multivariate analysis while uncovering the nanoscale heterogeneity in these natural materials. References: 1. Bechtel, H. A.; Muller, E. A.; Olmon, R. L.; Martin, M. C.; Raschke, M. B. PNAS, 111, 7191 (2014) 2. Muller, E. A., Pollard, B.; Bechtel, H. A.; van Blerkom, P.; Raschke, M. B. Science Adv. 2, e1601006 (2016). 3. Shi Z.; Bechtel, H. A.; Berweger, S.; Sun Y.; Zeng B.; Jin C.; Chang H.; Martin M. C.; Raschke M. B.; Wang, F. ACS Photonics, 2, 790-796 (2015). 4. Liu, M. et al. Phys. Rev. B, 91(24), 245155 (2015). 5. Johns, R. W.; Bechtel, H. A.; Runnerstrom, E. L.; Agrawal, A.; Lounis, S. D.; Milliron, D. J. Nature Commun., 7, 11583 (2016). 6. Wu, C.-Y.; J. Wolf, W.; Levartovsky, Y.; Bechtel, H. A.; Martin, M. C.; Toste, F. D.; Gross, E. Nature, 541, 511–515 (2017).

Nanoscale Chemical and Electronic Mapping of Carboxyl Graphene Oxide using Combined Tip-enhanced Raman Spectroscopy & Kelvin Probe Force Microscopy

Marc CHAIGNEAU, Weitao SU, Naresh KUMAR, Andrey KRAYEV

Visualizing chemical group distributions on GO surface challenges the sensitivity and spatial resolution of many metrological methods. In this study, for the first time, an in-situ tip enhanced Raman spectroscopy (TERS) and kelvin probe force microscopy integrated system was used to image chemical and electronic properties of Carboxyl Graphene Oxide (GO-COOH) samples with nanoscale spatial resolution. The chemical groups on GO-COOH were imaged by TERS with a spatial resolution of 14nm. TERS and KPFM integrated measurement indicates Fermi level on GO-COOH flakes increases with increasing ID/IG (intensity ratio of the D and G Raman bands), which enables to correlate the chemical groups distribution to local potentials in nanoscale.

Polymeric nanoparticle chemical analysis using Tapping AFM-IR

Alexandre Dazzi, Jérémie Mathurin, Ariane Deniset-Besseau, Elisabetta Pancani, Ruxanda Gref, Kevin Kjoller, Craig Prater

The AFM-IR technique is a very powerful and efficient technique to chemically characterize materials at the nanometer scale [1]. Recent improvements in mid IR lasers and AFM techniques have further extended the capabilities of this technique. AFM-IR is based on coupling a tunable infrared laser and an AFM (Atomic Force Microscope). This coupling allows performing ultra-local infrared spectroscopy and chemical mapping at the nanometer scale. The principle is based on the detection of the sample thermal expansion, irradiated at the wavelength of its absorption bands. The applications of this infrared nanoscope are numerous from biology to polymer science [1]. In this talk, we will present a new mode of imaging and spectroscopy allowing analysis of soft and or adhesive samples: the tapping AFM-IR. This approach combines the resonance AFM-IR [2], with tapping mode imaging to provide a new system with high sensitivity and lateral resolution able to probe a broad range of samples, including nanoparticles or soft samples. To exemplify the capabilities of this new technique, we will present results from the study of biodegradable polymeric nanoparticles. These nanoparticles are mainly used in nanomedicine as a drug delivery system. They appear to be a good alternative to overcome the problems encountered by antibiotics penetrating the cell membrane and increase the targeting of resistant bacteria. Unfortunately, there is a lack of knowledge about the chemical structure of these nanoparticles and the estimation of the amount of antibiotic inside the particles. The tapping AFM-IR technique is perfectly suited to answer these open questions and will help to fully characterize the polymeric nanoparticles. [1] A.Dazzi, C.B. Prater, Chem. Rev. (2016). [2] F. Lu, M. Jin, M.A. Belkin, Nat. Photon. 8, 307–312 (2014).

A Statistical Investigation of the Properties of Plasmonic Single Particles

Alexandre Brolo, Regivaldo Sobral Filho, XiaoYing Zhang

Single particle spectroscopy is well-established in the field of plasmonics. For instance, dark field methods allow the evaluation of the extinction of individual particles in an ensemble and those can be correlated with the geometric characteristics of the nanostructure. In this work, we report a novel synthesis of monodisperse metallic nanoparticles. The differences in particle to particle geometry was found to be negligible. On the other hand, interesting differences in the characteristics of the dark field images were observed for the first time. The extinction properties of individual particles were then correlated to other methods, such as surface-enhanced Raman scattering (SERS). The variation in optical properties from virtually identical nanoparticles were then interpreted as contributions from small scale roughness and defects. This work provides then an unique on the characteristics of plasmonic single particles that should be considered in the development of future applications, such as in the area of biosensors.

Plasmonic Nanostructures for Trapping and Sensing Single Molecules
Sang-Hyun Oh

Recent technological advances have enabled high-throughput fabrication of plasmonic nanogap structures with critical dimensions approaching 1 nm. Extreme confinement of optical energy in such single-nanometer gaps can lead to a series of new applications in biosensing and spectroscopy. This presentation will outline approaches to design and fabricate nanogap plasmonic sensors with precisely tuned optical resonances in the visible, near-infrared, or mid-infrared regime. Among various options, we use atomic layer deposition (ALD) to create a dense array of ultra-narrow slits and rings in metals, which is made possible by the Angstrom-scale thickness resolution of ALD. Our scheme – atomic layer lithography - has enabled rapid wafer-scale production of various nanostructures that are nontrivial to produce using any other methods. For example, now it is a routine task to produce a dense wafer-scale array of 1-nm-wide and 1-centimeter-long metal slits. Metasurfaces made with such nanoslits and rings that are resonant in the visible, infrared, and even terahertz regime have been produced over an entire 4-inch wafer. Since the nanogap size in our scheme can be precisely controlled with ALD to match the size of target biomolecules, our nanogap sensors can be used for trapping and detection of single molecules and nanoparticles. High-throughput production of these nanostructures allows researchers to perform such challenging experiments in a more reproducible manner. We will present experimental results on how to combine optical and electronic trapping techniques with these nanogap plasmonic sensors to accomplish single-molecule trapping and detection.

Synthesis of Copper-Silica Core-Shell Nanostructures with Sharp and Stable Localized Surface Plasmon Resonance
Jingyi Chen

Copper nanoparticles can exhibit intense and sharp localized surface plasmon resonance (LSPR) in the visible region; however, the LSPR peaks become weak and broad when exposed to air due to the oxidation of Cu. To overcome the instability issue of Cu nanoparticles, this work demonstrates the use of silica coating to protect the Cu nanoparticles in aqueous suspension against oxidation. The Cu-silica core-shell structures were synthesized by employing trioctyl-n-phosphine (TOP) capped Cu nanoparticles for the sol-gel reaction in a basic microemulsion containing NaOH. The resulting Cu-silica core-shell suspension possesses stable and well-preserved LSPR properties of the Cu cores. With the TOP capping, the oxidation of the Cu cores in the microemulsion was significantly reduced, thus allowing the Cu cores to sustain the sol-gel process used for coating the silica protection layer. It was found that the self-assembled TOP-capped Cu nanoparticles were spontaneously disassembled during the sol-gel reaction, thus recovering the LSPR of individual particles. During the disassembling progress, the extinction spectrum of the nanocube agglomerates evolved from a broad extinction profile to a narrow and sharp peak. For a mixture of nanocubes and nanorods, the spectra evolved to two distinct peaks during the dissembling process. The observed spectra match well with the numerical simulations. These Cu-silica core-shell nanoparticles with sharp and stable LSPR may greatly expand the utilization of Cu nanoparticles in aqueous environments.

Structural Analysis by enhanced Raman Scattering
Steven Demers, James Matthews, Cyna Shirazinejad, Grace Isakson, Jason Hafner

Gold nanostructures focus light to the nanometer scale at their surface, creating the possibility to analyze molecular structure. When illuminated at the plasmon resonance, the high optical intensity leads to surface enhanced Raman scattering (SERS) from nearby molecules. SERS spectra contain information on molecular position and orientation relative to the surface but are difficult to interpret quantitatively. We have developed a ratiometric analysis method that combines SERS and unenhanced Raman spectra with the theoretical calculations of the optical fields and polarizabilities. When applied to a decyltrimethylammonium bromide (DTAB) surfactant layer on gold nanorods, the alkane chain is found to be tilted 25 to the surface normal, which matches previous reports on the surfactant layer thickness by x-ray and neutron scattering. The analysis was also applied to fluid phase phospholipid bilayers that contain tryptophan on the surface of gold nanorods.The lipid double bond was found to be oriented normal to the bilayer and 13 Å from the nitrogen atom. Tryptophan was found to sit near the glycerol headgroup region with its indole ring 43 from the bilayer normal, in agreement with previous reports. The new method can determine specific interfacial structure under ambient conditions, with microscopic quantities of material, and without molecular labels. Ongoing analysis of the position of tryptophan in gramicidin channels in lipid bilayers will be discussed.

Photoexcitation of on-surface core-shell fluorescent plasmonic nanoparticles by grazing waveguides
Alexandre Grégoire, Jean-Philippe Bérubé, Réal Vallée, Denis Boudreau

Many researchers in photonics have been interested to confine and to propagate light over long distances with limited loss and divergence. Femtosecond laser direct inscription has emerged as a powerful technique for the fabrication of tridimensional photonic devices in transparent materials. One of the most promising applications of these buried dielectric waveguides is to take advantage of the evanescent wave to probe the external medium. For example, combining core-shell sensors (metal@dielectric fluorescent nanoparticles) with such near-surface waveguides allows innovative designs for detection and imaging using microscopy techniques. In this work, we present a wrinkled PDMS stamping method to produce large linear nanoparticle assemblies. By tuning the Ag@SiO2 concentric architectures, interparticle separation and plasmonic coupling can be adjusted in the final one-dimensional structures. Using a combination epifluorescence, dark-field scattering, and fluorescence lifetime microscopy with scanning electron microscopy, we were able to gather optical and structural data from isolated nanoparticle assemblies to describe interesting structure-properties relationships. Moreover, we also used the buried dielectric waveguides to launch plasmons into such linear architectures and collect the resulting fluorescence signal by a near-field excitation and detection technique. This method has the advantage to suppress medium fluorescence due to the distance-dependent near-field coupling of the evanescent field.

Coupling Ion Channels to Mobile Nanofluidic Devices (Nanopipettes)
Lane Baker

Integration of dual-barrel membrane patch-ion channel probes (MP-ICPs) to scanning ion conductance microscopy (SICM) holds promise of providing a revolutionized approach of spatially-resolved chemical sensing. However, to fully exploit the MP-ICP platforms for concurrent topography and chemical sensing, a comprehensive characterization of the analytical performance of this newly-developed platform is needed. A series of experiments were performed to further our understanding of the system and to answer some fundamental questions, in preparation for future developments of this approach. First, we constructed MP-ICPs that contained different types of ion channels including TRPV1 channels and BK channels to establish the generalizability of the methods. Next, we proved the capability of the MP-ICP platforms in single ion channel activity measurements, and demonstrated that the channel behaviors can be faithfully obtained with this approach. In addition, we studied the interplay between the SICM barrel and the ICP barrel. An interesting phenomenon that's related to the charge of the ligand molecules was discovered, for ion channels gated by uncharged ligands, channel activity at the ICP barrel is unaffected by the SICM barrel potential; whereas for ion channels that are gated by charged ligands, enhanced channel activity can be obtained by biasing the SICM barrel at potentials with opposite polarity to the charge of the ligand molecules. Finally, a proof-of-principle experiment was performed and site-specific molecular flux sensing was demonstrated at single-ion-channel level.

Chromatographic Properties of Ordered Carbon Nanomaterials

Susan Olesik

Carbon surfaces used for chromatographic purposes, such as glassy carbon, contain both basal and edge-plane reactive sites. These two different types of surfaces are known to have different sorption selectivity. The heterogeneity of surface sites often limits chromatographic efficiency due to different desorption kinetics. Ordered carbon nanomaterials that exclusively contain only basal or edge plane surfaces can be created. The homogenized surface sites of these carbon nanomaterials should allow for more efficient chromatographic separations which also have increased selectivity. This talk will illustrate ordered homogeneous carbon surfaces with high surface area. The chromatographic properties of these surfaces will be compared to publish results using commercially available carbon media. Column efficiency and kinetic performance will be highlighted. The separation of highly polar compounds for bioanalytical chromatography will be illustrated.

Exploring media for electrophoretic separation of DNA by sequence in microfluidic chips
Linda McGown, Wyatt Stevens, Jia Zhao

DNA analysis has widespread applicability in biology, medicine, ecology, agriculture, biodefense, biotechnology and forensics. DNA separation by length is readily achieved using sieving gels in electrophoresis. Separation by sequence is more difficult, generally requiring adequate differences in native or induced conformation or differences in thermal or chemical stability of the strands that are hybridized prior to measurement. We have discovered a more general approach that allows ssDNA to be separated by sequence using high concentrations of salts in simple buffer solutions. The approach does not require prior knowledge of the sequences or persistent conformational differences among the strands. The separation mechanism is currently under study. Here we describe progress toward a two-dimensional microfluidic platform that will use conventional sieving gels to separate ssDNA based on length in the first dimension and the new approach to separate DNA by sequence in the second dimension. This talk will focus on challenges we have encountered in implementation of the sequence-based separation and approaches we are taking to overcoming them. In particular, we have been exploring the use of gels in the separation channel to minimize diffusion, slow down the electromigration and protect the channel surface from the high ionic strength buffer.

Single Molecule Biopolymer Analysis Using Interface-Tailored Nanopore Sensing
Jason Dwyer, Buddini Karawdeniya, Nuwan Bandara, Jonathan Nichols, Robert Chevalier

Nanopores are a powerful tool for single-molecule and –particle analysis consisting, in their simplest format, of a ~10 nm-diameter channel through a <100 nm-thick film. The most common approaches electrophoretically force the passage of analyte molecules (or particles) through electrolyte-filled nanopores sized to allow transport of a single molecule at a time. The disruption of ionic current flow carries information determined by the analyte size, shape, and chemical properties; by the solution properties; and by the nanopore surface chemistry. Thus, nanopore sensing can be used for chemical analysis of solution-phase analytes (e.g. for chemical structure determinations and for molecular fingerprinting); to explore intermolecular interactions; and to probe solution-phase and interfacial chemical processes. Key to such efforts is the ability to fabricate nanopores with the desired sizes and shapes and—no less importantly—the desired surface chemical properties. We will demonstrate how to chemically tune nanopore function, and then how suitably tailored nanopores can yield insights into the structure and chemical properties of a variety of biopolymers—especially polysaccharides—significant for biological function and medical therapeutics.

LA-ICP-MS Analysis and Characterization of Adhesive Tapes as Forensic Evidence

Jose Almirall, Claudia Martinez-Lopez

Adhesive tapes are commonly found as physical evidence in cases involving violent crimes and national security threats. LA-ICP-MS was used for the characterization of chemical signatures of electrical tapes for forensic comparison and provenance purposes. The backings of 90 black electrical tapes, previously characterized by microscopical examination, Scanning Electron Microscopy-Energy Dispersive Spectroscopy (SEM-EDS), Fourier Transform Infrared Spectroscopy (FTIR) and Pyrolysis-Gas Chromatography-Mass Spectrometry (Py-GC-MS) were analyzed by LA-ICP-MS to evaluate the ability of the technique to discriminate samples originating from different sources and to associate pieces of tapes originating from the same roll. The results showed that LA-ICP-MS is a useful tool that complements current protocols for the organic and inorganic characterization and comparison of electrical tapes and results in improved discrimination and superior characterization. The developed LA-ICP-MS method alone provided 94% correct discrimination of the tapes known to originate from different rolls and 100% correct association of the tapes known to originate from the same roll. Moreover, LA-ICP-MS captured a large amount of compositional information, as the use of elemental profiles alone provide similar discrimination and classification capabilities as all the conventional methods together. The analytical sensitivity of LA-ICP-MS provides a means for the classification of tapes to support investigations as well as the potential for database searching capabilities in the future.

LA-ICP-MS imaging for anthropological samples
Matthieu Baudelet

The study of anthropological evidence has seen a new dimension with elemental analysis. For a long time, it relied on XRF and/or ICP-MS. The introduction of laser-ablation has given the opportunity to reduce sample preparation while having full access to the whole periodic table with limits of detection from ppb's to %. Laser-Ablation ICP-MS (LA-ICP-MS) for anthropological materials, from the cultural to the forensic point of view. We consider anthropological evidence in a general sense (not only bones but also hair and teeth). Teeth are great indicators in the human body for growth, evolution of diet and even geographical movements. This information can be extracted from LA-ICP-MS elemental maps. Information on the history and cultural distribution of an archaeological population will be discussed. DNA is not easy to recover from hair, preventing sex determination in some cases. Using elemental analysis of hair, we identified different minor elements of interest that can reflect the metabolism change in the body that can be different between male and female individuals. This study substantiates the potential of the elemental analysis of hair as a proxy for human sex determination.

Single-shot laser-ionization mass spectrometry for direct atomic analysis
Jose M Vadillo, J. Javier Laserna

Direct laser ionization of solids (LIMS, laser-ionization mass spectrometry) represents a powerful way to get full atomic information about sample microlocations. Spatial resolutions as low a few microns are easily attainable with conventional optics, with vaporized mass in the low femtogram per pulse range. Considering the lack of limitations with respect to sample type, it is clear that the only limitation of the technique relies on his poor reproducibility and moderate sensitivity. The talk talk will cover several fundamental and applied aspects of LIMS to fit the technique in a proper frame of comparison with other techniques.

Effect of Nanoparticles on Laser sampling and plasma emission
Alessandro De Giacomo, Rim Alrifai, Gabriele Valenza, Marcella Dell'Aglio

Recently we proposed the Nanoparticle Enhanced Laser Induced Breakdown Spectroscopy (NELIBS) for enhancing the LIBS performances in terms of signal enhancement in several applications, spacing from metal sample analysis to biological samples. In the present work we discuss some new fundamental aspects concerning the laser ablation mechanisms when NPs are deposited on the sample surface as well as the NPs effect on the dynamics and on the emission of the plasma produced by laser-matter interaction. For discerning the contribute of NPs enhanced ablation from the effect of NPs on the plasma features, the ablation has been carried on at different wavelength and the analyte signal has been detected both with LIBS and with LA-ICP-MS. At the same time spatial resolved imaging has been used for studying the distribution of emitters in the plasma with three different set-up. The results show that NELIBS allows a more efficient ablation as a consequence of the coupling between the laser and the electromagnetic field induced on the NPs, but a further improvement is due to the stagnation of emitters in a smallest volume during NELIBS with respect of conventional LIBS. Finally a general overview of the future perspectives and applications of NELIBS will be critically discussed in order to outline the advantages and the limitation of this recent technique.

Effect of Topological Charge on Femtosecond Laser-Induced Breakdown Spectroscopy with Optical Vortex Beams
Jason Becker, Xianglei Mao, Richard Russo, Costas Grigoropoulos, Vassilia Zorba

Laser-induced breakdown spectroscopy (LIBS) of a muscovite sheet (mica) was studied using an 800 nm Ti:Sapphire laser with vortex half-wave retarders to create optical vortex beam profiles in open air environment. LIBS serves as an efficient, non-invasive method for material analysis through laser ablation of target materials followed by investigation of the spectral characteristics of the materials plasmas. We present characteristics of spectra acquired using optical vortex beams with three different nonlinear polarization states and a conventional Gaussian beam with linear polarization. Pertinent plasma properties including excitation temperature, electron density, and persistence were derived from raw data collected in temporally resolved measurements. White-light interferometry was used to obtain crater depth and volume for the studied beam polarizations. The influence of laser polarization on ablation and resulting plasma are discussed. This work demonstrates the ability to use nonconventional beam shapes and polarizations for spectroscopic applications.

Teaching Students to Navigate Design Problems: Translating Fundamental Tools for Real-Time Learning
Barbara Smith, Emma Frow

Typical engineering courses focus on providing students with fundamental tools with which to build and develop. Frequently lacking in these courses are open-ended problems that allow students to grapple with the challenge of real-world scenarios. In the discussed course, we teach students how to identify global health problems, and to successfully formulate solutions that best fit the problems and needs articulated by end users. We use an iterative process that teaches students to work as part of teams to produce relevant designs, useful for a real setting. Within this course, we aim to provide the students with necessary tools for competence in: i) key processes within engineering design, ii) team-based learning, iii) understanding of the global healthcare market, and iv) presentation skills. This large-scale, project-based course provides a first introduction to engineering design and development.

ENGAGING LEARNERS OF ALL AGES WITH INQUIRY BASED ENVIRONMENTAL SCIENCE EDUCATION THROUGH INSCIED OUT

Seth K. Thompson, Christopher Pierret
Many of the world's most serious problems are tied to complex environmental issues. For example, the spread of vector-borne diseases will be influenced by a changing climate. Availability of drinkable water to communities across the globe will be influenced by land use and development of technologies to improve water treatment. An informed citizenship that is knowledgeable about environmental issues and actively engages in behaviors that promote environmental sustainability and ecosystem health is needed to solve such global problems. Environmental education (EE) is a powerful tool for creating informed citizens and imperative for promoting ecosystem health on a global scale, yet there is still no consensus on Environmental Education best practices. Previous work has identified several desirable characteristics of an EE initiative: meaningful experiences in nature, opportunities for students to take ownership over an issue, and empowerment of the student through belief that their actions can make a real difference. We incorporate these characteristics into an inquiry-based framework called the Integrated Science Education and Outreach (InSciEd Out) program to create EE opportunities for learners across the K-20 spectrum as well as adult learners. Fundamentally, InSciEd Out is a science education program built on developing strong scientist-teacher partnerships to provide cutting edge and authentic science experiences for all students. The program has been shown to be effective in promoting students' knowledge and interests in a variety of scientific content areas. Here, we present the design and outcomes of our programming on environmental chemistry and toxicology.

Where is reflection's place in the chemistry classroom?
Anna Donnell

Reflection is a powerful tool that can be used in teaching and in learning, but may not be commonly used in high content STEM courses such as chemistry. While certain concepts must be covered in courses, it can be easy to focus solely on covering content and assessing for student understanding in order to get through the semester. By incorporating small reflective activities and assignments, students can engage with the content to create stronger connections between concepts, become more conscious of their own learning, and build study skills to become independent learners. In a similar way, faculty can use critical reflection to examine their own teaching, identify areas of growth, and make informed decisions about changes to their teaching. This process often occurs informally and/or infrequently for faculty, but following a framework for faculty reflection can lead to significant change in one's teaching. Engaging in reflection as teachers and as students provides valuable formative feedback to both parties which can ultimately lead to greater student success and stronger relationships between faculty and students. In this session, techniques for incorporating reflection into the curriculum as well as faculty self-reflection will be discussed.

Experimenting with "build your own instrument" kits: A lab for an Analytical Chemistry course
Ingeborg Petterson

I will describe how I incorporated open source instrument designs that are available online into the chemistry curriculum. This includes building and testing an instrument as part an analytical chemistry lab course as special project that students worked on over a period of a few weeks. Students worked very independently to built their own potentiostats from a kit (WheeStat from Smokey Mountain Scientific) while being guided by online instructions provided for this open-source instrument. Students then tested the performance of their instruments with a simple electrochemical experiment. Purchasing "build your own instrument" kits provides a meaningful way to incorporate hands-on instrumentation skills into a course without having to design and build an instrument from scratch.

Research-based experiments in teaching-labs: how the chemistry of food can engage students in analytical and physical chemistry
Natasja A. Swartz, Paige W. Hall, Rob Jensen, Sam Danforth

This year, the American Chemical Society chose “Chemistry Helps Feed the World” for its Earth Day theme. The concept of food is an appealing framework for appreciating science in the undergraduate curriculum because everyone eats. Issues of abundance and variety aside, the quantification of ingredients that make up our food, desirable (vitamins, nutrients) or undesirable (pesticides, toxins), is one of the most practical applications of chemistry. Experiments related to food offer a unique and practical platform for increasing access to research experiences for undergraduates in teaching labs, which remains a high priority for educators in STEM fields. Traditional techniques and instrumental methods were used to identify and quantify relevant constituents of foods were used in upper-division teaching labs such as analytical and physical chemistry. Kombucha was investigated for simple acid content as well as its fermentation kinetics by volumetric titration and HPLC-UV. Another fermented product more universal in interest is beer, which makes for a unique study in vitamin analysis and quenching by protein binding with a fluorometer. The amount of caffeine in popular beverages was compared with that extracted from urban water ways and quantified by HPLC-UV. Uptake of systemic insecticides into honey was explored by HPLC-MS.

Exposing the Hidden Dangers of Dietary Supplements
Connie Ruzicka

Dietary supplements and herbal remedies are increasing in popularity all over the world due to the widespread belief that natural products are safer and healthier than synthetic ingredients. One of the major concerns related to the safety of dietary supplements is their adulteration with undeclared pharmaceutical substances in order to enhance the claims stated on the label. Supplements claiming to enhance weight loss and sexual performance are more commonly found to contain illegal adulterants and consumption of these adulterated products poses serious health risks to consumers. In response, the FDA has developed a rapid screening program to identify undeclared substances in imported dietary supplements and herbal products. The agency is currently piloting the use of ion mobility spectrometry (IMS) instruments at US ports of entry to screen weight loss and sexual enhancement products labeled as dietary supplements suspected of containing undeclared active pharmaceutical ingredients. Samples failing IMS screening are collected and sent to FDA laboratories for confirmatory testing using liquid chromatography-tandem mass spectrometry (LC-MS/MS). This presentation will describe the IMS method used by the FDA for the analysis of dietary supplements marketed for weight loss and sexual enhancement and discuss efforts to remove these harmful products from the marketplace. Several case studies will be presented highlighting the various types of products screened at import facilities for the presence of undeclared pharmaceutical substances.

LIBS and Raman Applications for Food Safety and Quality

Ismail Boyaci

Consumers must be informed about the composition, origin and quality of food and they should choose their diet according to this information. For this reason, labelling of foods should be honest and accurate. However, food fraud is worldwide problem due to economic gain and demand of analytical tools is increasing day by day to investigate the foods. Many methods have been developed for determination of food safety and quality but there are main technological challenges. Many of them is based on DNA and isotope analysis, proteomic methods and they are time-consuming, troublesome, lab-scale methods and require pre-processes and specialist. On the other hand, recently developed spectrometric methods provide rapid, practical, on-line analysis in foods. From these methods; LIBS and Raman systems which have ability to determine elemental composition and molecular structures respectively have high potential for rapid food analysis. In our research group, it was shown that potential usage of Raman spectroscopy for analysis of different food groups such as direct quantification of ethanol and methanol in distilled alcoholic beverages, rapid analysis of sugars in honey, label free bacteria detection and determination of tea varieties for their purities. On the other hand, LIBS is also versatile tool for determination of food adulteration and quality parameters of foods. In this context, it was shown that applicability of LIBS in identification of meat species, determination of whey adulterated milk powder, determination of ash and protein contents of cereals and NaCl analysis in bakery products. In addition to the promising results of LIBS and Raman spectroscopies on food applications, future respects of these methods and ongoing studies for portable systems were evaluated and discussed in the study.

Rapid Isolation of Pathogenic Listeria Serovars

Claire Crowther, Mark Hayes

The ability to rapidly detect foodborne pathogens, such as *Listeria* and *Salmonella*, has been the focus of many new detection tests. Specifically for *Listeria*, there is one species, *Listeria monocytogenes*, that has three pathogenic serovars. Traditional detection methods can take up to five days for identification of the bacteria, with newer techniques able to achieve detection in about twenty-four hours. Nevertheless, a test that could provide results in minutes would be preferable, as the earlier the bacterial contamination can be identified the easier it is to contain. The possibility for rapid separation and identification of cells is possible using microfluidics and dielectrophoresis (DEP), a force induced by non-uniform electric field. The slight differences in the intrinsic cell properties interacting with the electric field allows DEP to achieve the rapid and highly selective cell separations. This work will explore the differences and potential separation capability for the three pathogenic strains of *Listeria*. These isolations take only a few minutes and are highly sensitive and specific. Further development of this system, could result in the implementation of this rapid separation based detection method for easy identification of *Listeria* or other potential contaminants.

Data-Driven Strategies to Focus Food Fraud Prevention Resources
Karen Everstine

Development of methods to ensure food authenticity is a challenging problem for analytical chemists due to the unexpected nature of many adulterants. Regulatory agencies and the food industry need strategies for focusing these limited resources on food ingredients that are the most vulnerable to food fraud and that pose a health hazard to consumers. The United States Pharmacopeia (USP) Food Fraud Database (FFD) is tool designed to prioritize ingredients vulnerable to fraud and identify potentially hazardous adulterants. Information in the FFD includes public documentation of incidents of food fraud, evidence of potential or unreported fraud, analytical methods research, and surveys of fraud prevalence. The database contains more than 6300 records, including 855 incident records, 2116 inference records, 2985 method records, and 377 surveillance records. Fluid cow's milk is associated with the highest number of records (209) and the highest number of distinct adulterants (129), but also exhibits large geographic variability in incidence. Olive oil and honey are also associated with a high number of records (110 and 94, respectively) and chili powder is associated with a high number of distinct adulterants (59). Potentially hazardous adulterants are identified in the database using a framework developed by a USP Expert Panel. Of the 1300 adulterants in the database, approximately 25% have a documented history of causing illnesses or deaths in consumers, have resulted in safety-related regulatory action, or are allergens as defined by Codex Alimentarius. This challenges the conventional wisdom that most adulterants do not pose a health risk to consumers. This presentation will provide an in-depth overview of the data in FFD, challenges encountered in constructing and maintaining the database, and case studies for use of the database by industry and regulatory agencies. Food fraud continues to be a global challenge with economic and food safety implications. Making effective use of data from historical food fraud reports is an important component of risk mitigation.

Authentication of Turkish Honeys Using a Portable Raman System Spectrometry
Mei-Ling Shotts

Honey is one of the leading exports in Turkey behind olive oil. As a high commodity product, honey has been prone to adulteration. Adulteration of honey is completed by substitution of high fructose corn syrup (HFCS), maltose syrup (MS), beet sugar and other cheap invert sugars. This is a critical issue in the honey industry. Our objective was to develop a rapid untargeted approach to authenticate high-value honey by combining Raman spectroscopy and pattern recognition analysis. Samples (n=85) were collected from Turkish markets. The composition of samples were characterized by reference methods to identify potential adulteration by using sugar analysis by high performance liquid chromatography (AOAC Official Method 977.20).The reason may be that the major content in authentic honey is monosaccharide, e.g., glucose and fructose. Adulterated samples contained various disaccharides like maltose and sucrose. Spectra was collected by using a portable Raman system, analyzed by pattern recognition techniques that included Soft Independent Model of Class Analogies (SIMCA) to develop classification algorithms. Authentic honeys formed distinct clusters allowing the evaluation of commercial samples from local markets in Turkey showing some prevalence of adulteration. Spectral differences responsible for the separation of classes were associated with the signals of authentic and adulterated honey showed characteristics bands around 351, 425, 517, 592, 629, 705, 778, 824, 865, 915, 981, 1065, 1127, 1264, 1373 and 1461 cm?1.New generation portable Raman devices provided a viable tool for chemical profiling of Turkish honey samples allowing for the rapid, "in-field", and reliable authentication of food ingredients, making it a great alternative to time-consuming traditional testing methods.

Complex Electron Transfer Pathway at a Microelectrode Captured by in situ TERS
Ivan LUCAS, Thomas Touzalin, Suzanne Joiret, Emmanuel Maisonhaute

In the quest for analytical tools which enable the characterization of materials at the nanoscale under the condition of their operation (in situ, operando), the emerging Tip-Enhanced Raman Spectroscopy (TERS) now enters the spotlight. We demonstrate in this work the ability for TERS to capture the electrochemical transformation of a molecular layer self-assembled on a tapered gold microelectrode. The proposed set-up and protocol opens new perspectives in the characterization and development of redox architectures for molecular devices.

Applications of Nanoscale Chemical Imaging to Polyolefins

Mark Rickard, Gregory Meyers, Carl Reinhardt, Jamie Stanley

Atomic force microscopy coupled to infrared spectroscopy (AFM-IR) is a relatively new technology that provides the spatial resolution of AFM with chemical information from IR spectroscopy. It has the potential to connect the nanoscale chemical structure of a material to its macroscopic properties and performance. AFM-IR is based on the rapid thermal expansion of a material, which is detected by an AFM tip, when it absorbs IR light. This approach provides spatially-resolved chemical information at a length scale of < 50 nm, which greatly exceeds the diffraction limit of light. Applications of AFM-IR to polyolefins will be discussed. Studies of polymer blends are often hampered by the chemical and spectral similarity of the components. A method to overcome this challenge and its application to polyolefin blends will be discussed.

Tip-enhanced Raman spectroscopy as a new platform for plasmon-induced polymerization

Marie Richard-Lacroix, Zhenglong Zhang, Volker Deckert

Tip-enhanced Raman spectroscopy (TERS) is a technique that enables coupling the rich structural information available in vibrational spectroscopy with the spatial resolution and imaging capabilities of an atomic force microscope, thanks to the electric field enhancement caused by the formation of a localized plasmon at the tip apex. The plasmons generated have also shown their ability to induce the so-called "hot electrons" that catalyze various types of chemical reactions. In particular, the coupling of nitro- and amino-terminal phenol molecules into diazo have been shown to be efficiently plasmon-catalyzed in conditions where the reaction is otherwise impossible.[1] Here, we investigate the potential of this technique for producing polymer materials in a highly spatially controlled matter. As proof-of-concept example, we use surface-enhanced Raman spectroscopy (SERS) and TERS to investigate the 2D in-plane polymerization of dibenzo-1,2-dithiine3,8diamine (D3ATP), a bifunctional molecule, organized as self-assembled monolayers on gold and silver films. The use of TERS enables to simultaneously spectroscopically probe and induce the polymerization reaction, and with a nanometer scale resolution. Results show that both the diazo formation and the polymerization can be probed distinctively by the appearance of new bands in the spectra. [2] The conclusions drawn are further extended for the demonstration of other complex and commercially critical reactions such as diene epoxidation. [1] Z. Zhang, D. Kinzel, V. Deckert. J. Phys. Chem. C, 2016, 120, 3097-20983 [2] Z. Zhang, M. Richard-Lacroix, V. Deckert, Faraday Discussion, Accepted 09/05/2017

Recent Advances in Nanoscale IR Spectrosopy: Hyperspectral and Tapping AFM-IR Imaging

Curtis Marcott, Eoghan Dillon, Kevin Kjoller, Craig Prater

Nanoscale infrared spectroscopy has been successfully demonstrated in an expanding range of applications in recent years due to significant increases in capability. One method of nanoscale infrared spectroscopy, atomic force microscope based infrared spectroscopy (AFM-IR) directly detects IR radiation absorbed by the sample using the AFM probe tip to sense thermal expansion [1-4]. This thermal expansion depends primarily on the absorption coefficient of the sample and is largely independent of other optical properties of the AFM tip and the sample. The use of a quantum cascade laser (QCL) as the excitation source has dramatically improved the sensitivity of AFM-IR [4] and enabled spectra and absorbance images to be collected much more rapidly. Single AFM-IR spectra can now be acquired at speeds as fast 100 ms/spectra over the complete range between 1850 and 800 cm^{-1} at a spectral resolution of 0.5 cm^{-1}. This makes it possible to collect pointby-point hyperspectral data cubes of 50 x 50 pixels per image plane in well under an hour. A tapping AFM-IR methodology that enhances the spatial resolution of the technique will also be described. This new capability enables us to chemically differentiate features down to 10 nm spatial resolution. Example applications of hyperspectral AFM-IR, tapping AFM-IR imaging, and other new nanoscale spectroscopy and chemical imaging capabilities in the material and life sciences will be presented. [1] Dazzi, A.; Prazeres, R.; Glotin, F.; Ortega, J. M. *Opt. Lett.* **2005**, 30, 2388-2390. [2] Dazzi,

A.; Prater, C. B.; Hu, Q.; Chase, D. B.; Rabolt, J. F.; Marcott, C. *Appl. Spectrosc.* **2012**, 66, 1365-1384. [3] Lu, F.; Jin, M.; Belkin, M. A. *Opt. Express* **2011**, 29, 19942-19947. [4] Dazzi, A.; Prater, C. *Chem. Rev.* **2016**, DOI: 10.1021/acs.chemrev.6b00448.

Enabling Discoveries: TERS Imaging of the Grain Boundaries and Unexpected Nanoscale Heterogeneities in 2D Semiconductors.
Andrey Krayev

In recent years tip enhanced Raman scattering (TERS) imaging made a dramatic progress from "once in a lifetime experience" to a routine everyday characterization tool for nanoscale Raman imaging of various materials. This progress was mainly determined by the advent of new generation of TERS instrumentation, development of advanced TERS imaging modes and availability of highly reproducible commercially available TERS probes with high enhancement factors. Since TERS is intrinsically a hyperspectral imaging tool, it does not rely on apriori knowledge of the spectral properties of the investigated materials and therefore allows discovery of unexpected Raman peaks and mapping the distribution of their intensities. Cross-correlating the nano-Raman maps with other channels provides by scanning probe microscopy such as the topography, surface potential, conductivity, photocurrent, friction etc, is another powerful tool allowing comprehensive characterization of novel materials at nanoscale. We'll illustrate usefulness of TERS imaging and cross-correlation of nano-Raman with other SPM channels by the discovery of intrinsic nanoscale p- and n-doped domains in WSe2 flakes exfoliated to gold as well as the presence of interesting and potentially useful defects in CVD-grown flakes of this materials. We'll also discuss rather unexpected location of the grain boundaries in CVD grown flakes of MoSe2 and their evolution in the course of ageing of these flakes and in the course of their transfer from the original substrate.

Laser Ablation DART Imaging Mass Spectrometry (LADI-MS)—Applications to Forensics

Rabi Musah, Kristen Fowble

Laser ablation direct analysis in real time imaging mass spectrometry (LADI-MS) is a MS imaging technique in which a laser, interfaced with a DART ion source, is integrated with a mass analyzer to generate small-molecule spatial distributions for a variety of sample types of varying shape and size. The utilization of a DART source provides significant advantages including the ability to: (1) avoid sample pre-treatment steps; and (2) analyze samples under ambient conditions. The technique also enables the analysis of irregularly shaped samples. A 213 nm Nd:YAG laser coupled to a DART-TOF-HRMS system was used. The DART ion source was operated in positive and negative ion modes. Samples were affixed to the sample stage using adhesive carbon tape. They were then rastered beneath the laser as it fired at the surface, creating an ablation plume that was subsequently transported to the TOF-HRMS inlet by flow of helium. The applicability of LADI-MS in a forensics context was demonstrated through analysis of inks on paper, and the illustration of the content of psychoactive drugs and gunshot residue components in latent fingerprints, among other experiments.

Pyrolysis -Vacuum Assisted Plasma Ionization Ion Mobility-Mass Spectrometry for Insoluble Polymer Analysis.
Stephen Zambrzycki, Matthew Bernier , James Bradshaw, Facundo Fernandez

Characterization of insoluble and high melting point polymers such as nylon is difficult using conventional soft ionization techniques. Pyrolysis-GC-MS, with its use of electron ionization (EI) and gas chromatography (GC), has been the standard method of analyzing these types of polymers; however, its significant ion fragmentation is one of the major limitations of this technique. Using the same sampling technique of pyrolysis to break the polymeric chains into their monomeric and dimeric components, we present the coupling of a pyrolysis device to our recently described

Vacuum-assisted Plasma Ionization (VaPI) source (Keelor, 2016). Coupling a pyrolysis device to VAPI (Pyro-VaPI) leverages VaPI's enhanced ion collection efficiency, the relative softness of plasma ionization, the ability to probe samples remotely and the use of ion mobility to reduce spectral complexity. Using a home built pyrolysis device that reaches temperatures up to 1000 C, the sample polymer in question is placed into a stainless steel cup that is then inserted into the pyro device. Using a Waters Synapt G2s with a modified DART Vapur® interface to attach VaPI to, a sampling tube is connected via Swagelok® fittings to the VaPI source and the other end of the tube is placed over the outlet of the pyro device. As the temperature ramps on the pyro device, the polymer breaks apart and enters the vapor phase. These vapors are pulled through the transfer tube into the VaPI source to be ionized for MS analysis. Using this approach, we have been able to detect the protonated dimers for nylon 11, nylon 12, and nylon 6/6, the protonated monomers for nylon 6/12, nylon 6/6, and nylon 6/10, and for nylon 6, the protonated dimer and trimer. Further development of the device is required including enhancements in temperature control, testing other polymers such as polytetrafluoroethylene or polymers with dyes/pigments in them, and further leveraging the ion mobility stage to decrease spectral complexity and enhance peak capacity.

Temperature Programmed DART-MS Analysis of Fire Debris
Edward Sisco, Thomas Forbes

Analysis of ignitable liquid residues, commonly referred to as fire debris, represents a substantial area of forensic science that has remaining relatively unchanged for the past decade. While gas chromatography mass spectrometry remains the gold standard in this field, there lies potential for the use of other technologies to provide complimentary information. Once such area which has begun to explored is ambient ionization mass spectrometry. The work highlighted in this paper outlines recent efforts to utilize direct analysis in real time mass spectrometry (DART-MS) for the analysis of fire debris without sample preparation. This technique allows for near-instantaneous analysis of the volatile components found in fire debris. Coupled to a novel independent thermal desorption mechanism, temperature programmed desorption of the volatiles in fire debris can be achieved providing a GC-like chronographic fingerprint of different ignitable liquids. This talk will provide a detailed looking into this novel DART-MS configuration along with work that has been done to optimize the system, evaluate the effect of substrates, determine limits of detection, and highlight come of the advantages, and disadvantages this technique has over existing technology.

Linking smokeless powder residues to pre-burn smokeless powders using DART-TOTMS and GC-MS
Emily Lennert, Candice Bridge

After this presentation, attendees will understand how smokeless powders (SP) and residues may be analyzed pre- and post-burning using direct analysis in real time – time-of-flight mass spectrometry (DART-TOFMS), and how these results compare to traditional gas chromatography-mass spectrometry (GC-MS). Following an explosive event, such as discharge of an improvised explosive device (IED) or firearm, unburned or partially burned SP particles and burned SP residues may be recovered. Therefore, unburned and burned samples were analyzed to simulate potential real world evidence, as well as to examine the effect of burning on the identifiable organic compounds in SPs. Unburned SP extracts were prepared by following a simple extraction procedure: the National Center for Forensic Science's extraction procedures using dichloromethane. Extraction is not necessary for DART-TOFMS analysis of solid SP kernels; however, extractions were performed to allow for GC-MS testing and direct comparison of samples. Each SP was subsequently burned and the residue was extracted in dichloromethane. Pre- and post-burn samples were analyzed by DART-TOFMS in positive and negative mode, with helium carrier gas at 250?C, for the presence of several organic compounds of interest: diphenylamine, nitroglycerin, ethyl centralite, dinitrotoluene, phthalates, and others. Sample extracts were then analyzed by GC-MS to allow for comparison of results obtained by DART-TOFMS and GC-MS. Following characterization of organic compounds in each sample, classification models were created to determine if each burned SP sample would classify with its respective pre-burn group,
i.e how similar the pre- and post-burn samples were. Classification of SPs based on the organic components may aid investigators in determining the brand or origin of a suspect SP, burned or partially burned, recovered at a crime scene. DART-TOFMS allowed for rapid identification of compounds of interest within the samples with ease; sample runs may

be completed within minutes, and mass spectra may be observed and evaluated almost instantly. GC-MS may require over 10 minutes for one sample injection, and the sample run must be completed for mass spectra to be observed or evaluated. DART-TOFMS is a viable method for rapid analysis of SPs, which may be applied to explosive and GSR analysis.

The development of a Lab-on-a-Chip device for the track side detection of equine performance enhancement

Kim Quayle, Egan H. Doeven, Richard Alexander, Yi Heng Nai, Giorgio M. De Guzman, Paul S. Francis, Xavier A. Conlan, Stephen J. Haswell

The monitoring of metabolic performance manipulation within equine sports continues to prove highly challenging due to the sheer range and variability of prohibited substances. Performance modification in horses is defined as the "illegal application of any substance, except normal diet, that might modify the natural and present capacities of the horse at the time of the race" and is carried out with the aim of either increasing or reducing the mobility of a competing horse.[1] Over 500 substances have been identified in the Fédération Equestre Internationale (FEI) communication, which covered the following functional groups, including substances that have a similar chemical structure and biological effect: hormones, anabolic steroids, beta blockers, bronchodilators, non-steroidal anti-inflammatories, anticoagulants, analgesics, haemostatics, narcotics, diuretics, bronchodilators and many more.[2] The long term effects of equine performance modification varies from case to case, but it is not uncommon to result in the long term injury and death of many thoroughbred race horses. Modern methods of chemical analysis used for equine doping cases, such as Ultra Pressure Liquid Chromatography - Tandem Mass Spectrometry (UPLC/MS/MS), require complex sample extractions and bulky equipment that is unsuitable for field based applications. Advances in portable devices capable of low cost multiplexed analysis, with the easy interface with miniaturised sensors and smart technology, has led to the development of called Lab-on-a-Chip devices. This research showcases the development of a Lab-on-a-Chip device that utilizes non-invasive rapid sample collection with minimal processing, as a method for the detection of equine metabolic performance manipulation. Within this study we investigate the use of chemiluminescence and bioluminescence as methods for the targeted drug detection and metabolic biomarkers, inclusive of novel water soluble luminescent complexes. References [1] F. R. Ungemach, Doping control in racehorses, Tierarztl Prax, 1985, Vol 13 (1), pp 35-53 [2] FEI Equine Prohibited Substances List, Fédération Equestre Internationale, 2016, pp 1-15

Building on a Real-time Multivariate Monitoring Platform for Biopharmaceutical Manufacturing Operations: How to Teach an Old Dog New Tricks

Christopher Garvin, Cenk Undey

Biologics process development and manufacturing involves complex process dynamics that result from interactions between, and variability within, chemical, physical, and biological components. To improve process understanding and control, it is important to extract and analyze, in real-time, the large quantity of data generated by these processes toward the goal of managing variation, reducing risk, increasing process yields, and improving product quality. At Amgen, we employ a real-time multivariate monitoring platform to assess the behavior of processes and equipment relative to historical or expected performance. Our models can help detect weak signals in the data that indicate atypical performance of one or a combination of variables prior to an alarm being triggered in the control system, thereby providing an opportunity to prevent an issue before it has occurred. In this talk, we will discuss our past success deploying the platform in GMP environments and recent work to apply the platform to less traditional multivariate monitoring applications, such as new product introductions, high-mix/low-volume manufacturing, equipment preparation steps, and packaging operations.

Validation of Control Strategies Based on Advanced Sensors
Saly Romero-Torres

The use of advanced analytical techniques to control processes is becoming a reality. Nevertheless the notion that process sensors figures of merits should be comparable to those expected from quality measurements or lab based equipment still persistent. During this talk we will discuss the validation approach of a Raman based glucose control strategy for a production bioreactor. The Raman based control strategy reduces the glucose variation which mechanistically translates to better metabolite control.

Process analytical technology (PAT) in continuous bioprocessing
Edita Botonjic-Sehic, Steven Harris

In recent years biotech companies have been emphasizing on the implementation and utilization of process analytical technology (PAT) in continuous bioprocessing. An overview of PAT concepts for biotech processes with emphasis on integrated continuous bioprocessing and production of monoclonal antibodies will be presented. Biotech processes consist of several unit operations, with each unit operation serving a defined purpose. For some of the product quality attributes, a certain level of redundancy in the process is expected. The need for appropriate control of each step and continuous evaluation of its performance is of great importance. The objective of commonly used unit operations in downstream processing of biotech products are to isolate the product of interest from the process stream that comes from the harvest unit operations and contains a variety of impurities. These impurities include product-related variants that might closely resemble the desired product and have an equivalent potency and safety profile. Implementation of PAT for continuous measurement and monitoring of numerous critical quality attributes (CQAs) is being evaluated. UV spectrometer is being proposed as an on-line PAT tool for continuous measurement of protein concentration. In addition, real-time data collection helps in development of chemomtric models used for predictive estimation of properties of a process and hence help in process analysis, optimization and control. Pall Life Science will present data from a specific unit operation in the integrated continuous bioprocessing. By utilizing PAT into integrated continuous bioprocessing; better process understating is gained, production cycle time is reduced, improvement of yield is achieved, efficiency is improved from converting of the batch process into a continuous process, and cost reduction is achieved by the reduction of waste and energy consumption resulting from the real-time release of a batch with improved quality.

Development of Generic Raman Models for Process Monitoring
Thaddaeus Webster, Colin Jaques, Carrie Mason

The manufacture of safe and consistent bio-therapeutic products requires robust process monitoring and control of the production bioreactors. Historically, the Lonza bioreactors that operate platform processes have been monitored by measuring critical process parameters offline. This requires sample removal, expensive and sometimes toxic reagents and skilled operators to analyze the state of the bioreactor. The implementation of novel process analytical technologies for continuous monitoring has been proposed by regulatory agencies as a means of improving process control. To accomplish this an ideal PAT tool should be capable of monitoring several critical process parameters without sample removal from the bioreactor. Inline Raman spectroscopy coupled with multivariate software to develop predictive models for critical process parameters has been proposed to substantially improve control and reduce offline sampling by providing near continuous monitoring of the bioreactor's contents. To this end, generic, cell line-independent Raman models were developed to continuously monitor critical process parameters in bioreactors operating Lonza's platform processes. Generic Raman models were developed for key metabolites, viable and total cell concentration, and product concentration by culturing two different GS-CHO cell lines using a platform process in 12 five-liter bioreactors. Raman spectra and offline measurements were collected twice daily during the course of cell culture in order to construct predictive models. Metabolites, VCC & TCC, and titer were measured offline using a Nova Bioprofile 400, Vi-Cell XR Analyzer, and Protein-A HPLC respectively. Projections on latent structure models for each parameter of interest were created in SIMCA v13.0.3 by regressing Raman spectra with their corresponding offline measurements. Developed models were validated against an independent cell culture round using a GS-CHO cell line not used for model development. Developed predictive models were capable of monitoring changes in the concentration of glucose, lactate, ammonium, viable cell concentration, and total cell concentration with prediction errors of 0.44 g/L, 0.24 g/L, 0.028 g/L, 1.92 x 106 cells/mL, and 1.89 x 106 cells/mL respectively for cells cultured on the platform process. Moreover; the ability to continuously monitor these process parameters using generic models, independent of the cell line, potentially enables feedback control of Lonza's platform processes in the future.

Raman-based bioreactor control: building sustainable applications

John Bobiak, Dimuthu Jayawickrama, Nobel Vale, George Armenante, Greg Lane, Matthew Rehmann

Robustness and sustainability are goals for any analytical method. As Raman techniques are adopted across different bioprocessing locations, a number of spectral artifacts could enter the historical library of spectra. Such sources of spectral variation may lead to prediction error, and ultimately constrain feedback control applications. In this presentation, spectral residuals and background simulation exercises are discussed as the technical foundation for developing robust methods. A bioprocess control loop is described which uses a combination of paced feeding with small, on-demand boluses to maintain a desired glucose concentration. Looking forward, control applications will combine the strengths optical methods with those from other physicochemical techniques like capacitance and microscopy.

Combined Raman and Morphological Analysis of Fluorescent Propellants and Organic Gunshot Residue

Steven Bell, Yen Cheng Ho, Wendy Lee

Gunshot residue (GSR) is typically identified from its inorganic components by a combination of particle morphology with EDX analysis which searches for the characteristic combination of lead, antimony and barium. However, the development of lead-free ammunition undermines this well established procedure. An alternative approach is to investigate the organic particulate residues (principally nitrocellulose) which are also created during discharge. There are several reports in the literature describing the use of Raman spectroscopy for identifying organic components in GSR but there is some disagreement about the extent to which fluorescence is a problem. In addition, the possibility of introduction of nitrocellulose from other sources in the environment has not be considered in detail although this is clearly important if the results are to form the basis of criminal prosecutions. Here we describe an investigation which addresses both these issues by using sample pre-treatment to reduce fluorescence and includes particle morphology as an important parameter in identification of the samples as GSR. A set of ten smokeless propellants and three brands of 9 mm ammunition were tested along with common propellant additives. The majority of the samples showed strong background fluorescence but holding the particles on a microporous support allowed the samples to be rinsed with alcohol solvent mixtures which reduced the fluorescence to acceptable levels in all but one case. The process was designed specifically to retain the original particle morphology so that both chemical and morphological characteristics could be used for identification as GSR. Moreover, these measurements were recorded in a single process which is very useful for evidential purposes. The method was tested on both unfired propellant and GSR collected from test firings using the different brands of 9 mm ammunition. The results showed excellent identification of nitrocellulose in the GSR and discrimination of organic GSR from particulate environmental contaminants. In addition, some discrimination between different brands of ammunition before and after firing was found to be possible.

Standoff Explosive Detection Using an Eye-Safe, Wide-Area Hyperspectral Raman Sensor

Nathaniel Gomer, Charles Gardner, Matthew Nelson

Raman spectroscopy is a valuable tool for the detection and identification of explosive materials because it provides a distinct molecular fingerprint which allows for unambiguous target identification. Raman is most advantageous when utilized with UV excitation, due primarily to the increase in generated Raman scattered light at shorter wavelengths. Current generation UV Raman systems have numerous limitations that hinder their performance and integration into mobile or field portable configurations. These limitations include low signal throughput due to narrow spectrometer slit width requirements, large size and increased weight, and severely diminished area search rate. These limitations are directly related to a system's spectrometer, which is typically dispersive grating based and requires a very narrow slit width and long focal length optics to effectively resolve spectral bands. ChemImage is developing a novel, standoff wide-area Raman hyperspectral imaging system capable of providing high resolution measurements with greatly increased throughput in a small form factor that utilizes defocused laser excitation. This innovation combines a spatial heterodyne spectrometer (SHS), a slit-less spectrometer that operates similar to Michelson interferometer without requiring moving parts, with a fiber array spectral translator (FAST) fiber array, a two-dimensional imaging fiber that provides spatial information of the target area. This combination of technologies (termed FAST-SHS) creates a high throughput Raman hyperspectral imager capable of yielding very high spectral resolution measurements while simultaneously providing expanded area coverage and a faster search rate than traditional Raman systems. This presentation will provide background on the sensor and its underlying technologies and present initial results of the sensor's wide-area detection capabilities.

See_Through Raman Spectroscopy

Jun Zhao, Jack Zhou

A new Raman measurement method is presented that expands the applicability of Raman to See_Through diffusely scattering media such as opaque packaging materials and biological tissues, as well as to thermolabile, photolabile, or heterogeneous samples. Traditional Raman fiber optics probes employ a focused design similar to confocal microscopes. This confocal design has the advantage of maximum throughput. However, when the container diffuses the light strongly, light can no longer be focused inside the container, the confocal approach loses its effectiveness. Spatially Offset Raman Spectroscopy intentionally offsets the excitation beam and collection beam, and can be used effectively to collect Raman signal generated underneath diffusive top layers while largely avoiding their signature from overwhelming that of the sample. However, probes of such design cannot be easily used in a confocal arrangement and are generally inefficient for direct sample measurements. The See_Through Raman configuration illuminates and collects the Raman scattered light from a large sample area with enhanced throughput. This greatly increases the effective sampling depth, allowing the measurement of material inside visually opaque containers via a similar mechanism as SORS. The large area also has the additional advantage of preventing sample damage by reducing the power density, as well as improving accuracy by eliminating heterogeneous effect. With its high throughput design, the See_Through analyzer provides identity of common chemicals through thick packaging material in seconds. For normal measurements, a range of accessories can be used in the confocal configuration for different applications including microscopy and stand-off Raman. This flexibility makes the See_Through Raman a suitable technique for a variety of applications, from material ID to demanding research.

Silver-Gold Bimetallic Nanostructure Incorporated Nickel Foam as a Rugged and Reliable SERS Substrate
Tung Duy Vu

A three-dimensional (3D) nickel foam (NF) incorporating Ag-Au bimetallic nanostructure has been developed for a surface-enhanced Raman scattering (SERS) substrate. The adoption of NF as a substrate frame was to utilize its merits such as high porosity for easy access of analytes and superior chemical/physical stability for ruggedness. Next, the incorporation of Ag-Au bimetallic nanostructure in NF was to achieve optimal SERS measurement by combining the individual advantages of each SERS-active metal, the high SERS efficiency of Ag and good surface stability of Au. For the fabrication, Ag-Sn alloy was initially constructed on NF by electrodeposition and the incorporated Sn was then de-alloyed from the Ag-Sn alloy by free corrosion in NaOH solution. The de-alloying of Sn was to make the surface considerably roughened and improve SERS efficiency in a final substrate. Finally, a galvanic replacement reaction was performed to include Au into the Ag framework to produce NF/Ag-Au substrate. Raman intensity of R6G acquired using the NF/Ag-Au substrate approximately 20 and 76 times greater than those of employing the

NF/Ag-Au substrate without the incorporation of Sn and NF/Au nanoparticle substrate, respectively. In addition, the sensitivity was improved by 2.6 times in comparison with the structure built on a Ni foil. The porous structure of NF was attributed to the improvement. Further, since the inner Ag-Au structures are protected by the outer layer, SERS intensities were not significantly degraded even after considerable polishing of the substrate with a sand paper. Currently, additional modifications are underdoing to improve the sensitivity for sensitivity-demanding applications.

N-Heterocyclic Carbenes as a Surface Modifier for Analysis by Surface-Enhanced Raman Scattering

Michael Trujillo, Joseph DeJesus, Chaoxiong Ma, David Jenkins, Jon Camden

Surface-enhanced Raman spectroscopy (SERS) has developed into a rapidly growing field within analytical chemistry. A fundamental challenge in SERS-based analytical methods is to develop a way to bring a target molecule close enough to a noble metal surface where the scattering signal can be greatly enhanced by localized surface plasmon resonances. Ideally the molecule of interest would adhere to the surface where the enhancement is strongest. Amines and thiols are two functional groups known to have strong affinity for gold surfaces and have therefore been used extensively to functionalize metal surfaces, with thiols dominating because of the stronger binding to the surface. While thiols can be applied in many experiments, they are prone to oxidation and often unable to survive extreme environments such as high/low pH or temperature. A new class of molecules therefore needs to be developed which can rival the binding strength of thiols, yet are stable enough to survive extreme conditions. Herein, we present the development of N-heterocyclic carbenes as a new class of molecules which easily functionalize SERS-active gold surfaces.

Electrophoretic Separation of Single Molecules using Thermoplastic Nanocolumns: Applications in Single-Molecule Sequencing

Steven Soper

In this presentation, we will discuss a novel separation technique for single-molecule analysis, which utilizes capillary electrophoresis (CE) incorporating nanometer-scale columns (width and depth) that are tens of microns in length (nano-CE). Due to scaling-effects, nano-CE offers some unique separation characteristics not associated with micro-scale separations due in part to partial overlap in the electrical double layer (EDL). This results in non-plug like flow and coupled with surface charge effects, can give rise to transverse electromigration. Indeed, free solution separation of DNA molecules are possible in nano-CE that cannot be realized in micro-CE. The columns are produced in thermoplastics using Nanoimprint lithography (NIL) from resin stamps and Si masters. The masters are made using a combination of conventional optical lithography (microchannels) and focused ion beam milling (nanochannels). Thermoplastics can serve as viable alternatives to glass and quartz for nanofluidic devices because of their diverse surface chemistry, optical transparency and the ability to be manufactured in a high production mode using NIL. NIL has proven to be successful in patterning structures to sub-10 nm dimensions with the ultimate resolution determined by the minimum feature size associated with the imprinting tool. In fact, we have made columns as small as 18 nm (width x depth) using NIL. We will discuss the electrically driven transport properties of both small (deoxynucleotides monophosphates, dNMPs) and large (DNAs) molecules traveling through thermoplastic nanocolumns. The effects of scale (i.e., cross-sectional area of polymer nanocolumn), electrical double layer overlap producing non-plug-like flow profiles, surface chemistry, surface charge, transverse electromigration, nano-scale surface roughness and electroosmotic flow (EOF) will be discussed and the effects of these parameters on separation efficiency for nano-CE will be presented. We will also discuss the use of nano-CE as an identification technique for single-molecule sequencing of both RNA and DNA molecules.

Infrared Spectral Histopathology Using (H&E) Stained Glass Slides: A Route to Clinical Translation
Peter Gardner, Michael Pilling, Alex Henderson, Jonathan Shanks, Michael Brown, Noel Clarke

Infrared spectral histopathology (SHP) has shown great promise as a diagnostic technique, with the potential to compliment current pathological methods. However clinical translation has been hindered by the limited options for practical substrates. There are a number of impracticalities of using infrared transmissive substrates such as CaF2 which are both fragile and prohibitively expensive. The alternative method of infrared reflecting slides is controversial since this tansflection sampling geometry results in significant spectral distortion dependent upon sample thickness [1]. Recently, glass has been proposed as a potential replacement which, although largely opaque in the infrared, allows unrestricted access to the high wavenumber region (2500–3800 cm-1) [2]. Recent studies using unstained tissue on glass have shown that despite utilising only the amide A band, good discrimination between histological classes could be achieved, and suggest the potential of discriminating between normal and malignant tissue. However unstained tissue on glass has the potential to disrupt the pathologist workflow, since it needs to be stained following infrared chemical imaging. In light of this, we report on an SHP study utilising coverslipped H&E stained tissue on glass using samples as received from the pathologist. We present a rigorous study using results obtained from 182 prostate tissue cores obtained from 100 different patients on 18 separate H&E slides. Utilising a Random Forest classification model we show that four classes of histology of an independent test set can be classified with a high degree of accuracy (>90%). We investigate different degrees of staining using nine separate prostate serial sections, and demonstrate that we discriminate tissue types based on inherent spectral-biomarkers rather than the presence of the stain. We investigate different degrees of staining using nine separate prostate serial sections, and demonstrate that we discriminate on biomarkers rather than the presence of the stain. Finally, using a four-class model we show that we can discriminate normal epithelium, malignant epithelium, normal stroma and cancer associated stroma with classification accuracies over 95% [3] These results demonstrate that rapid IR imaging could be employed in the clinic without the need to use any specialist substrates. This overcomes one of the major obstacles to clinical adoption and could pave the way for a commercial IR screening system.
[1] Bassan P.; Lee J.; Sachdeva A.; Pissardini J.; Dorling K. M.; Fletcher J.; Henderson A.; Gardner P.; Analyst 2013, 138, 144 – 157 [2] Bassan P.; Mellor J.; Shapiro J.; Williams K. J.; Lisanti M.; Gardner P.; Anal. Chem. 2014, 86, 1648–1653 [3] Pilling M. J.; Henderson A.; Shanks J. H.; Brown M. D.; Clarke N. W.; Gardner P.; Analyst 2017, 142, 1258–1268

Designing a 3D culture platform for scalable quantitative drug screens
Amanda Hummon, Gabriel LaBonia, Matthew Boyce, Matthew Lockett

Bringing a new cancer therapeutic to market is challenging and expensive. Before a new drug can be brought to market, it must be rigorously evaluated to determine its efficacy and toxicity. Most initial drug testing is performed on monolayer cultures, which lack most of the complexity of a tissue or organ structure. 3D cell cultures recapitulate some of the complexity of in vivo tissues, and can be more easily manipulated than animal models to explore a range of chemical perturbations. While 3D cell cultures are increasingly popular laboratory models, a content-rich chemical analysis platform has not been available to evaluate the effect of drug candidates with the ease and throughput of many currently used monolayer culture assays. Using a blend of spheroids and paper based cultures, we are developing a drug screening platform to examine the response of hundreds of cultures simultaneously. Our platform enables both chemical and phenotypic readouts of the cells. The chemical readouts include quantitative proteomic profiling of the cells and imaging mass spectrometric analysis of the drugs and their metabolites. Viability, proliferation, and other phenotypes are evaluated by flow cytometry and fluorescent microscopy. With our paper based systems we can control the oxygen concentration, simulating the hypoxic microenvironment of a tumor. Using this platform, we are examining the response of colon carcinoma cells to a range of therapeutics under different oxygen concentrations.

After the diagnosis: monitoring cancer therapies with SPR sensors
Jean-Francois Masson

Significant focus is rightfully devoted to cancer diagnostics in the community. However, sensors can have a broader reach in cancer management beyond providing sensors for diagnostic purposes. Monitoring the therapeutic effect of drugs or dosage can have significant benefits to the patients. A suite of plasmonic instrumentation, surface chemistry and nanomaterials is thus reported for the quantification of a chemotherapeutic drugs, anti-drug antibodies and for HER-2. A fully integrated, multi-channel and small instrument has been developed for surface plasmon resonance (SPR) with several accessories such as a microdialysis for whole-blood sensing. Monitoring the therapeutic drug methotrexate (MTX) was achieved in actual clinical samples. The developed method consisted of employing folic acid functionalized gold nanoparticles (Fa-AuNP) in competition with MTX for a specific number of its targeted enzyme, human dihydrofolate reductase (hDHFR) immobilized on the sensor surface. Preliminary data with actual clinical samples from a local hospital shows good agreement between the SPR method and LC-MS/MS and Fluorescence Polarization Immunoassay (FPIA). In a second example, the detection of anti-asparaginase antibodies in patients during ongoing leukemia treatment was performed with clinical samples. This study showed the challenge of working with undiluted clinical samples in SPR sensing. Lastly, we have developed new surface chemistry for sensing of protein markers found in biopsied tissues. This was exemplified with HER-2 detection in HER-2-positive cell lysate. These examples with an integrated SPR instrument demonstrate the use of portable nanobiosensors in different clinical applications.

Mechano-NPS for Cancer Diagnostics
Lydia Sohn

We have developed a label-free method of screening cells for their phenotypic profile, which we call Node-Pore Sensing (NPS). NPS involves measuring the modulated current pulse caused by a cell transiting a microchannel that has been segmented by a series of inserted nodes. When a “contraction” channel through which cells can squeeze is inserted between the nodes, mechanical phenotyping is possible. In this talk, I will show how our method, which we call “mechano-NPS”, can screen cells multiparametrically, from determining cell size to measuring resistance to deformation and transverse deformation to quantifying cell recovery from deformation. I will show how mechano-NPS can distinguish malignant from non-malignant epithelial cells, discriminate between sublineages and chronological age groups of primary human mammary epithelial cells, and finally track malignant progression. Finally, I will discuss mechano-NPS’s potential clinical application in early breast cancer detection.

Bioanalytical and Nanomedicine Applications of Gold Nanorods, Nanostars, and Magnetic Nanoclusters
Alexander Wei

Plasmon-resonant nanomaterials with near-infrared (NIR) activity continue to provide outstanding opportunities in biophotonics research. In particular, coupling plasmon resonance with magnetomotive properties can introduce new twists to popular methods for biological imaging and chemical sensing, with application toward current objectives and challenges in the bioanalytical sciences. Topics to be discussed include magnetomotive modalities for biological imaging and surface-enhanced Raman scattering (SERS), improvements in surface chemistry permitting greater analyte specificity, the controlled delivery and intracellular release of siRNA using NIR pulsed laser excitation, and efforts to characterize components of the protein corona on nanoparticles in physiologically relevant media.

Gold Nanostars for Promoting Molecular Adsorption of Non-thiolated Molecules
Amanda Haes

Assessing and controlling solution-phase nanoparticle stability is important for exploiting the chemical and physical properties of nanomaterials but must be balanced with molecular adsorption in localized surface plasmon resonance (LSPR) and surface enhanced Raman scattering (SERS) spectroscopies. As such, trace molecule signals may be undetectable if they exhibit poor affinity to SERS-active surfaces. In this presentation, gold nanoparticle architecture and solution composition are considered as they relate to plasmonic detection of non-thiolated small molecules. Gold nanostars with varying sizes are used. Surface morphology and roughness, nanostar stability, as well as solution conditions are used to identify key features that rationally promote adsorbate-surface interactions without compromising the physical stability of the nanostructures. Systematic studies using a series of non-thiolated molecules are evaluated so that impacts of molecular adsorption and orientation can be learned. Solution composition induces changes in nanoparticle architecture and/or adsorption processes, yet systematic responses are observed. As a result, we expect these studies will broaden the scope of SERS and plasmonic-based assays as small molecules with weak affinity to metal nanostructures will be more readily detected.

Gold Nanostars: Can Cinderella Become a Princess?
Laura Fabris, Supriya Atta, Ted V. Tsoulos

Gold nanostars are powerful nanostructures when intense near field enhancement is required or hot carriers are involved. However, despite the wealth of applications reported in the literature, synthetic protocols ensuring sample monodispersity and batch-to-batch reproducibility, and detailed, shape-dependent characterization still appear to elude us. As a result, the reliable applicability of these nanomaterials has been drastically hampered and their use in device implementation has been extremely limited. To address these issues, our group has initiated a two-pronged experimental and computational effort to characterize, understand, and predict the properties of these nanoparticles, while developing novel and improved synthetic strategies that afford high monodispersity and reproducibility. For instance, we have recently implemented a novel experimental/computational method to determine volume, surface area, and shape-dependent extinction coefficient of nanostars (and in principle of any non-spherical nanoparticle) by first estimating them via a semi-empirical model and then confirming them with high-resolution TEM and STEM experiments, inclusive of tomography. We have employed the reconstructed 3D nanostar topography to calculate intensity and distribution of local scattered electric fields for the experimentally synthesized nanostars using finite element modeling, thus providing, for the first time, a practical, yet quantitative method to estimate their fundamental morphology-dependent physical and optical parameters. Furthermore, we have devised a novel synthetic strategy that allows us to obtain nanostars with finely tunable spike lengths and narrow plasmonic resonace bands that can be exactly reproduced with our 3D E-field finite element modeling. The ability to exactly reproduce computationally the plasmonic properties of the nanostars affords us the opportunity to use them as test beds to understand how plasmonic resonances in highly anisotropic nanoparticles may differ from those of their more characterized and understood spherical counterparts. Taken together, we believe that these results may be important to exploit the exceptional features of these nanoparticles and thus to render them truly applicable.

Beyond the Nanostar's Plasmon
Sean Burrows, Lixia Zhou

Nanostars are a relatively new geometry to the nanoparticle family. There are a variety of nanostar synthesis approaches that produce a large range of nanostar geometries. These geometries range from those with short stubby spikes to long protruding spikes that each have unique plasmon absorption and scattering components.

Reproducibility in plasmon spectra and signal generation has been a persistent challenge with analytical applications of nanoparticles and especially nanostars. We previously discovered that synthesis at temperatures less than room temperature produced more reproducible plasmons. We then became interested in two-photon induced Metal Enhanced Fluorescence on nanostars because single-photon studies did not observe enhancement through typical near and far field mechanisms. We recently found the nanostar plasmon's proximity to the fluorophore's excitation and emission spectra played a role in the observed fluorescence intensity. We are now interested in the possible mechanisms and causes of quenched or enhanced fluorescence. Our current focus is on the relationship between the nanostars optical density profile, scatter, and observed enhancement.

Chiro-optical activity of symmetric and asymmetric dimer plasmonic nanocrescents

Jennifer Shumaker-Parry, Peter Stevenson, Mark Swartz, Caitlin Coplan, Venkata Ananth Tamma, Vartkess Ara Apkarian

Nanosphere template lithography (NTL) is a simple, versatile fabrication method for preparing plasmonic nanoparticles with controlled structural and optical properties. When extending NTL to aluminum, the rather robust native oxide layer poses a challenge for the Ar ion milling step, thus limiting the scope of shape control. In order to overcome this challenge, a sacrificial copper layer was incorporated into the NTL fabrication process, ultimately eliminating the need for a milling step. Asymmetric dimer nanocrescents fabricated using this copper mask NTL process elicit giant circular dichroism (CD) responses over broad wavelength ranges. These responses potentially represent a simple approach in controlling desired handedness in chiral nanomaterials, irrespective of the inherent structural handedness. Orientation chiro-optical response dependencies are observed relative to in-plane as well as out-of-plane rotation. Optical responses are compared between symmetric nanocrescents and asymmetric dimer nanocrescents (i.e., nominally achiral and chiral nanostructures). Controlled chiro-optical response of plasmonic antennas could have impact on multibillion dollar endeavors in pharmaceutical, agriculture, food and specialty materials industries. It is anticipated that the understanding of the chiro-optical properties of the dimer nanocrescents relative to structural considerations as well as orientation-dependent handedness will facilitate chiral plasmon-driven chemistry, such as enantiomeric detection or enantioselective photochemical synthesis.

Infrared Response of Sub-Micron-Scale Structures of Poly(Oxymethylene): Surface Polaritons in Polymers

Naoto Nagai, Makoto Okawara, Yuta Kijima

An investigation of the infrared (IR) spectra of poly(oxymethylene) (POM) mold plates was undertaken to determine the sub-micron-scale morphology and molecular orientation. The nest-structured cells concerned with the orientation were observed from secondary electron microscope (SEM) measurements with the aid of Raman spectroscopy. The intensity of the anomalous IR reflectance peak of the C-O stretching A_2 mode

depends on the widths of the POM layers in the SEM image along the orientation direction. The results suggest that the spectral features originate from the Berreman effect of the bulk polaritons and the radiative surface polaritons. These polaritons will appear or disappear according to the intensity of the dipole moments of the C-O moiety (molecular orientation) and the characteristic morphologies, depending on the melt flow and/or preparation conditions. Moreover, the IR spectra of certain treated samples suggest that enhancement of the electromagnetic fields from the gap modes and transition dipole-dipole coupling influence the spectral shapes. If we accept the proposed interpretation, it suggests that IR spectroscopy has the advantage of investigating higher-order structures by using the various electromagnetic effects. The aforementioned phenomena should be universal effects in certain polar polymers with large dipole moments. It should be noted that such anisotropic (surface) polaritons are frequently transformed from "pure vibrations" by grinding (KBr or Nujol technique) and/or cutting in polymers. Therefore, we should carefully analyze such polar polymers when we use samples prepared via destructive techniques.

The wonderful world of high-precision isotopic analysis using multi-collector ICP-MS
Frank Vanhaecke

Originally, ICP-mass spectrometry (ICP-MS) was developed with the aim of combining the detection power of graphite furnace atomic absorption spectrometry in trace element determination with the multi-element capabilities of ICP-optical emission spectrometry and it can be stated that ICP-MS has more than fulfilled this expectation. Additionally, as a mass spectrometric technique, ICP-MS also provides information on the isotopic composition of the elements. This capability can, e.g., be relied on for elemental assay using isotope dilution as a highly reliable calibration approach or for tracer experiments with stable isotopes. Multi-collector ICP-MS (MC-ICP-MS) is a variant of ICP-MS specifically designed for isotopic analysis with a precision high enough to reveal and quantify natural variations in the isotopic composition of metallic and metalloid elements. For this purpose, an MC-ICP-MS instrument is equipped with a double-focusing sector-field mass spectrometer of Nier-Johnson geometry and an array of Faraday collectors, permitting the simultaneous monitoring of the intensities of the ion beams of interest. Under ideal circumstances, an isotope ratio precision down to 0.001% RSD can be realized. Elements with one (e.g., Sr) or more (e.g., Pb) radiogenic isotopes show a quite pronounced natural variation in their isotopic composition. Isotopic analysis of these elements is carried out for the purpose of geochronology (dating) or of provenance determination (i.e. determination of the geographical origin) of a variety of materials, ranging from raw materials used in the manufacturing of (pre)historic artefacts and objects of art, over atmospheric dust and agricultural products to human remains. Isotopic analysis of elements (only) showing natural variation in their isotopic composition as a result of isotope fractionation effects can be relied on for provenance determination purposes as well, but also for a multitude of other application types. Isotope ratios can be used as a paleoproxies for revealing the conditions prevailing at the time a process or reaction occurred, for obtaining an improved insight into the risks of environmental contamination and are currently even explored as novel biomarkers signaling the occurrence of diseases that otherwise can only be established at a later time or via a more invasive approach. Real-life examples from the author's lab will be discussed.

CHEMICAL RESOLUTION IN TANDEM ICP- MASS SPECTROMETRY: APPLICATION FOR MULTI-ELEMENT ANALYSIS IN MICROSAMPLES OF BIOLOGICAL FLUIDS
Eduardo Bolea-Fernandez, Kim Phan, Lieve Balcaen, Martin Resano, Frank Vanhaecke

Inductively coupled plasma – mass spectrometry (ICP-MS) is the most powerful technique for (ultra)-trace element determination in the most diverse sample matrices. However, since its commercial introduction in 1983, the scientific community has also been aware of its drawbacks, the occurrence of spectral overlap being the major one. Throughout the "ICP-MS history", several improvements have been made, permitting the user to tackle the problem of spectral interferences, although truly simultaneous (i.e. within a single measurement and under identical conditions) interference-free multi-element determination at ultra-trace levels via ICP-MS still remains challenging task. The use of high-resolution sector-field ICP-MS (HR-SF-ICP-MS) operated at higher mass resolution is an elegant approach that preserves the multi-element capabilities, although at the cost of sensitivity (1 - 2 orders of magnitude decrease). On the other hand, the use of collision-reaction cell technology in quadrupole based ICP-MS (ICP-CRC-QMS) instrumentation typically makes use of different methods and reaction gases for different analytes, thus requiring sequential measurements. In 2012, tandem ICP-mass spectrometry (ICP-MS/MS) was introduced onto the market. This relatively novel setup enables double mass selection by means of the introduction of an additional quadrupole located before a CRC system (Q1-CRC-Q2). By means of Q1, all concomitant (matrix) ions with a mass-to-charge (m/z) ratio different from that of the target nuclide can be easily removed, thus improving the control over the in-cell chemistry.[1] In this presentation, a general overview of the operating principles of ICP-MS/MS will be followed by a description of some key applications carried out at Ghent University in the context of the development of multi-element methods for the determination of biologically relevant metals at ultra-trace concentration levels. The use of chemical resolution, relying on the use of a highly reactive gas (CH_3F) in the CRC of an ICP-MS/MS instrument, offers capabilities for interference-free multi-element analysis at concentration levels beyond those of other present-day ICP-MS instruments[2] and enables reliable determination of various elements in very low sample amounts, as is the case in microsampling.[3] [1]Bolea-Fernandez et al., J. Ana. At. Spectrom., 2017, DOI:10.1039/c7ja00010C [2]Bolea-Fernandez et al., Anal. Chem., 2014,86,7969 [3]Bolea-Fernandez et al., Anal. Chim. Acta, 2016,941,1

Transient signal approaches in sector-field ICPMS for fast and sensitive trace element and isotopic analysis.

Christophe PECHEYRAN, Sylvain BERAIL, Oriol BALTRONS, Martin RESANO, Maité ARAMENDIA MARZO, Loic MARTIN, Chantal TRIBOLO, Norbert MERCIER, Bénédicte LELIEVRE

Developing reliable methods is a common thread in analytical chemistry. Facing the need to analyze more and more samples, the development of fast methods is now one of the driving forces in this field in order to improve the sample throughput and lower the price of analysis. In the last decade, multielemental and isotopic analyses at the trace level have received a growing interest and here again the speed of analysis if of great importance. In some extent, laser ablation, coupled to ICPMS fulfills these requirements. Gas chromatography can also be used to speed up the sample preparation processes in isotopic analysis. In addition, the arrival on the market of microsamplers allowing introduction of volume as small as few microliters in a short period of time (typically few seconds) opens new perspectives for fast and sensitive trace element and isotopic analysis, provided that appropriate data processing and time lag detector correction are used. We will present here some results based two analytical approaches first in a context of medical diagnosis and then in a context of archeological dating of ancient objects. The first approach relies on dried drop ablation by high repetition rate femtosecond laser ablation coupled to HR or MC-ICPMS and the second relies on fast introduction of low sample volume (microliter range) in a total consumption nebulizer used with HR or MC-ICPMS.

A high temperature torch integrated sample introduction system for characterization of nanoparticles in biological samples

Martin Resano, Esperanza García-Ruiz, Maite Aramendía, Diego Leite, Águeda Cañabate, José Luis Todolí

The performance of a high temperature torch Integrated Sample Introduction System (hTISIS) using inductively coupled plasma mass spectrometry (ICP-MS)1 is evaluated in the context of characterization of nanoparticles (NPs) in biological matrixes. The advantages offered by such a device maybe significant, not only because sensitivity can be usually improved, which is not significant at all in case of using single particle (SP)-ICP-MS, but because a higher, closer to 100% efficiency may help in diminishing any bias introduced when calibrating with non-matrix matched standards. Different analytical parameters are investigated in detail, including different sample introduction procedures (continuous aspiration and air-segmentation), and the combined the effect of the TISIS temperature on sample efficiency, NP characterization and the presence of different matrixes, in order to find conditions in which accurate results can be achieved for different types of metallic NPs. 1. A. Cañabate, E. García-Ruiz, M. Resano and J.L. Todolí, J. Anal. At. Spectrom., 2017, 32, 78-87.

Advances for the Determination of Elemental Impurities in Pharmaceutical Products

Erico Flores

Nowadays, there is a concern to the development of reliable methods for the determination of inorganic impurities in pharmaceuticals. It is due to the recent changes in regulatory requirements and the need for updating the current limit tests used in pharmacopeias for routine quality control. However, even considering the recent advances in instrumentation, non-specific methods still have a spread use in pharmacopeias for elemental impurities evaluation, especially for the determination of toxic elements, such as Cd, Hg and Pb. The use of inductively plasma-based techniques as ICP-MS and ICP-OES is a suitable alternative for this purpose and pharmacopeias are changing the official methods to comply with the new requirements. Strategies used for sample preparation including wet digestion and closed vessels combustion methods will be presented and the respective feasibility for further ICP-analysis will be discussed as well as the current challenges and trends for elemental impurities determination in raw materials and pharmaceutical products.

Role of Plasma Techniques in Petroleum Business

Francisco Lopez-Linares, Jenny Nelson, Estrella Rogel, Cesar Ovalles , Laura Poirier

The assessment of trace metals in petroleum feeds and derivatives is vital information required for running sustainable and daily petroleum operations. Spectrochemistry techniques such as Inductively Coupled Plasma Optical Emission Spectroscopy (ICP-OES), Inductively Coupled Plasma Mass Spectrometry (ICP-MS) and more recently Microwave Plasma-Atomic Emission Spectroscopy (MP-AES) using nitrogen as sustained gas are used as primary detection methods in different petroleum facilities due to their ability to perform multielement analyses, covering a broad range of concentrations as well as being robust and reliable. While ICP-OES is considered the workhorse technique, ICP-MS is becoming more integrated into petroleum laboratories due to its maturity and potentiality. The possibility to couple separation techniques with ICP-MS has been envisioned as a potential tool to go beyond of total metal determination. More recently, MP-AES is starting to call the attention to different petroleum lab as a possible alternative for elemental analysis, due to its ability to be independent of argon as the sustained plasma gas. In this talk, we will present some examples how the metals present in petroleum feedstock's can be determined by total concentration using different sample preparation methods as well as plasma techniques. Furthermore, the metal distribution along the boiling point, by using High-Temperature Gas Chromatography Coupled with ICP-MS (HT-GC-ICP-MS) or by solubility class in the non-distillable fractions by combining the asphaltene determination by in-line filtration technique with ICP-MS (ILF-ICP-MS) will also be covered.

Application of 2D correlation spectroscopy in characterization of thin film of biodegradable polymer

Young Mee Jung, Yujing Chen, Yeonju Park, Isao Noda

It has been well-documented that the physical properties of polymer thin film are significantly different from those of bulk sample. Proper understanding of the thermophysical properties and crystal structure of polymer thin film under spatial confinement is crucial in the field of polymer science and nanotechnology engineering. Biodegradable poly(3-hydroxybutyrate) (PHB) based copolymer poly(3-hydroxybutyrateco-3-hydroxyhexanoate) (PHBHx) has recently received considerable attention owing to its promising environmental, pharmaceutical and biomedical applications. We have recently investigated thermal behavior of spin-coated films of P(HB-co-HHx) copolymer and its blends with polyethylene glycol (PEG) by using 2D correlation spectroscopy. 2D correlation spectroscopy (2DCOS) is a well-established analytical technique that provides considerable utility and benefit in various spectroscopic studies of polymers. 2DCOS is now accepted as a powerful analytical technique that can provide the improved characterization of structural changes of materials, which is not readily observed in the conventional spectroscopy. 2DCOS opens a way to monitor the evolution of spectral changes of system of interest obtained as a function of a perturbation. This presentation reports on the potential, utility and versatility of 2DCOS for polymer. Details of 2D correlation spectra of P(HB-co-HHx) and P(HB-co-HHx)/PEG blends will be discussed.

Two-Trace Two-Dimensional (2T2D) Correlation Spectroscopy – a Method for Extracting Useful Information from Only a Pair of Spectra

Isao Noda

Two-dimensional correlation spectroscopy is a popular technique for analyzing systematically changing spectral intensities of a system under the influence of an external perturbation. It has traditionally been believed that at least three and preferably many more spectral data are required for the meaningful 2D correlation analysis. A newly introduced technique called two-trace two-dimensional (2T2D) correlation spectroscopy actually requires only two raw spectra to yield some useful information. In 2T2D correlation spectroscopy, bands originating from the same species are synchronized, while those from different constituents of the system are asynchronously correlated. 2T2D spectra may also be used in a form of multivariate curve resolution analysis to estimate sometimes even more than two traces of pure component spectra from the original two measured spectra of multiple component mixtures.

Vibrational Spectroscopy with Handheld Instruments: Recent Advances and Future Aspects

Heinz Siesler

Although the last four decades are characterized by a multiplicity of hard- and software developments for vibrational spectroscopy such as hyperspectral imaging, introduction of light-fiber optics, special probes and multivariate, chemometric evaluation routines the techniques always remained a domain for scientists. In contrast, the recent developments of miniaturized, handheld instruments show promise that in the future their use may be expanded to non-expert user environments. Inside the molecular spectrometer market, the fastest growing segment is miniature spectrometers and Tematys (Paris) expects the market of miniature spectrometers to reach $ 300m by 2021.This significant growth will not only be based on a wider adoption of spectrometers for industrial process control, in-the-field testing and on-site measurements but also on point-of-care tests and every-day-life consumer applications like food testing. The presentation will provide an overview on the building principles of the most recently introduced handheld spectrometers and highlight their performance by means of selected application examples.

IR spectroscopy: a powerful tool for analysis of perfluroalkyl compounds

Takeshi Hasegawa

A fully fluorine-substituted alkyl group is called a "perfluoroalkyl (Rf)" group. Since the chemical structure looks similar to that of a normal alkyl group, the spectral analysis has long been conducted on an extended way of normal alkyl groups even for an Rf compound. For example, $?_s CF_2$ is often considered to have an analogy of $?_s CH_2$. As a matter of fact, however, this mode appears at totally different positions at ca. 1150 and 720 cm^{-1} for IR and Raman spectra, respectively, which means that a largely different physical scheme is necessary to discuss IR spectra of Rf compounds. In this talk, fundamental matters of IR spectroscopy specific to Rf compounds will comprehensively be presented as well as another matter of the Berreman shift, which are both quite useful for quantitatively discussing a material of an Rf compound.

Investigation for coordination bond in Li+-Poly(ethylene glycol) complex by attenuated total reflectance spectroscopy in the FUV region.

Yusuke Morisawa, Nami Ueno, Tomonari Wakabayashi

Investigations for electronic states of polymers in FUV region have still been rare because it is difficult to observe variety of samples systematically. Recently, ATR-FUV was developed, and this technique is suitable for observing FUV spectra of liquids and solids easily and feasibly. Using ATR-FUV systematic measurement of FUV spectra of PECs composed of Li electrolytes and poly (ethylene glycol) (PEG) for dependence on variation of anions and their concentration were carried out and the spectra were compared with quantum chemical calculations. ATR-FUV

spectra of PEG shows 3 Rydberg (Ryd.) transitions in the region of 145-300 nm. Observed bands peaked at 155, 161 and 175 nm are assigned to n (OH)-3p Ryd., n (CH_2OCH_2)-3p Ryd., and n (CH_2OCH_2)-3s Ryd. For all the Li^+/PEG composites, ATR-absorbance decreases in the region of 140-200 nm where PEG has absorption bands. The band at 175 nm in PEG shift to shorter wavelength to 170 nm in the Li^+/PEG composites, while the other bands did not show thc pcak shifts. The blue shift in 170 nm
band caused forming the coordinated bond between Li and oxygen atom at the ether of PEG. To elucidate origin of these changes, we calculated simulation spectra with 2 coordinate structures, such as crown ether like structure (CE-like) and intercalation between 2 PEG (intercalation). They are well known structure about Li/PEG composites and suggested by vibrational spectroscopy. Both the results of quantum calculations at the level RCAM-B3LYP-TDFT 6-311++G (2d, p) indicated the blue shift in 170 nm band. The simulation spectrum for CE-like structure shows blue shift that for intercalation. The concentration dependence of the blue shift can be explained by the conversion structure from intercalation to CE-like one with increase the $LiNO_3$ concentration.

Reflectance confocal and fluorescence lifetime endoscopy in the oral cavity

Kristen Maitland, Javier Jo, Yi-Shing Lisa Cheng

Although the oral cavity is a relatively accessible organ, mortality of oral cancer remains regrettably high. Delayed diagnosis is believed to be one of the main causes for the low survival rate, compared to other cancers. Precancerous lesions can be diffuse, heterogeneous, and multifocal, complicating accurate diagnosis of pre-malignancy. Therefore, development of noninvasive clinical tools that can help the clinician identify the most accurate sites of precancerous lesions is of great importance and can potentially increase the diagnostic yield of the overall screening process. In an effort to improve early detection, we have developed a multi-modal, multi-scale optical imaging system based on macroscopic fluorescence lifetime imaging (FLIM) and high resolution reflectance confocal microscopy (RCM) for imaging of the oral mucosa in humans. FLIM provides biochemical screening by probing the endogenous fluorescence of structural proteins (collagen) and metabolic cofactors (NADH, FAD), and RCM provides information on nuclear morphology and overall tissue architecture. We have developed handheld endoscopes for FLIM and RCM designed to access the oral cavity. Following informed consent, patients presenting with oral lesions are imaged with the FLIM and RCM endoscopes prior to excisional tissue biopsy, which is processed for histopathology for diagnostic purposes. Preliminary results from a pilot clinical study will be presented. If this technology proves successful, it may enable real-time detection of premalignant and malignant oral mucosa, surgical margin detection, and treatment monitoring.

Develeopment of CARS system with dual-wavelength oscillation electronically tuned Ti:sapphire laser

Hidetoshi Sato, Bibin Andriana, Hiroko Matsuyoshi, Yasuhiro Maeda, Satoshi Wada

The second wave seems to be coming to the biomedical Raman spectroscopy, which is more extensive than the first wave that opened a gate of biomedical field for Raman spectroscopy in 90th. The driving power of the first wave was the development in near-infrared (NIR-) Raman system and application of chemometrics. The NIR-Raman system with CCD detector, NIR- diode laser and high throughput dispersion polychromator allowed one to obtain good spectra of biological sample and tissue. Chemometrics brought incredible progress to discrimination and quantitative analyses of complicated and transient biological data. The second wave seemed to be driven by imaging technology, based on non-linear Raman techniques, SERS, Raman-Tag, etc. In the present study, we focus into non-linear Raman technique, especially CARS imaging. Although CARS spectroscopy has been developed more than 20 years ago, it was in the hands of laser physics researchers for long time because the CARS system consisted of complex tunable lasers and pulse control instruments.

Although the development of multiplex CARS as well as commercial product of OPO system has taken the hurdle down much lower for biologist to use them recently, the problem has been remained; how to use the CARS spectroscopy with chemometrics. Stability of the present CARS system seems not enough high to extract trace changes from biological data, which arise from difficult controlling of the complex system. We are developing a simple CARS system based on a dual-wavelength oscillation electronically tuned Ti:sapphire laser (2W-ETL) for biomedical applications. 2W-ETL is able to emit 2 different wavelengths of laser pulses simultaneously. Since these 2 pulses are generated in single cavity, their build up times are equally modulated by outer stimulation, theoretically. A CARS spectrum is easily obtained by merely using this laser for excitation. We would like to show applications of chemometrics in biomedical hyperspectral image analysis and empirical study of the 2W-ETL CARS system in our talk.

A systems pathology approach to spectroscopic colon cancer diagnosis
Saumya Tiwari

Cancer is a complex set of diseases, presently characterized largely by the primary site of origin and/or average expression of biomarkers. The diagnostic gold standard, and for much of research, is the appearance of tissue in microscopy images. In colon cancer, minute changes need to be identified in epithelium as it begins to mutate before the disease has spread further. This was done by utilizing infrared spectroscopy which probes biochemical changes in tissue much before they are visible through chemical staining. In this project, we imaged colon biopsy samples by high definition infrared imaging microscopy and tested its viability in the accurate differentiation of different histological classes in anonymized colon biopsy samples collected from patients. Using computational models with IR spectroscopic imaging data, analysis of microenvironment and spatio-chemical features was performed which is an important component of systems pathology. Stromal features were used to stratify outcome of patients. Deep learning approaches further enabled us to separate normal cases from abnormal cases, which could potentially reduce the workload on pathologists.

Synergy of Multispectral Fiber Spectroscopy
Viacheslav Artyushenko, Urszula Zabarylo, Olga Bibikova, Andrey Bogomolov, Francesco Bianco, Iskander Usenov, Tatiana Sakharova, Olaf Minet, Svetlana TonevitskaiaHans Joachim Eichler

Key spectroscopy methods: Fluorescence, Raman and diffuse linear scattering, Mid & Near InfraRed and UV/Vis-absorption, -enable reaction monitoring in real-time collecting molecular composition changes with fiber probes in critical points in-line - and providing thereby an immediate feed-back for process control and its optimization. Synergy of the best combinations for spectral methods can provide a wealth of information that can improve sensitivity, speed, quality, and accuracy in such measurements. Nowadays this advanced concept is under development for industrial process control, but can be also used for so complicated bioreactor as human body - where the spectral fiber sensing used for label free analysis of tissue composition helps to differentiate malignant and normal tissue. Such spectroscopy guided approach helps to realize less invasive, but complete tumor removal or treatment -in contrast with the current statistics when in 25-30% of cases malignant tissues are left after operation (even after excessive resection of normal tissue). Results of multispectral cancer diagnostics by all key methods of Raman, fluorescence, diffuse reflection & MIR-absorption spectroscopy will be compared when used for the same spot of tissue - either for selection of the most specific, sensitive and accurate method or for their combination resulting in the synergy enhanced effect. The most informative spectral features defined for a distinct organs/ tumor allow to design customized portable spectral fiber sensors - with reduced cost and modern IT-features: with IP-address for each one and real time data transfer to the relevant iCloud for professional data treatment. These innovative features of diagnostics sensors will be close to IoT platform -to enable fast clinical data accumulation from the similar operations worldwide and fast guidance for surgery and therapy in-vivo to enhance its efficiency. Examples of multi-spectral tissue diagnostics will be presented toward the future clinical applications of tumor margin sensors. Translation of described spectral methods into clinical practice will be discussed in comparison with the other method of optical diagnostics which should enhance modern medicine by less invasive, more precise and more effective methods of clinical fiber spectroscopy.

Decoding breast cancer-induced stromal adaptations in pre-metastatic lungs with label-free Raman spectroscopy

Santosh Paidi, Asif Rizwan, Chao Zheng, Menglin Cheng, Kristine Glunde, Ishan Barman

Metastatic disease remains the largest cause for breast cancer-related mortality. Extending life expectancies demands a better understanding of metastatic progression and the underlying molecular adaptations at the distant site. Emerging reports increasingly suggest an active role of pro-metastatic environments created at secondary organs (prior to arrival of tumor cells) by primary tumors in supporting cancer cell infiltration in the future. However, a comprehensive understanding of such 'pre-metastatic niches' remains elusive owing to the limited set of tools that can map subtle differences in molecular mediators in organ-specific microenvironments. Novel non-perturbative tools that can provide quantitative insights into subtle changes in biochemical composition of secondary sites prior to morphological manifestations are urgently required. To address this unmet need, we have developed an approach to detect pre-metastatic changes in the lung microenvironment, in response to primary breast tumors, using a combination of metastatic mouse models, Raman spectroscopy and multivariate analysis of consistent patterns in molecular expression. Specifically, we employed orthotopic breast cancer xenografts comprised of high (MDA-MB-231) and low (MCF-7) metastatic tdTomato fluorescent protein expressing cells. Label-free Raman spectroscopic mapping was employed to record the molecular content of pre-metastatic lungs. Our study reveals discriminative Raman features, characteristic of the collagenous stroma as well as proteoglycans, that uniquely identify the metastatic potential of the primary tumor by recapitulating the compositional changes in the lungs. The spectroscopic findings agreed with Masson's trichrome staining observations and gene expression analysis of microarray data of pre-metastatic lung samples from mice harboring breast tumor xenografts of varying metastatic potential. Furthermore, the findings suggested a continuous pre-metastatic niche formation model from primary tumors with low and high metastatic potential, thus implying that these pre-metastatic adaptations are a necessary condition for further progression but not sufficient for eventual success of metastases. The presented data are both unique and complementary to that acquired using other analytical tools, notably downstream genomic tests, fluorescence-tagged cell tracking and emerging mass spectrometric methods. Overall, our findings create a new landscape for spectroscopic monitoring of pre-metastatic niche by uncovering discriminative stromal spectroscopic features in secondary sites.

Making the Jump from Graduate School to the Workforce

Anthony Stender, Alex Nemiroski, Karen Esmond-White, Barbara Smith, Alex Scheeline, Katelyn Mason, Colin Ingram

The purpose of this roundtable discussion is to incorporate multiple perspectives from scientists working in industry, government, and academia on how to navigate the early stages of a professional career. The panelists will share how they made the jump from school to the workforce, and then the floor will be opened up to questions from the audience. How do I land a position in academia? How about in Industry? Do I need a postdoc? Are there alternative options? How do I make myself more employable in a difficult job market? These are just a few of the many typical questions that graduate students and early career scientists find themselves asking while occupied with a job search. Scientists often approach decision making in a methodical manner. However, career decisions can be very complicated due to financial and other personal considerations. The timing of openings can seem mysterious or even cruel to job seekers. Many people tout the benefits of using social media or search engines to find employment, but even those approaches seem to have their shortcomings. In today's challenging job market, it is important to not become discouraged during the job hunt. Postdoctoral positions may seem like a natural first step after graduate school, but not every path requires one. Furthermore, postdocs can be difficult to secure without a fellowship to support your research. Alternative options do exist, including adjunct professorships, joining a startup, or stepping directly into an industrial position. Temporary early-career positions can provide you with opportunities to grow as a scientist and to forge new work-related relationships, but they are indeed meant to be temporary stops in a career. Once you finally make the jump into a more secure position, other career-related concerns can arise, such as job satisfaction, climbing the career ladder, and achieving balance between career and personal interests. These topics and more will be addressed as part of this special session.

Monolayer Dynamics at the Air/Water Interface
Michael Fayer

Monolayers and very thin films are fundamentally interesting and play important roles in naturally occurring phenomena and technological processes. In bulk materials ultrafast two dimensional infrared (2D IR) spectroscopy has developed into a powerful method for studying structural dynamics and molecular interactions. The advent of acousto-optic pulse shaping methods has greatly expanded the applicability of 2D IR spectroscopy to an increasing range of important materials. However, it has still been very difficult to study a monolayer or a very thin film because of the small number of molecules that comprise the sample, resulting in very weak signals. We have recently developed a new method for performing 2D IR experiments on monolayers and thin films that greatly increases the signals. The method is briefly described. The new method has been applied to study the dynamics and structure of a monolayer at the air/water interface. The surface pressure-area isotherm of tricarbonylchloro-9-octadecylamino-4,5-diazafluorenerhenium(I) (TReF18) monolayer has been reported in the literature. The symmetric CO stretch serves as an excellent vibrational probe. 2D IR measurements of spectral diffusion and chemical exchange were conducted at several surface densities. In addition, a water soluble version of the vibrational probe, ReCl(CO)3bipydicarboxylate (TReBDC), was studied in bulk water for comparison. In bulk water, TReBDC yields a fast spectral diffusion component of 1.5 ps, the same as the previously measured time for water hydrogen bond rearrangement. At the interface, the value is 2.9 ps, suggesting a substantial slowing of hydrogen bond rearrangement in the interfacial water layer. At low monolayer density, there is a single band in the 2D spectrum. At high density, three bands are observed, two "dimer" bands in addition to the main monomer band. Each dimer band has chemical exchange peaks with the monomer, showing that the monomer and dimers are in equilibrium with an interconversion time of 40 ps. Structural fluctuations in the monolayer give rise to configurations that permit the dimer formation. By taking thousands of 5 second data sets, the appearance and disappearance of the dimers were tracked, demonstrating that the monolayers undergo very slow structural evolution for many hours after they are formed.

Spectroscopic observation of triplet separation as a driving force of singlet fission
Daniel Turner

Singlet fission, a multi-step molecular process in which one photon generates two triplet excitons, holds great technological promise. Here, by applying a combination of transient transmittance and two-dimensional electronic spectroscopies with 5 fs laser pulses and high-sensitivity data-acquisition methods, we resolve the full set of fission steps before the onset of spin dephasing in rubrene. We selected single-crystalline rubrene because of its role as a viable singlet fission material and because its energetics and transition dipole alignment uniquely allow for the unambiguous identification of the various fission steps through their contributions to distinct spectroscopic features. Our measurements reveal that the neighboring correlated triplet pair achieves its maximum population within 20 fs. Subsequent growth of the triplet signal on picosecond timescales is attributable to triplet separation, proceeding nonadiabatically through weakly coupled but near-resonant states. As such, we provide evidence for a singlet fission step that, until now, has not been convincingly observed.

Fully Coherent Electronic-Vibrational Spectroscopy of Transition Metal Complexes

John Wright, Jonathan Handali, Nathan Neff-Mallon, Erin Boyle

The application of CMDS to dynamical investigations of complex biological samples requires high selectivity for extracting the multidimensional spectra of specific moieties within the samples. The requisite selectivity can be achieved with multiple quantum coherences (MQCs) containing distinctive vibrational and electronic states. This paper reports the use of two different fully coherent methods that form MQCs containing two vibrational states and an electronic state. The MQCs are formed by resonance with three independently tunable optical parametric amplifiers. The emission from the MQCs is cooperative and multiplicatively enhanced by the three resonances. The directional emission is defined by either a $\boldsymbol{k}_4$ = $\boldsymbol{k}_1$ + $\boldsymbol{k}_2$ + $\boldsymbol{k}_3$ or a $\boldsymbol{k}_4$ = $\boldsymbol{k}_1$ + $\boldsymbol{k}_2$ – $\boldsymbol{k}_3$ phase matching geometry. The output frequency is spectrally resolved from the excitation frequencies so scattered light is not important. Since the selection rules for the vibrational transitions are the same as those in infrared spectroscopy, these fully coherent pathways provide a resonance infrared absorption spectroscopy that is analogous to resonance Raman spectroscopy. The methods are applied to benzene [1], a Styryl 9M dye [2], copper phthalocyanine tetrasulfonate [3], chlorophyll, and cobalamin.

4D Coherent Electronic-Vibrational Spectroscopy: Theory and Applications
Elad Harel

The correlations between different quantum-mechanical degrees of freedom of molecular species dictate their chemical and physical properties. Generally, these correlations are reflected in the optical response of the system but in low-order or low-dimensionality measurements the signals are highly averaged. Here, we describe a novel four-dimensional coherent spectroscopic method that directly correlates within and between the electronic and vibrational manifold of states. This method, which we call GRadient-Assisted Electronic Raman Spectroscopy or GAMERS uses a combination of resonant and non-resonant interactions to control quantum coherence pathways. For instance, GAMERS can isolate coherences corresponding to specific electronic states, thereby directly distinguishing dynamics on the ground- and electronically-excited potentials. Critically, GAMERS is free from lower-order signals that have plagued other electronically non-resonant vibrational spectroscopies. Both the theory and experimental implementation of the method are presented. We highlight specific means by which non-trivial effects such as anharmonicity (diagonal and off-diagonal), mode-specific vibronic coupling, and curvature of the excited states manifest in different projections of the 4D spectrum. Further, we show preliminary results on molecular systems including conjugated organic molecules in solution and semiconductor colloidal quantum dots.

Cancer Diagnosis using LIBS and Machine Learning Tools: Progress and Challenges
Noureddine Melikechi, Rosalba Guadiuso, Ebo Ewusi-Annan

Cancer refers to a class of diseases that is triggered by abnormal cell growth in one organ with the potential to spread other parts of the body. Despite numerous research and development efforts that provide important tools to combat, cancer remains a challenging disease to diagnose and treat. World wide, it is the second leading cause of death and was responsible for nearly 1 in 6 deaths, or 8.8 million deaths, in 2015. Early diagnosis is crucial to increase the survival rate and quality of life of cancer patients. Thus, developing non-invasive screening methods would represent a key step towards point-of-care large scale screening and prevention of asymptomatic tumors such as Epithelian Ovarian Cancer (EOC) and others. By combining femtosecond LIBS with unsupervised classification techniques, we have shown that it is possible to discriminate blood samples extracted from healthy and diseased mice with an accuracy that approaches 80%. We will present our most recent results and will report on the effects of various substrates on which LIBS measurements are performed on the classification accuracy of blood samples extracted from various cancerous and healthy mice.

Using Portable LIBS to Inform Field Archaeology

Mary Kate Donais, Luke Douglass, David George

After a number of years developing protocols for and applications of portable X-ray fluorescence (XRF) spectroscopy for field archaeology, portable laser induced breakdown spectroscopy (LIBS) was used for the first time during our summer 2017 excavation season. Data collection was focused toward ceramics and construction materials. Considering our previous XRF work as a starting point, this talk will address the advantages and disadvantages to using LIBS for field work. Data formats, data export, and data analysis approaches will be discussed. Applications of multivariate statistics for informing the archaeology will be emphasized.

Spatio-temporal mapping of laser ablation plumes using laser-induced fluorescence

Kyle Hartig, Mark Phllips, Sivanandan Harilal

Laser ablation (LA) is a powerful technique that has been implemented into a number of material processing and analytical applications. Understanding the physical conditions of the plasma is utmost important for optimizing laser ablation for various applications. Most previous diagnostic tools developed for steady state plasmas can be used to diagnose and optimize laser induced plasmas (LIP), however, ablation plume properties change significantly with space and time. Hence it is very important to characterize the plume properties with high space and time precision. Active optical diagnostics are, perhaps, the best tools for measuring the fundamental properties of laser-induced plasmas with high accuracy and precision. Laser absorption spectroscopy (LAS) and laser-induced fluorescence (LIF) provide improved accuracy and sensitivity for laser ablation diagnostics compared to optical emission spectroscopy (OES). Moreover, both LIF and LAS monitor ground state populations which persist for a prolonged time in comparison with excited atoms, which contribute emission from the plasma. Hence, LIF and LAS are capable of providing physical conditions of the LIBS plumes at very late times. In this work, we present, for the first time, spatio-temporal mapping of LIPs using LIF. A frequency-doubled continuous-wave Ti:Sapphire laser is used as a LIF excitation source for plasma plumes generated by a Nd:YAG laser on a metal target. The LIF signals from various points in the plasmas are recorded with high time precision. Plasma temperatures are measured from the spectral line shapes. This work is supported by the DOE/NNSA Defense Nuclear Nonproliferation Research and Development Office. Pacific Northwest National Laboratory is operated for the U.S. DOE by the Battelle Memorial Institute under Contract No. DE-AC05-76RLO1830.

Argon Fluoride LA-LEAF for Arsenic: Toward Selective Measurements in Rice

Jonathan Merten, Patrick Tribbett, Christopher Jones

Rice is known to contain the carcinogen arsenic, at concentrations of 0.08-0.2 ppm, though Zavala and Duxbury have found samples containing as much as 800 ppb.[1] While LIBS is ideal for rapid screening, e.g. at rice mills during the busy harvest season or for process control during milling to remove the arsenic-rich bran, it does not provide adequate limits of detection. We have used laser ablation laser-excited atomic fluorescence (LA-LEAF) with a simple ArF laser to lower arsenic detection limits to 1 ppm in steel and 0.15 ppm in copper, a significant improvement over previously LIBS measurements, particularly since these LODs required only 10 shots.[2] Although there is a strong arsenic ground state absorbance under the 0.5 nm profile of the ArF laser's ~193 nm emission, Cheung et al. have shown such deep-UV reexcitation to be non-specific under certain conditions of LA-LEAF This leads to substantial background in ArF-induced LEAF, particularly in steel. We elaborate on the conditions for maximizing selectivity of ArF-induced LA-LEAF in various matrices using a standard broadband ArF excitation laser. Additionally, we report on the improved selectivity and limits of detection achieved after modification of the excitation laser to provide a tunable, line-narrowed (~100x) excitation pulse. These results provide insight into the origin of the nonspecific plume laser-excited atomic fluorescence technique of Cheung et al.[3] Finally, we present data from arsenic LA-LEAF measurements in rice as well as work to simplify the signal detection scheme. 1. Y. J. Zavala and J. M. Duxbury, Environmental Science & Technology, 2008, 42, 3856-3860. 2. J. A. Merten, C. A. Jones and P. D. Tribbett, Journal of Analytical Atomic Spectrometry, 2017. 3. X. Wang, Z. Huang, P.-C. Chu, Y. Cai, K. S. Y. Leung, J. T. S. Lum and N.-H. Cheung, Journal of Analytical Atomic Spectrometry, 2016.

Analysis of high silicon content samples by laser induced breakdown spectrometry: fusion process to match the sample matrix.

Alexandrina Carvalho, Maciel Luz, Cassiana Nomura

Laser induced breakdown spectrometry (LIBS) has become an attractive and much explored technique in chemical analysis. Despite all positive characteristics associated to solid sampling techniques, quantitative analysis is not an easy task and has been considered the "Achilles heel" of LIBS technique. Considering, most of calibration standard is used as pressed powder pellets, particle size and matrix effects can affect the precision of the measurement as well as its accuracy. In this context, the aims of this work were the evaluation and comparison of using glass beads and pressed powder pellets in quantitative measurements of Al, Fe, Si, Ti, Ca and Mg in high silicon content samples by LIBS. A LIBS system (J200 Tandem, Applied Spectra) with a Nd:YAG laser operating at 266 nm was used to carried out the experiments. Calibration curves were prepared from ten certified reference materials: SRM 679 (Brick Clay), SRM 688 (Basalt Rock), SRM 694 (Western Phosphate Rock), SRM 695 (Mult-Nutrient Fertilizer), SRM 1648a (Urban Particulate Matter), SRM 1886a (Portland Cement), SRM 2703 (Sediment) and SRM 2709 (San Joaquin Soil) from NIST; BCSS-1 and MESS-2 (Marine and Estuarine Sediments) from NRCC. The linearity of calibration curves (R^2) using samples in the form of pressed powder pellets was worse (R^2= 0.0254 – 0.9276) than that observed for those made with samples in the form of glass beads (R^2=0.5714–0.9620), which still show lack on the linearity. To improve the results using samples in the form of glass beads, B and Li were used as internal standard. With this strategy, coefficient of determination for Fe (R^2=0.9655), Mg (R^2=0.9661) and Si (R^2 = 0.9639) showed better results when B was used as internal standard while Al (R^2=0.9825), Ca (R^2=0.9999) and Ti (R^2=0.9589) worked better with Li as internal standard. Using these conditions, a certified reference of soil (NIST SRM 2711) was analyzed and relative errors between -8 and +8% were found, indicating the possibility to make quantitative analysis, since internal standardization improves the precision and accuracy of the method, minimizing instrumental fluctuations and matrix effects.

In-situ Ultrasonic Imaging of Dynamic Process Streams: System Calibration and Image Processing

Marcus Ingram, Anthony Gachagan, Alison Nordon, Anthony Mulholland, Carmelo Mineo, Martin Hegarty, Edo Becker

Ultrasonic imaging is a well-established technology that provides a safe method for imaging into optically opaque structures. These images can be generated by mechanically scanning the section of interest through a linear path with a single transducer element, using tomography or by using a multiple-element transducer array, where the latter provides the most focussed images due to the precise positioning of the transmitting and receiving elements. An ultrasonic array comprises a number of individually addressable active piezoelectric elements and is controlled by a Phased Array Controller instrument to manipulate the ultrasonic pressure field within the load medium. Such arrays are typically used for sonar, medical diagnosis and non-destructive evaluation, however they have not yet been applied in the field of industrial process analysis. This paper aims to highlight how ultrasonic transducer arrays can be used to acquire in-situ images of dynamic process streams and how these images can provide valuable spatial information about the process, such as particle size distribution. The impact of this means that visualisation of a process stream will no-longer be limited to those systems that are optically transparent, therefore creating a greater breadth of spatial measurements within process analysis. This paper is divided into two parts: image acquisition optimisation and image processing. The first part outlines the method used to determine the optimum number of array elements (active aperture) within the transducer array to acquire stable images of a dynamic system. For this work, the size of the active aperture ranges from 16 to 128 elements, corresponding to an image region width of 11 to 90 mm. However, larger images require a longer data acquisition time and therefore produce a poorer image resolution when the target is dynamic in nature. The second part shows how relevant information about the process can be extracted from these ultrasonic images and how these measurements can be calibrated using a static system. The conclusion of this paper discusses how an ultrasonic array image acquisition and processing system has the potential to become a new class of process analytical technology.

Towards Quantitative Process Monitoring Using SABRE Enhanced Benchtop NMR
Andrew Parrott, Peter Richardson, Meghan Halse, Alison Nordon, Simon Duckett

Nuclear magnetic resonance (NMR) spectroscopy is an extremely powerful analytical technique which can provide quantitative and structural information about chemical mixtures. However, typical high-field research instruments use very strong superconducting magnets. These superconducting magnets often need to be cooled with cryogenic fluids. This makes the overall NMR instrument expensive, large, and non-portable. Consequently, compared with other spectroscopic techniques such as infrared or Raman, NMR is rarely used for industrial process analysis [1]. Benchtop NMR instruments instead use permanent magnets. As such, these instruments are much smaller, cheaper, and more portable. Although, as the signal intensity of the NMR measurement is related to the magnetic field strength, benchtop instruments have much lower sensitivity than high-field instruments [1,2]. The signal intensity limitation of NMR can be overcome by using hyperpolarization, where the nuclear polarization is increased beyond that dictated by the strength of the detection field [2]. One such hyperpolarization method is signal amplification by reversible exchange (SABRE), where polarization from parahydrogen is transferred to a target molecule via a metal catalyst without the need for hydrogenation of the target [3]. In this work, the SABRE approach has been used to enhance the signal from a series of model compounds on a 1 T benchtop NMR spectrometer. When using at-line analysis the NMR signal was increased by a factor in excess of 10000 compared to the normal NMR signal. This technique was combined with an automated flow cell setup that allowed the repolarization of the sample many times. The level of hyperpolarization was found to be highly reproducible; such reproducibility allows for quantitative trace analysis. This paper discusses the implications of the combination of SABRE and benchtop NMR to make a cheap, highly specific and highly sensitive quantitative technique for industrial process monitoring. 1. L. A. Colnago et al., Chem. Eng. Technol. 37, 191–203 (2014). 2. M. E. Halse, TrAC 83, 76–83 (2016). 3. R. W. Adams et al., Science 323, 1708–1711 (2009).

Implementation of Terahertz Spectroscopy as a PAT Tool for Powder Compact Density Assessment
Shikhar Mohan, Md. Anik Alam, James Drennen, III, Carl Anderson

The use of Process Analytical Technology (PAT) in the pharmaceutical industry has increased significantly over the past decade due to its enhanced capability to monitor product CQAs, ensure product quality and minimize time and cost associated with product manufacturing. Several spectroscopic techniques have been utilized as PAT tools for their rapid and non-destructive mode of analysis. This work further investigates Terahertz spectroscopy as a PAT tool for monitoring powder compact density. Density of pharmaceutical compacts is a critical quality attribute due to its effect on tablet disintegration and dissolution time. Implementation of an on-line measurement method for monitoring compact density during pharmaceutical unit operations such as roller compaction and tableting can facilitate in improving quality and cost efficiency of the final dosage form. Terahertz spectroscopy is a potential PAT tool to measure compact density by using its ability to extract refractive indices from the instrument outputted time-domain waveforms. In this study, a quantitative method is developed to measure tablet density. The refractive indices obtained from terahertz spectroscopy were used to build a density model on a five-component pharmaceutical tablet. The calibration model included several tableting parameter variations including compression force, compression speed and tablet weight without any variation in the chemical composition. The robustness of the calibration model against chemical composition variations was also tested by predicting density of tablets with variation in chemical composition. Analysis of the prediction errors showed that the terahertz model had desired predictive ability for tablet density measurements and appropriate robustness against variation in chemical composition of the tablets. These results further enhance the prospect of terahertz spectroscopy as a PAT tool for pharmaceutical compact density measurements

Robustness Evaluation of Transmission and Backscattered Modalities for Measuring Content Uniformity of Pharmaceutical Tablets with Raman Spectroscopy

Md. Nayeem Hossain, Md Anik Alam, Douglas Steinbach, James Drennen, Carl Anderson

Process analytical technology (PAT) has elevated the role of sensors in pharmaceutical manufacturing. Often the ideal technology must be selected from many suitable candidates based on their model performance; in particular, the accuracy, precision, and robustness of the model must meet set validation criteria. For quantitative analysis of solid oral dosage form, raman spectroscopy (transmission and backscattered modes) is a widely used PAT tool. To obtain fit for purpose prediction performance, various multivariate raman calibration models have been used to extract relevant chemical and physical information contained in the spectra. However, this multivariate modeling performance can be affected by physical factors such as tablet density, as well as chemical interferences. This work demonstrated how different physicochemical variabilities, not related to the parameter of interest, affect the predictive ability of different modalities of raman calibration model for active pharmaceutical ingredient (API) in solid oral dosage form. This study also discussed about the direction of the development of robust calibration model. A four-factor full-factorial design was used to generate calibration samples where the four factors were high dose API (APAP) at five levels, low dose API (Caffeine) at five levels, MCC:Lactose ratio at three levels and compaction pressures at three levels. The amounts of Crosscarmellose-Na and Mg-stearate were kept constant in the formulation. A set of 225 design points were created for calibration. An independent test set of 27 samples was prepared using three levels of APAP, Caffeine and compaction pressures at a fixed MCC: Lactose ratio. Materials for each design point were weighed in a scintillation vial, blended in bin blender and mixed using a magnetic rotor. Tablets were compressed using a Carver Automatic Tablet Press equipped with 13 mm flat-faced punches and die. Transmission raman instrument and a backscattered raman system were used to scan both calibration and test set. Partial least squares regression and Net analyte signal analyses were performed for each system, which allowed for the computation of FOM and evaluation of robustness. Results indicated the importance of accounting the variances which are not related to parameter of interest to achieve robust calibration model performance for transmission and backscattered mode.

Approaches to Multiblock Modeling and Dimension Reduction for Combining Disparate Data Types

Heather Brooke, Frank Westad

Continuous manufacturing, which is being strongly promoted by the FDA, is an innovative approach to pharmaceutical processing that utilizes information from multiple sensors throughout the entire process. Due to the numerous factors that need to be monitored and controlled at the same time, a multi-step approach is necessary for optimizing individual unit operations as well as the overall process. In this presentation we will show some approaches to combining process and spectroscopic data, which can be applied to process monitoring for continuous manufacturing and other applications.

Fourier-transform Infrared spectroscopy for characterization of enzymatic protein chain reductions in industrial bioprocesses

Sileshi Gizachew Wubshet, Ulrike Böcker, Diana Lindberg, Kenneth Aase Kristoffersen, Ingrid Måge, Nils Kristian Afseth

For several decades, Fourier-transform Infrared spectroscopy (FTIR) has been an established method for characterization of protein and polypeptide structures. One of the intriguing applications, built on extensive research in the field, is the use of FTIR for probing protein modifications induced by enzymes. In enzymatic protein hydrolysis, proteases cleave peptide chains at the expense of a water molecule. Enzymatic protein hydrolysis (EPH) of food processing by-products is a valorization technology gaining a large industrial interest. By-products, e.g., from meat processing, are notoriously heterogeneous, hence, maintaining stable product quality requires continuous monitoring of the hydrolysis process.

Currently, there is no available means of monitoring EPH process in an industrial setting. In this study, we showed the potential of a "dry-film" FTIR approach as a process monitoring tool for EPH of complex raw materials, i.e., salmon and chicken processing by-products. FTIR spectra acquired at different time points during the course of hydrolysis were studied to qualitatively characterize the changes in the protein structure. The overall process dynamics were also studied by applying principal component analysis on the time-resolved spectra. Molecular size distribution of peptides is one of the most important parameter explaining the state of a given enzymatic protein hydrolysis at a given time during the process. In an effort to quantitatively predict this parameter, we have developed FTIR based partial least square calibration models. Average molecular weight values derived from chromatographic analysis were used as a reference measurement for the calibration. This has resulted in the first application of FTIR spectroscopy for monitoring average molecular weight of peptides during enzymatic protein hydrolysis of industrially-relevant raw materials.

Near Infrared Solutions for Biopharmaceutical Development and Manufacturing
Adam J. Hopkins

Near infrared (NIR) spectroscopy has long been used in the traditional pharmaceutical industry for applications such as wet-granulation and tablet analysis. However, the emergence of biopharmaceuticals and the emphasis by the FDA on both continuous manufacturing and Quality by Design to reduce out-of-spec products are important to minimize overall production cost. These approaches aim to increase efficiency in the development and production of drugs. The complexity of the products and production matrices are difficult to analyze by traditional methods and nearly impossible to measure in real time. NIRS is a technique that makes this possible. This talk demonstrates the capabilities and benefits of NIR spectroscopy for biological and biosimilar production. From materials inspection, to process development, and implementation as process analytical technology we will explore applications at each phase of the biopharmaceutical production cycle.

Analytical tools for physicochemical characterization of biopharmaceuticals
Sergey Arzhantsev

Biopharmaceuticals are complex macromolecules with secondary and tertiary structures that may be easily disturbed, potentially resulting in loss of therapeutic efficacy. The characterization of higher (secondary and tertiary) order structure is still challenging due to technical limitation of current analytical methods. Recently, new commercial spectroscopic instruments have become available to researchers. Raman optical activity (ROA) spectroscopy allows the measurement of vibrational optical activity in chiral molecules. Because of its sensitivity to chirality, ROA is an incisive new probe of higher order structure of biomolecules including peptides, proteins, carbohydrates, nucleic acids and viruses. Picosecond time-resolved fluorescence spectroscopy allows the measurement of the lifetime and anisotropy time of proteins to determine higher-order structure and aggregation state. Deep ultraviolet resonance Raman (DUVRR) spectroscopy is capable of characterizing the higher order structure of proteins in the presence of excipients in aqueous media with minimal sample preparation. An overview of these advanced spectroscopic tools for biopharmaceuticals characterization will be presented.

Designing a Calibration Set in Spectral Space for Efficient Development of an NIR Method for Tablet Analysis

Md Anik Alam, Md Nayeem Hossain, Douglas Steinbach, James Drennen, Carl Anderson

Designing a calibration set is the first step in developing a spectroscopic calibration method for quantitative analysis of pharmaceutical tablets. This step is critical because successful model development depends on the suitability of the calibration data. For spectroscopic-based methods, traditional concentration based techniques for designing calibration sets are prone to have redundant information while simultaneously lacking necessary information for a successful calibration model. The traditional method also follows the same design approach for different spectroscopic techniques and different formulations, thereby lacks the optimizing capability to be technique and formulation specific. A method for designing a calibration set in the Near Infrared (NIR) spectral space was developed for quantitative analysis of tablets. The pure component NIR spectra of a tablet formulation were used to define the spectral space of that formulation. This method minimizes sample requirements to provide an efficient means for developing multivariate spectroscopic calibration. Multiple comparative studies were conducted between commonly employed experimental design approaches to calibration development and the newly developed spectral space based technique. The comparisons were conducted on single API (Active Pharmaceutical Ingredient) and multiple APIs formulation to quantify model drugs using NIR spectroscopy. Partial least square (PLS) models were developed from respective calibration designs. Model performance was comprehensively assessed based on the ability to predict API concentrations in independent prediction sets. Similar prediction performance was achieved using the smaller calibration set designed in spectral space (11 for single API and 33 for multiple APIs tablets), compared to the traditionally designed large calibration sets (45 for single API and 225 for multiple APIs tablets). An improved prediction performance was observed for the spectrally designed calibration sets (33 tablets for spectral design) compared to the traditionally designed calibration sets of equal sizes (33 tablets for D-Optimal, CCD and I-Optimal designs). This study demonstrates a new and efficient technique for designing calibration set in spectral space for spectroscopic multivariate calibration of tablet analysis. This study also provides opportunity to design formulation and technique specific calibration sets to optimize calibration capability.

A Novel Fingerprint and High Wavenumber Raman Spectroscopy System for Hydration Quantification
Laura Masson, Anita Mahadevan-Jansen

Combining fingerprint and high wavenumber Raman bands in the analysis of tissues for biomedical application provides valuable information corresponding to molecular vibrations of lipids, proteins, and water. Previous dual fingerprint and high wavenumber Raman spectroscopy systems have utilized expensive alternatives to traditional silicon detectors or dual transmission gratings to capture both the fingerprint and high wavenumber regions on the detector simultaneously. We have developed a system capable of collecting information from either the fingerprint or high wavenumber region of the Raman spectrum using a single source with dual excitation wavelength and a single fixed grating spectrometer. This novel system allows sequential acquisition of both regions of the spectrum without moving the sample or the probe, and can be easily applied for analysis of chemical as well as biological specimens. Based on the broad peak corresponding to OH vibrations of water in the high wavenumber region, this system facilitates a quantitative approach to studying hydration in various tissues. We have demonstrated that this system is capable of obtaining hydration information with 1% accuracy, and as such is suited to a broad range of in vivo applications. Overall this system and approach provide the potential for using Raman spectroscopy to study tissue characteristics with increased sensitivity and accuracy.

Nanogap-Enhanced Raman Scattering (NERS)

Yung Doug Suh

Started from the simple wet-chemical nanogap generation induced by Coulomb aggregation among colloidal nanoparticles in aqueous solution after adding salt onto it shown independently by S. Nie and K. Kneipp in 1997, nanogap engineering to enhance Raman scattering signal to achieve single molecule sensitivity is getting more sophisticated. In this talk, single molecule Surface-Enhanced Raman Scattering (smSERS) field formed since 1997 will be briefly reviewed [1] and then different types of nanogap engineering strategy for smSERS developed in my lab will be discussed: single-junction 0-D external nanogap between two spherical nanoparticles connected with a double helix DNA [2-5], multi-junction 3-D spherical nanogap internally formed between spherical gold core nanoparticle and spherical gold shell nanoparticle connected by multiple single helix DNAs [6-7], and 2-D nanogap arrays formed on a 4-inch polymer wafer by simple two-

step process. Several reasons will be discussed why now SERS regime and NERS (Nanogap-enhanced Raman Scattering) regime should be separated on the enhancement factor (EF) distribution histogram. Recent result including direct near-field visualization of the nanogap field of the 2-D nanogap array, single molecule behavior of cytochrome C protein's Raman signal, and ultra-uniform distribution of SERS enhancement factor (EF) of benzene thiol molecule dispersed on this plasmonic 2-D nanogap array wafer will be presented. References 1. H. Lee, S.M. Jim, H.M. Kim, and Y.D. Suh*, Phys. Chem. Chem. Phys. (invited review article) 15, 5276 (2013) 2. D. Lim, K.-S. Jeon, H.M. Kim, J.-M. Nam*, and Y.D. Suh*, Nature Materials 9, 60 (2010)
3. J.-H. Lee, J.-M. Nam*, K.-S. Jeon, D. Lim, H. Kim, S. Kwon, H.M. Lee, and Y.D. Suh*, ACS Nano 6, 9574 (2012) 4. H. Lee, J.-H. Lee, S.M. Jin, Y.D. Suh*, and J.M. Nam*, Nano Letters, 13, 6113 (2013) 5. H. Lee, G.-H. Kim, J.-H. Lee, N.H. Kim, J.-M. Nam*, and Y.D. Suh*, Nano Letters, 15, 4628 (2015) 6. D. Lim, K.-S. Jeon, J.H. Hwang, H.Y. Kim, S.H. Kwon, Y.D. Suh*, and J.-M. Nam*, Nature Nanotechnology 6, 452 (2011)
7. J.-W. Oh, D.-K. Lim, G.-H. Kim, Y.D. Suh*, and J.-M. Nam*, J. Am. Chem. Soc. 136, 14052 (2014)

Fiber Sensors for Chemical and Biochemical Detection Based on Surface Enhanced Raman Scattering
Jin Zhang

SERS (surface-enhanced Raman scattering) sensors based on conventional optical fibers, D-shaped fiber, as well as hollow core photonic crystal fibers (HCPCF) have been developed to detect chemicals and biochemicals with molecular specificity and high sensitivity. We have demonstrated that the SERS fibers sensors are significantly more sensitive than traditional bulk detection using various samples including dye molecules, proteins, and bacteria. Several different sensing architectures have been developed and evaluated. Furthermore, we have demonstrated synergistically enhanced SERS activities in a double substrate "sandwich" structure using either two Ag nanoparticle substrates or a combination of Ag nanopartciles and nanowires to create "hot spots" for further electromagnetic (EM) field enhancement. In addition, we have shown that substrates with uniform structures, e.g. hollow gold nanospheres (HGNs) with unique structural and optical properties, can afford highly consistent SERS at the single nanostructure level. The HGNs are more sensitive for SERS in the near IR, which is desired for reducing possible fluorescence of target analytes. These sensors are very compact, highly sensitive, molecular specific, and low cost, and have remote detection capability and can be used in challenging environments. Our studies have demonstrated that fiber-based SERS sensors are promising for practical applications for chemical and biochemical analysis.

Novel Plasmonic Enable Solar-to-Chemical Energy Conversion Systems
Syed Mubeen, Wei-Chuan Shih, Wei Cheng, Jonathan Koonce, Abdul Sattar Al-Saedi

Environmental considerations require the world, to increase, ever more urgently, the renewables fraction of its energy usage. The focus on solar energy will certainly rise, and will include both photovoltaics and the much less well developed goal of using sunlight to convert available feedstocks such as water, CO2 and domestic and industrial waste directly to valuable chemicals and fuels. This talk will introduce our research efforts in collaboration with Dr. Shih's group at University of Houston to develop novel plasmonic systems to convert light energy directly into high free energy materials that can be used as fuels and fine chemicals, thereby storing the solar energy in a high energy density and transportable form. Specifically, here we report our efforts to develop a novel molecular level design and synthesis strategy for successfully utilizing hot electrons from plasmonically active metals to drive important fuel and chemical producing reactions such as water splitting and HMF oxidation reaction. The devices built using plasmonic metals demonstrated very stable operation and opens potentially dramatic new possibilities for renewable-energy powered electrochemical conversion systems.

Hot Electron Enhanced Thermionic Emission (HEETE) Converters for All-Metal Optical Power Generation

Matthew Sheldon

We report on our initial studies of photo-induced charge transport from metal nanostructures. In particular, we outline how the remarkable thermal and optical energy concentration provided by plasmonic resonances can enable a new thermodynamic power cycle whereby photo-excited 'hot' electrons and resonant photothermal heating provides a dual excitation mechanism for electron emission. Because this process is closely related to purely thermionic emission, we label an optical power-converting device based on this mechanism a Hot Electron Enhanced Thermionic Emission (HEETE) converter. The strong enhancement of both thermal and optical energy channels, may enable a significantly more efficient strategy for optical power conversion, and one that can theoretically out-perform traditional semiconductor-based solar sells. Our study emphasizes theoretical and experimental development of refined models for electronic distributions in metals, specifically when characteristic structural features are smaller than the mean free path of excited carriers. In addition, we have developed a mechanistic model of the photo-thermal response of optimized plasmonic absorbers accounting for factors such as spectral width of absorbance and emissivity, as well the role of thermal damping pathways such as conduction and convection. When radiative loss is the dominant cooling mechanism, as can be achieved with thermionic devices in vacuum, we anticipate solar-induced temperature increases over 900 K without additional optical concentration, and consequent photo-induced current densities with 10^15 enhancement compared with purely thermionic emission at the equivalent temperature. The optimal structures to achieve these temperatures have high absorption (> 90%) with spectral bandwidth that spans the visible up to 1100 nm, and emissivity of ~2% throughout the infrared. We show a variety of noble metal nanostructures (Au, Ag, Cu), and high melting point metals (W, Ir, Pt), combined with refractory dielectric cladding that can provide these ideal optoelectronic properties due to their thermal tolerance, highly absorbing and tunable plasmon resonances in the visible, and naturally low emissivity in the infrared. In combination with full-wave optical simulations that guide optimized designs, we have fabricated HEETE device test structures and measured the spectral and temperature performance, outlining opportunities for efficient power conversion.

Far-field and near-field plasmonic coupling in disordered nanoparticle arrays and applications in ultra-sensitive biosensing and super-resolution histopathology

Wei-Chuan Shih, Fusheng Zhao, Masud Arnob, Jingting Li, Camille Artur, Jason Eriksen, David Mayerich

Plasmonic nanoparticles (NP) permit light-induced free electron resonance, also known as localized surface plasmon resonance (LSPR). Recent advances in nanofabrication and synthesis facilitate the fine-tuning of LSPR properties in a wide spectral range for applications in sensing, imaging, energy conversion and catalysis. Ensemble plasmonic coupling (EPC) is an interesting phenomenon when plasmonic NP within a group interact with one another. Prominent examples to date include near-and far-field coupling, lattice plasmon resonance (LPR), Fano resonance, and etc. In existing literature, spatial regularity in highly ordered arrangements or arrays is considered a key prerequisite for EPC. Previous studies also suggested that far-field EPC is nil in disordered arrangements. In this talk, I will show that the demand on ensemble regularity can be significantly relaxed when plasmonic NP are strongly coupled. EPC-based plamonic NP array design enables wide tunability and versatility in many applications.1-7 In addition, the relaxation on ensemble regularity allows the use of disordered plasmonic arrays fabricated by extremely low-cost nanosphere lithography (NSL). In a second example, I will report our recent advances in using plasmonic NP as molecular labels in expansion microscopy (ExM) for superresolution imaging of fixed tissue specimen. In addition to the perceivable benefit of increased multiplexing capability from the inherent spectral diversity of SERS,3,5,7-9 we have discovered an intriguing phenomenon likely owning to distance-modulated EPC, which would potentially provide an additional mechanism for multiplexing as well as quantitation of tissue expansion. References: 1. Fusheng Zhao, Masud Arnob, Oussama Zenasni, Jingting Li, and Wei-Chuan Shih, Far-field plasmonic coupling in 2-dimensional polycrystalline plasmonic arrays enables wide tunability with low-cost nanofabrication, Nanoscale Horizons 2017. 2. Masud Arnob, Fusheng Zhao, Jingting Li, ans Wei-Chuan Shih, EBL-based fabrication and different modeling approaches for nanoporous gold nanodisks, ACS Photonics 2017. 3. Suyan Qiu, Fusheng Zhao, Oussama Zenasni, Jingting Li, and Wei-Chuan Shih, Nanoporous gold disks functionalized with stabilized G-quadruplex moieties for sensing small molecules, ACS Applied Materials & Interfaces 8(44): 29968-29976

2016. 4. Wei-Chuan Shih*, Greggy Santos, Fusheng Zhao, Oussama Zenasni, and Masud Arnob, Simultaneous chemical and refractive index sensing in 1-2.5 ?m wavelength range on nanoporous gold disks, Nano Letters, 16 (7): 4641–4647 2016. 5. Ming Li, Yong Du, Fusheng Zhao, Jianbo Zeng, Chandra Mohan, and Wei-Chuan Shih*, Reagent- and separation-free measurements of urine creatinine concentration by stamping surface-enhanced Raman scattering (S-SERS), Biomedical Optics Express 6(3): 849-858 2015. 6. Greggy Santos, Fusheng Zhao, Jianbo Zeng, Ming Li, and Wei-Chuan Shih*, Zepto mole cancer marker detection by surface enhanced fluorescence on nanoporous gold disk substrates, Journal of Biophotonics 8(10): 855-863 2015. 7. Fusheng Zhao, Jianbo Zeng, Md Masud Parvez Arnob, Po Sun, Ji Qi, Pratik Motwani, Mufaddal Gheewala, Chien-Hung Li, Andrew Paterson, Uli Strych, Balakrishnan Raja, Richard C. Willson, John C. Wolfe, T. Randall Lee, and Wei-Chuan Shih*, Monolithic NPG nanoparticles with large surface area, tunable plasmonics, and high-density internal hot-spots, Nanoscale 6, 8199-8207, 2014. 8. Ji Qi, Jingting Li, and Wei-Chuan Shih*, High-speed hyperspectral Raman imaging for label-free compositional microanalysis, Biomedical Optics Express 4(11): 2376-2383, 2013. 9. Ji Qi and Wei-Chuan Shih*, Parallel Raman microspectroscopy using programmable multi-point illumination, Optics Letters 37(8): 1289-1291, 2012.

Bridging the pressure and materials gaps: Methanol oxidation on perovskite thin-films and powders

David Mullins, Yafen Zhang, Michelle Kidder, Steven Overbury

Surface science has contributed fundamental insights into heterogeneous catalysis by studying reactions on model, single crystal surfaces under ultra-high vacuum (UHV) conditions. While these studies have provided precise mechanistic understanding their application to processes under "real world" conditions has been hampered by two so-called "gaps". The "pressure gap" refers to the pressure differential that occurs between studies conducted under UHV conditions and reactions conducted at atmospheric pressure. This difference can affect reaction mechanisms, kinetics or result in a reconstruction of the catalytic surface. The "materials gap" refers to the differences between precisely characterized, low-surface area materials used in UHV studies to isolate specific structural or compositional catalytic functions and high-surface area powdered catalysts which may have complex surface structures, defects and surface compositions. We are addressing these "gaps" by conducting parallel experiments on flat, highly crystalline $La_XSr_{1-X}MnO_3(100)$ thin films, and on the same materials synthesized by sol-gel techniques as moderate-surface area powders. The materials' structures are characterized by TEM, XRD, AFM, ion scattering and depth profiling XPS. A direct comparison of the oxidation of methanol, as a function of temperature and O_2 : CH_3OH ratio, was conducted at nominally
0.1 Torr using ambient pressure XPS (APXPS). The similarities and differences observed in the reactions on these various surfaces will be presented.

Using Tender Ambient Pressure XPS to probe Solid/Liquid Electrochemical Interfaces

Ethan Crumlin

Interfaces play an important role in nearly all aspects of life, and are essential for electrochemistry. Electrochemical systems ranging from high temperature solid oxide fuel cells (SOFC) to batteries to capacitors have a wide range of important interfaces between solids, liquids, and gases which play a pivotal role in how energy is stored, transferred, and/or converted. The ability to study these interfaces has proven to be difficult and is only further exacerbated by the limited number of techniques capable of operating under in situ/operando environments. To overcome these challenges, we use ambient pressure X-ray Photoelectron Spectroscopy (APXPS). Using synchrotron X-rays at Lawrence Berkeley Nation Laboratory, the Advanced Light Source has several beamlines dedicated to APXPS endstations that are outfitted with various in situ/operando features such as heating to temperatures > 500 C, pressures greater then 20 Torr to support solid/liquid experiments and electrical leads to support applying electrical potentials supports the ability to collect XPS data of actual electrochemical devices while its operating in near ambient pressures. A recent advancement has been to use tender X-rays between ~2 keV to 8 keV to probe solid/liquid electrochemical interfaces under in situ/operando conditions. This talk will introduce APXPS and provide several solid/gas and solid/liquid interface electrochemistry examples. Then I will focus on how we use tender APXPS to probe the electrochemical double layer (EDL) and our newest discoveries revealing metastable interfacial chemistries that are responsible for promoting titanium corrosion.

The Surface Chemistry of Water at Solid and Liquid Ionic Interfaces
John Newberg

Ionic interfaces are ubiquitous in physical, chemical, biological, environmental and technological processes. Herein we will present recent efforts examining metal oxide and ionic liquid interfaces in the presence of water vapor using ambient pressure X-ray photoelectron spectroscopy (APXPS). Building off the strong foundation of ultra-high vacuum studies of water interaction with metal oxides in the literature, we extend our understanding of surface thermodynamic and kinetic behavior of water dissociation under elevated pressures where adsorption and desorption occur concomitantly. The interaction of water with ionic liquids is known to affect various properties including density, viscosity, electrochemical window and surface tension. We have used APXPS to quantitatively assess the amount of water in the interfacial region, and the influence water has on the ionic liquid structure and electronic environment in the interfacial region.

XPS analysis of surface-bound biomolecules provides insight on their surface interactions and dissociative properties
Kenan Fears

The adsorption behavior of GG-X-GG and Xn ($n = 5, 10$) oligopeptides on inorganic substrates was investigated to elucidate the contributions of different amino acids (AAs) to peptide-surface interactions. Peptide adsorption was quantified by in vacuo X-ray photoelectron spectroscopy (XPS), the resulting systematic data revealed several trends in surface adsorption of oligopeptides as a function of their composition and length. Peptides adsorbed more readily on Au-coated Si wafers, on which the maximum surface coverage was ca. 3 times greater than that on the native SiOx. For a particular AA, adsorption tended to increase, sometimes dramatically, with increasing units of X (GG-X-GG < X5 < X10). In pairs of AAs having side chains that only vary by alkyl chain length (L and V, Q and N, R and K, E and D), the AA with the longer alkyl chain adsorbed more readily, although this trend diminished with the increasing number of X residues. The dissociation constants for these peptides were also obtainable via XPS by depositing drops of the peptides solutions (pH 1-12) on Au, and allowing them to dry at room temperature. Due to the ca. +1.3 eV shift in binding energy (BE) of protonated amines, pK values of basic amino acids were calculated by plotting the fraction of protonated amines as a function of solution pH. Similarly, the BE of carboxyl groups shifted ca. -1.3 eV upon deprotonation. While C 1s spectra were convoluted by the multiple chemical states of carbon present in the samples, the ratio of the C 1s components centered at $BE=289.0 \pm 0.4$ and $BE=287.9 \pm 0.3$ proved to reliably assess deprotonation of carboxyl groups. The pK values for the Asp (3.1 & 2.4), His (6.7), and Lys (11.3 & 10.6) containing peptides, were consistent with published values. Likewise, when this technique was applied to larger proteins, bovine serum albumin and fibronectin, the resulting isoelectric points, 4.8 and 5.7 respectively, were in agreement with literature values. These results, along with complementary circular dichroism spectroscopy, aided in the interpretation of XPS data collected from more complex peptide systems and biopolymers to discern surface-induced conformational changes.

Observation of Oxygen Binding on PGM-free Electrocatalysts by Ambient Pressure XPS and XAS

Kateryna Artyushkova, Elisabeth Weiler, Michael Dzara, Svitlana Pylypenko, Plamen Atanassov

The most promising class of PGM-free materials for oxygen reduction reaction (ORR) is based on graphene-like carbon containing nitrogen and transition metal (MNC). They show promise as replacement of Pt in two different technological platforms -alkaline exchange membrane fuel cells (AEMFCs) and proton exchange membrane fuel cells (PEMFC). It is well established that nitrogen coordination with metal in the carbon network of MNC materials is directly related to ORR activity; however, the exact nature of the active sites is still debated even after over 50 years of research. Understanding the specific roles of nitrogen and metal in the properties/activity/stability/durability of MNC-based catalytic materials is a prerequisite for the rational design of ORR electrocatalysts with improved performance. The key component in elucidating the relationship between the chemistry of active sites and activity is a better understanding of the formation of adsorbates, intermediates, and products during reactions occurring within the fuel cell. In situ monitoring reaction steps under realistic conditions in metal-free and metal-containing building blocks will shed light onto the reaction mechanism that is essential for developing active and durable PGM-free catalyst for ORR. [4] We will report on AP-XPS analysis for series of electrocatalysts belonging to Fe N carbon families based on sacrificial support method (SSM) [1, 2] and Metal-organic frameworks (MOF) [5]. The effect the nitrogen chemistry and the type of iron have on the oxygen binding was investigated by ambient pressure X-ray Photoelectron Spectroscopy (XPS) and X-ray Adsorption Spectroscopy (XAS) under an O2 environment at operating temperature of the fuel cell. The effect of the relative abundance of different types of nitrogens, such as pyridinic, coordinated to iron and hydrogenated nitrogens (pyrrolic and hydrogenated pyridine) on the preference of oxygen binding is studied by high-resolution nitrogen photoelectron spectra. The role of metallic and atomically dispersed iron is investigated by a combination of XAS and XPS.

Plasma Assisted Reaction Chemical Ionization using a Dielectric Barrier Discharge for High Sensitivity Elemental Quantification of Environmental Contaminants Separated by Gas Chromatography

Kunyu Zheng, Peter Haferl, Michael Dolan, Hamid Badiei, Kaveh Jorabchi

Atmospheric-pressure dielectric barrier discharge (AP-DBD) has recently attracted much attention as ion source for molecular mass spectrometry (MS) and excitation source for atomic emission spectroscopy. In this presentation, we evaluate the analytical performance of AP-DBD as an elemental ion source to quantify halogens and other non-metals at trace levels for analytes separated by gas chromatography. Halogen detection is particularly important because of prevalence of organohalogens among environmental contaminants and difficulties in detection of these elements with conventional elemental methods. Notably, we focus on detection of negative ions to enhance ionization of halogens. Furthermore, unlike conventional elemental ion sources where ions are directly sampled from the plasma, we allow extensive ion-neutral reactions in the plasma afterglow prior to ion sampling. This approach offers both fundamental and practical advantages. From fundamental perspective, the afterglow reactions open avenues to detect elements as polyatomic ions as well as elemental ions, expanding the range of analytically useful ionization reactions. From practical perspective, incorporation of afterglow reactions allows direct coupling of DBD to atmospheric sampling mass spectrometers with no modifications, offering the possibility to conduct both elemental and molecular mass spectrometry on one instrument. Using the above approach, we demonstrate high sensitivity detection for chlorine with a detection limit of 0.7 pg on-column. Further, we show that 14 chlorinated pesticides spiked into concentrated matrices (prepared by Quick Easy Cheap Effective Rugged and Safe method combined with dispersive liquid-liquid microextraction) from orange, apple, and mixed juices are quantitatively measured using a single chlorinated standard. These results illustrate that the Cl response factor is independent of the chemical structure of the compounds and that DBD provides sufficient power to quantitatively ionize chlorine in complex mixtures with minimal matrix effects. Performance characterization for other non-metals using DBD is under investigation and the results will be discussed in this presentation.

Determination of Magnesium, Aluminum, and Calcium in Red Spruce Foliage and Surrounding Soil from Alarka Laurel, North Carolina

David Butcher

Red spruce (*Picea rubens*) is a conifer that is found at high elevation sites in the Southern Appalachian Mountains. Along with the Fraser fir (*Abies fraseri*), they provide unique ecosystems in Western North Carolina. Red spruce decline has been observed since the 1960s that has partially been attributed to acidic deposition. The presence of acidic deposition has been shown to reduce calcium availability, enhance aluminum availability, and enhance the decline of red spruce. Here is reported the determination of magnesium, aluminum, and calcium in red spruce foliage and surrounding soil from Alarka Laurel in the Cowee Mountains of North Carolina. The goal of this study is to use the elemental analysis results to, at least in part, evaluate the relative health of red spruce in this unique ecosystem.

Optical emission hyperspectral imaging plasma diagnostics of atmospheric pressure ?DBD optimized for surface substrate erosion sampling

Songyue Shi, Xiaoxia Gong, Gerardo Gamez

An atmospheric pressure plasma jet (APPJ) based on a micro-capillary dielectric barrier discharge (DBD) has been recently proposed for depth profiling on metal substrates via mass spectrometry. Nevertheless, the underlying mechanisms of how the discharge interacts with the sample substrate surface to promote erosion are far from being well understood. Herein, optical emission hyperspectral imaging plasma diagnostics performed on the Helium-?DBD (in a 360 ?m o.d. and 150 ?m i.d. fused silica capillary), in the region where the discharge effluent interacts with the sample surface in the open air, will be presented. The lateral distribution of optical emission from different plasma species will be shown. A Fourier-based Abel inversion algorithm will be implemented to reconstruct the radial distribution of the plasma optical emission. The hyperspectral images from N2 and N2+ optical emission will allow spatially resolved vibrational and rotational temperatures to be calculated. This will allow us to get a better understanding of the ?DBD and its interaction with the sample surface during erosion.

Optimizing LA-LEAF for Arsenic in Metals Using a Tunable ArF Laser

Patrick Tribbett, Christopher Jones, Jonathan Merten

Laser Ablation Laser Excited Atomic Fluorescence (LA-LEAF) enables trace-level limits of detection (LOD) for arsenic in metal matrices through semi-selective excitation by a 193.3 nm standard broadband ArF laser. Our LA-LEAF arsenic data contrasts with that of Cheung's group, which has shown nonselective emission enhancement with ArF laser re-excitation involving the broadening of atomic energy states due to high atom number densities, a mechanism that he refers to as Plume Laser-Induced Fluorescence (PLIF).[1] Our work with arsenic demonstrated the important interplay of the plasma atmosphere and the inter-laser delay in maintaining adequate signal to background ratio, where the background is both thermal and laser-induced. The LODs for arsenic in a steel matrix under helium and argon atmospheres were 1.0 and 1.6 ppm, respectively, at appropriately long inter-laser delays.[2] The laser-induced emission from matrix atoms (presumably neutral iron) with absorption transitions near the arsenic absorption limits the selectivity of the ArF laser, though the exact mechanism of this background is unknown. The recent addition of a diffraction grating to the ArF laser in place of the HR mirror has narrowed the laser emission to ~70 pm FWHM and allows tuning to the arsenic absorption line at 193.76 nm. Preliminary results show improvements to the selectivity and background fluorescence, and indicate a single element-single transition mechanism rather than a PLIF mechanism for the matrix and arsenic fluorescence. We report improved As detection limits with the narrowband laser and studies of the limiting noise with an eye toward single-shot measurements. 1. Xiaochun Wang, Zhengyu Huang, Po-Chun Chu, Yue Cai, Kelvin S.Y. Leung, Judy T.S. Lum, Nai-Ho Cheung, " The mechanism of ArF laser-induced fluorescence of dense plume matter," Journal of Analytical Atomic Spectrometry, (2016), DOI: 10.1039/c6ja00290k. 2. J.A. Merten, C.A. Jones, P.D. Tribbett, "Argon fluoride LA-LEAF for rapid arsenic quantitation," Journal of Analytical Atomic Spectrometry, (2017), DOI: 10.1039/C7JA00092H.

Agglomeration of nanoparticle structures in the stagnation of colliding carbon plasma

John Oliver, Tatyana Sizyuk, Prasoon Diwakar

It has been demonstrated that the hot stagnation region formed during the collision of laser-produced carbon plasmas is rich with carbon dimers and trimers which have been shown to synthesize large carbon macromolecules such as soot and even carbon fullerene onions and nanotubes. In this study, we developed and integrated experimental and multidimensional modeling techniques to access temporal and spatial evolution of colliding plasma characteristics to elucidate the mechanism for early carbon dimer formation. Fast-gate monochromatic imaging, optical emission spectroscopy (OES), and laser-induced fluorescence (LIF) of graphite-produced carbon plasma were performed to obtain a spatiotemporal distribution of carbon radicals in the developing stagnation plume. Using laser-induced incandescence (LII), in-situ determination of particle concentration and size is made with excellent time resolution. Together, LIF and LII findings are married to create a robust theory of nanoparticle growth in the stagnation layer. Particle collections examined under scanning electron microscopy (SEM) were used to corroborate LII results and confirm carbon nanotube synthesis. Experimental results were compared with results of the 3D comprehensive modeling using our HEIGHTS simulation package. The results are explained based on a fundamental analysis of plasma evolution, colliding layer formation, stagnation and expansion. The precise mechanisms of the plasma collision, plume propagation, and particle formation are discussed based on the experimental and modeling results.

Benefits of ICP-MS with 10 Times Higher Sensitivity and 1/2 of Argon Consumption

Oliver Büttel, Oliver Büttel

Unique specifications like LOD, precision, cost and speed of analysis or productivity in the laboratory make ICP-MS a performance driven technique. When looking at any ICP-MS instrument's technical specifications one of the main parameters – sensitivity – is often under-estimated. While at the same time sensitivity in particular is a feature that directly affects LOD(s), speed and cost of analysis, and overall instrument productivity. Another important parameter is argon consumption. The Analytik Jena ICP-MS runs with ½ the argon consumption of competitive systems, thus allowing a significant reduction in running cost per sample.

Simulation of argon velocity distribution directly in front of the skimmer cone of an ICP-MS

Ross Spencer

In 2012 Taylor and Farnsworth made an intriguing measurement of the argon velocity distribution (along a single axis) just in front of the skimmer cone of an ICP-MS. The properties of this distribution function are influenced by the expansion of the argon gas and by its interaction with the surface of the skimmer cone. This makes their data particularly valuable as a standard against which flow simulations, like the DSMC code FENIX, can be checked. Some parts of their data are easily simulated. For instance, the part of the distribution function that is unaffected by collisions with the skimmer surface agree very well with the simulation. But finding a good model for the interaction of particles with the wall and having enough particles in the simulation to be sure that the simulation is converged, has proven difficult. The reflection, thermal (Maxwell), and CLL collision models as well as the details of the argon-argon collision cross-section will be discussed and compared to the data of Taylor and Farnsworth.

HIGH TEMPERATURE TOTAL SAMPLE CONSUMPTION COUPLED TO INDUCTIVELY COUPLED PLASMA MASS SPECTROMETRY FOR THE MULTIELEMENT ANALYSIS OF CEREBROSPINAL FLUID

Esperanza Garcia-Ruiz, Agueda Cañabate, Martin Resano, Jose Luis Todoli

For the first time, a total consumption low sample introduction system has been applied to the analysis of non-diluted cerebrospinal fluids (CSF) by means of inductively coupled plasma mass spectrometry (ICP-MS). A 2.5 µL sample volume was injected into an air carrier stream, following the air-segmented injection principle, and was later nebulized and introduced into a high temperature torch integrated sample introduction system (hTISIS) coupled to an ICP-MS, leading to transient signals. Under optimized conditions, the hTISIS provides peak areas around four times higher than those provided by a conventional double pass spray chamber. Additional advantages provided by the hTISIS include limits of detection up to 6 times lower much narrower peaks, and absence of matrix-effects. It was verified that, using the hTISIS recoveries for spiked real samples were close to 100% for a set of 14 different elements (V, Cr, Mn, Co, Ni, Cu, As, Se,

Mo, Cd, Sb, Ba, Tl and Pb).

HYDRODYNAMIC CHROMATOGRAPHY COUPLED TO ICP-MS: RETHINKING THE TECHNIQUE FOR THE ANALYSIS OF NANOMATERIALSCHNIQUE

Francisco Laborda, Maria S. Jimenez, Daniel Isabal, Maria T. Gomez, Juan R. Castillo

Engineered nanoparticles (ENP) are present in a broad range of scientific and industrial applications and consumer products. The consumption of ENP-based products is growing year by year. This fact requires the adaptation of existing techniques and methods, or the development of new ones to monitor their ocurrence, fate and transformations. In order to understand the environmental impact of ENP, it is critical to discriminate among dissolved and particulate forms. Different techniques and methodological approaches for the characterization and quantification of ENP and its derivatives in complex samples have been recently reviewed [1]. Hydrodynamic Chromatography (HDC) and Asymmetric Flow Field Flow Fractionation (AF4) and are commonly coupled to ICP-MS as element specific detector for the separation and determination of inorganic ENP in a variety of samples. Although HDC is a robust and versatile separation technique, its resolving power is much lower than AF4. However, separations take place in a shorter time and dissolved low molecular mass species are not lost, as in AF4 due to the ultrafiltration membranes used in its separation channel. Thus HDC-ICP-MS can provide simultaneous information about dissolved and particulate species of an element in less than about ten minutes, which is not the case for AF4. Most applications of HDC-ICP-MS involve the use of the commercially available PL-PSDA HDC column and a proprietary mobile phase, comprised of a salt mixture and surfactants. In this work, chromatographic parameters as well as mobile phase compositions have been systematically studied with the column cited above, to evaluate the performance of HDC-ICP-MS for the simultaneous determination of gold dissolved species and gold nanoparticles. Special attention was paid to the resolution achieved between dissolved species and nanoparticles and their column recovery. [1] F. Laborda, E. Bolea, G. Cepriá, M. T. Gómez, M. S. Jiménez, J. Pérez-Arantegui, J. R. Castillo, Analytica Chimica Acta, 904 (2016) 10 This work was supported by the Spanish Ministry of Economy and Competitiveness and the European Regional Development Fund, project CTQ2015-68094-C2-1-R (MINECO/FEDER).

DETECTION, CHARACTERIZATION AND QUANTIFICATION OF TITANIUM DIOXIDE NANOPARTICLES IN COMPLEX SAMPLES BY AF4-ICP-MS

Francisco Laborda, David Ojeda, Vanesa Taboada-Lopez, Eduardo Bolea, Antonio Moreda, Pilar Bermejo, Juan R. Castillo

One of the most representative cases of the current trend in the field of nanotechnology is titanium dioxide (TiO_2), mainly due to the increase in its use in a large variety of consumer products. Because of its refractive capacity at nanometric scale (1-100 nm), TiO_2 nanoparticles can be found in sunscreens, cosmetics, food packaging and surface coating, being photocatalysis another main property. These nanoparticles are also used in environmental remediation, as food additive or as whitening pigment. Considering its relevance and current commercial exploitation, the assessment of its potential effects, not only on the environment but also on human health, should be also considered. Due to the disparity of products and possible scenarios where these nanomaterials can be present, it is evident the need for the development of new methodologies which can contribute to obtain additional information respect to that provided by more conventional techniques, namely electron microscopies or light scattering-based techniques. This work is focused on the use of asymmetrical flow field flow fractionation (AF4) coupled to UV-Vis spectroscopy and inductively coupled plasma mass spectrometry (ICP-MS) for the detection and characterization of dioxide titanium nanoparticles (TiO_2 NPs) in complex samples. With this purpose, those variables that determine the separation process in AF4 have been analysed. Nanoparticles stability, mainly related to the carrier composition, or the resolution for NP suspensions of several sizes, by using different crossflow programs, have been selected in the first stage of the study. The use of TiO_2 standard suspensions of known size for size calibration and for the characterization of TiO_2 NPs in several samples, as an alternative to the use of other standard materials (such as latex or polystyrene, commonly used in AF4) will be discussed. Relatively high recovery rates (around 80% for NP standards), with limits of detection of 0.88 μg/L for Ti in AF4-ICP-MS under optimal separation conditions, have been achieved. Finally, this methodology has been applied to the detection and characterization of TiO_2 NPs in different types of samples containing TiO_2. This work was supported by the Spanish Ministry of Economy and Competitiveness and the European Regional Development Fund, project CTQ2015-68094 C2-1-R (MINECO/FEDER).

Role of Clusters for Non-metal Ionization in ICP-based Plasma Assisted Reaction Chemical Ionization (PARCI)

Joseph Lesniewski, Kaveh Jorabchi

Our recent ICP-based atmospheric pressure PARCI interface has shown excellent sensitivity for chlorine detection as well as compatibility with organic solvents for gradient liquid chromatography. In this interface, ICP is coupled to a cooled aluminum orifice followed by a stainless steel afterglow reaction tube where negative mode chemical ionization is promoted. Methanol and sodium have each been shown to improve chlorine sensitivity considerably. Here we provide further insights into the afterglow ionization mechanism and effects of sodium and methanol via investigation of a series of clusters formed upon introduction of sodium and methanol into the plasma. Soft sampling of ions from the plasma afterglow in our experiments confirm the presence of clusters composed of sodium, formate, and nitrite ions within the afterglow. Identities of clusters were confirmed through exact mass measurements by coupling the atmospheric pressure PARCI source with a time-of-flight (TOF) mass spectrometer. Importantly, the intensities of the clusters positively correlate with chlorine sensitivity at various operating conditions, implying that these clusters may play a role in ionization. We have also observed chlorine-containing clusters upon injection of chlorinated compounds, further strengthening the evidence for the role of these clusters in ion formation. Inspired by these observations, we discuss new ionization routes for elemental detection of chlorine and other non-metals.

Ultrasonic Autosampler to Reduce Nanoparticle Aggregation for ICPMS Analysis

Derrick Quarles, Jared Kaser, Mason Spilinek, Kyle Uhlmeyer, Daniel Wiederin

Many metallic nanoparticle solutions suffer from particle aggregation over time after the solution is produced. This can cause particle size misrepresentation upon analysis. Ultrasonication has been documented as an effective method to dis-aggregate and disperse these particles back into solution. An ultrasonication autosampler has been designed and implemented to keep nanoparticles in suspension while awaiting analysis after production. The autosampler has an integrated ultrasonic bath with a unique sample holder that keeps the samples suspended in the bath. Multiple sample rack designs allow for a variety of test tubes to be used with this autosampler. The work presented here will demonstrate an accurate method for long analytical runs of nanoparticle solutions without an impact on particle size distribution. Furthermore, a comparison of the traditional autosampler and the ultrasonic autosampler will be presented for Si, Fe, Au, and Ti based nanoparticles.

Stable isotope amount ratio analysis by using high-resolution continuum source graphite furnace molecular absorption spectrometry

Carlos Abad, Stefan Florek, Helmut Becker-Ross, Mao-Dong Huang, Hans-Joachim Heinrich, Sebastian Recknagel, Jochen Vogl, Norbert Jakubowski, Ulrich Panne

Analysis of stable isotopes has been used as proof of provenance of mineral and biological samples, to estimate a contamination source and to determine geological processes. This kind of analysis needs high accuracy and precision for reliable conclusions. Currently, stable isotope analysis is dominated by mass spectrometric techniques that are time consuming and expensive. Here we present a fast and low cost alternative for isotope analysis of boron and magnesium: high-resolution continuum source graphite furnace molecular absorption spectrometry (HR-CS-GFMAS). Two stable isotope systems were evaluated separately: boron (^{10}B: ^{11}B) and magnesium (^{24}Mg: ^{25}Mg: ^{26}Mg). Their isotope amount ratios were estimated by monitoring their absorption spectra *in-situ* generated monohydrides. The molecular absorption spectrum of a XH molecule (X= B or Mg) with *n* isotopes would be a linear combination of *n* isotopologue spectra and the amount of each component (isotope) could be calculated by a multivariate regression (*n*= 2 and 3 for B and Mg respectively). For the analysis of boron certified reference materials (CRM), the band 1?1 for the electronic transition $X^{1}?^{+}$? A^{1}? was measured around wavelength 437.1 nm. Since boron has a memory effect in graphite furnaces, a combination of 2 % (v/v) hydrogen gas in argon, 1 % trifluoromethane in argon, an acid solution of calcium chloride and mannitol as chemical modifiers were used during the BH vaporization at 2600
C. Partial least square regression (PLS) for analysis of samples was applied. For this, a spectral library with different isotope ratios for PLS regression was created. Magnesium does not have memory effect. Therefore, only 2 % of hydrogen in argon as gas modifier during vaporization at 2500 C was employed for analysis of magnesium CRM. Absorption spectra of MgH for the X^{2}??A^{2}? electronic transition (band 0?0) were recorded around wavelength 513.45 nm. A similar PLS procedure to the BH was applied. Results for B and Mg CRM are metrologically compatible with those reported by mass spectrometric methods. An accuracy of 0.08 ‰ for B and 0.1 ‰ Mg was obtained as the average deviation from the isotope CRM. Expanded uncertainties with a coverage factor of k = 2 range between 0.10 - 0.40 ‰.

Towards Quantitative Analysis of Metallic and Metal Oxide Nanoparticles Using Microsecond Time-Resolved Single-Particle ICP-MS

Ingo Strenge, Antonio R. Montoro Bustos, Karen E. Murphy, Michael R. Winchester

Production and use of engineered nanoparticles (NP) are steadily increasing. In Europe, first regulations regarding the labeling of cosmetics, food, and biocide products became effective. [1] However, to date a thorough characterization of these materials requires time- and cost-intensive techniques such as scanning or transmission electron microscopy. One emerging technique for the rapid analysis of inorganic NP is single-particle inductively coupled plasma mass spectrometry (spICP-MS). Briefly, time-resolved measurements of a single isotope are performed using short, consecutive, measurement periods in the milliseconds to tens of microseconds [2,3] range. Individual NPs appear as discrete spikes in the time-resolved signal intensity profile. The advantage of spICP-MS, compared to commonly used ensemble screening techniques, is its ability to provide mean size, size distribution and number concentration based on measurement of discrete particles. However, the development of spICP-MS towards a quantitative analysis technique remains challenging. Accurate determination of the transport efficiency is required if sample introduction is carried out using conventional solution nebulization. Other difficulties include a lack of reference materials (RM) for the various NPs used nowadays, and the limited suitability of available RMs with respect to the diversity of NP properties (e.g. surface chemistry, density, elemental composition) for NPs in consumer products. In this work, recent accomplishments in the analysis of metal (Ag, Au) and metal-oxide (TIO2) NPs using microsecond time-resolved spICP-MS are presented. In addition, efforts and achievements toward establishing a quantitative spICP-MS method are critically discussed. Utilization of measurement periods as short as ten microseconds allowed the acquisition of ion cloud profiles for each detected particle. With this, measurement artifacts from split-particle events and coincidence from multiple particles occurring in quick succession could be virtually eliminated, and the NP signal vs. background noise ratio could be improved. The availability of individual time-resolved signals allowed the implementation of a refined dead time correction approach, intended to meet the requirements of short transient signal measurement. In combination with a suitable data processing scheme, accuracy and precision of captured particle events was improved, and lower size detection limits were achieved. Furthermore, the possibility to increase the applicable size range using different degrees of signal attenuation was explored. [1] H. Rauscher, K. Rasmussen, B. Sokull-Klüttgen, Chem. Ing. Tech., 2017, 89, No. 3, 224–231. [2] I. Strenge, C. Engelhard, J. Anal. At. Spectrom., 2016, 31, 135-144. [3] M. D. Montaño, H. R. Badiei, S. Bazargan, J. F. Ranville, Environ. Sci.: Nano, 2014, 1, 338-346.

Validation of ATR Correction and Reverse ATR Correction Algorithms, Improved by Optimized Corrections

Gregory M. Banik, Michelle D'Souza, Keith Kunitsky, Robin O'Connor

To address wavelength-dependent differences in intensity between ATR and transmission FTIR spectra, a mathematical ATR correction can be applied to an ATR spectrum to provide a better match when searching against a transmission FTIR reference database. Conversely, the reverse ATR correction can also be applied mathematically to a transmission FTIR spectrum to provide a better match when searching against an ATR reference database. An original study validated the ATR correction and the reverse ATR correction algorithm in Bio-Rad's KnowItAll Informatics System spectroscopy software. This study updates and compares search results from the original study to search results using a patent pending Optimized Corrections technology recently added to the software that iteratively optimizes the query and reference spectra for each comparison by automatically applying multiple corrections to compensate for differences between spectra caused by the variability of different instruments and accessories as well as other factors, including human error. This technology solves many of the issues surrounding spectral search that cannot be adequately addressed by traditional algorithms and manual methods. Corrections include: baseline correction, clipping, horizontal shift, vertical shift, intensity distortion, and ATR correction.

Eliminating Fringes from Hyperspectral Data to Localize Chemically Distinct Macromolecules

Ghazal Azarfar, Ebrahim Aboualizadeh, Nick Walter, Achim Kohler, Carol Hirschmugl

Fourier Transform Infrared(FTIR) microspectroscopy is a powerful technique for chemical imaging of micrometer size samples. Employing a bright synchrotron source, a microscope coupled to a FTIR spectrometer, and replacing a single detector with an array of detectors (128 x 128 detectors) enables collecting combined spectral and spatial information simultaneously. Fringes arise in infrared spectra and images of thin films due to multiple internal reflection (similar to that encountered with soap bubbles or oil slick on water for visible wavelengths) and hinder the quality of the measured hyperspectral data. In this work, an extended multiplicative signal correction method based on Fourier analysis of the spectra removes the fringes in the spectral domain that enhances micro-spectral images. The method is applied to 24800 spectra with fringes that are up to 20% of the intensity. Spectra are extracted from 1037 images covering spectral range between 950 cm^{-1} to 2950 cm^{-1} (3.4 ?m to 10 ?m) from a live algal cell (~ 40x70 $?m^{2}$) maintained in a liquid flow cell. The dimension of the improved hyperspectral information is reduced using multidimensional Principal Component Analysis to extract the most important components (spectrally and spatially). Comparison of the first three principal components (raw and after correction) of the data show a high degree of improvement in visualization of spatial distribution of macromolecules after the correction, this drives the ability to localize the important biochemistry.Similar spatial distributions are observed for control and stressed cells, and from spectral analysis, lipids, and carbohydrates are responsible for these changes. This work was supported by UWM Physics department and NSF Grants CHE-1508240 and CHE-1112433.

Quantitative analysis of water content in polymer samples by terahertz spectroscopy
Hiromichi Hoshina, Yoh Iwasaki, Eriko Kometani, Makoto Okamoto, Chiko Otani

Due to the recent development of the terahertz (THz) wave technologies, their spectroscopy and imaging applications have attracted keen interests in applied physics and chemistry. For the sensing and imaging applications in THz frequencies, liquid water is one of the significant targets, because of their high absorbance in THz region (~250cm^{-1} at 1THz). THz images are quite sensitive for small amount of water and show clear contrast in the images of human tissues, plants and industrial products such as papers, plastics and cloths. Also, THz spectra contain rich information of water dynamics. At lower frequency region below 1THz, spectra reflect picosecond rotational relaxations. On the other hand, intermolecular vibrational motion shows absorption bands at higher frequencies such as intermolecular vibrations observed around 6 THz. In this paper, we demonstrated quantitative measurement of water content in Poly(ethylene-vinylalcohol copolymers) (EVOH) by THz absorption spectroscopy. Polymer containing hydrophilic group adsorb water molecules and change their physical and chemical properties. In the case of EVOH, it has been known that mobility of gas molecules in the polymer changes with the humidity, which result in instability of the oxygen permeability of EVOH products. We have prepared 60 ?m thick EVOH films in different humidity conditions and measured THz absorption spectra in 1-9 THz by a FT-FIR spectrometer. In the observed spectra, absorption of lattice modes of crystalline EVOH and intermolecular vibrational motion of water molecules were observed. By subtracting spectra with those measured in vacuum dried conditions, spectral feature of water in EVOH samples were obtained. The spectra of the adsorbed water show main band around 6 THz suggesting that the major part of water is in similar environment as bulk water. Simultaneously, molar content of water was obtained by measuring the weight of samples. The plot between water content and THz absorption intensity at 6THz shows clear linearity, that demonstrates the potential of THz spectroscopy as a quantitative analytical method for bound water in the polymer products.

Structural analysis of the bound water in Poly(ethylene-vinylalcohol copolymers) by terahertz two-dimensional correlation spectroscopy

Hiromichi Hoshina, Yoh Iwasaki, Eriko Kometani, Makoto Okamoto, Chiko Otani

Since bound water in the macromolecules is a key to understand their properties and functions, their structure and dynamics has been studied in various methods. The amount of bound water can be estimated by DSCs; however, their structural information is hard to obtain. There are various studies observing bound water by infrared spectroscopy, observing shift of OH stretching mode, however, information is indirect and sometimes concealed by the intense main band. Terahertz (THz) spectroscopy may have a potential to observe bound water because intermolecular vibrational band of liquid water is observed around 6 THz, which gives direct information of intermolecular structure. However, analysis of the THz spectra is not a straightforward task, because vibrational bands in THz region are broad and overlapped. In the case of water adsorbed polymer, lattice modes, hydrogen bonds in polymer, hydrogen bonds between polymer and water, and intermolecular vibration of water molecules are overlapped. For such complicated spectra, two-dimensional correlation spectroscopy (2DCOS) works as a power tool for analysis. In this study, sheets of Poly(ethylene-vinylalcohol copolymers) (EVOH) prepared in different humidity condition. THz spectra were measured by a FT-FIR in the frequency region in 1-9 THz. In the observed spectra, absorption of lattice modes of crystalline EVOH and intermolecular vibrational motion of water molecules were observed. By subtracting spectra with those measured in vacuum dried conditions, spectra of adsorbed water were obtained. The series of spectra were analyzed by 2DCOS using humidity as parameter. The spectra of the adsorbed water show main band around 6 THz and weak features at lower frequencies. 2DCOS shows that lower frequency vibrations arise at lower humidity condition and main band arise at higher humidity suggesting that low-frequency vibration is originated in the bound water close to the polymer chains.

2D correlation analysis of surface reaction on LiFePO4 cathode during charging-discharging process using in-situ Raman spectroscopy

Young Mee Jung, Yeonju Park, Su Min KIm, Sila JIn

As increasing the demands on the Li-ion battery in hybrid electric vehicles (HEVs), electric vehicles (EVs) and grid energy storage, it become more important issue to enhance the performance and safety for Li-ion battery. However, electrochemical reactions involved in the degradation process are very complicated and the full scenario of the relationship between the surface structural degradation and degradation of devices is still not well understood. Therefore interesting for development of in-situ techniques is getting growth. In this study in situ Raman spectroscopy was used for real time monitoring the change of composite electrode during electrochemical performance. To deeper understand the electrochemical reaction in Li-ion battery, 2D correlation spectroscopy was applied to in situ Raman spectra obtained during charge-discharge processes. In this presentation, the result of 2D correlation in situ Raman spectroscopy in Li-ion battery during charge-discharge processes will be discussed in more details. Acknowledgement This work was supported by the National Research Foundation of Korea (NRF) grants funded by the Ministry of Science (2015R1D1A3A01015885).

Theoretical Modeling of Vibrational Coupling between Cyano- and Azido- Reporters

David Hogle, Ryan Gustafson, Matthew Tucker

Two-dimensional infrared (2DIR) spectroscopy can report structural information based upon the coupling between selected vibrational modes in a target molecule. The use of site-specific reporters enables global structure determination and observation of local dynamics. We have investigated the interaction between two potentially useful reporters: the cyano- and the azido- moieties. We are developing theoretical maps that relate the magnitude of coupling between these reporters to the spatial orientation and distance. By building different scaffolds and developing coupling models, these reporters will provide a means for monitoring structural changes within oligonucleotides and peptides.

Spectroscopic Methods for Quantifying Diesel Particulate Matter

David Parks, Arthur Miller, Peter Griffiths

Diesel engines are one of the primary contributors to the presence of ultrafine and nano-aerosols in ambient air and occupational environments. Exposure to diesel emissions has been shown to contribute to various adverse health outcomes of the pulmonary system and cardiovascular system. Currently diesel emissions in the workplace are monitored by way of collecting the aerosol onto filters, which are then sent to a lab for thermal-optical analysis, also known as the NIOSH 5040 method. The NIOSH 5040 method measures elemental and organic carbon explicitly. This process can take days or even weeks, and workers can potentially be exposed to excessive levels of diesel particulate matter (DPM) before the problem is identified. To remedy this, NIOSH researchers are seeking to develop a field-portable method for measuring elemental and organic carbon in DPM aerosols. In the current study we investigated the use of FT-IR spectrometry, which lends itself more readily to implementation in a real time system than do the thermal-optical analysis methods. Further, it is anticipated that FT-IR spectroscopy will prove to be superior to current field-portable methods based on single channel absorption or optical scattering, which cannot differentiate between elemental and organic carbon. DPM was collected onto filters and the infrared spectra measured in transmission mode. We demonstrated good linear correlation between the integrated area of a selected peak and the mass of total carbon on the filters (R2 value of 0.95). The chosen peak was ascertained based on previous published results for analyzing carbon compounds, and lay in the wavenumber range of 1675 to 1748 cm-1 on the IR spectrum. Additionally a principal component analysis was performed to identify functional groups that may facilitate the sensitivity and accuracy of the anticipated real-time method.

Enhancement of spectral discrimination among samples using spectroscopic mapping over a drop-and-dry spot
Daun Seol, Eunjin Jang

When aqueous samples are measured with IR spectroscopy, a drop-and-dry method analyzing dried samples formed on a substrate after solvent evaporation is advantageous since it enables to highlight spectral features of analytes by avoiding strong water absorption. When an extract from a solid sample is dried on a substrate, the formation of successive ring patterns, the so-called coffee ring effect, occurs due to the result of dissimilar interactions of components with surfaces over the course of solvent evaporation. When many spectra are collected over the dried spot by mapping, relatively diverse chemical information of individual components separated on the substrate can be obtained and the use of all spectral information for analysis, such as discrimination analysis in this study, could be helpful to enhance accuracy. For the evaluation, first, IR spectra of the extracts from red pepper powders were measured by drop-and-dry method, and the corresponding mapped spectra over the dried spot were used for discrimination of geographical origins. To collecte IR spectra of extract samples, Si wafer with hydrophobic coating was prepared to make the formation of sample droplets as consistent as possible in shape. In parallel, NIR spectra of the same red pepper powders were collected and the result of discrimination was compared. Second, using the same method, the identification of gallbladder (GB) polyp and GB stone was attempted using bile juices as specimens. IR mapping spectra of the dried bile juices on the substrate were acquired as the same fashion above. In both cases, several feature selection methods with different classification tools were evaluated to choose optimal variables out of the mapping dataset to eventually improve the discrimination accuracy.

Mass Balance and Unit Process Efficiency for the Refinement of Natural Products are Revealed via Spectroscopic Chemical Imaging

Mark Boatwright, David Wetzel

The objective of industrial wheat milling is to isolate the fine, starchy endosperm portion of the wheat kernel as flour. This is accomplished by sequential, selective grinding steps followed by physical segregation of particle size via sifting. Quantitative spectroscopic chemical imaging enables discrimination based on molecular structural differences between the botanical parts of the wheat kernel that are mixtures of organic compounds. Large spectral libraries have been developed that provide standards for pure endosperm vs. various forms of nonendosperm. Application of partial least squares regression enables reliable quantitative determination of endosperm vs. nonendosperm. Typically, thousands of individual analyses are produced by the pixel. The quantitative result of each image is a summation of the individual pixels and the chemical image enables visualization of the distribution of the analyte within the matrix. Select unit processes at various stages of the milling process were chosen to demonstrate the utility of this technique to highlight departures from ideal operation and to observe the effect of operational adjustments.

Determination of Total Petroleum Hydrocarbons (TPH) in Soil by Portable Mid -ATR System
Toni Miao, Natasha Sihota

The application of mid-infrared spectroscopy as a rapid screening tool for TPH concentrations in contaminated soils is presented. Current literature reports use of Near and Mid DRIFT (Diffuse Reflectance Infrared Fourier Transform) spectroscopy with some success. In this study, we describe an ATR (attenuated total reflection) mid-infrared technique for determining TPH in soil using a handheld spectrometer with no solvent extraction required. Our data show that this technique is extremely robust and even could be used for neat whole soil samples with 15-30 wt% water content. The advantage of using ATR is that the Reflectance IR spectra can be acquired easily and rapidly, in addition, high sensitivity of alkyl-CH_3 vibrational group makes the TPH measurements robust and reliable for quantitative studies. Partial least squares (PLS) and cross validation were used to develop correlations for determination of TPH. Our study includes soil samples from different contaminated sites and comparison ATR with the DRIFT techniques. From our preliminary results, we conclude that Mid IR-ATR can deliver promising results for field screening, providing real time in-situ TPH analysis in minutes.

Application of Infrared Microscopy to the Forensic Examination of Automotive Paint Smears

Barry Lavine, Undugodage Perera, Kaan Kalkan, Linqi Zhang

Modern automotive paints typically use thinner undercoat and color coat layers protected by a thicker clear coat layer. All too often, a clear coat is the only layer of paint left at the scene of a "hit-and-run" incident where damage to a vehicle and/or injury or death to a pedestrian has occurred. In these situations, clear coat formulations are too similar for automotive paint databases or commercial library search algorithms to generate accurate hit-lists of potential suspect vehicles by searching the clear coat infrared (IR) spectra alone. Smears represent another type of automotive paint sample found at a crime scene that is also problematic for forensic paint examiners as there are no reference materials available in forensic automotive paint databases to generate hit-lists of potential suspect vehicles from paint smears. A procedure based on a Charpy impact tester will be developed to generate paint smears (e.g., abraded or deformed clear coats, clear coat and color coat layers mixed together, or clear coat, color coat, primer, and e-coat layers mixed together) due to paint transfer from an automotive substrate to any surface of interest under conditions simulating those of a real collision. As revealed by an ATR-IR microscope and multivariate curve resolution, there is separation of the automotive paint layers in certain regions, whereas in other regions the paint smear is intact. In contrast, the paint smear appears intact everywhere using transmission IR microscopy. The unmixing of the layers is evident only with ATR-IR microscopy because of its superior spatial (x, y, and z) resolution. Because the smeared paint in all likelihood undergoes melting due to the heat generated by the impact, a separation phenomenon may occur during fluid flow due to differences in viscosities and/or affinities for the substrate. This result indicates that paint smear standards can be developed for proficiency testing using the approach described here. The proposed technique can simulate an automobile collision with controllable collision speed and direction as well as momentum/energy transfer. The long term goal of this proposed paint transfer effort is the development of forensic standards for automotive paint transfer and paint smear analysis.

Differentiation of Bovine, Porcine, and Fish Gelatins by Attenuated Total Reflectance Fourier Transform Infrared (ATR-FTIR) Spectroscopy coupled with Pattern Recognition

Ahmet Kemal Aloglu, Peter de B. Harrington

Bovine, porcine, and fish gelatins have been differentiated based on their spectra collected by attenuated total reflectance Fourier transform infrared (ATR-FTIR) spectroscopy and coupled with pattern recognition. Three tree-based classification methods, a fuzzy rule-building expert system (FuRES), support vector machine classification trees (SVMTreeG and SVMTreeH), and one reference model, super partial least squares discriminant analysis (sPLS-DA) were evaluated with and without two preprocessing techniques, namely standard normal variate (SNV) and principal component orthogonal signal correction (PC-OSC). Validation of these methods was obtained with 95% confidence intervals by 10 bootstraps and 4 Latin partitions (BLPs) (10:4). The ATR-FTIR spectra were used with four different ranges at full spectra [4000-650] cm-1, the fingerprint region [1731-650] cm-1, specified spectra [4000-800] cm-1, and narrow fingerprint region [1731-800] cm-1. Classification rates for the methods were improved with SNV and PC-OSC when they were used separately or together. The highest classification rates were obtained from the narrow fingerprint region with SNV and PC-OSC at 97.4±1.6% for FuRES, 100±0% for sPLS-DA, 99.3±0.5% for both SVMTreeG and SVMTreeH. ATR-FTIR spectroscopy combined with pattern recognition could be a potential analytical technique for differentiating the sources of bovine, porcine, and fish gelatins with fast and reliable results.

Water solubility measurements with FTIR for carbon capture and sequestration relevant gas mixtures
Christopher Wiseall

With carbon dioxide levels rising in the atmosphere there is a significant effort across all areas of industry to cut emissions by reducing energy use. Carbon capture and sequestration (CCS) involves capturing the carbon dioxide produced during fossil fuel combustion and pumping it underground. Once the CO_2 is removed from the atmosphere it cannot contribute to the greenhouse effect[1]. For CCS to become a viable technology option the costs need to fall; the knowledge of these limits will help to reduce costs for modelling and designing the network as well as any future legislation. As the CO_2 has been captured from the combustion of fossil fuels there will be a wide range of impurities present such as, N_2, H_2 or H_2O. These impurities can have a large effect on the physical properties of the CO_2 rich fluid. For the most efficient transport network, the CO_2 is expected to be transported in the supercritical phase[2]. There is limited knowledge of supercritical CO_2 containing impurities. Knowing the water solubility limit of these mixtures is vital as liquid water causes increased risk of corrosion in pipelines. A method for measuring water solubility limits of gas mixtures at high pressures has been developed in the George/Poliakoff group at Nottingham[3]. The method tracks the evolution of the 1605 cm^{-1} water band with increasing volumes of water. This method has been used for a variety of CCS relevant gas mixtures, which will be shown by this poster. 1 M. Z. Jacobson, Energy Environ. Sci., 2009, 2, 148–173. 2 Z. X. Zhang, G. X. Wang, P. Massarotto and V. Rudolph, Energy Convers. Manag., 2006, 47, 702–715. 3 S. Foltran, M. E. Vosper, N. B. Suleiman, A. Wriglesworth, J. Ke, T. C. Drage, M. Poliakoff and M. W. George, Int. J. Greenh. Gas Control, 2015, 35, 131–137.

Quest ATR Spectroscopy: Fast and repeatable analysis of pharmaceuticals, drugs and forensic samples
Jeff D'Agostino, Fawzi Abou-Chahine, Mia Abbott, Ben Schazmann

Infra-red transmission spectroscopy is regularly used in academia and industry to analyse a variety or liquid and solid samples. This effective and non-destructive method can detect contaminants and can quantify sample purity for quality control purposes. This paper shows how ATR spectroscopy is particularly important for the healthcare and security industries. Focus is given to narcotic analysis and discrimination, and a calibration plot of a test drug sample is presented. The user-friendly design of the Quest ATR makes this technique easier and faster than traditional transmission spectroscopy with KBr matrix windows.

Measurement of Crystalline Silica Aerosol Using Quantum Cascade Laser–Based Infrared Spectroscopy
Kevin Ashley, Shijun Wei, Pramod Kulkarni, Lina Zheng

Exposure to airborne crystalline silica poses major health risks in many industrial workplaces. Over 2 million workers are exposed to respirable crystalline silica (RCS) in the United States alone. Sensitive methods for in-field or near real-time measurement of crystalline silica aerosol are necessary to promote improved compliance and to support epidemiological studies. The objective of this study was to develop a quantum cascade laser (QCL) based infrared spectroscopy (IR) method for quantification of airborne concentrations of RCS. Three aerosol collection methods were investigated for effective coupling with QCL: i) conventional total filter collection; ii) focused spot collection directly from the aerosol phase; and iii) dried spot analysis after deposition using liquid sample slurries. Spectral analysis methods were developed to obtain QCL-IR spectra from the collected particulate samples in the range 750-1030 cm-1, with measurement of the quartz (and/or cristobalite) { <--Si-O-Si--> } band at 800 cm-1. The system was calibrated using primary standard respirable quartz and QCL data were compared with results from standardized methods using Fourier transform infrared (FTIR) spectrometry. Results show that sufficiently low detection limits for RCS (?330 ng) can be achieved with effective microconcentration and careful coupling of the particulate sample with the QCL beam. These results offer promise for further development of sensitive filter-based laboratory methods and direct-reading QCL portable monitors for near real-time measurement of crystalline silica aerosol.

Analysis of Diffuse Reflectance Spectra of Dye in Milk Using Representative Layer Theory

Akihiro Nojima, Takuya Kambayashi, Shin-ichi Taniguchi

The nonlinear relationship between absorbance and concentration of substrates due to multiple scattering events was observed in diffuse reflectance spectra. In this study, we investigated a method for increasing the linearity based on representative layer theory. The sample consisted of Rhodamine B dissolved in milk. The thickness of the sample was greater than 20 mm, so the intensity of the detected transmission light was less than 1/1000 of the incident light. The white light was incident normal to the surface of the sample, and the diffuse reflected light was collected at 45° from surface normal. From this measurement, the fraction of the remission of the whole sample (R) is determined. Because the sample is thick enough, the fraction of the transmission light (T) that passes through the whole sample can be ignored. Therefore, the fraction of absorption of the light (A) is nearly equal to 1-R. In representative layer theory, the sample is divided into hypothetical thin layers, which are thin enough that the multiple scattering events can be ignored. The fraction of absorption (a) and remission (r) of the thin layer can be related to the fraction of absorption (A) and fraction of the remission (R) of the whole sample as follows: (1-R)2/R=(a/r)(2-a-2r). Here, r can be estimated from the scattering coefficients of the milk. The Pearson correlation coefficient between the absorbance (log (1-a) and log (1-A)) and the concentration of Rhodamine B was determined to be 0.99 for log (1-a) and 0.88 for log (1-A). This result clearly shows that the linearity can be improved by the method presented above by using representative layer theory. In addition, the effect of the scattering coefficients of the milk on the spectra will be discussed in the presentation.

Precise Recovery of Coordinates by Image Processing Algorithm for FTIR Micro-Spectroscopy Limited Angle Computed Tomography

Sugato Ray, Nicholas Walter, Alexander Schofield, Ghazal Azarfar, Reinhold Blumel, Achim Kohler, Carol Hirschmugl

Limited Angle Computed Tomography (LACT) to extend the capability of FTIR spectromicrotomography applications is presently being developed. LACT requires perfectly aligned projections to reconstruct the object-of-interest with high contrast and accurately determined edges. However, experimental sources of error can overwhelm such reconstruction. Such sources of error could stem from translational or rotational errors associated with the mechanical motion of the stage that the object-of-interest lies on. The stage for such imaging must have at least five degrees-of-freedom (DOFs); three translational and two rotational DOFs. The approach described here is designed to compensate for any misalignment of projections due to stage inaccuracies. Four fiducials are used as visible-anchors around the sample to recover the exact coordinates of each field-of-view (FOV) on the sample plane by means of post-image collection image-processing. Micro-grids with strong infrared absorbing vibrational excitations are used as the fiducials. These fiducials are placed around the sample, on the sample plane, prior to FTIR imaging for LACT. Images obtained for each of these fiducials can then be processed to recover the translational and rotational coordinates for each micro-grid, which are used to determine precise coordinates for the FOVs. The four-fiducial system provides four concurrent triangulation systems for a given sample plane. The coordinate-recovery algorithm from image-processing ensures high accuracy when implemented with multiple fiducials. This presentation will focus on a demonstration of this approach, using the image processing of the fiducials to accurately correct the coordinates of the FOVs and reconstruct the chemical 3D Images of the sample of interest. Further, this approach will be used to calibrate the range of positional errors induced by the motions of the sample stage. This work was supported in part by the National Science Foundation CHE-1508240 and CHE -1112433.

Mid-Infrared Dispersion Spectroscopy for Trace Gas Sensing
Jakob Hayden, Pedro Martin-Mateos, Pablo Acedo, Bernhard Lendl

Laser absorption spectroscopy is extensively used and studied for sensing trace amounts of molecular species in air and exhaust gasses due to short response times, high sensitivity and selectivity and moderate instrumental complexity and cost. However, absorption spectroscopic sensing, due to the measurement of intensities, is susceptible to intensity noise and fluctuations caused by the laser source or any non-absorptive attenuations, e.g. scattering at particles, in the beam path. To overcome this limitation while maintaining the mentioned advantages of laser spectroscopic sensing, dispersion, rather than absorption based spectroscopic schemes were proposed. These techniques sense the characteristic refractive index profile associated with a molecular resonance via a measurement of light's phase, rather than absorption via intensity. We demonstrate heterodyne phase sensitive dispersion spectroscopy (HPSDS) in the mid-infrared spectral range for sensing trace amounts of CO.[1] HPSDS is an instrumentally particularly simple technique that recovers the sample's refractive index from the phase of an intensity modulation imposed on the light emitted from a continuous wave laser.[2] In our implementation, we use a current modulated quantum cascade laser emitting at 4.59 μm to target the fundamental CO vibration. After passing through the gas sample, the modulated infrared light is detected by a fast detector and the phase of the modulation is recovered by a phase detector. By slowly scanning the laser wavelength across the molecular resonance, a refractive index spectrum is recorded and from it, the concentration of CO is retrieved. In this contribution, we discuss the most significant characteristics of HPSDS as compared to absorption spectroscopy, thereby focusing on wavelength modulation spectroscopy. We present experimental results on the limit of detection and linear dynamic range under different points of operation of the laser. The modulation frequency and the laser bias current are identified as critical parameters as they determine the extent of undesired frequency tuning associated with the direct current modulation. In addition, we demonstrate the technique's insensitivity to the intensity reaching the detector. 1. P. Martín-Mateos, J. Hayden, P. Acedo, and B. Lendl, Anal. Chem. 89, 5916 (2017). 2. P. Martín-Mateos and P. Acedo, Opt. Express 22, 15143 (2014).

Environmental and Industrial Gas Applications of Integrated Cavity Output Spectroscopy
Diane Errigo, Frédéric Despagne, Steven M. Barnett

The Off-Axis Integrated Cavity Output Spectroscopy (OA-ICOS) patented technology for gas analysis provides a unique combination of sensitivity, linearity, portability and robustness that makes it a tool of choice for trace gas quantitation and isotopic ratio assessment in various locations. Multiple OA-ICOS instruments have been deployed in the past 15 years for various applications in the fields of environmental and industrial analysis. The presentation will provide an overview of the basic principles and features of the technology. It will be illustrated with application examples related to greenhouse gas analysis in several challenging environments, and also some industrial applications.

Determination of domain stability using High Throughput-Developability and Comparability Assessment (HT-DCA) platform and 2D IR and Co-distribution correlation spectroscopies
Belinda Pastrana

High throughput-Developability Comparability Assessment (HT-DCA) platform technology has proven useful in the determination of domain stability for an array of proteins under thermal stress. Hyperspectral Imaging (HSI) using quantum cascade lasers were collected in the MID IR fingerprint region, thus rendering enhanced IR SNR. The spectral data were subject to 2D IR and Co-distribution correlation analysis. This innovative method is direct, free of labels and tags providing the dynamic behavior of an array of proteins under the exact same stressor conditions in one experimental run. The novel Co-distribution correlation analysis provides molecular changes due to the stress that is representative for the population of proteins in aqueous solution. Also, 2D IR correlation analysis of the amide I and II bands within the spectral region of 1700 - 1450 cm^{-1} provides molecular information crucial to the design of therapeutic proteins. Furthermore, the approach allows for the evaluation of side chain vibrational modes serving as internal probes that validate the domain stability of the protein of interest. The sequential order of molecular events for three distinct proteins: **(1)** NIST mAb standard **(2)** bovine serum albumin (BSA) and **(3)** lysozyme with molecular weights of 147, 67 and 14 kDa, respectively will be discussed. These results provide initial proof-of-concept for further development of our HT-DCA platform for the screening of protein therapeutic candidates.

Fano resonances in infrared spectroscopy

Alex Schofield, Reinhold Blumel, Achim Kohler, Rozalia Lukas, Carol Hirschmugl

We show that mid-infrared transmission spectra from typical dilute biological media can be modeled by Fano theory. In particular, we demonstrate that materials with a constant refractive index cause scattering that leads---to a very good approximation---to Fano resonances when the imaginary part of the complex refractive index is small and Lorentzian in nature, as is the case for many biological media. We demonstrate the effect for two geometries: thin films, and small spheres; and provide a general reasoning for arbitrary geometries. Furthermore, an algorithm for fitting a Fano line shape to transmission spectra is presented and used to demonstrate that the shift in peak frequency can be recovered with high accuracy for general, isolated Fano line shapes. In addition, fitting Fano line shapes in the context of small and optically soft spherical scatterers, recovery of the true absorption strength, and one or the other of constant refractive index or radius (given approximate knowledge of the other), is possible. We believe that understanding scattering phenomena in FTIR as Fano resonances may provide new insight and foster the development of new corrective techniques, and the improvement of spectral analysis.

Determination of water and glycol concentration in engine oil via induced emulsification

Torrey Holland, Ali Mazin Abdul-Munaim, Dennis Watson, Poopalasingam Sivakumar

We investigated the glycol and water content of engine oil at various levels of contamination using Fourier transform infrared (FT-IR) spectroscopy and laser-induced breakdown spectroscopy (LIBS). The amount of water and glycol in SAE 15W-40 diesel engine lubricating oil is highly dependent on the timing of the measurement since the initial contamination when using conventional sampling methods. Ultrasonic probe emulsification is employed to disperse the water and glycol molecules into the oil homogeneously which significantly reduces the influence of sampling time and location. Analysis of variance (ANOVA) modeling and calculation of detection limits are applied to determine the contamination levels. The results show that the ultrasonic probe emulsification is a very effective technique to determine the degree of contamination in engine oil with greater accuracy.

Ultrafast Angle-Resolved Photoemission Studies of Cuprate High-Temperature Superconductors
Christopher Smallwood

The origin of high-temperature superconductivity at critical temperatures T_c > 40 K in cuprate ceramics, and more recently in iron-based pnictides and chalcogenides, is one of the great unsolved mysteries in condensed matter physics. Among the most powerful tools for solving this mystery is the technique of angle-resolved photoemission spectroscopy (ARPES), in which ultraviolet photons are exposed to a sample to produce photoelectrons that directly map out the material's occupied-state band structure. Significant discoveries by ARPES in facilitating a better understanding of the cuprates throughout the course of the past 30 years have included the discovery of an anisotropic d-wave symmetric superconducting gap, and the discovery that this gap persists in cuprate band structure at temperatures well above the superconducting critical temperature. Nevertheless, mysteries remain, including the roles that the superconducting gap and the normal-state "pseudogap" play in relation to each other, the role that phonons and/or magnetic phenomena play in facilitating superconductivity in high-T_c materials, and the interplay of phase fluctuations and intertwined orders in the materials. Here I report on studies of the cuprate superconductor Bi2212 using a variation of the ARPES in which a material is exposed not only to an ultraviolet probe, but rather to a short succession of two pulses&#8212an infrared pump followed by an ultraviolet probe [1]. Time-, momentum-, and energy-resolved spectra reveal a transient population of quasiparticles and electron-boson coupling effects that strongly depend on both crystal momentum and pump pulse excitation density [2,3]. The results offer a glimpse into the dynamics of unoccupied states in cuprate superconductors, and they hint at the possibility of quasiparticle recombination "hot spots" at specific points in momentum space. Such hot spots, if present, suggest a connection between the physics of superconductivity and the physics of antiferromagnetism and/or charge density wave phenomena in cuprate superconductors.
[1] C. L. Smallwood, C. Jozwiak, W. Zhang, and A. Lanzara. Review of Scientific Instruments 83, 123904 (2012)

[2] C. L. Smallwood et al. Science 336, 1137 (2012)

[3] W. Zhang et al. Nature Communications 5, 4959 (2014)

Fluorescence Optical Rotary Dispersion for Surface-Specific Chiral Analysis
Garth J. Simpson, Fengyuan Deng, James R. W. Ulcickas

A new interface-specific spectroscopic method is proposed based on optical rotary dispersion in fluorescence and Raman measurements from monolayer films, with sensitivity to chirality that is orders of magnitude greater than typically encountered in analogous chiral-specific measurements of isotropic assemblies. The symmetry requirements underpinning the surface specificity of second harmonic generation (SHG) and sum frequency generation (SFG) are shown to have analogs in four-wave mixing measurements, which include both conventional Raman and fluorescence spectroscopies. Furthermore, the remarkable sensitivity of SHG and SFG to interfacial chirality is also predicted to be present in fluorescence measurements of oriented surface assemblies. Since the effects are fully electric dipole allowed in both cases, fluorescence measurements of chiral surface films are predicted to produce chiral-specific fluorescence optical rotary dispersion (F-ORD) with predicted chiral-specific effects up to 25% of the achiral contributions from a single monolayer film. Previous single-molecule chiroptical fluorescence studies by Barnes and coworkers (Science 2006, 314, 1437) provide compelling evidence supporting these theoretical predictions.

Exciton Correlation in Monolayer MoS2 revealed by 2D Electronic Spectroscopy
Liang Guo, Daniele Monahan, Graham Fleming

Monolayer transition metal dichalcogenides exhibit strong potential as new optoelectronic materials and in many other applications. Due to quantum confinement and reduced dielectric screening in 2D materials, excitons, which play an important role in energy conversion and transfer in such materials, are featured by large binding energy and affected by

strong many-body effects. In this study, two-dimensional electronic spectroscopy (2DES) is used for exploring many-body effects between exction A and B in MoS_2 monolayer,
the most important two exciton modes in the visible range. The details of many-body effects are reflected in the dynamics of the cross peaks in 2DES plot. This work reveals fundamental correlation between the two exciton modes, which is a important concern in design of future optoelectronics.

CHARACTERIZATION OF IONIC LIQUIDS USED IN LIQUID-LIQUID EXTRACTION PROCESSES BY RAMAN SPECTROSCOPY.

Teresa Alejandra Razo Lazcano, María del Pilar González Muñoz, Mario Ávila Rodríguez, Mercy Sugey Dzul Erosa

Ionic liquids have had a great attention in the search of new materials for a variety of applications to industrial level and in technology, because of their physicochemical properties. Some of these properties are: very low vapor pressure (because of this, they are a potential substitutes of organic solvents), low melting point, high chemical and thermal stabilities (350 – 400 C) and they are nonflammable compounds. By the above, it is possible to performance several mixtures of anions and cations to make a specific ionic liquid for a particular application, as the liquid-liquid extraction processes. Therefore, it is necessary to perform the characterization of these compounds and Raman Spectroscopy is a technique very useful to this purpose. In this work, the characterization and analysis of two ionic liquids of phosphonium salt's class by Raman Spectroscopy, had been performed. Ionic liquids studied were: Cyphos IL-101 (trihexyl (tetradecyl) phosphonium chloride) and Cyphos IL-104 (trihexyl (tetradecyl) phosphonium bis-2,4,4-(trimethylpentyl) phosphinate), mixture of Cyphos IL-101 and Cyanex 272 (derivative of phosphonic acid); and mixtures of Cyanex 272 (0.1 and 1 mol.L-1) and Cyphos IL-101, made in laboratory, were studied too. It is important to note that the characterization of these compounds doesn't exist recorded. The Raman spectrum of Cyanex 272 exhibited characteristic bands at 1068 and 747 cm-1 assigned to the symmetric stretching vibrations P-O, at 873 cm-1 to the stretching and wagging vibrations P-O and at 472 cm-1 to the wagging vibration O-P-O. The main bands for the Raman spectrum of Cyphos IL-101 were assigned at 1109, 1070, and 887 cm-1 to the rocking vibrations P-CH3 and stretching vibrations P-C at 663 cm-1. The Raman spectrum of Cyphos IL-104 showed characteristic bands of Cyanex 272 and Cyphos IL-101, with assignments at 1110 cm-1 for the rocking vibrations P-CH3, at 1060 and 871 cm-1 to the asymmetric stretching vibrations P-O, symmetric stretching vibrations at 926 and 746 cm-1 P-O, at 656 cm-1 stretching vibrations P-C and wagging vibrations at 475 cm-1. On the other hand, it was possible to observe that at higher concentration of Cyanex 272 in the mixtures, the probability of found the specific band of this compound increases.

Interactions between ionic liquids and liposomes studied by differential scanning calorimetry

Antti Rantamäki, Suvi-Katriina Ruokonen, Alistair King, Susanne Wiedmer

Differential scanning calorimetry (DSC) is a valuable tool for the analysis of interactions between bioactive compounds and bilayers of biomimetic liposomes in aqueous environment. The technique is based on the detection of an endothermic phase transition taking place in the lipid bilayer. The bilayer undergoes a transition from an ordered gel phase into a fluidic L-alpha phase when the temperature of the system is raised above the lipid-specific phase transition temperature T_m. Dipalmitoylphosphatidylcholine liposomes (T_m of
41.3 C) have been used as a reference biomimetic membrane in our studies. However, other saturated lipids with suitable Tm such as dimyristoylphosphatidylcholine can be used as well. When liposomes and surface-active molecules are mixed in aqueous solution, the surfactants interact with the liposome bilayer to a differing extent depending on the surfactant properties. Typically, depending on the nature of the interaction, the Tm shifts towards lower temperature as a function of the surfactant concentration. This is caused by the decreased order of the lipid bilayer. Based on the changes observed in the endotherms, the character of the bilayer reorganization can be assessed. The described method is a powerful tool for measuring the interactions between ionic liquids (IL) and liposomes. ILs are organic salts, utilized for instance as solvents and co-solvents in cellulose and petrochemical industry. They have been considered as greener alternatives for organic solvents. However, since ILs are typically readily water soluble, they create a potent threat to a variety of organisms from bacteria to humans. From a structural point of view, ILs are a very diverse group of compounds, and therefore, also the toxicities vary from very toxic to harmless. The liposome

interaction studies described here are utilized to assess the toxicity and the toxic mechanism of ILs in parallel to toxicity measurements with living organisms. Particularly, the interactions between the surface-active ILs and liposomes are observed as permeation of the surface-active components into the bilayer. This is typically due to the long alkyl chains bound to cation and/or anion. High tendency to permeate into the liposome bilayer very often correlates with the high toxicity of the IL.

Spectrophotometric determination of the concentration of Zn (II), Eu(III), Lu (III), and La (III) with eriochrome black T

Daniela Tapia Pitzzu, Jorge H.S.K. Monteiro, Ana de Bettencourt-Dias

A complexometric titration between eriochrome black T and Zn^{II}, Eu^{III}, Lu^{III}, and La^{III} standards was used to develop calibration curves for the determination of metal concentration in aqueous solution at pH 10.00. The detection limits for the systems ranged from 1.27 to 11.8 ?M at 500 nm and from 0.922 to 1.05 ?M using a ratiometric method (650 nm/ 500 nm).

Multiple Simultaneous Spectroscopic Measurements with a Single Spectrometer
Claudio Egalon

We demonstrated a new type of device that uses a side illuminated optical fiber, a single spectrometer and three LED sources to analyze multiple samples distributed along the fiber length. The heart of this system lies in our side illuminated optical fiber that allows the multiplexing of multiple signals from different sources of non–overlapping spectral distribution. The combination of this fiber illumination technique with the linear array of sources generates a system with multiple spectral signals that can be easily separated by a single spectrometer simultaneously. Such a technique, also known as wavelength division multiplexing, have been performed in the past with axially illuminated fibers however we report its use with side illuminated fibers for the first time. There are two ways an optical fiber can be illuminated: axial illumination and side illumination. Axial illumination, involves the illumination of the fiber end face. Side illumination, pioneered by this author, consists in illuminating the cylindrical surface of the fiber. Although axial illumination is widely used by the scientific community, it can accommodate only one source at a time because of the small size of the fiber end face: 2 mm of diameter or less. In contrast, side illumination can be accomplished with several sources deployed along the fiber length to generate multiple signals. This allows the probing of the optical fiber at different positions along its length increasing system redundancy, the output signal and sensitivity. We demonstrated a proof of concept using the following: 1. A tapered optical fiber. 2. Three LEDs: blue, green and red LEDs. 3. A multichannel LED driver. 4. Red, blue and green food dye solutions with concentrations between 0 and 5 drops per 0.5L of water to emulate three sample solutions and 5. An Ocean Optics USB-2000 mini spectrometer. The red, blue and green solutions were simultaneously illuminated by the blue, green and red LEDs, respectively. Our data indicates that the combination of red LED and green solution had the greatest sensitivity of all. We are now working to extend this concept to N samples using N sources of non-overlapping spectral distribution.

Molecular Recognition of Muramyl Dipeptide Occurs in the Leucine-rich Repeat Domain of Nod2
Mackenzie Lauro, Brian Bahnson, Catherine Grimes

Genetic mutations in the innate immune receptor nucleotide-binding oligomerization domain-containing 2 (Nod2) have demonstrated increased susceptibility to Crohn's disease, an inflammatory bowel disease that is hypothesized to be accompanied by changes in the gut microbiota. Nod2 responds to the presence of bacteria, specifically a fragment of the bacterial cell wall, muramyl dipeptide (MDP). The proposed site of this interaction is the leucine-rich repeat (LRR) domain. Surface plasmon resonance and molecular modeling were used to investigate the interaction of the

LRR domain with MDP. A functional and pure LRR domain was obtained from Escherichia coli expression in high yield. The LRR domain binds to MDP with high affinity, with a KD of 212 ± 24 nM. Critical portions of the receptor were determined by mutagenesis of putative binding residues. Fragment analysis of MDP revealed that both the peptide and carbohydrate portion contribute to the binding interaction.

ICP-MS based methods for the analysis of nanomaterials

Francisco Laborda, Eduardo Bolea, Maria T. Gomez, Maria S. Jimenez, Juan R. Castillo

The need of reliable methods to determine nanoparticle identity, characteristics and concentrations, as well as their transformations in complex systems (e.g., environment, foods, biological) is one of the most challenging problems in nanosciences. In the case of environmental nanosciences, analysis at environmentally relevant concentrations adds an extra level of difficulty [1]. Inductively coupled plasma mass spectrometry (ICPMS) is a multielemental-specific technique, which is used routinely for the quantification of the elemental content of nanoparticles and nanomaterials. However, novel approaches based on the use of ICPMS are emerging and consolidating during the last years. Direct analysis based on the detection of individual nanoparticles (single particle-ICPMS) and hyphenation of asymmetrical flow field flow fractionation (AF4) to ICPMS are two of the most promising ones. Whereas AF4-ICPMS allows the separation and quantification of nanoparticles according to their size, single particle detection ICPMS provides information about dissolved and nanoparticle forms of an element, size distributions, and number and mass concentration without previous separation. Other hyphenated techniques, like hydrodynamic chromatography and capillary electrophoresis, are being used to separate nanoparticles according to their size or their charge-to-size ratio, respectively. On the other hand, the use of ultrafiltration in combination with ICPMS analysis allows to fractionate an element in a suspension as dissolved and nanoparticle forms, whereas methods based on cloud point extraction are being used to separate and concentrate nanoparticles from dissolved species. The complexity of the analytical problems to be solved in the realm of nanosciences involves the use of multi-method approaches. In the case of inorganic nanomaterials, the use of ICPMS alone and in combination with different separation/fractionation techniques provides a platform of analytical methods capable of facing these problems. State of the art of the cited above methodologies will be presented, discussing their performance, challenges, limitations and complementarity. Application to selected cases will also be presented. This work was supported by the MINECO/FEDER project CTQ2015-68094-C2-1-R. [1] F.Laborda, E. Bolea, G.Cepria, M.T. Gomez, M.S. Jimenez, J. Perez-Arantegui, J.R. Castillo. Analytica Chimica Acta, 904, 10 (2016).

New Nanometrology Approaches for examining nanomaterials released from products undergoing weathering

James Ranvile, Katie Challis, Ronald Lankone, Jing Jing Wang, Howard Fairbrother, Paul Westerhoff

As the use of nano-enabled products rises, the potential for engineered nanoparticles (ENPs) release to the environment increases. One process involves product exposure to precipitation and sunlight. New nanometrology that quantifies and characterizes (size, composition, physical form) ENPs are needed. A methodology using a combination of single particle ICP-MS (spICP-MS) and field flow fractionation (FFF) was developed to facilitate the investigation of ENP release during weathering. The spICP-MS method provides the inorganic NP size and particle number concentration based on elemental mass. Two FFF techniques, asymmetric flow field-flow fractionation (AF4) and centrifugal field-flow fractionation (CFFF), separate and size particles based on their hydrodynamic size and buoyant mass, respectively. Combing these methods leads to insights into the form of the released nanomaterials. The release of single-wall carbon nanotubes (SWCNTs) from a carbon nanotube polymer nanocomposite (CNT-PNCs) provides an example of the need for improved methods for characterizing released ENPs. Pulses in 89Y that each result from the detection of SWCNTs in a suspension generally are < 50 counts. sp-ICP-MS analysis of a suspension generated by 10 days of UV irradiation of a CNT-PNC show larger pulses (averaging 100-200 counts) and demonstrate that SWCNTs are released as multiples/bundles embedded within photolyzed polymer fragments.

The environmental fate and impact of SWCNTs are likely highly influenced by their form. The talk will highlight the details of combining sp-ICP-MS and FFF to fully characterize ENPs and to address the issue of background nanoparticles, which create a challenge to ENP quantification. Other studies to be reported on include the release of TiO2 from concretes and sunscreens. Further refinements of the nanometrology will lead to improvements in the life cycle assessment of nanomaterials, particularly in the use and end of life phases.

ACHIEVING SI TRACEABILITY FOR NUMBER CONCENTRATION OF INORGANIC NANOPARTICLES USING SP-ICPMS

Susana Cuello-Nunez, Dorota Bartczak, Heidi Goenaga-Infante

Nanoparticles are increasingly used in innovative products manufactured by advanced industries and provide enhanced, unique properties of great commercial and societal value. The demand for high performance materials places increasingly stringent tolerances on the properties of nanoparticles to ensure reliable performance and to meet regulation. Measuring the number concentration of particles in colloidal suspension is a major commercial interest for a large range of industries, such as personal care, fine chemicals, pharmaceutical and biomedical industries. It enables compliance with potential EU regulation linked to the EU definition of a nanomaterial (2011/696/EU)[1], this is based upon the number concentrations of particles within size limits. However, there is currently a lack of validated methods that can be used to determine nanoparticle number concentrations. SP-ICP-MS has been proposed as a powerful technique for the determination of size distribution and number concentration of inorganic nanoparticles[2]. One of the drawbacks of SP-ICPMS is the need for reference materials to determine the transport efficiency, which represents a key factor to obtain the number based concentration of nanoparticles in the sample. Reference nano-materials are very scarce and the few existing ones are not like-for-like with nanoparticles used in real sample matrices. The aim of this study is to develop a SI traceable method to measure nanoparticle number concentration in liquid suspensions with no need to use a reference material (RM) to calculate the aforementioned transport efficiency. In order to do that, commercial citrated capped Au spherical particles were chosen as 'model material' due to the availability of like-for-like reference materials for spherical gold nanoparticles (NIST SRM series). This offered the possibility to evaluate the impact of key parameters on the accuracy and uncertainty of the number concentration data using the newly developed method since confirmatory RM-derived data is available. Results obtained by SP-ICPMS using the methodology presented here were compared with an independent technique (Nanoparticle tracking Analysis, NTA) in order to obtain additional confirmatory data. [1]EC Commission Recommendation, 2011/696/EU; European Union: Luxembourg, 2011 [2]F. Laborda, E. Bolea, J. Jiménez-Lamana, Anal. Chem 84 (2014) 2270-2278

Metrological Traceability and Validation of Single Particle ICP-MS Measurements of Nanoparticle Size and Number Size Distribution using High -Resolution Scanning Electron Microscopy as Reference Method

Antonio R. Montoro Bustos, Kavuri P. Purushotham, Natalia Farkas, Antonio Possolo, András E. Vladár, Karen E. Murphy, Michael R. Winchester

Single particle inductively coupled plasma-mass spectrometry (spICP-MS) is considered an emerging and promising analytical technique for the ultrasensitive detection and characterization of metal-containing nanoparticles (NPs). Particularly, spICP-MS offers exceptional potential for providing information about particle size and size distribution, agglomeration/aggregation state, and particle number concentration, at environmentally relevant concentrations (mass concentration levels down to ng L-1). Additionally, simultaneous detection of dissolved, pristine NPs, and their agglomerated/aggregated species can be carried out in a single analysis by using spICP-MS. However, despite its highly valuable utility, some important limitations remain and for spICP-MS to be considered an established method, the metrological quality of spICP-MS measurements must be thoroughly proven. This communication presents for the first time, a rigorous validation of spICP-MS capabilities for measuring mean NP size and number size distribution of gold nanoparticles (AuNPs) and establishes metrological traceability of spICP-MS results to the International System of Units (SI). Metrological traceability is achieved by

i. calibration with a National Institute of Standards and Technology (NIST) Reference Material (RM) 8013 with a consensus particle size obtained by six different sizing techniques, ii. comparison with high-resolution scanning electron microscopy (HR-SEM) used as reference method, which is linked to the SI through a calibration standard characterized by the NIST metrological Atomic Force Microscope, and iii. deriving comprehensive uncertainty estimates associated with the measurement of the mean particle size to enable comparison of the spICP-MS and HR-SEM methods. Upon establishing HR-SEM and spICP-MS measurement protocols, both methods are used to thoroughly characterize commercial AuNP suspensions of three different sizes (30 nm, 60 nm, and 100 nm) with four different coatings and surface charge at pH 7. Single particle ICP-MS measurements (corroborated by HR-SEM analyses) reveal bimodal number size distributions for three of the 60 nm commercial suspensions that are not evident in the vendor supplied transmission electron microscopy (TEM) data, demonstrating the utility of spICP-MS for characterization of commercial NP suspensions regardless of size or coating.

Field-Flow Fractionation coupled to ICP-QQQ for characterization of functionalized inorganic nanoparticles

JOSE MANUEL COSTA-FERNANDEZ

Applications of functionalized inorganic nanoparticles (NPs) in life sciences involve dispersions of such materials in biological fluids, which typically contain high amounts of ions and biomolecules. Such conditions may lead to the formation of a protein adsorption layer (known as "protein corona") on the surface of colloidal nanoparticles, which plays an important role in their interaction with living matter. The assessment of the interactions between engineered NPs and biomolecules that can be present in biological fluids is of utmost importance for understanding how exposure to nanoparticles affects the biological responses of cells and organisms. On the other hand, the increasing use of NPs in imaging studies and in the design of platforms for controlled drug delivery typically requires the bioconjugation of nanoparticles to antibodies (against specific biomarkers) for active targeting. Though many approaches have been reported in this direction, the controlled bioconjugation of NPs, a key issue to achieve a success in such applications, is still a challenge [1]. In this context, on-line hyphenation of asymmetric flow field-flow fractionation (AF4) to an elemental detector (inductively coupled plasma-mass spectroscopy, ICP-MS) appears as a powerful diagnostic tool for studies involving interactions between NPs and biomolecules. For such purposes, the use of the novel ICP-MS instruments such as the triple quadrupole ICP-MS, enabling sensitive and reliable quantification of native elements present in the biomolecules (such as P or S) typically hardly to be detected by ICP-MS, turned out to be highly advantageous in order to investigate in depth interactions between the NPs and S-containing analytes. Moreover, a key issue in future development of analytical and bioanalytical NP applications, that is bioconjugation efficiency, can be easily attained with such hyphenated tool. This new approach to bioconjugation efficiencies evaluation opens the door to enable robust determinations of metal NPs: protein actual stoichiometries. References [1] J. M. Montenegro, V. Grazu, A. Sukhanova, S. Agarwal, J. M. de la Fuente, I. Nabiev, A. Greiner and W. J. Parak, Adv. Drug Deliv. Rev., 65 (2013) 677 – 688.

Frank Vanhaecke: The Man in the ICP
Martin Resano

On occasion of receiving the Lester Strock Award, this presentation discusses the relation of Dr. Vanhaecke with rock and roll music, food, beer, and with inductively coupled plasma mass spectrometry, not necessarily in that order, moreover providing a personal view on an almost 20 years of scientific cooperation.

Tracing mercury pollution in and unraveling exposure pathways for marine species via high-precision isotopic analysis with multi-collector ICP- mass spectrometry

Eduardo Bolea-Fernandez, Ana Rua-Ibarz, Amund Maage, Sylvia Frantzen, Jörg Feldmann, Eva M. Krupp, Frank Vanhaecke

Mercury (Hg) occurs in the environment as a result of natural and anthropogenic sources, and both wildlife and humans are exposed to the harmful effects of this highly toxic heavy metal and its compounds. Nowadays, Hg is recognized as one of the most important global pollutants due to its high toxicity (methyl mercury being the most toxic Hg species), capacity for traveling long distances in the atmosphere, and for its bioaccumulation in the food web. Thus, increasing our understanding of the biogeochemical Hg cycle, which involves the transport of and transformation pathways for the different Hg species, is of the utmost importance for assessing the risks of Hg contamination. The study of the natural variations in the isotopic composition of Hg can be seen as a key tool for unravelling the origin of Hg pollution and for obtaining a more profound insight into Hg exposure in marine ecosystems. However, the variations in the Hg isotopic composition are relatively small and high-precision multi-collector ICP-MS (MC-ICP-MS) is required.[1] In this work, mass-dependent (MDF) and mass-independent (MIF) isotope fractionation of Hg in different marine species at different levels within the tropic chain, i.e. crustaceans,[2] deep-water fish and marine mammals, have been used for tracing possible Hg sources and exposure pathways. Different degrees of exposure and accumulation mechanisms have been observed for the different species and tissue types, pointing to complex biological processes for Hg re-distribution and detoxification routes. [1] A. Rua-Ibarz, E. Bolea-Fernandez and F. Vanhaecke, Anal. Bioanal. Chem., 2016, 408, 417 – 429 [2] A. Rua-Ibarz, E. Bolea-Fernadez, A. Maage, S. Frantzen, S. Valdernes and F. Vanhaecke, Environ. Sci. Technol., 2016, 50, 10361 -10369

Green Sample Preparation Methods for Trace Element Determination by ICP-based Techniques

erico Flores

Classical methods for sample digestion and further determination by atomic spectrometry involve the use of a relatively high volume of concentrated reagents (inorganic acids in case of further metals determination) and a relatively long time for digestion. Moreover, the digestion efficiency of these systems presents some limitations for many matrices. In this sense, even using methods based on microwave-assisted closed vessels some drawbacks can be present for some matrices and incompleteness of digestion is frequently reported. Nowadays, there is a trend for the development of methods requiring lower reagent consumption, less analytical steps and lower waste generation combined with high efficiency of digestion. In addition, it is important that digests are suitable for the determination techniques avoiding excessive dilution or high blank levels. On this aspect, some trends for sample preparation will be presented in this lecture for further metals and also non-metals determination. The use of diluted solutions using oxygen pressurized systems with or without UV radiation, use of combustion systems, etc, will be covered. Recent applications will be presented showing the advantages of methods using diluted reagents (combustion, UV digestion) for metals and non-metals determination using ICP-MS and ICP-OES.

Characterization of SiO2 Nanoparticles by Single Particle – Inductively Coupled Plasma – Tandem Mass Spectroscopy (SP-ICP-MS/MS)

Maite Aramendía, Eduardo Bolea-Fernández, Diego Leite, Ana Rua-Ibarz, Lieve Balcaen, Martín Resano, Frank Vanhaecke

The increase in the use of SiO_2 nanoparticles (NPs) is raising the concern about their environmental and health effects, thus necessitating the development of novel methods allowing for their straightforward detection and characterization. Single Particle ICP- mass spectrometry (SP-ICP-MS) is able to provide information on the size of NPs and their particle and mass concentrations. However, the determination of Si via ICP-MS is strongly hampered by the occurrence of spectral overlap from polyatomic species present in the plasma (e.g., CO^+ and N_2^+). The use of tandem ICP-MS (ICP-MS/MS) enables interference-free conditions to be obtained, even in the case of the most demanding applications. Upon testing of several gases, the use of CH_3F (monitoring of SiF^+, mass-shift) and of H_2 (monitoring of Si^+, on-mass) were demonstrated to be the

most suitable to overcome the spectral interference affecting ultra-trace Si determination (LoD < 15 ng L^{-1}). By using these approaches, SiO_2 NPs (ranging between 80 and 400 nm) have been detected and characterized.
For SiO_2 NPs > 100 nm, it was possible to provide accurate results in a straightforward way, as the signals they give rise to are well resolved from that of the background. In the case of 80 and 100 nm, the use of a simple deconvolution approach following a Gaussian model has been successfully applied to characterize SiO_2
NPs apparently showing incomplete distributions as a result of the presence of the background signal. Overall, the methods developed using SP-ICP-MS/MS are sensitive and selective enough for interference-free determination of Si at ultra-trace levels, as proved for SiO_2 NPs.

LA-ICP-MS: theoretical considerations on and perspectives for its future development.
Stijn J. M. Van Malderen, Frank Vanhaecke, Olga Borovinskaya

This talk will discuss recent trends in the field of laser ablation-inductively coupled plasma-mass spectrometry (LA-ICP-MS) imaging in terms of instrumental development and data processing strategies. LA-ICP-MS systems are sensitive elemental microprobes characterized by their high selectivity, lateral and depth resolution on the single ?m to nm level, and wide dynamic range. The analysis cost per sample for this technique can be greatly reduced through modifications of the aerosol transport system, increasing the pixel acquisition rate (<10 ms per pixel), whilst simultaneously improving absolute limits of detection with a factor of 10^2. In parallel with these developments, significant progress has been achieved in image formation and data extraction approaches for LA-ICP-MS. One of the key strategies to faster imaging with sequential mass analysers is based on synchronising the detector acquisition cycle to the laser repetition rate to mitigate aliasing artefacts induced by resampling the varying peak signal intensity generated by low-dispersion ablation cells. Quasi-simultaneous mass spectrometers on the other hand, which provide a more comprehensive elemental profile, can be triggered by the laser for every laser shot to capture the signal peak induced by this laser shot in a single acquisition interval. This strategy uses a delay generator between the laser and the data acquisition card to start the acquisition at a fixed interval after the laser trigger has been engaged. This talk will also discuss a how these developments leverage 3D tissue mapping applications, which rely on image registration approaches based on computed tomography to reconstruct the 3D volume describing the elemental distribution. The essence of this talk is that LA-ICP-MS is evolving from an experimental, slow mapping technique into a full-fledged automated 2D and 3D elemental microscope.The UGent-A&MS team acknowledges Teledyne Photon Machines, Teledyne CETAC Technologies and TOFWERK AG for financial and/or logistic support.

Turning chemical imaging data to useful information
Rohit Bhargava, Shachi Mittal, Kevin Yeh

Infrared (IR) spectroscopic imaging offers significant potential for clinical translation due to its high sensitivity signal involving ~10-70% of photons and the ease of performing spectroscopic imaging. The parameters involved in a practical clinical protocol, however, are numerous and performance often has to be traded for practicality. Here we discuss developments that have provided a path to feasible, accurate and useful imaging that rely on advances in data processing to leverage advances in instrumentation. We first show that high-definition (HD) IR imaging can lead to accurate tissue histopathology. Spectral parameters and noise characteristics of the data are available that then form the requirements for translating IR imaging data to useful information. We used the information to develop a new IR imaging instrument that only records data that is needed. Using this discrete frequency IR (DFIR) imaging approach, we tuned a quantum cascade laser based instrument to take advantage of the limited features needed for classification while ensuring that the signal to noise ratio (SNR) Is high enough to ensure accuracy. We report on two configurations in IR imaging that provide information at different length scales and functionalities for digital molecular pathology. The results allow us to rapidly provide histopathologic information from high dimensional data. Here we report rapid imaging of human breast biopsy samples and integrated imaging of histopathology to provide clinically actionable information in minutes.

Does every image pixel matter in TB-sized images?

Peter Bajcsy

This talk will address the problems related to conducting microscopy research over terabyte (TB)-sized image collections and publishing traceable results derived from such image collections. While the motivation behind collecting TB-sized datasets ranges from population measurements to rare event detection and heterogeneity quantification, there are traceability challenges associated with derived statistics included in journal publications. The traceability challenges include (a) computational aspects (managing, analyzing, viewing, and summarizing TB-sized datasets), (b) user-driven aspects (interactive selection of data points and models), and
(c) publication aspects (verifying results during reviews and reproducing results based on a publication). The goals of our work are to overcome these challenges by (a) providing web image processing software to address computational aspects, (b) enabling interactive multi-scale measurements over Gigapixel images by designing OpenSeadragon extensions, and (c) delivering traceability of published measurement statistics to their persistent source by using web technologies. The talk will describe a prototype web system that is meeting the above goals and has been applied to analyzing a TB-sized collection of phase contrast and fluorescent microscopy images taken from live stem cell colonies. More information can be found at https://isg.nist.gov /deepzoomweb/resources/csmet/pages/web_image_pipeline/web_image_pipeline.html The persistent image collections are accessible from https://isg.nist.gov/deepzoomweb/data/stemcellpluripotency.

Chemometric Concepts Applied to Digital Signal Processing
Garth Simpson, Scott Griffin, Scott R. Griffin, Gregory Eakins, Fengyuan Deng, Atanu Sangupta

As dimension reduction methods continue to advance in mining data-intensive chemical measurements, so too grows the potential for expansion of this toolkit into mining other digital datasets. Specifically, the spectra most routinely used in chemometric analysis are generally produced from raw instrument signals that themselves can benefit from dimension reduction methods. In this work, chemometric tools greatly improved the temporal resolution of raw distorted electronic measurements. In one example, linear discriminant analysis (LDA) enabled the design of a digital filter to recover quality images in two-photon excited fluorescence microscopy from a raw signal distorted by impedance mismatch in the detection electronics. Expanding on this capability, adaptation of multivariate curve resolution (MCR) for digital filtering was employed to not only allow for resolution improvements in highly distorted waveforms, but also to recover the waveform responsible for the distortion self-consistently. Applications, implications, and limitations of these methods will be critically evaluated.

Rethinking the Classification Process with Data Fusion
John Kalivas, Brett Brownfield

Data fusion is considered as a "multilevel, multifaceted process dealing with the detection, association, correlation, estimation, and combination of data and information from multiple sources" [1]. Classifying samples into categories is a common problem in many fields and numerous methods are available. Most classification schemes require optimization of one or more tuning parameters for best classification accuracy, sensitivity, and specificity. To avoid these ambiguous decisions, multiple methods can be used as a data fusion approach. Unique to the presented fusion method is that specific tuning parameter values do not need to be selected for each method, e.g., the number of latent variables (LVs) for PLSA-DA. Instead a window of LVs is evaluated thereby removing subjectivity and avoiding local anomalies for a particular tuning parameter value. Also unique to the data fusion method is that raw values are used in the fusion rather than setting thresholds for each classification method. The approach is demonstrated with several data sets of multiple classes. Letting p and n denote the number of corresponding variables and objects, two data sets have $p >> n$ and a third data set has $p < n$. In all cases, data fusion without model selection significantly improves the classification results. [1] Klein LA. Sensor and data fusion concepts and applications, SPIE Optical Engineering Press (1999).

New methodolgy for finding optimal spectral matches in reference databases

Dr. Gregory M. Banik, Karl Nedwed, Ty Abshear

Optimized Curve Matching and Display is a novel and valuable curve matching and visualization methodology. It allows users to identify optimal spectral matches within reference databases and visualize the comparative results in a way that is more discernible to the human eye. We will discuss multiple corrections that can be applied automatically to compensate for differences between spectral instruments, environmental conditions, sample concentration, ATR correction, and others to optimize the match between spectral curves.

The FDA Food & Veterinary Medicine Science and Research Program: In Search of Partnerships to Develop Innovative Food Safety Applications
Palmer A. Orlandi, Janie Dubois

The Food Safety Modernization Act (FSMA) passed in 2011 reformed U.S. food safety law and enabled the FDA, the regulatory steward of the U.S. food supply, to shift from a strategy of reactive response to one of prevention. This shift in the FDA's regulatory paradigm and its successful implementation relies heavily on public–private partnerships to harness the expertise and resources outside the federal regulatory arena to achieve mutual food safety goals. Such partnerships would facilitate and promote cooperation and collaboration between the FDA, other Federal food safety entities, academia, regulated industry, international stakeholders and commercial technology sectors to leverage capacity and to foster technological advancements and develop innovative applications. The latter is of particular importance to the FDA's Food and Veterinary Medicine (FVM) Science and Research Enterprise. FDA FVM understands that such broad research partnerships are important to spur "outside-the-box" thinking and to avoid institutional myopia if it is to maintain an analytical portfolio capable of addressing evolving food safety challenges in the global market place. This presentation will detail how FDA FVM has embarked on a multi-faceted effort to take advantage of those administrative tools at its disposal to facilitate collaborative public-private research activities within the constraints posed by its regulatory responsibilities. Such mechanisms and examples to be discussed will include innovative challenge activities under the auspices of the America COMPETES Act and HHS Innovates; Broad Agency Announcements and The Reagan-Udall Foundation.

My Method is Valid" – Can You Prove That in Court?
DeAnn Benesh

You believe your method is "valid" because you have conducted a comparison study of your method with a reference method, and shown the results of your method to be "equivalent" to results with the reference method. But on what criteria was your "equivalency" based? How would you convince a jury that you have conducted your comparison in a non-biased manner? There are only a few Method Certification bodies that are globally recognized and accepted by competent authorities. These are: AOAC INTERNATIONAL, AFNOR Certification and MicroVal

Certification. Both US FDA and USDA FSIS accept methods validated through these Certification bodies - when they have been validated against the appropriate reference method, have been shown to be fit for purpose, and have been collaboratively studied. European Directive 2073/2005: Microbiology Criteria of Foodstuffs clearly states that alternative methods may be used to meet the requirements in the directive, if they have been validated as fit for purpose using the protocol outlined in ISO 16140-2. Requirements for method validation through both AOAC INTERNATIONAL and following ISO 16140-2 will be outlined and contrasted to provide a basic understanding of the criteria required for method validation and acceptance by "competent authorizes" worldwide.

Use of Chemometrics to Identify Marker Compounds for Food Authentication

Zhengfang Wang

Chemometrics is nowadays finding applications in a wide range of disciplines to improve analysis of compounds. This presentation aims to demonstrate how chemometrics based multivariate analysis is able to extract distinctive information from GC/MS by using mathematical and statistical methods to support food authentication. Extracts of basil plants cultivated by organic and conventional farming practices were tested with gas chromatography/mass spectrometry (GC/MS). Baseline correction and retention time alignment were performed prior to data processing. Partial least squares discriminant analysis (PLS-DA) was used to build classification models. Bootstrapped t-statistical weight feature selection method was used to select informative features (marker compounds) from GC/MS chemical profiles of organic and conventional basil extracts. Characteristic components in basil extracts were putatively identified. For method validation, selected maker compounds were used to construct economic classification models, e-PLS-DA. Both the PLS-DA and the e-PLS-DA classifiers were used in parallel to classify a new validation set collected 2.5 months later with no parametric change to experimental procedure. Classification accuracy of e-PLS-DA and PLS-DA classifiers was compared. Results showed that marker compounds have been successfully identified from the gas GC/MS data of basil extracts. Feature selection proves valuable as a powerful and cost-effective approach for identifying marker compounds in chemical data. As all algorithms can be implemented in any numerical computing environment such as MATLAB, C, C++, Python, and R languages, expensive specialized data processing software is unnecessary.

Pathway selective CMDS for revealing weak interactions in complex systems
Jeffrey Davis, Jonathan Tollerud, Fabio Novelli

Multidimensional coherent spectroscopies are designed to identify and quantify interactions between quasi-particles. These interactions include population transfer, coherent coupling, correlations, and many-body effects. With broadband excitation, data can be acquired and spectrally resolved in terms of both absorption and emission energies in a single experiment, without having to change the excitation wavelength. In this case all available quantum pathways are excited and probed. In contrast, experiments with narrowband pulses resonant with specific transitions can be used to selectively excite specific quantum pathways. This can become particularly useful when the sample under investigation has broad overlapping peaks; when the interactions between states is weak and the relevant signals are swamped by larger contributions; or when states that do not contribute to the signal are excited by the laser and interact with the states that do, thereby changing the nature of the multidimensional spectrum measured. We have developed an approach to acquire pathway selective coherent multidimensional spectra by independently controlling the wavelength and polarization of the excitation pulses, while maintaining phase stability between all pulses. Using this approach, we have been able to reveal weak interactions in complex systems that are otherwise hidden[1-4]. We first demonstrated this by isolating coherent superpositions between excitons in different semiconductor quantum wells, showing a substantial enhancement in signal to noise over broadband measurements[1]. A similar approach in double quantum 2D spectroscopy enables control of the pathway by which the double quantum coherence is excited. With one pathway selected, overlapping peaks can be removed, allowing analysis of the relevant peak-shape and thus details of interactions with the environment and other carriers. An alternative pathway identifies a double-quantum state associated only with GaAs defects, the signature of which has previously been confused with other interaction induced effects [2]. Finally, we have recently applied this approach to probe the low-energy states (<60meV) of high-Tc cuprate superconductors. The results reveal coherent dynamics and details regarding the nature of the states in this energy range, which are believed to play a crucial role in the mechanisms of superconductivity in these materials. [1] J. Tollerud et al., Optics Express 22, 6719 (2014). [2] J.O. Tollerud, J.A. Davis, Laser & Photonics Reviews, 11, 1600249 (2017) [3] J.O. Tollerud, J.A. Davis, Progress in Quantum Electronics, 55, 1-34 (2017) [4] J.O. Tollerud, ST Cundiff, J.A. Davis, Phys. Rev. Lett. 117, 097401 (2016)

2D Spectroscopy of 2D Materials: Insight into Exciton Dynamics in Atomically Thin Semiconductors
Galan Moody

Recent advances in isolating monolayers of van der Waals materials have enabled the exploration of fundamental physics and ultrathin devices at the atomic limit. Within the family of 2D materials, monolayer semiconductors, such as transition metal dichalcogenides (e.g. WSe2, WS2, MoSe2, and MoS2), exhibit remarkable optical and electronic properties that emerge from their unique band structure and exceptionally strong light-matter interaction. The former endows electrons with coupled spin and momentum (valley) degrees of freedom, while the latter is a consequence of strong valley-dependent Coulomb interactions associated with neutral and charged excitons, which govern the material's optical response. In this talk, I will discuss our efforts in understanding the complex optical properties of these excitonic states using polarization-resolved 2D coherent spectroscopy. One-quantum and zero-quantum 2D spectral maps reveal previously unknown properties and unexpected phenomena that are not readily accessible with conventional spectroscopy methods, including: (1) contributions to the optical transition linewidth from disorder and dissipative exciton-exciton and exciton-phonon coupling; (2) oscillations of cross peaks associated with coherent and incoherent energy transfer; (3) identification of neutral and charged biexcitons; and (4) ultrafast optical manipulation of the valley quantum degree of freedom as a novel information carrier analogous to electronic spin. These results provide insight into possible strategies for developing high-bandwidth and low-power opto-electronic devices, ultrathin and efficient photonic platforms, and novel quantum technologies that leverage the unique valley physics in these materials.

2D Transient Absorption Spectroscopy of Transition Metal Dichalcogenide Heterostructures
Kyle Czech, Zach Matusinec, John Wright

Transition metal dichalcogenides (MX_2), such as MoS_2 and WS_2, have sparked interest within the semiconductor physics community due to their unusual optical properties and potential use in spintronics/valleytronics applications. As a result of lattice inversion asymmetry and strong spin-orbit coupling, MX_2 exhibits spin-asymmetric excitonic states at conjugate K and K' points of the Brillouin Zone. Structurally, MX_2 forms in hexagonally packed layers, each layer held to the next by van der Waals forces. This van der Waals interface enables $MX_2/M'X'_2$ heterojunctions to form with greatly reduced lattice-matching constraints, as compared to more traditional epitaxial heterostructures. The exotic physics characteristic of MX_2 and the enhanced stability of van der Waals heterojunctions, highlight MX_2 as a promising platform for investigating the semiconductor physics of complex heterostructure architectures.

Here, we conduct 2D transient absorption spectroscopy of MoS_2/WS_2 layered heterostructures. Metal evaporation and a chemical vapor deposition tube-furnace synthesis was used to create MoS_2/WS_2 samples of two distinct morphologies: **(1)** single-crystal/thin-film MoS_2/WS_2 and **(2)** thin-film/thin-film MoS_2/WS_2.

The focus of this talk will be on the use of 2D transient absorption spectroscopy to selectively excite excitonic states within MoS_2 and probe the presence of ultrafast carrier transfer to WS_2. Charge carrier tranfser dynamics will be evaluated for each morphology and will aim to address the influence of defect density and heterojunction uniformity on the excited state evolution of MoS_2/WS_2 layered heterostructures.

Global Analysis of Transient Grating and Transient Absorption Spectra of PbSe Quantum Dots
Daniel Kohler

We examine the non-linear response of PbSe quantum dots about the 1S exciton using two-dimensional transient absorption and transient grating techniques. The combined analysis of both methods provides the complete amplitude and phase of the non-linear susceptibility. The phased spectra reconcile questions about the relationships between the PbSe quantum dot electronic states and the nature of non-linearities measured by two-dimensional absorption and transient grating methods. The fits of the combined dataset reveal and quantify the presence of continuum transitions.

Ultrafast laser-induced plasma diffusion and mixing processes at interfaces
Vassilia Zorba, Ran Hai, Xianglei Mao, Rick Russo

Understanding the role of sample homogeneity on Laser Induced Breakdown Spectroscopy (LIBS) analysis is of critical importance in chemical analysis and chemical imaging applications. For example, when the lengthscale of laser sampling is commensurate to that of the sample homogeneity, the laser-material interaction, plasma expansion and subsequent spectral emission detection may be influenced, especially at the interface between media with different optical, thermal and physical properties. In this work we study of the role of spatio-temporal plasma separation, mixing and diffusion processes at the interface between two model systems in the femtosecond time regime. The extend to which these processes affect femtosecond LIBS emission is also investigated. These results underline the importance of carefully selecting the independent laser sampling and detection parameters in chemical imaging applications with LIBS.

Fundamental aspects and applications of Laser Induced Plasma under water in a range of pressure between 1-150 bar.
Alessandro De Giacomo, Marcella Dell'Aglio, Antonio Santagata, Gabriele Valenza

In this work the fundamental aspects of Laser Induced Plasma (LIP) under water are discussed from a theoretical point of view in a range of pressure between 1 – 150 bar. In general, LIP in bulk liquids is characterized by the following stages: laser-ablation, plasma expansion, cavitation bubble. A good knowledge of these processes allows to exploit LIP underwater for different application, including Nanoparticles (NPs) production and submarine Laser Induced Breakdown Spectroscopy. With the aim of investigating the plasma characteristics and how the LIP parameters evolve in a water at different pressure, Optical Emission Spectroscopy and spectral resolved imaging have been employed; on the other hand, for the study of laser induced bubble dynamics, high temporal resolved shadowgraphic images and laser scattering have been used. Although the laser-matter interaction is not affected by the external conditions of the liquids, the dynamics of the plasma and the spectral features are strongly dependent on the hydrostatic pressure. In this view, while the formation of NPs, that occurs in the first few hundreds of ns, is not affected by the pressure, the spectroscopic signal is strongly degraded in terms of number and shape of emission lines. For this reason a DP approach has been implemented, where the second pulse LIP acts like a light source for inducing the spectral absorption of the species generated by the first pulse with the main result that at high pressure the analytical signal is an absorption spectrum. The material after the plasma's and the bubble's extinction has been characterized with TEM and SPR with the aim of correlating the spectroscopic observation to the produced particles and of giving a complete description of the LIP under water.

Isotopic analysis of uranium using laser induced fluorescence of LIBS plumes
Sivanandan Harilal, Kyle Hartig, Brian Brumfield, Igor Jovanovic, Mark Phillips

Currently isotopic analysis of nuclear materials is predominantly laboratory based (mass spectroscopy) which is time consuming and not useful for field applications. Optical spectroscopic methods (emission, absorption, fluorescence) in conjunction with laser ablation sampling have the potential to overcome these limitations with capabilities for in-field, rapid (< 1 sec), and non-contact analysis. LIBS is a proven technology for stand-off detection of all elements in the periodic table; however, spectral and instrumental broadening affect its performance for isotopic analysis from atomic emission for materials of interest to nuclear forensics and verification. Laser-absorption spectroscopy (LAS) provides the necessary spectral resolution for isotopic analysis in laser ablation plumes, but in a measurement geometry that is not compatible for stand-off measurements. Laser-induced fluorescence (LIF) spectroscopy incorporates aspects of both atomic absorption and emission and is capable of overcoming limitations of both emission and absorption spectroscopy methods for stand-off isotopic analysis in laser ablation plumes. In LIF spectroscopy, like atomic absorption, the ground state atoms in the plume are excited by a tunable laser. However, instead of directly monitoring the amount of light absorbed by the atoms, the emission resulting from the decay of the excited atoms is measured. In this work, we demonstrate a uranium isotope ratio measurement employing 2D fluorescence spectroscopy (2D-FS) of femtosecond laser ablation plumes up to atmospheric pressure in a stand-off configuration. For performing LIF of LA plumes, we used a narrow-band CW laser to probe absorption from selected U atomic transitions. The high spectral resolution provided by the LIF measurement allows isotopic emission to be easily resolved regardless of their possible overlap present in the LIBS emission spectrum. This work is supported by the DOE/NNSA Office of Nonproliferation and Verification Research and Development (NA-22). Pacific Northwest National Laboratory is operated for the U.S. DOE by the Battelle Memorial Institute under Contract No. DE-AC05-76RLO1830.

Comprehensive Ultrashort Laser -Metal Ablation at Terawatt Laser Power
Ahmed Elsied, Prasoon Diwakar, Ahmed Hassanein

The fundamental properties of ultrashort laser interactions with metals using up to terawatt power were comprehensively studied, i.e., specifically mass ablation, nanoparticle formation, and ion dynamics using multitude of diagnostic techniques. Results of this study are useful in many fields of research including spectroscopy, micromachining, thin film fabrication, particle acceleration, warm dense matter, and equation of state determination. A Ti:Sapphire femtosecond laser system (100 mJ maximum energy, 40 fs, 800 nm, P-polarized, single pulse mode) was used, which delivered up to 3 terawatt laser power to ablate 1 mm tin film in vacuum. The experimental analysis includes the effect of the incident laser fluence on the ablated mass, size of the ablated area, and depth of ablation using white light profiler. Atomic force microscope was used to measure the emitted particles size distribution at different laser fluence. Faraday cup detector was used to analyze the emitted ions flux by measuring the velocity, total charge. The study shows that the size of the emitted particles ranges from tens to hundreds of nanometers with a peak at 10 nm. The size of the ablated area tends to increase slightly with laser fluence. Laser ablation depth was found to increase logarithmically with laser fluence then leveling off at laser fluence > 300 J/cm^2. The total ablated mass tends to increase logarithmically with laser fluence up to 400 J/cm^2 while, increases gradually at higher fluence due to the increase in the ablated area. The measured ion emitted fluxes shows a linear dependence on laser fluence with two distinct regimes. Strong dependence on laser fluence was observed at fluences < 300 J/cm^2. Also, a slight enhancement in ion velocity was observed with increasing laser fluence up to 300 J/cm^2.

Plasma dynamics following femtosecond filament interactions with solids in single and multiple filament regimes
Patrick Skrodzki, Milos Burger, Igor Jovanovic

High-peak-power ultrafast laser filaments allow for extended energy delivery to a target advantageous for standoff ablation of solids considering applications in remote sensing and remote laser-induced breakdown spectroscopy (R-LIBS). Prominent challenges addressed in this work include: poor laser-material coupling efficiency following filamentation resultant from low contribution of the energy reservoir surrounding the filament core toward ablation; notable background congesting optical emission spectra originating from breakdown of atmosphere around the filament-produced plasma and white-light emission from self-phase modulation of the laser pulse undergoing filamentation; and non-uniform ablation due to sporadic formation of multiple filament cores seeded by beam instabilities and noise for peak-powers greatly exceeding the threshold peak-power for filamentation. We employ a dual-imaging system in order to observe the various components of filament-produced plasma in the time-scales relevant to R-LIBS including emission from the target and surrounding atmosphere as well as expansion of the laser-supported detonation wave (LSDW) front. Filtered spectral imaging reveals space- and time-resolved emission from neutral and ionic species from the target as well as emission from neutral nitrogen originating from the atmosphere; furthermore, we probe the shock front produced from filament-target interaction using time-resolved shadowgraphy and infer laser-material coupling efficiencies for both single and multiple filament regimes through analysis of shock expansion with the Sedov model for point detonation. The results provide insight into internal plasma structure for the range of peak-powers up to 30 times the threshold power for filamentation in air. Significant bifurcation of the plume in both single and multiple filament regimes occurs several hundreds of ns after the laser pulse potentially hindering plasma diagnostics. Despite stochastic nucleation of multiple filaments at peak-powers greater than 16 times the threshold for filamentation, emission of ionic and neutral species reveals a monotonic increase with pump beam intensity, and short-lived nitrogen emission originating from the ambient is consistently observed. Ultimately, results suggest favorable scaling of emission intensity from target species to laser pump energy relevant to applications including remote sensing and R-LIBS.

Development of Continuous Flow Chemistry Using Online PAT
Eric Fang

Continuous manufacturing technology, or flow chemistry, provides a potentially highly controlled and data rich environment for the development of robust manufacturing routes to complex molecules. Further, the technology can be leveraged to deliver low cost manufacturing plants and significantly simplified tech transfer. Realizing the benefits from continuous manufacturing technologies requires chemical, analytical and engineering expertise. The development of appropriate scale-down models for continuous manufacturing further complicates the efficient development of continuous manufacturing processes. The utility of the application of various process analytical technologies to the lab scale design and developing continuous processes for the pharmaceutical and fine chemical industry will be discussed. Case studies highlighting solutions and challenges of developing flow chemistry processes will be presented

Continuous Processing Strategies for Global Health Targets
Katherine Belecki

In the pharmaceutical industry, potential drugs are usually first synthesized by a medicinal chemist using a diversity-oriented approach. In this “discovery” synthesis, the goal is to make a wide variety of structurally-related target molecules, in which only small quantities are required. The commercial synthesis of an active pharmaceutical ingredient (API) often evolves directly from these medicinal chemistry routes because for many therapies, this approach would maximize the marketable patent lifetime by minimizing the time-tomarket after FDA approval. Moreover, legacy API routes often continue to be used post-patent expiration. For generics and many global health drugs, the API manufacturing cost represents a significant portion of the total cost for a drug. By streamlining the chemical synthesis of an API, medicines can be manufactured more economically, allowing for the same financial input to treat more patients. In the case of global health targets such as HIV, a more efficient chemical synthesis can make the difference between life and death. Moreover, in developing countries, novel manufacturing platforms such as continuous chemical processing can potentially be implemented more readily than in countries which already have an intensive capital investment in traditional batch manufacturing infrastructure. Through a series of vignettes about important global health targets including HIV and malaria, a comprehensive and generalizable strategy to develop streamlined process routes to molecular targets will be presented, and the success of this multifaceted approach to addressing API process inefficiencies will be demonstrated. The implementation and influence of real-time monitoring techniques will be discussed, along with opportunities and challenges in the incorporation of continuous flow technology.

Advanced Process Control and PAT – Criticality of attribute measurement on a Continuous Tableting line – Feed monitoring example.
Benoit Igne

The development of continuous manufacturing for pharmaceutical oral solid dosage forms brings unique challenges. To ensure product quality, developing the ability to directly measure product Attributes at multiple points on the line is an important factor to ensure consistency of the product characteristics along the line, be it for confirming the product control strategy or to provide direct process control of the process. In-line analytics provide means to measure product attributes in a relevant time frame to the process, potentially allowing for feed-back and feed-forward control to ensure product quality. In this presentation, the attribute monitoring of feeder output quality will be presented. Using Near-Infrared Spectroscopy and a custom feeder interface, the homogeneity of a blend was monitored in-line and in real-time.

Integrating Sensors for Monitoring Blend Content in a Pharmaceutical Continuous Manufacturing Plant
Savitha Panikar, Jingzhe Li, Varsha Rane, Sean Gilliam, Gerardo Callegari, Fernando Muzzio

In a pharmaceutical manufacturing process, Critical Quality Attributes (CQAs), as defined by the FDA, need to be monitored not only for the final product but also for intermediates. Blend uniformity of powders is one such attribute that needs to be measured to ensure the quality of the final product. In-line sensors, to monitor the blend content of the powders, were implemented within a Direct Compaction (DC) continuous tablet manufacturing line. In most cases, since the primary ingredient of interest is the active pharmaceutical ingredient (API), the amount of the API was monitored/predicted over the course of manufacturing. For the calibration model building process, a unique setup involving dynamic powder spectral acquisition method was used. This setup was aimed at mimicking the powder flow characteristics within the manufacturing line. A Raman probe and a portable NIR were used as sensors and they were placed at the exit of the blending process before the tableting stage. The performance of the two sensors and their respective models were evaluated in terms of accuracy, precision, operating range, measurement frequency, placement, reliability, robustness. Additionally, their performances were validated by off-line traditional analytical measurements.

Identification of subvisible particles in biopharmaceutical formulations: Raman spectroscopy and more
Miguel Saggu, Ankit Patel

We exploit the potential of Raman microscopy for chemical identification of particles in monoclonal antibody formulations. The classification of particles as extrinsic, intrinsic, and inherent is a critical step during the development process as well as during investigations into unexpected appearance of particles. Particles in the subvisible size range are particularly difficult to identify and methods are evolving for particle identification. Here, we will discuss several methodologies for determining the identity of subvisible particles including Raman spectroscopy as well as flow microscopy. Through a combination of Raman experiments and density functional theory (DFT) calculations, we identified unique spectral marker bands for insoluble degradation products of polysorbate 20. Raman spectroscopy is used in other examples to identify particles arising from disposable equipment as well as in-use administration items. In addition, we also discuss the use of flow microscopy methods for better classification and counting of subvisible particles based on particle morphology.

Analytical Methods and Characterization Techniques for Drug/Device Combination Products
Sundaravel "Ananth" Ananthavel

Drug/Device combination products ranges from simple prefilled syringes to much complex drug eluting stents. Development and validation of characterization techniques for Drug/Device combination products are challenging and time consuming and unique requirements for various regulatory agency such as FDA, EMA (European Medicine Agency). This talk will outline some of the unique methods that were developed and validated for drug/device combination products (such as Inhalation devices, Patient-controlled analgesia (PCA) through different phases of the development and how it is implemented for commercial production at multiple countries.

Monitoring Oxidation of Methionine Residues in Biotherapeutic Proteins using Raman and FTIR Spectroscopy
Gurusamy Balakrishnan, Gregory Barnett, Tapan das, Sambit Kar

Protein oxidation is a degradation pathway that commonly results in chemical modification of methionine, tryptophan, histidine, cysteine, or tyrosine residues in a protein. For biotherapeutic proteins, oxidation of susceptible residues may compromise product quality or efficacy. Therefore, characterization of protein oxidation in biopharmaceuticals is essential for developing robust control strategies to prevent, detect, and monitor oxidation during manufacturing, storage and handling. Oxidation of biotherapeutics may occur rapidly if proteins are stored improperly, such as if they are left exposed to visible or UV light, or in the presence of oxidizing agents. This work demonstrates the application of Raman and FTIR spectroscopic methods to monitor the oxidation of methionine residues in biotherapeutic proteins, and discusses advantages over traditional analytical methods such as peptide mapping.

Raman spectroscopy for monitoring bispecific antibody assembly
Andrew Maier

Recent advances in biologic drug development have led to creation of new formats of biologics to target diseases in new ways. A large and rapidly growing area of new biologic formats is bispecific antibodies, with over 80 formats currently in development. One approach for generating bispecific antibodies is assembly of two half antibodies into a bispecific. This approach can result in unique impurities which may require multiple orthogonal assays to quantify. New analytical tools are needed to increase process understanding, reduce analytical burden, enable real-time monitoring of process dynamics, and reduce yield loss to unwanted byproducts. Raman spectroscopy is an emerging technique in biopharmaceutical industry and has demonstrated value for fermentation process monitoring. This talk will cover application of in-line Raman spectroscopy to monitoring assembly of a knob-hole bispecific antibody. Partial Least Squares (PLS) modeling was used to predict assembly products from multiple assays (HPLC-SEC, HPLC-RP, mass spec, Ellman's assay). Raman spectral preprocessing and region selection increased model accuracy. Model strength was confirmed via predictive validation. Analysis of prediction errors suggests that formation of bispecific antibodies, aggregates, and homodimers as measured by HPLC-SEC are well modeled across a range of temperatures, pH, and reductant concentrations. Accurate predictions were also made for system free thiols as measured by the Ellman's assay.

Field flow fractionation characterization of the NIST monoclonal antibody standard RM 8671

Robert Reed, Soheyl Tadjiki, Thorsten Klein

The NIST monoclonal antibody (mAb) standard reference material provides a platform for evaluating methods used to characterize physicochemical and biophysical attributes of other mAbs. In this study, the NIST mAb was used to evaluate separation parameters for asymmetrical flow field flow fractionation (AF4) analysis of mAbs, especially in comparison to size exclusion chromatography (SEC). Results will be shown for AF4 and SEC, both coupled with refractive index and multi-angle light scattering detectors for measuring molecular weight. For AF4, both a standard (33.5 cm) and micro channel (8 cm) were used, to evaluate separation parameters such as analysis time, resolution, and recovery. The NIST mAb was agitated at room temperature to induce aggregation, and the ability of both AF4 and SEC to measure the full distribution of aggregated species was evaluated.

Use of Portable Raman Spectroscopy in Low Resource Settings
Matt Keller, Changwon Lee, Wenbo Wang, Jim Stafford, Ben Wilson

Raman spectroscopy possesses many characteristics that have made it attractive as a potential alternative to traditional chemical analyses. These advantages include the ability to perform in vivo or in situ non-destructive measurements, which can produce objective results in near real time. These qualities could be particularly advantageous in low-resource settings that lack the infrastructure and skilled labor to carry out many traditional lab-based analyses. To date, a major limiting factor in its implementation has been the cost of devices; however, recent advancements have brought the cost down to levels that are approachable by potential end-users in low-resource settings, including health clinics and government or private agricultural extension programs. At Intellectual Ventures Laboratory, we have investigated several potential applications of Raman in low resource settings, including examining the quality and makeup of fertilizers that are often adulterated or mislabeled, analyzing matrices like grains and milk for both nutritional information and the presence of adulterants or toxins, and determining the authenticity of pharmaceutical drugs. In all of these applications, Raman has demonstrated its advantages over near infrared (NIR) and mid infrared (MIR) devices that have been tested in parallel – namely, better specificity than NIR and better ability to work with liquid and complex layered samples than MIR. The shortcomings of current Raman approaches have been highlighted as well, including significant fluorescence interference from certain samples in every application, as well as higher device costs compared with novel NIR devices now on the market. Thus, before Raman can truly emerge as a promising tool for use in low resource settings, the Raman community must seek to further reduce hardware costs and to come up with reasonable strategies to avoid or to mitigate fluorescence interference.

Extending Raman
David Creasey

Over the last decade, miniature fiber optic spectrometers have greatly expanded the ability of Raman spectroscopy to tackle practical applications in the field, from mobile pharmaceutical ID to hazardous material assessment in remote locations. There remains a gap, however, between the typical diode array spectrometer and their more sensitive benchtop analogs. High sensitivity, cooled Raman spectrometers have the potential to narrow that gap by providing greater sensitivity, better SNR, and faster measurement times. In this presentation, we'll look at the key factors in the design of high sensitivity miniature Raman spectrometers and their associated accessories, as well as the key metric for direct comparison of these systems - limit of detection. Having established this baseline, we'll present some applications that are now becoming practical thanks to high sensitivity Raman systems operating at wavelengths from the UV to NIR. UV and NIR Raman are showing great promise for fluorescence suppression in the study of biological samples, enabled by more readily available excitation sources and high throughput spectrometers that cut acquisition times by more than 10x. Handheld explosives detection surfaces and in the field are facilitated by SERS, but as the demands on trace level detection become more stringent, the need for faster acquisition times, greater sensitivity, and through barrier measurements have increased. These same benefits are also fostering an increasing number of applications in the study of transient events and high-throughput monitoring. But the proof is in the data come see the performance difference that is closing the gap with benchtop Raman.

Dual Wavelength Applications in Portable Raman Spectrscopy
Robert Chimenti

As portable Raman spectroscopy systems are gaining more and more acceptance in the field, users are expecting higher and higher levels of performance. This is in large part due to the drive to expand the application space to include applications such as medical diagnostics, gemology/minerology, and even laboratory research in addition to traditional applications such as incoming inspection and unknown material identification. In this presentation, we will discuss a novel method that shows a great deal of promise for increasing the spectral range, resolution, sensitivity, and background removal in portable and handheld Raman systems. In this approach, two individually controllable diode lasers are integrated into a single package with either a single collinear output beam or a single fiber optic allowing them to be sent through the same optical path. By employing this dual wavelength approach, it enables systems designers to utilize a wide variety of techniques that are not possible with a single laser. For example, they can utilize two different diodes separated by an optimal wavelength (determined by the spectral range of the spectrometer) in order to double the total spectral range of the system while maintaining spectral resolution through spectral stitching. Additionally, with this platform technology, a variety of different shifted excitation techniques can be performed by utilizing two diodes with similar center wavelengths. Lastly, with the same dual laser approach, orthogonal polarizations can be utilized allowing for applications ranging from crystallography to chirality.

Ocean Optics Miniature Raman Spectrometer Development

Joseph Bonvallet, Bryan Auz, Doug Anderson, Ty Olmstead

The need for techniques to rapidly identify samples, ranging from molecule and particle imaging to detection of high explosive materials, has surged in recent years. Raman spectroscopy gives a molecular fingerprint, with no sample preparation, and can be done remotely. Raman spectroscopy offers a technology that can meet the unmet needs for rapid sample identification. These systems can be small, compact, lightweight, and with a user interface that allows for easy use and sample identification. Ocean Optics Inc. has developed several systems that would leverage Raman spectroscopy for rapid identification of samples. This talk will describe the development of different Ocean Optics Inc miniature Raman spectrometers and their capabilities to identify samples. Three separate techniques will be presented: SOAP (spectrometer on a phone), Accuman (a bench top portable Raman spectrometer) and the HRS-30 (a handle held Raman system.) The spectrometer on a phone (SOAP) system was designed using commercial off the shelf (COTS) components. The footprint of the system measures 40x40x14 mm (LxWxH) and was coupled directly to the cell phone detector camera optics. However, it gets roughly only ~40 cm-1 resolution. The Accuman system is the largest (210x200x100 mm) of the three, but uses our QEPro spectrometer and get ~7-11 cm-1 resolution. Finally, the HRS-30 measuring 165x85x40 mm is a combination of the other two systems. This system uses a miniature Ocean Optics Inc. spectrometer and gets ~7-12 cm-1 resolution. Each of these units uses a peak matching algorithm that then correlates the results to the pre-loaded and customizable spectral libraries.

Raman Imaging for Drugs & Explosives Particle Detection with a High Throughput Virtual Slit

Edward Gooding, Erik Deutsch, Joseph Huehnerhoff, Jason Lozo, Courtney Johnson, Arsen Hajian

Proximal and standoff Raman-based detection of narcotics and explosives is becoming widely accepted by first responders and forensic investigators. Handheld units suitable for field use typically use point rastering with 785 nm excitation to create an averaged Raman spectrum over a small sample area. While this approach works well with homogeneous samples, it is less effective when the substance of interest is present as trace particles in an inhomogeneous mixture. Point rastering NIR Raman instruments focus the diode laser beam to a small spot at the sample using a spherical lens, then image or fiber-couple the excitation spot onto the entrance slit of a mini-spectrometer. In both cases, the mismatch between the circular excitation profile and the narrow slit required to produce adequate spectral resolution results in inefficient photon collection. Further, it is often necessary to attenuate the laser in order to avoid photobleaching sensitive samples. A small excitation spot thus delivers high spatial and spectral resolution and good photon collection efficiency, but requires an unacceptable amount of time to map the sample area. A large spot can map an area quickly, at the cost of poor spatial resolution, spectral dilution of trace components, and inefficient photon collection. Hindsight Imaging, Inc. uses a different approach. Our portable Raman spectral imager creates a slit-shaped excitation profile from a high-power diode laser. We image the illuminated area onto the slit, enhance the resolution using High Throughput Virtual Slit technology, and collect a 2D spectral slit image using an array detector. We then sweep the excitation across the sample using pushbroom scanning. This reduces mapping times significantly, while detecting trace particles, as each sample location is mapped to a specific row of the detector. It also enables the use of inexpensive, lightly cooled CMOS detectors.

We will discuss how spectral maps created by our instrument are used to detect trace particles in forensic, narcotics enforcement, security and defense settings.

REAL TIME MEASUREMENTS OF POLYMER PROPERTIES BY RAMAN SPECTROSCOPY

Patrice Bourson, David Chaprob, Elise Dropsit, Isabelle Royaud , Marc Poncot, Abdessalam Dahoun, Alain Durand, Sandrine Hoppe, Sarah Saidi

Raman spectroscopy is a technique particularly rapidly evolving in spectral qualities, quality of instruments ... and transportability of the new equipment. This allows uses in-situ or real time using of these devices and the possibility of their use for monitoring response and industrial flows directly into the plant or reactor We show, in this presentation, two examples of real-time monitoring by Raman spectroscopy. The first one is the control in real time of industrial flows using Raman spectroscopy combined with chemometrics to monitor very effectively an industrial polymer production. The second example is an original coupling of a Raman spectrometer with a rheometer for monitoring in real time of the polymerization of acrylic acid.

Chemometric Model Development and Maintenance – an Industry Case Study

Michael F Roberto, Randy J Pell

A project has been undertaken to assess how to reduce the amount of effort devoted to producing, maintaining, and stabilizing Raman model performance in support of refinery and chemical plant labs. Over the last year, Infometrix has examined an unprecedented historical collection of spectra spanning 1-5 years from three oil refineries with the goal of developing and maintaining stable models for long-term deployment. Technologies utilized follow a pattern of Best Practices, including the use of Robust outlier diagnostics, local and hierarchical modeling, and model augmentation. The effort has resulted in significant progress towards automation of model creation, stability, and maintenance in an industrial process. This presentation will highlight recent efforts to streamline model assessment and deployment, with examples from chemometric application in motor fuel property assessment. It will also include a live demonstration of data visualization, with interactive comparisons of different approaches investigated through the length of the project.

Crystallinity and Density Assessment of Layered Polyethylene Using Confocal Raman Microscopy

Mohammed Ibrahim

Crystallinity is an important characteristic of polymers, and affects their optical, mechanical, thermal and chemical properties. Most methods of evaluating the polymer crystallinity assume a three-phase morphological structure: a crystalline region, a disordered region, and a transition region. These methods include DSC, XRD, IR, NMR, and density measurement. Crystalline areas are generally more densely packed than amorphous areas; hence, higher density. For example, the density of crystalline polyethylene (PE) is 1.0 g/cm^3 whereas the density of amorphous PE is 0.85 g/cm^3. Currently, there are several standard techniques for PE density measurement, including ISO 1183/ASTM D1505, ASTM D792, and ASTM D4883. While relatively simple and straightforward, these methods can be labor-intensive and time-consuming, and are limited to bulk PE. Furthermore, they can't be easily adapted to measure PE density in polymer laminates or multilayer polymer films. To that end, confocal Raman microscopy provides an attractive alternative by enabling direct measurements of the PE layer in polymer laminates and multilayer films without a need for sample preparation. Strobl et al. first proposed using Raman spectroscopy to characterize the morphological structure of PE. Fractions of the three phases could be derived directly from the integrated intensities of characteristic bands. The method has since been extensively improved by several researchers, and is now widely applied in the structure analysis of different PE's and their blends. To our best knowledge, however, existing work are limited only to measurements of bulk PE (resin, pellet etc.). There has been no report on extending this method to PE films. In this work, we will demonstrate the feasibility of using confocal Raman microscopy to assess the density and crystallinity of PE films. The densities of a series of PE films will be measured and compared with those of the starting PE pellets. The density variation along the film thickness will also be evaluated. Our results indicate that the confocal Raman microscopy can provide facile and *in situ* measurements of the density and crystallinity of PE films in polymer laminates and multilayer films.

Raman Analysis of Multiple Melting Peaks of Polyethylene
Young Jong Lee, Ying Jin, Anthony Kotula, Chad Snyder, Angela Hight Walker, Klaman Migler

We investigate multiple melting peaks of high density polyethylene (HDPE) with Raman spectroscopy and differential scanning calorimetry (DSC). The origin of the observed multiple melting peaks are studied by quantitative analysis of Raman spectral change associated with each melting peak with respect to temperature. A least squares moving-window (LSMW) method is used for effective suppression of derivative noise. The conformational similarity among multiple melting peaks strongly suggests that dominant melting occurs from the crystalline phase directly to the amorphous phase at all transitions. The observed multiple melting transitions are consistent with the multimodal lamellar thickness distribution model as the dominant origin. However, a two-dimensional Raman correlation spectroscopy suggests that non-crystalline trans-rich conformations appear during the melting process, whose mass fraction is estimated as 10% or less.

Plasmon Driven Reactivity
Zachary Schultz, Darby Nelson, Zhicong Zeng, Hao Wang

The excitation of plasmon resonances results in electric fields that can be used for a variety of applications from sensing to catalysis and photovoltaics. The intense electric fields are often associated with surface enhanced Raman scattering (SERS). High sensitivity SERS sensing often requires plasmon coupling between nanostructures, to form hotspots or junctions with the greatest electric fields. In addition to these high electric fields, plasmon resonances have been associated with the production of large temperature gradients on the surface of nanostructures. Experiments utilizing SERS, photothermal heterodyne imaging (PHI), electron microscopy, and optical microscopy illustrate how these different affects can result in conditions that alter chemical reactivity in the vicinity of plasmonic nanostructures. Experimental results we have obtained indicate that the intense electric fields on nanostructures can give rise to second order nonlinear phenomena, such as optical rectification. The presence of a large dipole moment in gap between nanostructures is shown to break the symmetry of the interface, such that electron tunneling in biased between metal structures giving rise to a DC bias with optical excitation. This rectified plasmonic field is evident as a shift in the vibrational frequency of reporter molecules on the nanostructures. The resulting bias is shown to alter electrochemical activity of reactions on the nanostructure surface.

Biocompatible, Liposome-Based Surface Enhanced Raman Spectroscopy (SERS) Substrates
Laura Sagle, William Lum, Ian Bruzas, Zohre Gorunmez

Surface enhanced Raman spectroscopy (SERS) is a powerful technique capable of probing single molecules with unprecedented detail for biophysical studies. Unfortunately, most currently used SERS substrates contain <10 nm gaps between gold or silver particles, in which biological probe molecules are inserted. Although this strategy often leads to single molecule detection of the biological probe molecule, it also results in denaturation and inactivation of these molecules, which severely limits the use of SERS for biophysical measurements. In the Sagle group, we are working with novel liposome-based SERS substrates, in which probe molecules are encapsulated in the liposome component, rather than tethered to or sandwiched between metallic surfaces. Initial characterization of a liposome-based gold nanoparticle on mirror substrate shows high field enhancement of 107-108, stability to laser illumination, and good measurement reproducibility. We have recently extended this idea to include silver particles and nanoshells around the liposome component, both of which exhibit increased SERS enhancement over the gold nanoparticle on mirror substrate and improved biocompatibility over traditional SERS substrates.

Plasmon-enhanced fluorescence and energy transfer in composite nanoparticles and applications to sensitive chemical detection

Denis Boudreau, Jeremie Asselin, Nicolas Fontaine, Rihab Bouchareb, Mazeyar Parvinzadeh Gashti, Simon Labrecque, Paul De Koninck, Jesse Greener, Patrick Mathieu

The use of fluorescent probes is extremely popular in bioanalytical chemistry and cellular imaging, offering numerous advantages including non-invasiveness and high detection sensitivity. A plethora of fluorophores are already commercially available and well-characterized for a variety of applications, but they often suffer from low photostability and limited brightness on a molecular basis. In this presentation, we will discuss the use of Metal-Enhanced Fluorescence (MEF) and Förster Resonant Energy Transfer (FRET) to improve the analytical properties of molecular fluorophores. Our architecture of choice is designed around a plasmonic metal core (Ag, Au, In) surrounded by one or more silica shells acting as spacers or incorporating dyes at optimal positions in the local electric field. Furthermore, these nanocomposites may in turn be immobilized on solid optical substrates such as microscope coverslips or waveguides. The impact of particle geometry on detection sensitivity, dynamic range and photostability, as well as the efficiency of energy transfer between donor and acceptor molecules immobilized within distinct layers can be evaluated by time-resolved and steady-state fluorimetry. Using these strategies, nanosensors have been designed for, e.g., detection of DNA and quantification of pH and halides (Cl^-, Br^-, I^-) for various applications, notably for the investigation of local chemical changes near excitable cells and bacterial biofilms.

Plasmonic Nanoparticle Probes for Optical Spectroscopy

Andrea Tao, Tyler Dill, Yuan Zheng

Many optical spectroscopy and imaging techniques are limited by the ability to fabricate nanoscale probes that are robust, reproducible, and support high quality optical resonances. I will describe how colloidal plasmonic nanoparticles overcome these obstacles and serve as enabling materials for spectroscopy and imaging. First, I will present our recent work on the synthesis and self-assembly of colloidal nanoparticles for the fabrication of resonant optical nanojunctions. Previously, we demonstrated that shaped colloidal nanoparticles can be organized into nanojunctions that possess intense “hot spots” due to electromagnetic field localization. Here, I will describe how colloidal nanoparticles can be assembled and used as scanning near-field optical probes for tip-enhanced Raman spectroscopy. Nanoparticles are assembled onto atomic force microscope cantilevers that can then be raster scanned across a surface during optical measurements. Nanoparticle size and shape can be used to modulate the optical response of the scanning probe. Second, I will present spectroscopic data for plasmonic nanoparticles that exhibit coupling to excitonic systems, such as fluorophores and quantum dots. These coupled probes have the potential to be used for distance-sensing and nonlinear optical spectroscopies.

Plasmonic enhancement of fluorescence as a tool in single-particle analysis of ion concentrations

Jeremie Asselin, Nicolas Fontaine, Simon Labrecque, Paul De Koninck, Denis Boudreau

Using a wide range of surface functionalities and chemical compositions, recent nanotechnologies have demonstrated a tremendous potential to probe individual components of interest in cellular biology and medicine. Plasmonic materials can be adapted to improve efficiently the analytical properties of common fluorophores with regards to spatial and temporal resolutions, non-invasiveness, photostability, sensitivity, and signal-to-noise ratio. Moreover, integrating multiple moieties in an architecture allows to encode multiple signals simultaneously in single-particle imaging. In this study, we present a methodology to apply metal-enhanced fluorescence (MEF) in controlled core-multishell nanosensors for detection in spectrofluorimetry and microscopy experiments. The particles consist of a metallic core coated with multiple shells which incorporate dyes at optimal positions in the plasmon-induced electric field. The fundamental influence of the plasmon on the MEF phenomenon was studied by introducing a non-fluorescent dielectric layer in the general design. Then, sensitive molecules like fluorescein, rhodamine, quinolinium and seminaphtharhodafluor (SNARF) are grafted covalently in silica and used with improved brightness, dynamic range of detection and photostability. Such nanoprobes were characterized thoroughly in time-resolved and steady-state measurements before their demonstration for the quantification of pH, halide concentrations (Cl-, Br-, I-) and temperature in different environments.

Integrating Novel Microscopy into Battery Research: From Atomic Resolution to In Situ and Functional Imaging
Miaofang Chi

Despite their different structural configurations and chemistries, current energy-storage systems (e.g. Li-ion, Li-S, Li-air, and all-solid-state Li batteries) rely on the use of solid-state ion conducting materials as electrodes and/or electrolytes. The high interfacial resistivity within solid electrolytes and between electrolytes and electrodes, is currently the major concern associated with the application of solid electrolyte materials. However, the root origin is unclear. To understand the interfacial mechanisms at play, correlated studies of the potential structural and chemical heterogeneity, chemical, electrochemical, and mechanical stability, and space charge, must be conducted and ideally, under simulated operation. In this talk, I will show several examples of elucidating the ion conduction behavior at solid electrolyte interfaces, including both electrolyte-electrolyte and electrolyte-metallic lithium interfaces, primarily using atomic-scale and in situ electron microscopy. The urgency of developing new microscopy techniques for investigating solid-solid ion conducting interfaces will also be discussed. The integration of new microscopy techniques, such as atomic resolution differential phase contrast imaging and valence electron energy loss spectroscopy, into battery research will be discussed.

Recent developments in AFM analysis of soft materials at the nanoscale

Igor Sokolov, Maxim Dokukin

Here I present recent development of using atomic force microscopy (AFM) technique for quantitative analysis of soft materials. I will describe the mandatory steps to image mechanical properties of a sample. Furthermore, I overview two novel and quantitative modes of AFM operation, which were recently developed in my lab. The first one, FT-NanoDMA AFM mode is a combination of three different methods: the quantitative dynamic mechanical spectroscopy (DMS), AFM indentation, and Fourier-transform spectroscopy. This novel mechanical spectroscopy mode is fast and sensitive enough to allow DMS imaging of nanointerfaces and single cells. Compared to the existing state-of-the-art nanoindentation, FT-NanoDMA demonstrates ~ 100x improvements in both spatial (down to 10nm) and temporal resolution (down to 0.7 sec/pixel). The second mode (we call this "ringing" mode) is an extension of popular sub-resonant oscillatory AFM modes (for example, PeakForce QNM by Bruker). This mode utilizes the signal information from free resonance oscillations of the cantilever, which occur after detaching the AFM probe from a sample surface (ringing). Processing the ringing signal allows deriving five new channels of information, such as disconnection energy loss, sample deformation at disconnection, size of possibly stretchable molecules, etc.

Materials Science at Surfaces - 2D materials and Nanospheres
Petra Reinke, Cameron Volders, Ehsan Monazami, Gopalakrishnan Ramalingam, John Brandon McClimon

This presentation will discuss the study of complex surfaces, and the discovery of new 2D and nanomaterials using STM and STS in combination with in-situ materials synthesis. This work extends from the manipulation of graphene to silicene synthesis, control of electronic properties in molecules, and design of self-repairing carbide films. In the first part I will focus on a new pathway to the synthesis of silicene using the Si-terminated h-MoSi2 (0001) surface The silicene properties are strongly influenced by sheet buckling due to a partial sp3 bonding and control of buckling has been proposed as a pathway to the expression of quantum phases. The details of surface structure, weak buckling and synthesis pathways are derived from imaging (STM) and spectroscopy maps (STS). The silicene structure and a weak buckling state are confirmed and are examined in the context of substrate-silicene bonding [Nano Lett., **17** 299 (2017)]. The growth of silicene on MoSi2 opens new possibilities of monolithic integration in future devices, and the use of reconstructions as a platform for 2D materials synthesis. The second part will illustrate the synthesis of so-called nanospheres, which are based on fullerene molecules whose electronic structure is modulated by their reaction with the substrate. The reaction of C60 with a W-surface is controlled by annealing. The topography of the reacted fullerene molecules, respectively nanospheres, remains nearly intact while the reaction at the C60 –W interface drives the transition to a metallic material. The majority of studies on the interaction of large molecules with metal surfaces operate in a regime where the molecules integrity is conserved. However, the control of the interfacial reaction offers unique opportunities in the targeted manipulation of electronic structure [ACS Appl. Mater. Interfaces **8** 34854 (2016)]. It is relevant for the use of molecular electronics devices and polymeric solar cells (apart of some interesting science).

Analysis of Nanoscale Semiconductor Devices Using Advanced Scanning Probe Microscopy Techniques
Phil Kaszuba

The Atomic Force Microscope (AFM) has been in existence for approximately 30 years. Over the course of its lifetime, numerous analytical techniques have been developed and implemented on the AFM platform, evolving into what is collectively called the Scanning Probe Microscope (SPM). The SPM has since become firmly established as a mainstream analytical instrument with a continually increasing role across many disciplines in both academia and industry. SPM analyses have provided critical information in the evolution of leading edge nanoscale semiconductor technologies providing data from initial device concept to technology development to manufacturing to failure analysis. This presentation focuses on two scanning probe analytical techniques and their application to nanoscale semiconductor technology. Scanning Capacitance Microscopy (SCM) is a well-established workhorse technique used for nanoscale diffusion analysis applications in advanced semiconductor devices. Scanning Surface Photo Voltage Microscopy (SSPVM) is a recently invented technique with implications in analyzing material properties as well as stress in nanoscale structures. A structural and functional overview of two-dimensional and three-dimensional nanoscale semiconductor devices will be presented followed by a discussion of each technique (SCM, SSPVM) along with examples of how they're applied in the development of new technologies, the monitoring of manufacturing processes, and the failure analysis of nanoscale semiconductor devices.

Nanoscale Chemical Analysis with Photo-induced Force Microscopy
Sung Park

Infrared Photo-induced Force Microscopy (IR PiFM) is based on an atomic force microscopy (AFM) platform that is coupled to a widely tunable mid-IR laser. PiFM measures the dipole induced at or near the surface of a sample by an excitation light source by detecting the dipole-dipole force that exists between the induced dipole in the sample and the mirror image dipole in the metallic AFM tip. This interaction is strongly affected by the optical absorption spectrum of the sample, thereby providing a significant spectral contrast mechanism which can be used to differentiate between chemical species. Due to its AFM heritage, PiFM acquires both the topography and spectral images concurrently and naturally provides information on the relationship between local chemistry and topology. Due to the steep dipole-dipole force dependence on the tip-sample gap distance, PiFM spectral images have spatial resolution approaching the topographic resolution of AFM, demonstrating sub 10 nm spatial resolution on a variety of samples. PiFM spectral images surpass spectral images that are generated via other techniques such as scanning transmission X-ray microscopy (based on synchrotron source), micro confocal Raman microscopy, and electron microscopes, both in spatial resolution and chemical specificity. The breadth of the capabilities of PiFM will be highlighted by presenting data on various organic, inorganic, and low dimensional materials. By enabling imaging at the nm-scale with chemical specificity, PiFM provides a powerful new analytical method for deepening our understanding of nanomaterials and facilitating technological applications of such materials.

High Analytical Power Density with the Liquid Sampling-Atmospheric Pressure Glow Discharge (LS-APGD) MicroplasmaHigh Analytical Power Density with the Liquid Sampling-Atmospheric Pressure Glow Discharge (LS-APGD) Microplasma

R. Kenneth Marcus

We are all familiar with the term "get more bang for your buck". In the economics world, they refer to price-value analysis. To be sure, the primary advantage of the newly developed inductively-coupled plasma (ICP) in the 1970s and 80s was the ability to solve many of the current challenges involving elemental analysis. Immunity from the severe matrix effects experienced in flame spectroscopy allowed far greater analytical flexibility in terms of quantification, as well as great improvements in limits of detection. Beyond the ability to perform high quality elemental analysis of liquids, development of alternative sample introduction tools including furnace atomizers, laser desorption cells, and chromatographic couplings improved the analytical versatility of ICP-OES/MS. Given its power, there are indeed missions that the ICP cannot fulfill and even if they can, do not do so with sufficient efficiencies of process or cost. Applications that require "taking the instrument to the sample", are not in the wheelhouse of the ICP. It is too energy, support service, reagent, and space intensive to put in the back of a minivan. By the same token, while chromatographic systems and mass analyzers have gotten smaller, they do not couple effectively to the standard ICP torch. For these reasons, there have been many efforts to develop miniaturized plasma sources. This laboratory has developed the liquid sampling-atmospheric pressure (LS-APGD) microplasma for elemental analysis by OES and MS. Low power (<50 W), low sample load rates (<50 ?L/min) and a small footprint (<0.5 ft^2) are well in line for transportable systems. Perhaps a more impactful aspect is the fact that the source can be directly coupled to any of the common ambient-sampling interfaces that are common to electrospray MS. Thus, atomic/isotopic analyses can be affected on thousands of *organic* mass spectrometers. Taken a step further, the same ion source can be applied for the full complement of molecular mass spectrometry applications. Finally, the microplasma can also be applied in ambient desorption and laser ablation sample introduction modes for both OES and MS analysis. This presentation will provide examples of the high analytical power density of this microplasma source.

Chemical, electrical, and spectroscopic studies of a solution-cathode glow discharge.

Michael Webb, Denise Moon, Scott Crowe

Atmospheric pressure plasmas with liquid electrodes have received increased attention for trace metal analysis in the past few years. These plasmas offer advantages including size and cost over traditional plasmas used for atomic spectroscopy. They have been used to analyze such diverse samples as tea leaves, tuna fish, aquatic plant matter, oyster tissue, coal fly ash, hepatitis-B vaccines, lake water, soil leachates, spruce needle leachates, colloidal silica, and zirconium alloys. Although these applications are promising, greater understanding of the processes involved – including the mechanism by which analytes enter the plasma and the influence of instrument parameters on emission strength and stability – may lead to improved figures of merit. This presentation will describe studies of the solution-cathode glow discharge (one example of an atmospheric pressure plasma with a liquid electrode), with the goal of better understanding the discharge and thereby improving its performance. Studies will include spectroscopic and electrical studies of the plasma, as well as studies of chemical changes to the sample solution.

Charge Controlled Electrospray Ionization and Plasma Ionization via Triboelectric Nanogenerator (TENG)

Anyin Li, Yunlong Zi, Zhong Lin Wang, Facundo Fernandez

Ion sources are usually driven by electronic power supplies. Voltages applied to the ion source serve to initiate the process of converting neutrals into gas-phase ions. However, the number of generated ions does not depend on the applied voltage in a straightforward fashion. As a result, controlling the number of charges used in the ionization process is typically impossible. We recently demonstrated that the output of triboelectric nanogenerators (TENGs) can quantitatively deliver charge into the ionization process, yielding excellent sensitivity and reproducibility. The triboelectric nanogenerator (TENG) device was composed of three layers of thin film materials. The two outside layers serve as electrodes and the middle layers serves as triboelectric electret. Mechanical movement of these layers with respect to each other induces charge redistribution in the electrodes. The total charge number is dependent on the area and the charge density of the functional material, which is fixed and typically in the tens to hundreds of ?C m-2. The charges are supplied to either a nanoelectrospray emitter for electrospray ionization, or to a needle electrode for plasma discharge ionization. The amount of total charge in one ionization event was controlled by varying the output of the TENG. The duration, polarity and frequency of the ionization events were controlled via TENG device's on-demand actuation. For a given nanoelectrospray emitter, accurately controlled ion pulses ranging from 1.0 to 5.5 nC were generated. Spray pulses were generated at frequencies as high as 17 Hz, and the pulse duration could be adjusted on-demand between 60 ms and 5.5 s. Plasma discharge ionization was also generated using TENG. In air, this was achieved using a TENG of 140 nC capacity, a stainless steel needle-plate electrode pair with a distance of 1.5 mm. Highly repeatable plasma discharge ionization pulses with a relative standard deviation of 7% were generated. A variety of compounds such as dimethyl methylphosphonate, hexylamine, and trinitrotoluene, were readily ionized by this plasma discharge, and can be detected in either positive or negative mode mass spectrometry. Sub-ppb detection limits were achieved for the analysis of these explosive and chemical warfare simulant chemical compounds.

Soft argon-propane dielectric barrier discharge for mass spectrometry

Alexander Schütz, Felix David Klute, Sebastian Brandt, Joachim Franzke

Since about 10 years dielectric barrier discharges (DBD) are used as soft ion source for organic mass spectrometry (DBDI-MS). Helium based DBDI are often chosen because of a good ionization efficiency and low ignition voltage. Using the same geometry argon needs much higher ignition voltages compared to helium and the outcome could be a filamentary instead of a homogenous plasma which is not suitable for soft ionization. In this study a mixture of argon with propane was used as ignition aid to decrease the ignition voltage which leads to a homogenous argon based DBD for soft ionization. Custom-made gas mixtures enable to apply a DBDI in front of a mass spectrometer in an enclosed system. Usually the analyte has to be introduced into the plasma at the outlet of the capillary at ambient air. It is assumed that a huge amount of analyte is lost and will not be sucked into the mass spectrometer successfully. But an argon-propane DBDI can be operated at ambient air such as in an enclosed system. This is due to the fact that nitrogen is not necessary for the plasma anymore. This allows now to introduce the analyte in such an enclosed system either in front of the inlet of the mass spectrometer or through the discharge capillary itself. Furthermore argon-propane has some advantages concerning the vacuum: the vacuum pumps can deal much better with argon-propane than with helium. The high vacuum conditions are now comparable with ambient air and there is no degradation of resolution anymore. The novel setup could lead to a better limit of detection (LOD) because the probability of interaction between the reactant ions and the molecules is higher compared to an open system while a decrease of background signals is expected. This novel non-ambient DBDI-MS setup is presented and characterized which could advance novel hyphenation of DBDI with analytical separation techniques like gas chromatography (GC) and high pressure liquid chromatography (HPLC) in the near future.

Analysis of Nanomaterials using CE-spICP-MS with Microsecond Dwell Times
Carsten Engelhard

In this presentation, recent developments in inductively coupled plasma quadrupole mass spectrometry (ICP-Q-MS) instrumentation for nanomaterials research will be reviewed and an approach for direct, fast, and high-throughput nanomaterial characterization on a single particle level will be presented. In the first part of the presentation, the current state-of-the-art in single-particle (spICP-MS) for the detection and characterization of single nanoparticles (NP) will be reviewed. Also, a custom-built high-speed data acquisition unit with 5 μs dwell times (μDAQ) specifically tailored to the needs of spICP-MS [1] and capillary electrophoresis (CE) coupled to spICP-MS will be presented. The μDAQ was built in-house and enables continuous acquisition of signals from discrete ion clouds in the ICP with microsecond time resolution and 100% duty cycle. Fully time-resolved temporal profiles of individual single droplet or particle events can be recorded. The performance of this approach compared to conventional ICP-MS operation and selected applications will be discussed. Using a time resolution much higher than the typical duration of a particle-related ion cloud, the probability of measurement artifacts due to particle coincidence could be significantly reduced and the occurrence of split-particle events in fact was almost eliminated. Fully time-resolved temporal profiles of transient signals originating from single gold NP as small as 10 nm are presented. The advantages and disadvantages of millisecond versus microsecond dwell times are critically discussed including measurement artifacts due to particle coincidence, split-particle events, and particle number concentration. Also, it will be discussed how separation methods coupled to ICP-Q-MS can aid the detection of NPs in complex mixtures. Specifically, a method that we developed for NP separation, size characterization, speciation, and quantification of gold and silver NPs using capillary electrophoresis (CE) and ICP-MS [2] will be discussed and our recent efforts to expand this approach also to CE-spICP-MS [3] and different surface modifications of NP will be presented. [1] I. Strenge, C. Engelhard, J. Anal. At. Spectrom., 2016, 31(1), 135-144. [2] B. Franze, I. Strenge, C. Engelhard, Anal. Chem., 2014, 86(12), 5713-5720. [3] B. Franze, I. Strenge, C. Engelhard, J. Anal. At. Spectrom., 2017, under revision.

Analysis of lipoprotein and cholesterol content using chemometrics and deep-ultraviolet resonance Raman spectroscopy

Renee JiJi, Michael Eagleburger

Cardiovascular health is intimately related to ones levels of high-density (HDL) and low-density (LDL) lipoproteins and their respective cholesterol contents. Lipoproteins vary in cholesterol, protein and unsaturated lipid content. It is well known that the vibrational bands from the protein backbone are resonance enhanced in deep ultraviolet resonance Raman (dUVRR) spectra of soluble and lipid solubilized proteins. New studies show that the carbon-carbon double bond of unsaturated fatty acids, lipids, and cholesterol is strongly resonance enhanced in dUVRR spectra. Further, the dUVRR feature arising from unsaturated lipids (1655-1661 cm^{-1}) is sufficiently resolved from that of cholesterol (1675 cm^{-1}) to allow estimation of cholesterol content in intact LDL when combined with multivariate methods. The ability to separate and directly quantify lipoproteins, their associated cholesterol levels and size distributions would be a major step forward in understanding how lipoproteins relate to cardiovascular health. DUVRR spectroscopy coupled with size exclusion chromatography was evaluated as a potential method for direct separation and analysis of lipoproteins in blood serum samples. Spectral and chromatographic profiles were resolved for a lipid/cholesterol, protein and buffer features using parallel factor (PARAFAC) analysis and multivariate curve resolution (MCR).

Chemometric Applications of Walsh-Hadamard Filter Functions
Timothy Corcoran

A special situation exists in chemometric analysis when spectral loadings of the analytes are already known. In this case, the true scores of the analytes in a mixture may be estimated directly, in an on-the-fly fashion using spectral filter functions. Conceptually akin to optical filters which pass a desired portion of a spectrum, the dot product of these spectral filter functions on the spectrum yields information pertinent to a particular analyte, even if that information is spread widely across the spectral range. In some forms, filter functions can be used to apply a compressive detection strategy. Rather than collecting the entire spectrum over the relevant region for the mixture, a filter function may be applied within the spectrometer itself so that only the scores are recorded. Consequently compressive detection shrinks datasets tremendously. Filter functions may also be applied outside the context of compressive detection as a powerful and fast chemometric analysis technique. The Walsh functions, the binary basis used in Walsh-Hadamard transform spectroscopy, form a complete orthonormal set well-suited to forming filter functions via linear combinations. Computational methods of creating and evaluating optimal filter functions for both compressive detection and other analysis strategies in which the crosstalk between analytes is minimized are laid out. Monte Carlo simulations using mixtures of analytes with highly overlapped Raman and excitation-emission matrix loadings demonstrate the utility of the technique.

Multivariate Analysis and Raman Microspectroscopic Imaging: Implications for the Search for Life on Mars

Joseph Smith, Frank Smith, Karl Booksh

The search for evidence of extant or past life on Mars is a primary objective of both the upcoming Mars 2020 rover (NASA) and ExoMars 2020 rover (ESA/Roscosmos) missions. This search will involve the detection and identification of organic molecules and/or carbonaceous material within the Martian surface environment. For the *first* time on a mission to Mars, the scientific payload for each rover will include a Raman spectrometer, an instrument well-suited for this search. Hematite (?-Fe_2O_3) is a widespread mineral on the Martian surface.
The 2LO Raman band of hematite and the Raman D-band of carbonaceous material show spectral overlap, leading to the potential misidentification of hematite as carbonaceous material. Here we report the ability to spatially and spectrally differentiate carbonaceous material from hematite using multivariate curve resolution-alternating least squares (MCR-ALS) applied to Raman microspectroscopic imaging under both 532 nm and 785 nm excitation. For this study, a sample comprised of hematite, carbonaceous material, and substrate-adhesive epoxy in spatially distinct domains was constructed. Principal component analysis (PCA) reveals that both 532 nm and 785 nm excitation produce representative three-phase systems of hematite, carbonaceous material, and substrate-adhesive epoxy in the analyzed sample. MCR-ALS with Raman microspectroscopic imaging using both 532 nm and 785 nm excitation was able to resolve hematite, carbonaceous material, and substrate-adhesive epoxy by generating spatially-resolved chemical images and corresponding Raman spectra of these spatially distinct chemical species. Moreover, MCR-ALS applied to the combinatorial data sets of 532 nm and 785 nm excitation, which contain hematite and carbonaceous material within the same locations, was able to resolve hematite, carbonaceous material, and substrate-adhesive epoxy. Using multivariate analysis with Raman microspectroscopic imaging, 785 nm excitation more effectively resolved hematite, carbonaceous material, and substrate-adhesive epoxy as compared to 532 nm excitation. To our knowledge, this is the *first* report of multivariate analysis methods, namely MCR-ALS, with Raman microspectroscopic imaging being employed to differentiate carbonaceous material from hematite. We have therefore provided an analytical methodology useful for the search for extant or past life on the surface of Mars.

Variable Selection to Improve Reliability and to Reduce Cost of Ownership for Classifications and Calibrations

Undugodage Perera Perera, Matthew Allen, Joshua Ottaway, , Kristl Adams, Chance Carter, Steven Brown, Karl Booksh

Spectroscopic and sensor array data are commonly enlisted for disparate applications including routine analyses, process monitoring, and quality assurance. While there have been significant advances in measurement science during the past 20 years, computational analysis of responses from instrumental techniques has lagged. Two strategies have been prevalent to address the analytical modeling needs for chemical data. The first strategy has been to develop either "robust methods" immune to deviations from the assumed data structure or to develop complicated non-linear models that incorporate all deviations from the assumed data structure. The second strategy is to select subsets of data or even single variables for use in modeling with the goal of isolating highly informative variables in the data and to limit the subsequent analysis to only those variables. A hybrid approach that selects variables in a way that reflects the inherent structure of the overall data has been developed. Several projects demonstrating the utility of this approach are highlighted (e.g., analysis of edible oils for adulteration and detection of N-linked serum glycan biomarkers for esophageal cancer) in this presentation.

Modeling Microalgal Biosediment Formation based on FTIR-ATR Monitoring
Frank Vogt, Zachary Ogburn

With increasing amounts of anthropogenic pollutants being released into ecosystems, it becomes ever more important to understand their fate and interactions with living organisms. Microalgae play an important ecological role as they are ubiquitous in marine environments and sequester inorganic pollutants which they transform into organic biomass. Of particular interest in this study is their role as a sink for atmospheric CO2, a greenhouse gas, and nitrate, one cause of harmful algal blooms. Novel chemometric hard-modeling methodologies have been developed for interpreting phytoplankton's chemical and physiological adaptations to changes in their growing environment. These methodologies will facilitate investigations of environmental impacts of anthropogenic pollutants on chemical and physiological properties of marine microalgae (here: Nannochloropsis oculata). It has been demonstrated that FTIR-ATR spectroscopy can gain insights into both and this study only focuses on the latter. From time-series of spectra, the rate of microalgal biomass settling on top of a horizontal ATR element is derived which reflects several of phytoplankton's physiological parameters such as growth rate, cell concentrations, cell size, and buoyancy. In order to assess environmental impacts on such parameters, microalgae cultures were grown under 25 different chemical scenarios covering 200 ppm – 600 ppm atmospheric CO2 and 0.35 mM – 0.75 mM dissolved NO3-. After recording time-series of FTIR-ATR spectra, a MCR-ALS extracted spectroscopic and time profiles from each data set. From the time profiles, it was found that in the considered concentration ranges only NO3- has an impact on the cells' physiological properties. In particular, the cultures' growth rate has been influenced by the ambient chemical conditions. Thus, the presented spectroscopic+chemometric methodology enables investigating the link between chemical conditions in a marine ecosystem and their consequences for phytoplankton living in it.

Rapid and accurate peripheral nerve imaging by multipoint Raman spectroscopy
Yasuaki Kumamoto, Yoshinori Harada, Hideo Tanaka, Tetsuro Takamatsu

Raman spectroscopy allows label-free, minimally-invasive molecular analysis of biological tissue at microscopic spatial resolution (< 1 mm), and thereby is expected to detect in vivo an object that is not identifiable by its macroscopic morphology. One potential biomedical application of Raman spectroscopy is peripheral nerve detection in nerve-sparing surgery and surgical nerve repair, where accurately identifying peripheral nerves is the key to success but surgeons find it hard to identify a thin nerve by eye. A periprostatic nerve of a tissue dissected from a cancer patient was detected against adjacent tissues by Raman spectral analysis. The Raman-based nerve detection sensitivity and specificity were 94.4% and 92.0%, respectively, as evaluated with ex vivo rat tissues. Differences in lipid, protein, and collagen contents in peripheral nerves and other tissues allowed the nerve detection. To implement the Raman-based nerve detection in surgery, the technique needs improvement of the speed for imaging; the current Raman imaging technique takes minutes to hours for nerve imaging, which is too long a waiting time for a surgical procedure. Here we present a multipoint Raman measurement technique that has enabled acquisition of 32 spectra from rat ex vivo tissues in only 5 seconds. Individual spectra obtained from peripheral nerves and connective tissues at ? = 532 nm excitation were discriminated. The overlay of the tissue discrimination results with the corresponding bright-field image revealed the number of peripheral nerve-positive prediction spots (N_N) and the total number of detection spots (N_{total}) on given tissues. Of 158 peripheral nerves and 120 connective tissues measured, 152 nerves had values of N_N/N_{total} equal to or larger than 0.4, while 119 connective tissues had those smaller than 0.4. The sensitivity and specificity of the nerve detection by discriminating tissues at N_N/N_{total} = 0.4 were 96.2% and 99.2%, respectively. These results surpass the above-mentioned ones by singlepoint Raman spectroscopy. Additionally, the concurrent analysis of the tissue discrimination result with the corresponding bright-field image enabled visualization of the nerve running direction. The multipoint Raman measurement technique and following analyses will contribute to implementation of Raman-based nerve detection in surgery.

Spectrocopic evaluation of collagen fibrosis in post-myocardial infarction cardiac tissue
Kathleen Gough, Negar Atefi, Richard Wiens, Ian Dixon

Cardiac fibrosis is a primary contributor to heart failure; there is no cure for cardiac fibrosis in post-myocardial infarction (MI) hearts. We are studying the pathogenesis of cardiac fibrosis in in vivo mouse and rat models of cardiac fibrosis and heart failure, specifically following MI. Our study combines spectrochemical imaging of healthy and MI heart tissue with a long-term investigation of genes important to regulation of cells that influence cardiac fibrosis, particularly how fibroblast-specific gene expression contributes to heart failure in cardiac tissues. The latter research is focused on Ski protein as a negative regulator of Scleraxis, a driver for cardiac fibrosis. It is proposed that Ski plays a central role in modulation of global cardiac fibrosis in the heart following myocardial infarction. Far field polarization contrast FTIR imaging is being performed on tissue sections to assess amount, distribution and organization of scar deposited in the interstitium of failing hearts, at the micron scale. Near Field IR will permit unique spectrochemical assessment of collagen fibril organization at the nm scale in an effort to obtain information on quality and quantity of collagens in the fibrosing heart.

Challenges in Infrared Spectroscopic Data Collection and Processing From Collagen-Containing Tissues
Mugdha Padalkar, Rutvin Kyada, Farzad Yousefi, Ramya Ailavajhala, Jessica Falcon, Nancy Pleshko

Collagen is the most abundant protein in the body, and provides the structural basis for diverse tissue types. Accordingly, compositional changes in collagen with development and disease states is of great interest. Collagen molecular structure has been investigated by infrared spectroscopic techniques for over 50 years. Initially, transmission spectroscopy of solutions of collagen was performed, and assessment of structural changes was based primarily on the Amide I absorbance. However, the development of advanced spectroscopic technology, including fiber optics and hyperspectral imaging, have enabled more detailed investigations. Commensurate with technology development are challenges in understanding the nature of data collected from these novel infrared modalities. Hyperspectral Imaging allows evaluation of collagen structure from thin histological sections at pixel resolution from 3 to 6 microns. But, accurate results can be dependent on the data collection mode, i.e., transflectance vs transmittance, as well as the substrate utilized. For fiber optic data collection, attenuated total reflectance (ATR) mode is frequently used, and at times, reflectance mode. Peak shifts can occur which are very dependent on data collection mode, and thus comparison of data must be done with knowledge of potential shifts due to collection mode differences. Further, data processing techniques can vary significantly among investigations, and are not necessarily translatable from one experiment to another. Here, we will discuss some of the major challenges that present for spectroscopic investigations into collagen molecular structure.

Quantification of differences in bone mineral phosphate peak positions and relative intensities in transmission vs. ATR spectroscopic data collection
William Querido, Ramyasri Ailavajhala, Mugdha Padalkar, Nancy Pleshko

Shifts in peak position and relative intensities in the mid-infrared region can reveal many features of the poorly crystalline bone apatite mineral, and have been widely used to assess questions related to bone biomineralization and pathology. Peak assignment and ratio parameters of apatite have traditionally been established based on transmission spectroscopy using KBr pellets. Alternatively, the attenuated total reflectance (ATR) mode has been increasingly used in bone research. However, many relevant parameters of the apatite spectra acquired by ATR have yet to be compared to those established using KBr. In this work, we analyzed samples of bovine bone and of commercial bone mineral (Bio-Oss) by data collection in ATR vs. in transmission using KBr. We also performed ATR correction. We found several previously undocumented differences in peak position and relative intensities in the bone apatite spectra from ATR, ATR corrected, and KBr. Specifically, we found a significantly higher frequency shift and broadening of the $?_3$ phosphate peak

(900-1200 cm^{-1}) in KBr. The crystallinity assessment (1030/1020 cm^{-1} ratio) also showed significantly higher values in KBr, as did the assessment of acid phosphate content (1127/1096 cm^{-1} ratio). ATR correction was not sufficient to adjust ATR values to those of KBr. Importantly, in the analysis of 1030/1020 cm^{-1} ratio, a direct comparison of bone and Bio-Oss by ATR was not adequate to reveal the existing difference in sample crystallinity, which was seen clearly in KBr. In the second derivative spectra, the $?_3$ phosphate showed a striking difference between ATR and KBr: KBr spectra showed a split of the main peak into a second very clear component, around 1029 cm^{-1} in bone and 1039 cm^{-1} in Bio-Oss. These peaks are hallmarks of bone apatite and of more crystalline/stoichiometric apatite, respectively. However, in the ATR spectra, the primary absorbance was located at 1110 cm^{-1} for bone and 1019 cm^{-1} for Bio-Oss, with no secondary higher frequency peak. We highlight that although ATR is a very useful method to analyze bone mineral, care should be taken when comparing the obtained spectra parameters with reference values described using KBr pellets, and especially when analyzing the $?_3$ phosphate peak.

Carbonate in bone mineral: Discriminating the structural changes simultaneously imposed by carbonate in A and B sites of apatite

Mary Tecklenburg, Honey Madupalli

The mineral component of bone, teeth, and other biological calcifications is primarily a carbonate substituted calcium apatite. Integration of carbonate into two sites, substitution for phosphate (B-type carbonate) and substitution for hydroxide (A-type carbonate), influences the crystal properties which relate to the functional properties of bone. The extent of carbonation controls biomineral properties such as crystallinity and solubility. Synthesized carbonated apatite is commonly used to model the mineral phase of hard tissues and explore its properties as a function of substitution site and level of carbonate in apatite. Spectral methods (IR, Raman) and diffraction (XRD) are ideal for characterization of mineral phases, crystallinity, and interaction between ions, in bone or in model materials. A series of AB-type carbonated apatites (AB-CAp) having varying A-type and B-type carbonate weight fractions were prepared. A complete characterization and B-site carbonate assignment in the FTIR ?3 region is proposed. The mass fractions of carbonate in each site of AB-CAp correlate differently with crystal axis length and crystalline domain size. In this series of samples, reduction in crystal domain size correlates only with A-type carbonate which indicates that carbonate in the A-site is more disruptive to the apatite structure than carbonate in the B-site. This characterization is consistent with the dominance of B-type carbonate in biological apatites.

Proteomic Approaches to Increasing the Value of Hair Shaft Evidence

Robert Rice, Pei-Wen Wu, Tempest Plott, Zachary Goecker, Glendon Parker

Exploiting recent advances in mass spectrometry, the proteome of the human hair shaft has now been established. The ability to identify proteins from their tryptic peptides has circumvented the intractability of isolating hair proteins cross-linked by isopeptide bonding into large aggregates. From the aggregates, >350 proteins were first identified a decade ago. The hair was prepared by vigorous treatment with sodium dodecyl sulfate and dithioerythritol at elevated temperature before alkylation with iodoacetamide. The protein was then isolated by ethanol precipitation and digested extensively with reductively methylated trypsin, solubilizing nearly 90% of it. This procedure works well using milligram amounts of protein, three orders of magnitude larger than necessary in view of the sensitivity of mass spectrometry. Reflecting the desire to analyze much smaller hair amounts, a modified procedure is now in use that avoids ethanol precipitation. Rather, the hair is treated with sodium dodecanoate (instead of sodium dodecylsulfate), which is removed by ethyl acetate extraction after acidification. Further modifications of the procedure are being explored to minimize exposure to high temperature and subsequent tryptic digestion time. Two approaches to mining the information in hair tryptic digests have been employed. First, profiles of protein expression have been compared to distinguish among individual hair donors. Originally suggested by our finding that inbred mouse strains are distinguishable by their hair protein profiles, this approach has been applied to characterization of hair disease in humans and mouse models and has been extended to corneocytes of normal and diseased skin. Results with monozygotic human twins indicate primarily a genetic basis for the differences in hair shaft protein expression level among individuals. Supporting eventual application to forensic investigation, the expression levels appear not to change much with age. A second approach to mining the information in hair digests focuses on detection of polymorphic peptides. Peptides reflecting single nucleotide polymorphisms in the DNA encoding the protein of origin can be identified and used to impute the genotype of the hair donor. This approach is envisioned to permit distinguishing individuals and conducting paternity analysis from hair samples with high precision.

Combing mass spectrometry protein analysis and DNA PCR-STR testing for contact traces

Mechthild Prinz, Steven Kranes, Stacey Ann Sterling, Glendon Parker, Katelyn Mason, Deon Anex, Bradley Hart

Forensic applications of human protein assays include body fluid and tissue identification, as well as genetic individualization. This type of peptide analysis will benefit from being combined with DNA identity testing on the same item. To facilitate this, we developed a trypsin based protein-DNA co-extraction method for contact traces that is scalable for different volumes and sample sizes. Fingerprint samples were deposited on non-porous surfaces such a glass slides and metal cartridge casings. Nickel, aluminum, steel and brass cartridges were tested before and after being fired. The incubation buffer consists of 0.01% Protease Max (Promega), and 5mM DTT in 50mM NH4CO3. The extraction proceeds with a 56oC denaturation step, and a 3 hour 37oC trypsin digestion and may include a concentration step with Microcon 100 membranes (Millipore). PCR-STR testing was performed with the Identifiler Plus kit (Thermo Fisher Scientific). Peptides were separated by reversed-phase liquid chromatography using Easy-nanoLC 1000 HPLC fitted with a Q Exactive Orbitrap mass spectrometer (Thermo Fisher Scientific). STR results were similar to Wan et al. 2015, with full profiles on glass, and DNA degradation and mostly insufficient DNA on metal surfaces, especially brass. Mass spectrometry analysis was successful for samples negative for DNA and detected between 838 and 1903 unique peptides on metal, and 1227-2309 on glass. As with contact DNA evidence, the probative value of peptide results will depend on background, transfer and persistence of proteins on forensically relevant items. We have started to assess background on surface control samples and will present strategies for the evaluation of transfer scenarios. [1] Wan, T. C. R., MacDonald, L., Perez, Y., Bille, T. W., & Podini, D. S. (2015). Recovering Touch DNA From Cartridge Casings Using a Method of Tape Lifting. In AAFS (p. B135).

"NextGen Serology: Leveraging Mass Spectrometry for Protein Based Human Body Fluid Identification

Phillip Danielson, Kevin Legg, Heather McKiernan

DNA profiling is often considered to be the "gold standard" for modern forensic testing. The characterization and confirmatory identification of the tissue source of the DNA, however, still presents numerous challenges for forensic serologists. Current antibody and enzyme activity based assays used by forensic practitioners for biological stain identification yield only presumptive results. Positive results with non-target body fluids or cross-reactivity with non-human sources has been well documented. Some serological tests can consume unacceptable quantities of precious evidence while for some body fluids there are simply no available tests at all. Comprehensive proteome mapping and comparative analyses by multidimensional HPLC in combination with Q-TOF mass spectrometry has successfully identified and verified a series of protein "biomarker panels" six body fluids (i.e., peripheral and menstrual blood, vaginal secretions, semen, urine and saliva). Subsequent research and development activities have focused on the development of a multiplexed serological assay for the single-pass identification of these body fluids which all have clear forensic relevance. Using the Agilent Technologies Agilent® AssayMAP Bravo automation platform for front end sample preparation and an Agilent 6495 triple quadrupole mass spectrometer coupled to a 1290 UHPLC for mass analysis, the resulting workflow is designed to meet the demanding needs of a forensic operational environment. To date, a 23-protein, 51-peptide panel for fluid identification has been developed with a run time of only 10 minutes. In conjunction with the automation platform, several hundred samples can be analyzed per week. A series of developmental validation studies are currently in progress to establish performance limits of the assay (reproducibility, repeatability, sensitivity, case-samples, etc.) and to inform the creation of standard operating procedures and reliable interpretation guidelines. The data generated to date demonstrate the utility of mass spectrometry as a unified platform for high sensitivity confirmatory body fluid identification. This technology has the potential to replace the diversity of test methods currently employed with a simpler and more standardized work flow for forensic body identification.

High-throughput Screening for Lipid Biomarker Discovery in Novel Arabidopsis thaliana Signaling Pathway

Rebecca Hansen, Hongqing Guo, Yanhai Yin, Young-Jin Lee

FERONIA is a receptor-like kinase in *Arabidopsis thaliana* that has critical roles in controlling growth and biotic/abiotic stress responses, and so the study of this signaling pathway may allow greater insight into the coordination of these important functions. This protein is highly expressed in wild-type (wt) plants, and the loss-of-function mutant (*feronia*) displays significant growth suppression, is more susceptible to bacterial infestation, and more resistant to drought conditions. In this work we applied multiple mass spectrometry (MS) techniques to identify lipid ligands involved in the FERONIA signaling pathway and to understand their biological roles. A direct-infusion FTICR-MS approach was used for initial high-throughput screening of wild and *feronia* plant extracts. A series of related features discovered to be more abundant in *feronia* in statistically significant amounts were confirmed to be Arabidopsides through MS/MS. Arabidopsides are known wounding response lipids, but have unknown biological significance. Interestingly, these lipids contain oxo-phytodienoic acid (OPDA) and dinor-oxo-phytodienoic acid (dnOPDA) side chains which are precursors of jasmonic acid (JA), an important plant hormone in stress response. These Arabidopsides, OPDA, dnOPDA, and JA were further studied using direct-infusion and LC-QTOF with and without wounding. Before wounding, higher levels of OPDA, dnOPDA, and Arabidopsides were detected in *feronia*. In both genotypes Arabidopsides and JA significantly increase with wounding, however this increase is more evident in *feronia* than wt. Based on these results, we hypothesize that *feronia* is in a constant stress state and can more rapidly respond to biotic/abiotic stresses, but at the cost of plant growth. To ascertain biological activity, Arabidopside A was isolated and applied to wt and *feronia* plants growing in a petri dish. Inhibition of root growth was observed in wt plants, but there was no significant change in *feronia*, indicating biological activity for Arabidopsides. Finally, matrix-assisted laser desorption/ionization MS imaging (MALDI-MSI) was used to visualize the localization of Arabidopsides, and in both genotypes we confirmed Arabidopsides are highly abundant at wounding sites. More significantly, five micron high-spatial resolution MALDI-MSI was performed for cellular localization of Arabidopsides in *feronia* which were found to be co-localized with chloroplast-related compounds, but only near vascular bundles.

Analytical solutions to the finite-pulse Bloch model for multidimensional coherent spectroscopy
Christopher Smallwood, Travis Autry, Steven Cundiff

Multidimensional coherent spectroscopy (MDCS) is a nonlinear spectroscopy technique exhibiting a uniquely sensitive ability to characterize dephasing rates, coherent coupling effects, and many-body phenomena in optically sensitive materials. In recent years, the technique has improved significantly in both sensitivity and resolution at optical wavelengths, sparking a growing need for theoretical models. To date, however, most approaches to modeling MDCS have either incorporated computationally intensive numerical techniques, or have adopted an impulsive analytical model in which excitation pulses are treated as having infinite bandwidth and are assumed to be of infinitesimal duration. In this talk, I will present recent work showing that the impulsive analytical model can be extended, without a great deal of difficulty, to be able to incorporate pulses of finite duration and bandwidth if the excitation pulses are assumed to have a Gaussian profile, and if dephasing and decay phenomena are modeled using exponential decays [1]. The incorporation of finite-duration pulses leads to a number of important effects that are not captured using a model restricted to infinitely short pulses, and contains a particularly noteworthy ability to characterize the roles of bandwidth, resonance, and pulse chirp in MDCS spectra. [1] C. L. Smallwood, T. M. Autry, and S. T. Cundiff, J. Opt. Soc. Am. B 34, 419-429 (2017).

Simulating vibrational spectroscopic experiments with Wilson

Magnus Ringholm

Ab initio computational tools in spectroscopy offer valuable opportunities for analysis and corroboration of experimental results and may aid in the design of new experimental techniques. Our program Wilson can be used for frequency-resolved simulation of arbitrary-order wave-mixing experiments involving vibrational but not electronic resonance for laser pulses in the infrared or ultraviolet/visible range. Wilson, using a recursive approach, will take information about a vibrational spectroscopic experiment specified by the user, determine expressions for the spectroscopic intensites, evaluate them using ab initio data available from software developed by our group or other sources, and finally render the resulting spectra. The wide range of experiments that fall within the scope of Wilson - including 2D-IR techniques -show a variety of appealing features such as direct probing of vibrational anharmonic effects, interface-specificity and tailoring of spectral selection rules, and may thus permit modes of analysis that are unattainable by traditional techniques. Furthermore, the large number of effects that may contribute in these processes and the close relation to other categories of processes offer several opportunities for future development and richer analysis, such as improved description of lineshapes stemming from dephasing and pulse delay effects, inclusion of high-order anharmonic contributions and extension to other categories of experiments. In this presentation, we show how the user-specified information is processed by Wilson to arrive at the spectral result and present data for selected small molecular systems. We show that the working expressions identified by Wilson contain various geometrical derivatives of polarization properties which can be calculated in a fully analytic ab initio manner by the OpenRSP program library, which is also developed in our group. We proceed to describe opportunities to extend the present functionality, with particular attention given to improvements in the description of lineshape effects.

Many-body theory of quasiparticle-resolving spectroscopies
Mackillo Kira

Laser spectroscopy delivers perhaps the most accurate control and characterization of matter. At the same time, the related quantum properties remain mostly unexplored because many-body interactions scramble the information behind macroscopic processes. I will briefly overview the latest theory insights on how photons can resolve and excite diverse quantum processes relevant for nanotechnology. I will explain the principles how to convert ordinary laser spectroscopy into quantum-optical spectroscopy in order to directly access complicated many-body states and processes. I will also present how different combinations of optical and terahertz spectroscopy can be utilized in solids when the fields become extremely strong. In particular, I consider situations beyond the perturbative regime of typical 2D spectroscopies, when all wave-mixing signatures become completely nonperturbative. As a major benefit, a strong field can move quasiparticles on time scales much faster than typical phonon and Coulomb scattering times. In this ultrastrong excitation regime, new phenomena can be directly observed, ranging from electronic quantum interferences, quasiparticle collisions, and nonlinearities among Landau electrons.

Round Table Discussion

Kyle Czech

Ultrafast multidimensional spectroscopies have been applied to investigate the dynamics of molecular and semiconductor systems. Although multidimensional spectroscopies provide a wealth of information about the evolution of excited states, the field has primarily provided qualitative investigations of excited state dynamics. The goal of this session is to permit participants to openly discuss the current limitations of ultrafast multidimensional spectroscopies and prospective developments that will improve the quantitative power of these measurements.

Spectral identification in the attogram regime through laser induced breadown spectroscopy of single optically-trapped nanoparticles in air
Javier Laserna, Pablo Purohit

Current trends in nanoengineering are bringing along new structures of diverse chemical compositions that need to be meticulously defined to ensure their correct operation. Few methods can provide the sensitivity required to carry out measurements on individual nanosubjects without tedious sample pre-treatment or data analysis. In the present study, we introduce a pathway for the full elemental identification of single nanoparticles that avoids suspension in liquid media by means of optical trapping and laser-induced plasma spectroscopy. We demonstrate spectroscopic detection and identification of individual Cu nanoparticles of masses down to 73 attograms and report, for the first time, stable optical trapping in air and manipulation of Cu particles from 25 to 70 nm in diameter. We found an increase in the absolute number of photons produced as size of the particles decreased; pointing towards a more efficient excitation of ensembles of only 7 x exp(-5) Cu atoms in the onset plasma.

Femtosecond LIBS with Optical Vortex Beams
Vassilia Zorba, Jason R. Becker, Xianglei Mao, Rick Russo, Costas P. Grigoropoulos

The use of ultrafast lasers in plasma spectroscopy offers high-resolution three-dimensional chemical imaging capabilities and suppressed matrix effects, leading to improved analytical performance, such as precision, accuracy and limit of detection. The advent of optical vortex beam technologies over the past few years has enabled their use in femtosecond laser ablation and structuring applications. Vortex beams carry orbital angular momentum and topological charge, thereby demonstrating differences in their interactions with matter as compared to traditional Gaussian beams, and leading to specific advantages in laser-induced surface and in-volume material structuring and modification applications. In this work, we introduce the use of optical vortex beams in chemical analysis and imaging applications involving femtosecond plasma spectroscopy, specifically Laser Induced Breakdown Spectroscopy (LIBS) for elemental analysis of solid samples. We study the influence of vortex beam wavefront helicity and polarization on the spatio-temporal plasma emission intensity and the laser-induced plasma properties in different classes of materials, including metals, dielectrics and semiconductors. We find that laser ablation sampling as well as the emission signals vary significantly based on the independently variable vortex beam parameters, thereby opening the possibility of using vortex beams for selectively tailoring ultrafast laser ablation-based chemical analysis technologies.

Calibration-Free Monte Carlo Method for Laser Induced Breakdown Spectroscopy

Igor Gornushkin, Alexander Kazakov, Ulrich Panne, Norbert Huber, Johannes Pedarnig, Simon Eschlböck-Fuchs, Roman Rössler

The improved Monte-Carlo (MC) method for standardless LIBS-analysis is presented. Concentrations are found by fitting model-generated synthetic spectra to experimental spectra. The algorithm is designed for a graphic processing unit (GPU) and allows for fast parallel computing and short analysis time ranging from seconds to minutes depending on a number of elements and lines in spectra analyzed. The algorithm is characterized using synthetic spectra; it shows high accuracy, precision, and robustness toward noise. The procedure is maximally unsupervised; an operator needs only to introduce spectral lines and their corresponding spectroscopic characteristics. The knowledge of Stark broadening parameters is unnecessary; line widths due to both natural (Stark, Doppler, resonance, etc.) and self-absorption (optical thickness) mechanisms are determined automatically. The performance is tested on slag samples consisting of CaO, FeO, MgO, MnO, TiO2, Al2O3, Cr2O3, and SiO2 oxides, standard reference materials, and other samples. Experiments show that GPU-based MC LIBS is capable of rapid multi-element analysis with relative error between 1 and 10's percent that is satisfactory for industrial and other applications. The results are positively compared to that obtained with standard calibration-free (CF) LIBS methods and to that obtained with XRF and other analytical methods.

Exploiting UV-femtosecond LIBS in single- and double-pulse modes for the sensitive analysis of aerosols and thin films

Demetrios Anglos, Konstantinos Marmatakis, Nikos Giannakaris, Panayiotis Siozos

The use of femtosecond LIBS is investigated in the context of analysis of liquid samples in the form of aerosols and solid thin film samples. Emphasis is placed on the study of the plasma properties when aerosol ablation is carried out in atmospheric air or with argon as a carrier gas. Spectral data are collected either in the single-pulse mode or by use of appropriately delayed pairs of pulses, considering that the latter has been shown to lead to distinct enhancement in the emission signals. Results will be shown from the analysis of metal ions in aqueous and organic solutions, including metal containing biomolecules. In addition, thin solid films are examined by use of a femtosecond LIBS microscope exploring optimum conditions for performing minimally invasive analysis of very thin layers of materials, motivated by the demand to analyze multi-layered films. A main challenge in this application is to achieve efficient analysis with just a single pulse as often the thin layer probed is fully removed by the ablating pulse.

Advances in LIBS and Challenges

Mohamad Sabsabi

Advances in LIBS and Challenges M. Sabsabi, P. Bouchard, A. Harhira, J. El Haddad, K. Rifai, F. Vanier, A. Moreau, A. Blouin. National Research Council Canada (NRC), Boucherville (QC), J4B 6Y4, Canada Corresponding author email address: mohamad.sabsabi@cnrc-nrc.gc.ca Laser-Induced Breakdown Spectroscopy (LIBS) technique has the advantage to analyze the material without contact, thus making it suitable for in the field and real time analysis of any type of materials, solid, liquid, slurries or gas. Because of its unique features, LIBS is currently a subject of great interest by several industrial sectors where there is a need for fast elemental analysis that cannot be addressed by conventional analytical methods. The driving force for this growth appears to be the unveiling of significant technological developments in the components (lasers, spectrometers, detectors) used in LIBS instruments, as well as emerging needs to perform real time measurements under conditions to which conventional techniques cannot be applied. The LIBS principle of operation is quite simple, although the physical processes involved in the laser–matter interaction are complex and still not completely understood. In this presentation, we will outline NRC activities in the development of LIBS technique and its applications for on-line measurements in various fields. Based on these applications, we will discuss the LIBS instrumentations in terms of robustness, analytical performance for the mining value chain and comparison to conventional techniques. We will highlight a LIBS case study for gold detection in ore samples for the mining industry.

Chemical denaturation of bovine serum albumin observed by DUV photoelectron yield spectroscopy combined with IR-laser ablation of droplet beam

Hiroya Asami, Tomoko Hasegawa, Jun-ya Kohno

Protein molecules exhibit unique structures for the intrinsic function in aqueous environment, which have been generally investigated by using X-ray crystallography and NMR spectroscopy. The protein structures are constructed through self-assembly and hydration. In order to elucidate the key factors to determine the protein structure, comparison of the structure in aqueous solution with that in the gas phase is really significant. Studies on structural changes in protein, so called denaturation, facilitate the understanding of a fundamental role of hydration to build a high-order structure of protein. In gas phase, a UV photodissociation or photoelectron technique combined with ESI-mass spectrometry has been frequently used for the structural analysis of protein. This method is effective for measuring the gas-phase UV absorption, whereas several problems are still lying, such as (1) limit of sample concentration, (2) thermal degradation of huge biomolecule, (3) lack of low-valence ions. In order to overcome the problems, we have developed a novel UV photodissociation or photoelectron spectroscopy by using IR-laser ablation of a droplet beam, in which lower-valence ions are effectively generated than the typical ESI method. Herein we have successfully isolated the low-valance negative ions of bovine serum albumin (BSA) and first measured a UV photoelectron yield spectrum at a deep UV region (192-208 nm). Two absorption peaks of native BSA, which are assigned to ?-?* transitions of peptide bonds in the ?-helix structures, are reported in UV absorption and CD spectra in the aqueous solution. The UV photoelectron spectrum obtained in this study also exhibits the characteristic two bands, which is analogous to the results in the liquid phase. On the other hand, the photoelectron spectrum of a chemically denatured BSA with cetyltrimethylammonium bromide (CTAB) shows a single peak in the same wavelength region, which can be attributed to the ?-?* transition of peptide bonds in the random coil structures. This result suggests that the native and the denatured BSA are successfully distinguished by the photoelectron spectroscopy in the gas phase. In the near future, our own spectroscopic technique will open up new doors in structural chemistry of protein denaturation.

Vibrational Spectroscopy Studies of the Interaction of Cytochrome c with Cardiolipin in Phospholipid Membranes

Jay Kitt, David Bryce, Joel Harris

Outer mitochondrial membrane (OMM) permeabilization is an integral step in apoptosis which results in the release of pro-apoptotic signaling proteins and leads to destruction of the cell through activation of the cysteine-aspartic protease (caspase) cascade. Despite the importance of the permeabilization process, the mechanism of OMM permeabilization remains unclear. A key step appears to be the translocation of the electron-transport protein cytochrome c from the inner-mitochondrial membrane (IMM) to the OMM just preceding permeabilization. This observation, combined with recent work demonstrating cytochrome c can induce pore formation in cardiolipin-containing phospholipid membranes leading to large dextran and protein permeability, suggests that association of translocated cytochrome c with the OMM may lead to permeability. In this work, the interaction of cytochrome c with cardiolpin-containing phospholipid vesicles, which serve as models of the OMM, is investigated to probe the nature of cytochrome c-induced permeability. Lipid vesicles with bilayers composed of either pure 1,2-dipalmitoyl-sn-glycero-3-phosphocholine (DPPC) or mixedDPPC/cardiolipin leaflets and which contained a membrane-impermeable Raman tracer, 3-nitrobenzenesulfonate (3-NBS), were optically trapped, translated into a solution containing cytochrome c, and monitored for 3-NBS leakage. Cytochrome c-correlated leakage was observed only in vesicles with cardiolipin-containing membranes. Structural changes observed in the Raman spectra during permeabilization of cardiolipin-containing membranes indicated acyl-chain disordering and less intense cardiolipin cis-double-bond stretching modes. Additionally, when the cardiolipin-vesicle-associated cytochrome c Raman spectrum is compared with a spectrum in buffer, heme-resonance Raman bands are absent indicating loss of Met-80 coordination associated with a change in protein conformation. To verify interactions of cytochrome c with cardiolipin are selective, these experiments were repeated where the DPPC acyl-chains were deuterated (D62-DPPC), allowing spectral resolution of DPPC acyl-chain Raman scattering from that of cardiolipin acyl chains. Interestingly, D62-DPPC acyl-chains were unaffected by cytochrome c accumulation, while cardiolipin showed major changes in acyl-chain structure. These results suggest that cytochrome c-induced permeabilization of cardiolipin-containing membranes proceeds through selective interaction of cytochrome c with cardiolipin, resulting in protein unfolding, where the unfolded form interacts with cardiolipin acyl-chains within the bilayer to induce permeability.

Structure and Stability of BioPharmaceuticals with Raman and ROA spectroscopy

Rina Dukor, Juanita Sanchez, Carolina Carballo, Laurence Nafie

Pharmaceutical industry has been using FTIR and RAMAN for variety of tasks from identification of raw material, formulation, processing and quality control. However, in biopharma, and specifically in structure elucidation, where the drugs are based on either peptides, proteins, carbohydrates or nucleic acids these techniques have been taking much longer to gain ground. FT-IR has become almost routine but Raman is in its very early stages of implementation. Its advantage is ability to detect the conformation of disulfide bonds and gain information from side-chains, in addition to secondary structure. More recently, several studies have shown an enhanced sensitivity of ROA (Raman Optical Activity) with differences observed when none are seen with any other spectroscopic technique. Raman optical activity (ROA) is a highly sensitive technique that combines the structure-rich detail of conventional Raman spectroscopy with chiral sensitivity of electronic circular dichroism (ECD or CD) [1]. Just as CD enhances the biological structure content of its parent UV absorption spectra, so ROA tremendously enhances the structure-sensitive content of conventional Raman spectra. Since its discovery over 40 plus years ago, ROA is now recognized world-wide, at first in academic research laboratories and increasingly throughout the pharmaceutical industry as technique that is significantly more sensitive to protein structure than either CD or Raman [2]. . The first report on the sensitivity of ROA to structure of biological drugs was published by Dr. Li and co-workers at Amgen [3]. In more recent studies, other companies including Merck [4] and BMS [5] confirmed the unusual sensitivity of ROA to enable incisive characterization of even minor changes in biologics structure and/or early detection of structural degradation and/or aggregation. In this lecture, we will describe the fundamentals of ROA and compare the sensitivity of the four forms of vibrational spectroscopy as applied to structural studies of proteins, with emphasis on predictability (and thus sensitivity) nature of ROA. References: 1. L. A. Nafie, "Vibrational Optical Activity: Principles and Applications", 2011, Wiley & Sons, Ltd. 2. L. D. Barron, "The Development of Biomolecular Raman Optical Activity Spectroscopy", Biomed. Spectrosc. & Imaging, 2015, 4, pp 223–253
3. C. H. Li and T. Li, "Application of Vibrational Spectroscopy to the Structural Characterization of Monoclonal Antibody and Its Aggregate", Curr. Pharm. Biotechnol., 2009, 10 (4), pp 391-399 4. G. Thiagarajan et. al., "Use of Raman and Raman optical activity for the structural characterization of a therapeutic monoclonal antibody formulation subjected to heat stress", J. Raman Spectrosc., 2015, 46 (6), pp 531–536 5. Mary E. Krause et al "Imapct of Stress-Sensitivity and Batch Variability in High Concentration Antobody Formualtion Development", 2015 AAPS National Meeting

Deconvolution of Vibrational Spectra for Analysis of Protein Secondary Structure
John Wasylyk, Mary Krause, Daniel Fichana, Ming Huang, Robert Wethman

The increase in biopharmaceuticals has generated considerable interest in enhancing the understanding of protein stability especially during the process development cycle. The knowledge gained through comprehensive studies involving stress on bioactive proteins is key to delivering a stable product to the patient. The ultimate goal is to develop and implement ideal transportation, delivery, and storage conditions to ensure maximum product integrity. As a result, there is a need for rapid analytical techniques for the purpose of characterization of secondary structural changes during stability studies and processing. Fourier transform IR (FTIR) spectroscopy is a non-destructive technique that has been used to determine the secondary structure of proteins. This is largely based on the Amide I and Amide II IR absorption bands at ~1700-1600 cm-1 and 1500-1400 cm-1. Furthermore, additional information, based on fine structural perturbations of key amino acids in the fingerprint region, provides an insight into the initiation of aggregation. In this study, several biopharmaceuticals were stressed at varying conditions, and the resulting FTIR spectra provided valuable information on structural changes in the amide bond regions and the fingerprint region. Changes involving specific amino acids such as phenylalanine, tyrosine and tryptophan were thoroughly vetted. Other analytical techniques were utilized such as dynamic light scattering, differential scanning calorimetry, size-exclusion chromatography, and Raman spectroscopy in order to add complementary data.

Biopharmaceutical Applications of ICP-MS

Lydia Breckenridge, Masano Huang, Mary Krause

As the landscape of pharmaceutical sciences shifts in focus from small molecules to biological medical products the analytical techniques employed to ensure the highest standards of product quality must too evolve. Biopharmaceuticals present different analytical challenges relative to synthetic small molecules owing to the complexity of their varied matrices. Solubility, potential interferences, corrosive buffers, protein aggregation and potential contamination from processing equipment or packaging materials are just a few of the variables which may pose difficulties to sample preparation or instrumental analysis. Inductively-coupled plasma-mass spectrometry (ICP-MS) has been employed as a technique for the determination of trace metals in various biological compounds. Various metals can be introduced into a biopharmaceutical formulation by a number of sources, including cell culture media, excipients, pre-filled syringes, water or stainless steel reactors. In some cases, these levels may not be controlled if the metal is not of toxicological concern and therefore not subject to regulated specifications. However, even part-per-billion levels of metals such as iron or chromium may result in degradation of a protein and thereby a loss in efficacy of the pharmacological action, therefore it is imperative that a metal control strategy be incorporated in formulation development. ICP-MS is a highly sensitive, selective and robust technique that provides the necessary low detection limits required for metal analysis in a number of challenging biological matrices. A number of applications of ICP-MS for the analysis of biopharmaceutical samples will be presented, including its use to monitor low-level metal leaching from stainless steel processing equipment under different stresses, to assess the efficiency of metal chelators in different biological formulations and to identify potential glass delamination from various packaging equipment. Results generated via ICP-MS have had a direct impact on biological formulation development and demonstrate that the technique is a crucial component in any biopharmaceutical analytical strategy.

Process Optimization Strategies for Cell Culture Bioreactors using Raman Spectroscopy and Multivariate Analysis

Karin Balss, Ann Edwards, Wan Su, Carl Rafferty, Dan Trout, Steve Mehrman, Priya Ramachandrula, Raghunath Shivappa, Olav Lyngberg

Efforts are underway to employ PAT tools as part of an advanced feed control strategy for enhanced process monitoring, robustness, and titer for mammalian cell culture bioreactors. Advantages of feed control include enhanced productivity, inhibition of lactate, longer batch times, reduction in glycation, and improved glycosylation control. Real time process monitoring for example can lead to faster, more efficient technical transfer activities with fewer demonstration batches as well as provide valuable insight during manufacturing investigations. Real time monitoring also has the opportunity to allow for mid-batch corrections. One example of a real time process monitoring tool is Raman spectroscopy combined with multivariate modeling to describe nutrients, metabolites, and biomass. Raman has been shown previously to be a robust industrial tool in a variety of industrial processes such as fermentations and cell cultures. The technique is highly chemically specific and scalable from the lab to commercial scale. In this study, we present the utilization of in situ Raman spectroscopy to quantitatively monitor multiple components, including nutrients and metabolites, over the course of a 2L scale fed batch process. Multivariate Partial Least Squares (PLS) regression models were developed to correlate real-time Raman spectra to off-line reference values. The models were then implemented for use in real time feed control strategies and used to compare two processes for product quality and titer. We seek to establish this and other continuous process monitoring tools to ensure consistent product quality and provide feed control strategy options during product process development.

Use of PAT in the Selection of a Design Space based on Process Understanding: A Case Study
Antonio Ramirez, Daniel Hallow, Michael *Fenster, Sha Lou, Nathan Domagalski, Srinivas Tummala, Sushil Srivastava, Lindsay Hobson*

This presentation describes the development of a control strategy for impurities in the final intermediate step of the Asunaprevir drug substance utilizing the concepts outlined in the International Conference on Harmonisation guidelines (ICH Q8 (R2), Q9, Q10, and Q11). Detailed mechanistic understanding enabled the construction of a kinetic model that was used in conjunction with a process risk assessment and well-defined quality attributes to guide the selection of the reaction design space. Implementation of continuous monitoring of the reaction using PAT facilitated the expansion of the design space and provided suitable parameter ranges to enable a robust process for commercial manufacturing.

Aldehyde detection, speciation, and quantification using surface-enhanced Raman spectroscopy
Mark Peterman, Samuel Kleinman, Merwan Benhabib

Aldehydes are important contaminants found in environmental and beverage samples. One selectic colorimetric approach for the detection of aliphatic C1-C3 aldehydes relies upon their reaction of aldehydes with 3-methyl-2-benzothiazolone hydrazone (MTBH), which forms a blue dye in the presence of $Fe3+$. Although this method is specific to aldehydes, it is not specific within the aldehydes. Formaldehyde, acetaldehyde, and propionaldehyde all form the same blue compound. We demonstrate here the use of surface-enhanced Raman spectroscopy (SERS) for the quantitative analysis and speciation of these aldehydes in aqueous solution. This approach relies upon MTBH to bind the aldehyde, while using SERS to detect the resulting compound. The three compounds resulting from the aldehyde reaction have Raman spectra that differ, even though all three compounds have the same color. Furthermore, by including an isotopologue as an internal standard, we can quantify the aldehydes. We demonstrate the detection and quantification of acetaldehyde down to sub-ppb detection limits. We also extend this approach to the detection of metaldehyde, a pesticide used against slugs and snails, through hydrolysis into acetaldehyde. Our approach suggests a general approach to chemical identification and quantification relying upon tried-and-tested colorimetric reactions coupled with SERS analysis. Many colorimetric reactions rely upon amine- and/or thiol-containing molecules. These compounds lend themselves to SERS detection at very low detection limits while opening the possibility for additional structural identification.

Label free approach to monitor ultra-violet radiation induced changes in skin cells
Surya Singh, Jeon Woong Kang, Ramachandra Rao Dasari

Chronic ultraviolet radiation (UVR) exposure from the sun or other sources is a known cause for the development of photo-aging, actinic-keratosis and melanoma. Novel and alternate diagnostic strategies are being explored to assess photo-oxidative toxicity in early stages. In contrast to routinely employed methods, recently developed label free approaches can provide detailed analysis of single cell in an objective and quantifiable manner in their natural state. In the present study, we have used quantitative phase imaging and Raman spectroscopy as complementary live cell analysis tools to evaluate the efficacy in identifying changes induced by minimal time bound UVR exposure. Normal skin (Hs895.Sk) fibroblasts were used. Sub-lethal doses and optimal exposure time were determined using cellular viability assay. Spectra were acquired using a custom-built confocal Raman microscopy and phase microscopy system. Spectral differences in the Raman fingerprint region indicates differences in the protein and nucleic acid composition. These differences were successfully utilized to develop an automated classification model based on principal component analysis. In addition to this distinct changes in the cellular morphology were observed and validated through quantitative phase imaging. Significant dose dependent differences in different biophysical parameters such as dry mass, cell height and matter density were observed. Combination of these two techniques, one suited for detection of subtle morphological/biophysical alterations while the other appropriate for capturing molecular perturbations, could pave the way to address issues of label-free monitoring of cellular responses in response to an external stimulus. These findings can provide an accurate understanding of different markers associated with radiation damage and would assist in providing a quantitative tool to our future studies on designing alternate diagnostic tools.

Substrate Evaluation Using Single Molecule SERS
Alexandre Brolo, MuYang Zhang

Surface-enhanced Raman scattering (SERS) is a powerful spectroscopic method, capable of single molecule (SM) level detection. A large number of publications addressed the fabrication of a variety of SERS substrates, which are the platform that support the effect. Traditionally, the enhancement factor (EF) is the quantity used as an indicator of substrate quality. However, other parameters might also provide interesting insights into the characteristics of a particular substrate. In this work, we explored the stochastic characteristic of SM-SERS to propose alternatives approach to evaluate a SERS substrate. Six evaluation criteria was implemented and compared. Interestingly, some substrates exhibited better performance in terms of one criterion but worse in terms of others. Those variations in performance can be explained by the plasmonic characteristics of the surface.

Raman optical activity of biofluids: Molecular signature of health and disease?
Vladimír Setnicka *, Lucie Habartová*

Raman optical activity (ROA) has matured into a powerful tool for studying the structure, molecular geometry and conformation of chiral molecules (1). However, it has never been systematically employed for the analysis of complex biological samples as biofluids, except within our recent studies (2–5). The situation is similar for other methods of chiroptical spectroscopy – electronic (ECD) and vibrational circular dichroism (VCD). We address the practical aspects of measuring the ROA of human blood plasma, probably one of the most important biofluids for a future sensitive screening of large populations at risk of various diseases. However, human blood plasma contains hundreds of molecules, many of which often exhibit a spectrally broad fluorescence background of large amplitude, primarily below the ROA and Raman spectra. Thus, we developed a methodology to minimize the fluorescence effect and to improve the reliability of the ROA and Raman data (2). Using a combination of ROA with other chiroptical methods supplemented by unpolarized Raman and infrared spectroscopies, we monitor spectral pattern differences that reflect the conformational and stereochemical changes of chiral biomolecules circulating in the blood plasma of patients with serious diseases – cancer, neurodegenerative and metabolic (3–5). We illustrate our approach through multivariate statistical analyses of spectral data obtained in studies of pancreatic cancer and Alzheimer's disease. Our results suggest a promising perspective of ROA to reveal disease-induced pathological changes in the native states of chiral biomolecules. The financial support by the Ministry of Health of the Czech Republic (No. 16-31028A) and the Czech Science Foundation (No. 17-05292S) is gratefully acknowledged. The work was partially supported also by Specific University Research (MSMT No. SVV-20/2017), Operational Program Prague – Competitiveness (No. CZ.2.16/3.1.00/21537 and CZ.2.16/3.1.00/24503) and National Program of Sustainability (NPU MSMT No. LO1601 – 43760/2015). References 1. Berova, N.; Polavarapu, P. L.; Nakanishi, K.; Woody R. W. (eds.): Comprehensive Chiroptical Spectroscopy; Wiley: New Jersey, 2012. 2.Tatarkovi?, M. et al.: Anal. Bioanal. Chem. 2015, 407, 1335. 3. Š?oví?ková, L. et al.: Analyst 2015, 140, 2266.4. Tatarkovi?, M. et al.: Analyst 2015, 140, 2287. 5. Synytsya, A. et al.: Anal. Bioanal. Chem. 2013, 405, 5441.

Raman optical activity: From carbohydrates to pharmaceuticals
Václav Profant, Petr Bouř, Vladimír Baumruk

Methods of vibrational optical activity - Raman optical activity (ROA) and vibrational circular dichroism (VCD) -provide rich structural information and high specificity to the spatial arrangement of biomolecules and other biologically active compounds. In our contribution we focus on recent discoveries demonstrating the applicability of these methods to studies of various conformational related phenomena in solution. Chondroitin sulfate (CS), one of the most common glucosaminoglycans, is a linear carbohydrate polymer responsible for many of the important biomechanical properties of cartilage. So far, there is still uncertainty about nature of CS secondary structure in solution and its alteration by different type of sulfation. We have employed thorough experimental study of several natural differently sulfated CS samples and inspected changes in their Raman and ROA spectra caused by modifications of physico-chemical properties. Based on study of chondroitin basic monosaccharide and repetitive disaccharide units together with their sulphated analogs, we were able to describe the effect of diverse sulfation on chondroitin structure. Despite the recent improvements the simulations of VCD and ROA spectra of monossacharides remain challenging and usually require averaging of high number of explicitly solvated MD snapshots. We present an improved approach based on the initial cluster analysis of MD trajectory which leads to the substantial decrease of amount of necessary snapshots. On the model of dimethyltartarate we demonstrate comparison between experimental and simulated spectra in various solvents (carbontetrachloride, chloroform, water, DMSO, acetonitrile) inducing different hydrogen bonding. Taxol is one of the best-selling medicaments used in the treatment of ovarian, lung, and other cancers. In a low yield, it may be isolated from the bark of the Pacific yew tree; this is however insufficient for industrial production. We investigated the performance of ROA spectroscopy in distinguishing four possible stereoisomers of the synthetic precursor used for the production of taxol from baccatin III. It is shown that for reliable comparison of the experimental and computed Raman and ROA intensities a large number of conformers had to be averaged, to properly account for molecular flexibility in solution. If combined with simulations, ROA spectra provide convenient and economic means of absolute configuration determination.

Quantitative analysis of Raman optical activity spectra for the assignment of polysaccharide structure in the gel phase

Steffen Lüdeke, Anja Rüther, Aurelien Forget, Anjan Roy, Carolina Carballo, Florian Mießmer, Rina K. Dukor, Laurence A. Nafie, Christian JohannessenV. Prasad Shastri

The polysaccharide agarose forms stiff hydrogels. However, the introduction of carboxylic acid groups into the polysaccharide backbone leads to soft hydrogels. This change in mechanical properties was assumed to be associated with a change in higher order structure of the polysaccharide scaffold. While agarose is well known to form ?-helices and bundles of double ?-helices, until now the chain organization of carboxylated agarose in a hydrogel has remained unclear. We applied Raman optical activity (ROA) on gel samples of agarose and agarose with different degrees of carboxylation (28, 40, 50, 60, 80 and 93%). The observed carboxylationdependent spectral changes could be clearly attributed to higher order structure. We employed matrix least squares global fitting for the deconvolution of the spectra and identified two spectral species that interconvert into each other as a function of carboxylation. Comparison to ROA spectra from quantum chemical calculations allowed to assign them to double helical structure in agarose and ?-strand-like conformation in the carboxylated derivatives. Our results suggest that 40% of the polysaccharide are already single stranded in native agarose and that carboxylation fully inhibits chain organization as double strands. The interchaininteractions of ?-stranded polysaccharides seem to be much weaker, thereby leading to softer hydrogels.

Raman optical activity of the hydrogen out-of-plane mode is a sensitive probe of chromophore distortions in a photoreceptor protein

Masashi Unno

Raman optical activity (ROA) measures the different intensity of right- and left-circularly polarized Raman scattered light from chiral molecules and provides information on dynamic changes in stereochemistry associated with vibrational modes. We have constructed a ROA spectrometer using near-infrared laser excitation at 785 nm, which enables us to measure ROA spectra for a variety of biological samples, including photoreceptor proteins, without fluorescence from the samples. In previous works (1-4), we applied the ROA spectroscopy to some photoreceptor proteins and showed that this technique has a capability to detect structural deformations of light-absorbing chromophores embedded in a protein moiety. In the present study, we measured the ROA spectra of a bacterial blue-light receptor, photoactive yellow protein (PYP) that has p-coumaric acid as a chromophore. The ROA spectra of PYP exhibit several intense bands below 1000 cm–1, which are suggested to be hydrogen out-of-plane bending modes. We have assigned the characteristic ROA bands of PYP by using an isotopically labeled sample. We also examined structural factors that affect the ROA signals by quantum chemical calculations based on density functional theory. 1) Unno, M.; Kikukawa, T.; Kumauchi, M.; Kamo, N. J. Phys. Chem. B 2013, 117, 1321-1325. 2) Shingae, T.; Kubota, K.; Kumauchi, M.; Tokunaga, F.; Unno, M. J. Phys. Chem. Lett. 2013, 4, 1322-1327. 3) Kubota, K.; Shingae, T.; Foster, N. D.; Kumauchi, M.; Hoff, W. D.; Unno, M. J. Phys. Chem. Lett. 2013, 4, 3031-3038. 4) Haraguchi, S.; Hara, M.; Shingae, T.; Kumauchi, M.; Hoff, W. D.; Unno, M. Angew. Chem. Int. Ed. 2015, 54, 11555-11558.

Studying the various structural forms of alpha synuclein by means of Raman optical activity

Christian Johannessen, Carl Mensch

Alpha synuclein is a protein found in abundance in the human brain. Upon aggregating, the protein forms Lewy bodies associated with Parkinson's Disease. Thus, alpha synuclein falls in the category of neuropathological fibril forming proteins, together with for example Tau protein and amyloid beta. The protein is believed to exist in an intrinsically disordered protein (IDP) native form in the brain, but has been observed to adopt traditional secondary structure elements under different physiological circumstances, even earning the nick name "the chameleon protein". Thus, both with respect to understanding the native functions of alpha synuclein, which are still not fully understood and also with respect to disease pathology, it is essential to discern what structural motifs may lead to different behaviors. In this presentation, Raman optical activity (ROA) spectroscopy is employed to study the different structural motifs of alpha synuclein. ROA is a well-established tool for studying solution phase structures of proteins[1,2], and has previously been employed to study alpha synuclein in the IDP form[3]. Here, we present ROA spectra of alpha synuclein in the IDP form, as a mainly alpha helical protein on the surface of SDS micelles, and a beta sheet protein and molten globule in varying concentrations of trifluoroethanol (TFE)[4]. Furthermore, structural differences on full length and truncated alpha synuclein are identified using ROA. Thus, we show that ROA can be used as a stand-alone technique to study all the different forms of this flexible protein. References: 1. Barron, L.D. Biomed. Spectrosc. Imaging 2015, 4, 223-253. 2. Mensch, C., Barron, L.D., Johannessen, C. Phys. Chem. Chem. Phys. 2016, 18, 31757-31768. 3. Syme, C.D., Blanch, E.W., Holt, C., Jakes, R., Goedert, M., Hecht, L., Barron, L.D. Eur. J. Biochem. 2002, 269, 148-156. 4. Mensch, C., Konijnenberg, A., Van Elzen, R., Lambeir, A-M., Sobott, F., Johannessen, C. J. Raman Spectrosc. 2017, DOI: 10.1002/jrs.5149

Scalable nanofabrication of cost-effective 3D plasmonic chips for point of care applications

Wei-Chuan Shih

Abstract: Cost-effectiveness has been a key factor in point of care applications. In this talk, I will discuss our strategy to fabricate robust 3D plasmonic chips that feature intense and high-density hot spots, large surface area, and high structural integrity and reproducibility. Using localized surface plasmon resonance (LSPR) and coupling modes, enhanced light-matter interactions near the nanosurface can be realized. The surface enhancement results in stronger signal which reduces the burden on high-end optical detection systems. We have applied them to several analytical Chem/Biosensing platforms for a range of sensing targets by various spectroscopic and imaging techniques. Nanoplasmonic sensors appear to provide potential solutions in a range of applications from precision medicine to point-of-care diagnostics and wearable technologies.

Manipulating Plasmonic Hot Spots as a Strategy for Enhanced Clinical Diagnostics
Marc Porter, Nicholas Owens, Jason Beck, Michael Granger

Plasmonic hot spots play a vital role in the amplification of the signal strength in surface-enhanced Raman scattering (SERS) and various other forms of the so-called enhanced spectroscopies. This presentation investigates the role of nanoparticle-nanoparticle and nanoparticle-substrate interactions on the SERS response in immunoassays that capture biolytes on a smooth gold or carbon substrate which has been modified with a layer of antibodies and then tag the captured biolytes with a Raman-active gold nanoparticle also coated with a layer of antibodies. Work aimed at the development of an approach that extends this labeling approach to using surface-enhanced resonance Raman scattering (SERRS) in which the resonance species is immobilized as a coating between the substrate and capture antibody is also described. Finally, the advantages and limitations of both approaches as platforms for the low-level detection of biolytes will be analyzed and future prospects examined.

A biogenic silver nanoparticle based assay for improved cancer diagnostics
Mark McDermott, Sunil Rajput

Prostate cancer (PCa) is the most commonly diagnosed cancer in North American men; accounting for 11% of all cancer types. A prominent method to improve diagnosis and treatment of cancer is to target a reliable biological factor or marker that predicts a patient's response (sensitivity or resistance) to a specific treatment. Two major hormonal therapies for PCa are Enzalutamide and Abiraterone; both of which prevent the binding of a steroid (androgen) to its protein receptor (androgen receptor, AR). Recently, the presence of a receptor variant (AR-V7) has been linked to resistance of these hormonal therapies. We are developing a nanomaterial enhanced antibody sandwich assay for AR-V7 in the blood of prostate cancer patients. Highly sensitive detection will be driven by the optical properties of noble metal nanoparticles. The biomarker will be captured onto a chip containing specific antibodies and subsequently labeled for detection by antibody coated silver nanoparticles. The relationship between the number of nanoparticles on the chip and the presence and amount of AR-V7 will be established. We will describe our efforts to incorporate biogenic spherical silver metal NPs, produced by the fungus Fusarium oxysporum, as plasmonic labels for enhanced spectroscopy detection. Biogenic silver NPs are stabilized with a surface corona that is unique relative to their chemically prepared counterparts.. This method produces particles that contain a native biological scaffold surrounding the nanoparticles that provides enhanced stability and enables facile antibody attachment. Measurement of the number of nanoparticles that bind to the chip will be accomplished using surface-enhanced Raman scattering (SERS) which provides a unique chemical fingerprint of the label. Raman spectroscopy (e.g. SERS) is emerging as a technique that can be translated into the diagnostic clinic as a sensitive method that provides rich biochemical information in real-time, non-invasively, and in an automated manner. Furthermore, the biomarker will be measured in blood or urine and will not require an invasive tissue biopsy.

Field-deployed SPR sensors environmental applications

Jean-Francois Masson, Thibault Brulé, Geneviève Granger, Natalia Bukar

In this presentation, the successes and challenges associated with the deployment of SPR sensors for the analysis on site of environmental samples will be exposed through the example of a sensor for RDX environmental contamination of underground wells. We applied a portable SPR instrument for sensing on site the presence of RDX in underground water samples of contaminated wells near a Canadian military base. The SPR sensor was modified with a molecularly imprinted polymer selective for RDX with a detection limit in the low fM range. The sensor was thoroughly characterized with a series of SEM, AFM, and ellipsometry measurements. The instrument was deployed in winter conditions on a sled (lab-on-a-sled) and in summer conditions using a Jeep (lab-on-a-Jeep). The results obtained on site correlated with the accuracy of better than 1.6 ppb with the results obtained with the lab method of the EPA (EPA method 8330b). These results show the potential of SPR for monitoring environmental contaminants on site in actual environmental conditions.

Localized surface plasmon resonance for studying the immobilization of liposomes on sensor surfaces
Susanne Wiedmer, Joanna Witos, Giacomo Russo, Filip Dusa, Wen Chen

Indirect nanoplasmonic sensing (NPS) is a novel microanalytical technique based on the optical phenomenon of localized surface plasmon resonance of metallic nanostructures. In indirect NPS experiments the plasmonic particles are inert and adjacent to the material of interest to probe a process occurring either in the material investigated or on the surface of the material. Changes in the nanoparticle properties cause a shift in the maximum-extinction wavelength, which can be monitored and recorded in real-time. Among the notable advantages of NPS are the simple instrumentation and the robust and noninvasive methodology, which is very flexible regarding the type of material investigated. In addition, highly surface-sensitive measurements can be carried out with great temporal resolution. NPS has been successfully applied to the development of biosensors, biochemical processes, and to the binding characterization of polymers including kinetic measurements. We have studied the immobilization of lipid vesicles on various NPS surfaces. Unilamellar vesicles composed of mixtures of synthetic phospholipids or of lipids extracted from natural membranes have been compared. Vesicles adsorb and self-assemble to form different adsorption patterns on sensors, depending on the surface properties and the composition and size of the vesicles. The type of pretreatment of the sensor is highly essential for the adsorption process. The increasing use of ionic liquids in industrial processes has resulted in strong demands on achieving a better understanding of the toxicological effects of ionic liquids. Therefore, in this work we have studied the interactions between ionic liquids and lipid membranes, using a wide range of different ionic liquids. Both harmless and harmful ionic liquids, by means of their effective concentrations measured on Vibrio Fischeri bacteria, have been included. The effect of ionic liquids on lipid membranes is highly dependent on the hydrophobicity and molecular structure of the ionic liquid. In addition, the interaction of various surfactants with liposomes adsorbed on the surface was studied in detail. In general, this project will lead to a better understanding of the immobilization process of lipid vesicles on NPS sensor surfaces, as well as on the possible harmfulness of novel ionic liquids on lipid membranes.

Accessing the relevant temporal and lateral scale of interfacial electrochemistry through multimodal approach
Vijay Murugesan, Kee Sung Han, Karl T Mueller

In advanced battery systems, electrode-electrolyte interfacial processes span extraordinarily wide both spatially—ranging from one-tenth of a nanometer up to one-hundred nanometers—and temporally—lasting from nanosecond up to a few minutes. Gaining an in-depth understanding about these interfacial process under realistic battery operating conditions requires multi-modal probes that integrate theoretical computation methods with multiple analytical methods having the necessary spatial and temporal resolution. Joint Center for Energy Storage Research (JCESR) an energy innovation hub, has brought together a wide range of characterization capabilities into a focused, multi-modal characterization protocol through inter-laboratory collaboration that will enhance the accessing the wide span regions of interest in Li-S battery, which can accelerate the discovery and optimization of electrode and electrolyte materials. This presentation will highlight the development and implementation of multi-model characterization that includes advances and synergies in in-situ X-ray photoelectron spectroscopy (XPS), X-ray absorption spectroscopy (XAS) and nuclear magnetic resonance (NMR) combined with ab initio molecular dynamics (AIMD) computational modelling studies focused on complex interfacial interactions in Li-S batteries.

Liquid-jet X-ray Photoelectron Spectroscopy as a Probe of Liquid/Gas and Liquid/Solid Interfaces
John Hemminger

We have combined liquid-microjet technology with ambient pressure x-ray photoelectron spectroscopy (LJ-APXPS) to study the liquid/vapor interface of aqueous solutions. The use of a synchrotron x-ray facility provides us with the ability to obtain spectra of the liquid/vapor interface as a function of x-ray photon energy. Changing the x-ray photon energy allows us to obtain a depth profile from the surface into the bulk of the solution since the photoelectron kinetic energy increases with increased photon energy, thus increasing the experimental probe depth. We have shown that most solutes (even very soluble ones) segregate to the surface of a water solution at low solute concentrations. A range of molecular and ionic solutes will be discussed in detail. Molecular dynamics simulations of the solutions provide a molecular level picture of how the solvation differs between the bulk of solution and the liquid/vapor interface.

Spectroscopy and Microscopy Investigation of Interfacial Processes Affecting Uranium in Abandoned Mines

Jose Cerrato, Sumant Avasarala, Lucia Rodiguez, Adrian Brearley, Kateryna Artyushkova, Johanna Blake

The interaction between solid and water chemistry on the reactivity of metals in mine wastes from two sites in the Southwestern US was investigated using microscopy and spectroscopy. Metal release from these mine wastes could pose potential health risks for neighboring communities. Spectroscopy analyses on the first site located Northeastern AZ suggest that U-V phases are present in abandoned mine wastes from; the dissolution of these U-V phases is key to understand U transport. Electron microprobe and electron microscopy analyses on the second site located in Laguna, NM suggest that U-Si and U-P phases on mine wastes from which could be a source for U in neighboring surface waters. The presence of 80% U(VI) and 20% U(IV) was detected with X-ray photoelectron spectroscopy (XPS), and nano-particulate coffinite was identified in mine waste samples using X-ray diffraction (XRD). Accumulation of U in roots of salt cedar plants was also investigated. This presentation will emphasize the importance of interfacial interactions on the transport of metals in waters adjacent to abandoned mine waste sites.

EnviroESCATM – Routine surface analysis under environmental conditions

Thomas Schulmeyer

For many decades, XPS (or ESCA) has been the well-accepted, standard method for non-destructive chemical analysis of solid surfaces. To fulfill this task, existing ESCA tools combine reliable quantitative chemical analysis with comfortable sample handling concepts integrated into fully automated compact designs. Over the last few years, it has been possible to develop XPS systems that can work far beyond the standard conditions of high or ultrahigh vacuum. Near Ambient Pressure (NAP) XPS has become a fast-growing area of research, inspiring many scientists to apply the technique to new fields. Thus, by crossing the pressure gap, new insights in complicated materials systems have become possible using either synchrotron radiation or laboratory X-ray monochromators as excitation sources under NAP conditions. Based on this experience, SPECS Surface Nano Analysis GmbH has developed a revolutionary tool to bring to fruition the long existing dream in many analytical laboratories: reproducible chemical surface analysis under any environmental condition. EnviroESCATM allows for different applications such as extremely fast solid surface analysis of degassing (but also non-degassing) samples, ESCA analysis of liquids or liquid-solid interfaces, chemical analysis of biological samples, and materials and device analysis under working conditions. After introduction of the technological realization, a comprehensive survey of results will be given on standard solid conductive samples under different pressure conditions, bulk insulators with environmental charge compensation applied, high throughput analysis of batches of similar objects, geological samples, biological specimen from plants and animals, chemical analysis of pharmaceuticals, and the comparative analysis of ultrapure liquid water with different aqueous solutions. The influence of the ambient conditions on quantification will be demonstrated and discussed. Finally an outlook is presented on the application to electrochemical and other in-operando devices.

In Situ Chemical Imaging of the Evolving Material Interface in Liquids

Xiao-Ying Yu

A vacuum compatible microfluidic reactor, SALVI (System for Analysis at the Liquid Vacuum Interface) was employed to study the evolving material interface in liquids. Two examples will be presented. The first study is a model switchable ionic liquid (SWIL) system consisting of 1,8-diazabicycloundec-7-ene (DBU) and 1-hexanol. When CO_2 gas is added to the DBU and 1-hexanol mixture, the solvent polarity is known to change. A series of ionic liquids with different CO_2

loading was analyzed. Spatial chemical differences were observed within the same ionic liquid, indicating inhomogeneity of the ionic liquid. Spectral principal component analysis (PCA) was conducted. Clear distinctions were observed among SWILs with different CO2 loadings. The loading plots strongly indicate that fully loaded SWILs share similar spectral components as those of the non-loaded ILs. This finding confirms the hypothesis of the biphasic structure in the fully loaded IL predicated by molecular dynamic simulation and presents the first physical evidence of the liquid microenvironment of IL determined by liquid ToF-SIMS. In the second example, we investigated the formation of metal organic framework (MOF) using in situ imaging liquid SIMS. Zn-MOF-74 is the model system. Zn-acetate is the metal center and DHTA is the ligand linker in a DMF solvent. MOFs formed at different times were analyzed to ascertain the growth mechanism. Ex situ XRD and HeIM/SEM are also used to characterize MOFs formed over different time periods to complement the in situ analysis. The vacuum compatible microchannel in SALVI offers an innovative perspective to study the evolving liquid-liquid and solid-liquid interface in liquid. Our approach directly visualizes spatial and chemical heterogeneity in complex liquids by dynamic ToF-SIMS and provides new insights for improved understanding of the evolving material interface using imaging mass spectrometry.

Development of Ultraviolet Photodissociation Mass Spectrometry for Characterization of Proteins
Jennifer Brodbelt

Activation of ions by absorption of UV photons is an alternative to conventional collisional activation methods that affords fast, high energy deposition. Ultraviolet photodissociation (UVPD) results in broad sequence coverage of intact proteins via more extensive backbone fragmentation than can be obtained using other MS/MS methods, and ion activation/dissociation can be accomplished using a single 5 nanosecond laser pulse. This translates to an MS/MS technology that be used to identify proteins, including mapping post-translational modifications, and can be implemented for high throughput LCMS workflows. There has also been growing interest in employing top-down approaches to directly characterize native-like protein structures, primarily using MS/MS methods to disassemble the complexes, sequence the proteins, and draw conclusions about protein conformation based on fragmentation behavior. UVPD offers high levels of sequence coverage for native-like proteins in a manner that mirrors the B-factors of proteins, suggesting that fragmentation depends on the intramolecular and intermolecular interactions that stabilize particular regions of the proteins or protein-ligand complexes. Products retaining non-covalently bound ligands are also observed. For multimeric protein complexes, UVPD disassembles the complexes in a manner that reflects sub-unit architecture.

Using Microchip Electrophoresis and Electrochemical Detection to Investigate Cellular Communication
R. Scott Martin

Our lab has long been interested in the development of robust microfluidic devices that integrate cell culture and analysis, with electrophoresis being used to separate molecules of similar structure and electrochemistry utilized for detection of neurotransmitters/modulators in close-to-real-time. This includes the integration of PC 12 cell culture with microchip-based electrophoresis (to monitor stimulated exocytosis of multiple neurotransmitters) as well as creation of an endothelium mimic with integrated electrodes for nitric oxide detection. This talk will detail recent technology we have developed around these applications. This includes the fabrication and use of polystyrene-based microfluidic devices, where encapsulation of materials (tubing and electrodes) can be used to improve cell viability and separation performance. We will show that a unique feature of this encapsulation approach is the ability to create planar membranes from 3-dimensional pillar electrodes. Work towards using this type of device to monitor cell-to-cell interactions will be discussed. Finally, more recent work on the development of more realistic, 3-dimensionsal cell culture models for microfluidics will be presented. This includes the use of 3D-printing to create an air sheath device to dynamically focus electrospun fibers into a fully closed fluidic channel, with a fibrous scaffold being created on the inner channel wall. We will discuss these results and how electrospinning can be used with a modular microfluidics approach to create more realistic and robust approaches for investigating cellular communication.

Online Elemental Analysis of Airborne Nanoparticles
Murray Johnston, Justin Krasnomowitz, Andrew Horan

Understanding the formation, fate and impact of airborne nanoparticles requires analytical methods that can detect and quantify very small amounts of material with fast time resolution. Our group has developed a Nano Aerosol Mass Spectrometer (NAMS) for characterization of ambient particles between 10 and 100 nm in diameter. This instrument performs quantitative elemental analysis on a particle-by-particle basis by irradiating individual particles with a high energy laser pulse. Particles are disintegrated into electrons and positively charged atomic ions, characteristic of a plasma ionization mechanism. Quantification is performed by summing the signal intensities of all ions detected for each element. Particle-to-particle variations of the mass spectra appear to arise from plasma energetics – lower charge states are enhanced in some spectra while higher charge states are enhanced in others. The relative proportion of low vs. high energy spectra is particle size dependent. In quantitative measurements, precision is determined by the number of particle mass spectra averaged, and the time resolution is determined by the amount of time needed to acquire the requisite spectra. Accuracy appears to be dependent on the kinetic energy distribution of the ions produced, which determines how efficiently they are transmitted through the mass analyzer. In most cases, elemental mole fractions greater than ~1% can be determined within 10% of the true values, and the total amount of material analyzed is less than 1 femtogram. Fundamentals of particle sampling and analysis will be presented along with applications to laboratory and field measurements.

Airborne laser spark ionization

Jens Riedel, Andreas Bierstedt

A novel ionization scheme for ambient mass spectrometry is presented and discussed. Desorption and ionization are achieved by a quasi-continuous laser induced plasma (LIP) ignited in front of the atmospheric pressure interface of a time-of-flight mass spectrometer. This setup comprises the advantages of i) an ambient probe, ii) electro neutrality, iii) low power consumption, iv) a sufficient duty cycle, v) a ubiquitous plasma medium (air) and vi) high sensitivity caused by the high electron number densities. The plasma properties and operating conditions are investigated to understand the processes that lead to the formation of molecular ions. Comprehensive studies include optical emission spectroscopy, shadowgraphic shockwave visualization and time-of-flight mass spectrometry investigations. The observed MS signal (including reagent-ion signal, as well as analyte spectra) closely resembles the ionization behavior of other electrically-driven plasma-based ionization sources, such as DBD, LTP or DART. Conversely that means that although LIPs are commonly known to have temperatures way above 10.000 K and thus efficiently atomize and ionize molecules, MS spectra were obtained that showed the formation of intact molecular ions. For an insight into the LIP properties the optical emission of the LIP is examined and exhibits a pronounced degree of dissociation into atomic and ionic species, which reflects the higher temperature and electron density inside a LIP. Accordingly, rather elemental (ICP-type) than molecular spectra should be expected. However due to the absence of such ions, we conclude that the analyte does not enter the hot center of the plasma itself. Moreover, the initial, highly-excited plasma species react cascade-like to lower energetic species via reactive collision with the surrounding atmosphere, which can then subsequently ionize molecules of interest. The concept potentially involves charge transfer, proton transfer, electron impact and photoionization. These assumptions are confirmed via the shadowgraphs of the expanding shockwave around the LIP. An effective spatial separation between the two regions is maintained by concentrically expanding pressure waves resulting in a strongly unidirectional diffusion. Additionally, the strictly outbound matter transport suggests a rarefaction inside the plasma region for each plasma event.

The Liquid Sampling-Atmospheric Pressure Glow Discharge (LS-APGD): Adding Versatility to Your "Organic" Mass Spectrometer

R. Kenneth Marcus, Edward Hoegg, Htoo Paing, Tyler Williams

For over three decades, the field of atomic/elemental mass spectrometry has been dominated by the use of the inductively-coupled plasma (ICP). ICP-MS, most commonly practiced on quadrupole mass analyzer platforms, is far and away the most widely-applied quantitative, multi-element technique. That said, the platforms which are applicable for organic (molecular) mass spectrometry cross an incredibly diverse range of manifestations, from linear quadrupoles, to three-dimensional quadrupole traps, linear traps, quadrupole-time of flight, ion cyclotron resonance, and Orbitrap instruments. To this plethora of analyzers, an equally diverse plethora of ionization sources is available. The palette of sources and analyzers allows practitioners to tailor their instruments to their applications; pollutants, metabolites, proteins, etc. To be clear, though, any desire to perform "elemental analysis" required the purchase of a completely different instrument. This laboratory has developed the liquid sampling-atmospheric pressure glow discharge (LS-APGD) microplasma as a low power, low complexity *alternative* to the ICP source. The microplasma operates at dc powers of <25 W and solution flow rates of <30 ?L/min. The general operating parameters were first targeted for operations in a field-deployable mode, but the resulting figures of merit provide problem-solving capabilities that are applicable across many fields of application. Uniquely, the device can also be operated in a "molecular-MS" mode that is relevant to small molecule, protein, and organometallic species' studies. Thus, the LS-APGD is itself a versatile ionization source, but more versatile is the fact that the device can be applied on any ambient-sampling "organic" mass spectrometer having the basic capability to sample electrospray ionization (ESI) or atmospheric pressure chemical ionization (APCI) sources. In this laboratory alone, the microplasma has been interfaced to quadrupole ion trap, Orbitrap, and triple-quadrupole mass spectrometers, originally outfit for ESI mass spectrometry. As such, the operator can choose the mass analyzer system that provides the desired/required performance based on their specific analytical challenges. We will demonstrate here the performance of the LS-APGD on these diverse platforms in terms of elemental, isotopic, and molecular mass spectrometry applications.

Distance of Flight Mass Spectrometry for Atomic Spectrometry

Steven Ray, Andrew Schwartz, M. Bonner Denton, Gene Atlas, Jaime Orejas-Ibanez

In this presentation, a new type of mass analyzer known as the distance-of-flight mass spectrometer (DOFMS) will be evaluated for use in atomic spectrometry. The DOFMS concept is best explained by comparison with traditional time-of-flight mass spectrometry (TOFMS). TOFMS determines the mass-to-charge (m/z) of an ion by imparting the same energy to all ions and then measuring the time required for each m/z to traverse a known distance and arrive at a detector. In contrast, DOFMS imparts the same momentum to all ions and then measures the m/z of an ion by determining the distance the ion travels during a set time period. Essentially, Ions of lower m/z travel longer distances than ions of greater m/z, and their m/z can be calculated based upon their measured location. As a consequence of this strategy, the DOFMS can employ position-sensitive and array-based ion detectors to great advantage. Unlike TOFMS, which employs a single detector to detect all ions, a spatially-dispersive technique like DOFMS can employ an array of ion detectors, each of which can be tuned to best quantitate the ion flux at a given location (m/z). Thus, the full dynamic range of the detector can best be exploited, and ratio-based measurements obtained with greater fidelity. Here, both an electo-optic position sensitive ion detector and a new 1704-channel solid-state ion detector array will be examined for use within this role. Although DOFMS and TOFMS operate based upon very different concepts, both DOFMS and TOFMS can be made to employ the same instrumental architecture. Thus, both techniques are combined here into a single instrument to realize the advantages of both methodologies within a single instrument.

Atmospheric-Pressure Glow Discharges As Sources of Spatially Resolved Atomic and Molecular Information

Jacob Shelley, Sunil Badal, Montwaun Young, Jessica Hellinger, Garett MacLean, Courtney Walton

With recent advances in analytical instrumentation and data capture/processing, the ability to generate comprehensive chemical maps (or "images") of solid samples with high spatial resolution is being realized by a number of groups. In particular, unambiguous identification of non-targeted analytes within these complex samples necessarily requires multiple analytical approaches performed on the same sample; these types of methods have been termed 'multimodal chemical imaging' approaches. While tandem imaging methods provide a wealth of information, they often suffer from weak sensitivity, poor selectivity, or compromised spatial resolution as dictated by the spectroscopic method employed. Greater success in comprehensive chemical imaging has been achieved with mass-spectrometry-based instruments, due to the excellent sensitivity and selectivity. Here, we will present our recent work towards the development of a multimodal chemical imaging apparatus capable of providing simultaneous molecular and elemental information from the exact same spatial location (i.e. each pixel) at high spatial resolution (e.g., better than 100 microns). This dual imaging approach is achieved through laser ablation of samples at atmospheric pressure followed by simultaneous mass-spectrometric and optical-emission measurements. Aerosolized particles from the laser-ablation event were swept with a gas stream to a plasma-based molecular ionization source and, subsequently, mass-analyzed with an Orbitrap mass analyzer. At the same time, atomic emission from the laser-induced plasma (LIP) formed during the ablation event was recorded with a fiber-coupled optical spectrometer to provide elemental information on the laser-sampled area. By scanning the laser across the sample, followed by appropriate data processing, dual atomic and molecular chemical images were generated. The effect of various laser-ablation parameters, such as laser fluence and ablation-gas composition, in conjunction with ion source operating conditions were explored in detail for a variety of sample types including pharmaceutical tablets, biological tissues, and polymer films. Importantly, these parameters were found to influence atomic-emission and molecular-ion signals to different degrees. As such, approaches to obtain multimodal atomic and molecular images, as well as the data processing needed to generate and compare chemical images will be presented. Finally, approaches to enable quantitative information for both molecular and atomic analyses will be presented.

Dual Cleavable Crosslinking Technology (DUCCT): A New Strategy for High Confidence Identification of Crosslinked Peptides

Saiful Chowdhury, Jayanta Chakrabarty, Abu Hena Kamal

Chemical crosslinking and mass-spectrometry is now emerging as a confident technique for covalent labeling of nearby proteins. This produces a stable link so weak and transient interactions can be identified during pulldown or antibody-based experiments. Recently, mass spectrometry-cleavable crosslinkers are emerging as one of the most prominent method for crosslinking proteins. In this methods, inter crosslinked peptides can be cleaved into two signature peptides so their sequences can be identified by tandem mass-spectrometry. Unfortunately, confident identification of crosslinked products by cleavable crosslinkers is also challenging. Further MS3rd sequencing of the cleaved peptides is required, which are sometimes very ambiguous due to the fragmentation of the crosslinked residues attached to the peptides. In order to provide high confidence in crosslinked-peptide identifications, we introduced a novel crosslinker which we called DUCCT (Dual Cleavable Crosslinking Technology). This crosslinker contains two lysine reactive groups and two mass spectrometry-cleavable bonds. One cleavable bond is selective for CID-MS cleavage where as other one selectively cleaves by ETD tandem mass-spectrometry. This produces two signature mass spectra of the same crosslinked peptides after CID and ETD tandem mass-spectrometry. Crosslinked peptides can be identified with high confidence from these two signature mass spectra of the same crosslinked peptides. DUCCT was tested in peptide and protein level using neurotensin peptide as well as several standard proteins. In addition, its labeling efficiency was tested in macrophage immune cell upon exposures to microbial ligand lipopolysaccharide. Furthermore, a pulldown studies with a HA-TLR2 HEK 293 cells enriched two novel interactors of TLR2 after DUCCT crosslinking. Moreover, to study the crosslinking in a large-scale sample, a stand -alone software are also in progress. We believe this crosslinking approach will provide high confidence and reproducible identification of protein-protein interactions in large-scale samples.

Fourier Analysis Method for Analyzing Large, Polydisperse Membrane Protein-Lipid Complexes with Native Ion Mobility-Mass Spectrometry

James Prell, Sean Cleary

Native ion mobility-mass spectrometry (IM-MS) can be a powerful tool for determining the composition, oligomeric state, topology, and even the shape of biomolecular complexes that are challenging to study by conventional methods. In recent years, the application of native IM-MS to membrane proteins has provided tantalizing evidence that weak protein-lipid interactions can be preserved upon transfer of membrane protein-lipid assemblies, such as Nanodiscs, into the gas phase. Characterizing these interactions present an exciting challenge, due to the heterogeneity of these samples, their resistance to chromatographic separation, and the extreme congestion of their electrospray ionization mass spectra. We demonstrate that mass spectra for very large assemblies containing many non-covalently bound subunits can often be analyzed straightforwardly using a Fourier analysis method developed in our laboratory. The native stoichiometry, charge states, and lipid masses for small Nanodiscs containing approximately 150 lipids are analyzed using this method, and IM-MS data are used to obtain charge-state-specific mass spectra and collisional cross sections for these discoidal assemblies. We further illustrate that this method can be used to characterize large, intact Nanodiscs containing more than 300 lipids. Intriguingly, relative lipid polydispersity is observed to be higher for smaller Nanodiscs than for large Nanodiscs. The use of fundamental, 2nd, and 3rd harmonic peaks in the Fourier domain is compared and provides a means to quantify the uncertainty of the method. In a second set of studies, we show that samples containing mixed-lipid Nanodiscs can be distinguished from mixtures of pure-lipid Nanodiscs by this native IM-MS method, providing tantalizing evidence that this technique can be used to study non-ideal lipid mixing at the nanoscale. Together, these experiments indicate that the Fourier analysis method can dramatically facilitate interpretation of extremely complex native IM-MS spectra that often arise in the study of membrane protein complexes and other samples with repeated subunits.

Characterization of Glycan and Carbohydrate Structures and Conformations using Hydrogen/Deuterium Exchange-Mass Spectrometry

Elyssia Gallagher, Tara Liyanage, Jamie Kim, Marina Mulenos, Matthew Brantley, Touradj Solouki

Glycosylation is a post-translational modification that changes during cellular differentiation, immune response, and in metastatic cancerous tissue. Characterizing glycan structures is analytically challenging since glycans are heterogeneous, branched polymers with different monosaccharide components, linkages, and stereochemistries. Currently, glycans are analyzed by tandem mass spectrometry and ion mobility (IM). Electron transfer dissociation (ETD) and electron capture dissociation (ECD) result in different fragments for the same oligosaccharide adducted to different metals (Huang et al., *J Am Soc Mass Spectrom* 2014, 25, 1451-1460). Additionally, different drift times are measured in IM-mass spectrometry (MS) for oligosaccharide isomers adducted to the same metal (Huang et al., *Anal Chem* 2013, 85, 9728-9735). These results indicate that the three-dimensional conformations of the metal-adducted carbohydrates are dependent on both the metal ion and isomeric structure of the glycan. Hydrogen/deuterium exchange (HDX)-MS has been used to analyze structures and conformations of biomolecules, including carbohydrates (Kostyukevich et al., *J Mass Spectrom* 2015, 50, 1150-1156). In HDX-MS, the mass increase of an analyte is measured as labile protons are replaced with deuterium (D) upon exposure to deuterated solvent (ND_3, D_2O, MeOD). The rate of exchange is dependent on the chemical functional group, the presence of hydrogen bonds, pH, temperature, and solvent accessibility. Here, we utilize rapid in-solution, in-ESI, and gas-phase HDX-MS to characterize the structures of metal-adducted oligosaccharides as model systems for glycans. Similar to previous studies (Zhang et al. *J Am Chem Soc* 2004,

126, 5906-5919), we observed minimal H/D exchange in the gas-phase for cationic metal-adducted oligosaccharides, even after hundreds of collisions. Conversely, D-uptake was observed for oligosaccharides following in-solution or in-ESI HDX, with different D-uptake patterns observed for each exchange method. For in-ESI HDX, the distribution of peaks resulting from HDX was dependent on the source conditions, oligosaccharide, and metal. The combination of HDX methods permits us to examine oligosaccharide conformations as analytes transition from solvated to gas-phase ions, allowing us to compare differences in the solution-phase and gas-phase structures of carbohydrates and glycans.

The PepSAVI-MS pipeline for natural product bioactive peptide discovery

Leslie Hicks, Christine Kirkpatrick, Nicole Parsley

The increase in multidrug-resistant pathogens demonstrate the immediate need for new antibiotic backbones with novel mechanisms of action. While natural products are a well-studied source of biologically active small molecules, peptidyl factors contributing to their medicinal properties remain largely unexplored. To this end, we have developed the PepSAVI-MS pipeline that employs bioactivity assays, statistics, and mass spectrometry to identify bioactive peptide targets from complex samples. To validate this pipeline, we have demonstrated successful detection and identification of a known peptide, cycloviolacin O2 (cyO2), from the botanical species *Viola odorata*. CyO2 is a small, cysteine rich cyclotide comprised of 30 amino acids, which has been shown to have diverse activity against many Gram-negative bacteria as well as several cancer cell lines. Following our pipeline, *V. odorata* peptide library was subject to an array of bioassays against *E. coli*, a panel of ESKAPE pathogens, and several cancer cell lines. CyO2 from *V. odorata* was within the top candidates when modeled against the averaged bioactivity data, and was characterized via a multi-step mass spectrometry-based approach and these results are in agreement with previously published findings. To allow for greater access to the scientific community, the statistical analysis methods established for this platform have been developed into an R package that will be made publically available through CRAN. Additionally, we have widened the known antimicrobial spectrum for *V. odorata* cyclotides, including antibacterial activity of cyO2. We further demonstrate the broad applicability of the platform through the identification of novel anticancer activities for cycloviolacins by their cytotoxicity against ovarian, breast and prostate cancer cell lines. The developed platform is highly versatile as it is adaptable to any natural product source of peptides and can test against diverse physiological targets, including bacteria, fungi, viruses, protozoans, and cancer cells for which there is a developed bioassay.

Microscale Mass Spectrometry Analysis of Live in vitro Tumors

Zhibo Yang, Xiang Tian, Mei Sun

The multicellular spheroids (i.e., *in vitro* tumors), which possess numerious characteristics similar to in vivo tumors, gained increasing interest as 3D tumor models in varies studies such as drug penetration and resistance. However, due to the lack of appropriate analytical techniques, the corresponding studies are largely limited. In our research group, we developed novel mass spectrometry (MS)-based techniques for microscale analysis of live *in vitro* tumors. (1) Depth profiling analysis. A microscale sampling and ionization device, the Single-probe, was coupled with MS to conduct depth profiling analysis of anticancer drug and metabolites distribution in live *in vitro* tumors. The HCT-116 cell line was used for 3D culture of *in vitro* tumors and treated using anticancer drug irinotecan. The Single-probe tip was gradually inserted into live in vitro tumors to extract biomolecules at different depths for real-time MS analysis. Results indicate that the concentrations of drug in the core regions are lower than those in the outer regions. The apparent diffusion coefficient of Irinotecan in *in vitro* tumor was obtained. In addition, we observed groups of cell metabolites regulated upon anticancer drug treatment. The accomplish of our studies will develop new analytical methods to study drug penetration and metabolism in live tissues, which is an experimental approach not possible before. (2) Analysis of extracellular molecules. Extracellular compounds in tumors play critical roles in intercellular communication, tumor proliferation, and cancer cell metastasis. We introduced a microscale collection device, the micro-funnel, fabricated from biocompatible fused silica capillary. With a small probe size (~25 ?m), the micro-funnel was implanted into live in vitro tumors to accumulate extracellular metabolites produced by cancer cells. Collected metabolites were extracted by the Single-probe for real-time MS analysis. We successfully detected the abundance change of anticancer drug irinotecan and its metabolites inside *in vitro* tumors. Moreover, irinotecan treatment dramatically altered the composition of extracellular compounds. We observed increased abundances of many lipids, which are potentially related to the drug resistance of cancer cells. The successful development of our technique can potentially benefit the research in multiple areas, including microenvironment inside live tissues, cell-cell communication, biomarker discovery, and drug development.

ComDim: from multiblock data analysis to path modeling

El Mostafa Qannari, Véronique Cariou, Douglas N. Rutledge, Evelyne Vigneau

ComDim: from multiblock data analysis to path modeling Keywords: Multiblock datasets, latent variables, PLS regression, canonical analysis, ComDim. The concept of data fusion is becoming increasingly important. The objective is to acquire knowledge about a system or a phenomenon by combining information from several sources. Naturally, the statistical analysis of multiblock datasets has become a central issue. A plethora of methods have been developed and, moreover, it is not infrequent to find several names and acronyms for the same method of analysis. In a first stage, we propose a unified framework that makes it possible to better grasp the similarities and differences among the various methods. In a second stage, we focus on a method of analysis called ComDim which has proven to be versatile and flexible. Initially called Common Components and Specific Weights Analysis, this strategy of analysis has been efficiently applied to various kinds of data ranging from sensory and preference data [1] to chemometrics data [2]. It has been extended to the situation where we aim at assessing the effects of several factors on a multivariate dataset [3]. We recall that for a set of multiblock data, ComDim aims at determining common components (or latent variables) and saliences, which reflect the importance of each block dataset in determining the underlying common components. We show how the method of analysis can be extended to the situation where we aim at explaining an independent dataset from K explanatory datasets. A more general situation consists in a multiblock setting with a specific pattern of directed relations among the blocks of variables. We discuss how ComDim analysis can be extended to investigate the relationships between these blocks taking account of the path diagram. The methods of analysis are illustrated on the basis of case studies and the outcomes are compared to those of standard methods. References: [1] E. M. Qannari, P. Courcoux, E. Vigneau, Common components and specific weights analysis performed on preference data. Food Quality and Preference, 12 , 365-368 (2001). [2] G. Mazerolles,
M. Hanafi, E. Dufour, D. Bertrand, E. M. Qannari, Common components and specific weights analysis: A chemometric method for dealing with complexity of food products. Chemometrics and Intelligent Laboratory Systems, 81, 41-49 (2006). [3] D. Jouan-Rimbaud Bouveresse, R. Climaco Pinto, L. M. Schmidtke, N. Locquet, D. N. Rutledge, Identification of significant factors by an extension of ANOVA-PCA based on multi-block analysis. Chemometr Intell Lab.;106(2):173–182 (2011).

BIVREGBLS : a new R package in method comparison studies with tolerance intervals and (correlated)-errors-in-variables regressions

Marion Berger, Bernard G. Francq

When comparing methods (say methods X and Y), the Bland–Altman plot (or (M,D) plot for Means versus Differences) with its Agreement Intervals (AIs) are frequently used. Another interesting approach is the use of errors-in-variables regressions in a classical (X,Y) plot. Francq and Govaerts (2014) demonstrated that the Bivariate Least Squares regression is the most suitable and general regression as it covers all the cases handled by the Deming regression, the orthogonal regression, the mean geometric regression, the vertical and horizontal ordinary least squares regressions. Francq and Govaerts (2016) also reconciled the (M,D) and (X,Y) approaches by proposing tolerance intervals and predictive intervals with good coverage probabilities to replace AIs. They also demonstrated that errors in (M,D) were correlated. Bias soar and coverage probabilities collapse if this correlation is ignored. They proposed a novel regression, CBLS regression (Correlated Bivariate Least Square) where predictive intervals and confidence intervals have excellent coverage probabilities whatever the design. A new R package BivRegBLS has been developed to perform BLS and CBLS under various designs (unreplicated data, replicated data with balanced or unbalanced design, variance homogeneity or heterogeneity) including a new generalized interval that combines the confidence and the predictive concepts. The use of BLS and CBLS should be encouraged as one allows visualizing the comparison in the original unit scale, while the second fulfills the deficiencies of the Bland-Altman approach.

Combining Independenat Components Analysis and multiblock data analysis: application to hyperspectral images
Benoit Jaillais, Eloise Lancelot, Douglas N. Rutledge

Combining Independenat Components Analysis and multiblock data analysis: application to hyperspectral images. Germination capacity of wheat is influenced by both genetic and environmental factors and the process of germination is complex [1]. The mechanism of germination is well-known and can be split into two different phases [2]. The first phase, called imbibition phase, consists in a rapid water absorption which induces perturbations in the cell membrane structures, and is followed by a constant uptake of water (the second phase). The tracking of the diffusion of water into single wheat kernel sections over time is very challenging and can be assessed by Magnetic Resonance Micro-Imaging (MRMI) [3] and by Near Infrared hyperspectral imaging (NIR-HSI) [4]. However, since the protocol classically used to produce imbibition [4] consists in immerging whole seeds in water for hours before analysis, we have proposed a different method [5] to provide a detailed picture of water uptake and distribution in wheat grain tissues in real time using NIR Imaging. To do that, the seed is transversely cut and the part containing the embryo is placed in a specially-designed sample holder, the cut surface oriented towards the imaging system. NIR images can be regarded as three-way data structures that can be unfolded into two-way data matrices which can then be analyzed by Independent Components Analysis (ICA) which has been recently applied on hyperspectral images to determine the distribution of peanut traces [6]. As the spectrum acquired for each pixel in an image can be considered as the weighted sum of unknown source signals resulting from chemical and physical phenomena that evolve during the imbibition, it was decided to apply Independent Components Analysis (ICA) to the unfolded 3D NIR hyperspectral images acquired for each seed over time. Common Components and Specific Weights Analysis (ComDim) was developed by Qannari et al. [7] for sensory profiling analysis where each of the judges has his own table of descriptors to describe the same set of samples. In order to determine which of these source signals are common to all images of all seeds and which are specific to particular times and seeds, the independent components were introduced as column vectors into the Common Components and Specific Weight Algorithm (ComDim). By applying ComDim, we obtained a set of global common signals and were able to detect specific ones related to the water uptake by different tissues of the wheat kernels.

The application of micro-Raman spectroscopy and micro-XRF to analysis of duct tapes
Sergey Mamedov

Raman and X-Ray Fluorescence (XRF) spectroscopies are useful tools for identification substances and confirming their identity with little or no sample preparation. Raman spectroscopy provides information about molecular structure of the material, whereas X-ray fluorescence spectroscopy gives elemental composition. Both techniques are able to record not only spectra of small particles, but also hyper-spectral image, as well as collect average spectrum over certain area. Multivariate analysis (MVA) can produce chemical distributions of elements and/or material classification based on Principal Component Analysis (PCA), in particular, with association between elements or structural motives that can aid in identification of bonded phases. The analysis of micro-XRF and Raman data of duct tapes can be used to identify source of the duct tape or the vendor of the product. X-ray fluorescence and Raman analytical microscopes were used in this study. X-ray fluorescence spectra of the materials were collected using 30 keV acceleration voltage and with X-ray spot size of 1.2 mm. Two excitation wavelengths (532 nm, 785 nm) were used to collect Raman spectra. The spectra of duct tapes from different sources were collected and analyzed by micro-XRF and Raman spectroscopy. XRF analysis was performed in the range of 1.00-40.96 keV. Some tapes contain elements Ti, Ca, S, and Al in a fiber substrate, which may be used for duct tapes differentiation. Classification of duct tapes based Principal Component Analysis of the spectra will be shown. Raman spectra of duct tapes were measured in the range of 100-3500 cm-1. PCA was applied to these spectra to extract differences in connection with different source of the tapes. Raman spectra of duct tapes show many common features with some differences, which may be originated from the filler. Data fusion technology was applied to the set of XRF and Raman data to create PCA model of fused data. Results of standard and data fused analysis are compared and discussed. In conclusion, this study provides methods that allow one to differentiate duct tapes based on spectra analysis of micro-XRF and micro Raman data.

Evidence of fibrin accumulation preceding altered collagen production in combat related heterotopic ossification identified via Raman spectroscopy

Katherine E Cilwa, Benjamin M Wheatley, Devaveena Dey, Thomas A Davis

Heterotopic ossification (HO) is the pathological formation of bone in soft tissue and is a debilitating complication of burns and traumatic injuries. Over 60% of injured service members with blast-related combat extremity injuries involving amputations develop HO; sometimes resulting in chronic pain, ulceration, tissue entrapment, and reduced mobility. Early radiographic evidence of immature bone is limited by current clinical imaging modalities. Raman spectroscopy has identified altered collagen formation as an early marker of disease in pre-HO tissue in ex vivo specimens within 1-2 weeks post injury and may serve as a non-invasive tool for guiding surgical excision at HO tissue margins. However, treatment development is prevention focused with intervention timed shortly after injury before surgical excision is required. Therefore, identification of the earliest detectable alterations in tissue pathology may guide clinical decision making and intervention selection. This study examines the ability of Raman spectroscopy to non-invasively detect changes in tissue associated with early HO development in a rat combat-trauma model of HO within the first week of injury. Spectroscopic signatures of tissue pathology and the effect of Palovarotene treatment, a retinoic acid receptor agonist and known potent chondrogenic HO inhibitor, are explored (n=5 animals per time point per treatment group). HO development occurred in 100% of vehicle control animals and 67% reduction in HO volume was observed in Palovarotene treated animals. Significant spectroscopic differences between vehicle control and Palovarotene treated animals are evident as early as post-operative day 1 (POD-1). For vehicle control animals, calculated protein disorder/order [1240/(1270+1300)] band area ratio, is significantly increased at POD-1 for vehicle control animals compared to Palovarotene treated animals ($p<0.001$) and is indicative of elevated ?-sheet/random coil protein content suggestive of fibrin accumulation. Spectroscopic evidence of fibrin accumulation within the first 48 hours following injury may serve as a potential early marker of HO formation prior to evidence of endochondral ossification and emphasizes the need to investigate the role of fibrin in altered collagen production post-traumatic injury. This study highlights the non-invasive ability of Raman spectroscopy to serve as a potential tool for identifying patients at risk of HO development.

Multi-Centre Raman imaging for pathology classification in Barrett's Oesophagus

Gavin Lloyd, Catherine Kendall, Nick Stone, Lauren Denton, Jennifer Dorney, Geaint Thomas, Riana Gaifulina, Aaran Lewis, Martin IsabelleIan Bell

Raman spectroscopy has been shown to be an accurate tool for cancer diagnosis, but it has not yet been implemented as a histopathology tool in a clinical setting. The majority of work to date has used a single instrument at a single centre. The Stratified Medicine through Advanced Raman Technologies (SMART) project, funded by Innovate UK, is an industry-academic collaboration too address some of the hurdles to clinical adoption including tissue sample preparation, data collection, pre-processing and model transferability. A multi-centre study consisting of three Raman RA800 series instruments (Renishaw, UK) at three sites (Gloucester Royal Hospital, University of Exeter and University College London) was developed. Raman spectra were collected across the same homogeneous regions of tissue pathology at all sites. A multivariate algorithm incorporating steps to minimise between-site variability was developed and applied to a database of >100 samples from multiple pathology groups relevant to oesophageal cancer diagnosis. These included samples of normal squamous, Barrett's oesophagus, low and high grade dysplasia and adenocarcinoma. Pathology classification models were trained at a single site and cross-validated to ensure a robust model. The model was then tested using data from the other sites to verify the applicability of the developed models to secondary sites. This study illustrates the potential of non-invasive rapid Raman spectral mapping measurements and development of a robust and validated oesophageal classification model, constructed from three remote instruments sites, able to classify tissue pathology.

Motor capsules for esophageal cancer screening and diagnosis

Rohith Reddy, Jing Dong, Michalina Gora, Metthew Beatty, Kanwarpal Singh, Tim Ford, Emilie Beaulieu-Ouellet, Catriona Grant, Mireille RosenbergGuillermo Tearney

Optical coherence tomography (OCT) is an imaging technology that provides depth-resolved images of a sample by measuring the optical delay of backscattered light using interferometry. OCT has been shown in prior studies to be capable of accurately diagnosing Barrett's Esophagus (BE), a metaplastic change that conveys an increased risk of developing esophageal adenocarcinoma (EAC). Tethered capsule endomicroscopy (TCE) using OCT is a new technology developed in our lab where a tethered, opto-mechanical pill is swallowed and obtains 10 µm resolution cross-sectional OCT images of the entire esophageal wall as it traverses the esophagus via peristalsis. Clinical studies using TCE were initially conducted using a probe that used a rotary junction (RJ) and a driveshaft to convey torque that rotated optics within the capsule to scan the beam along the esophageal wall. The cost of this driveshaft-based probe was high, potentially hindering widespread adoption of this technology for screening. We have recently developed a next-generation TCE device that is significantly less expensive in both fixed and disposable costs while producing better and more quantitative image data. In this new design, an inexpensive cell-phone micro-motor replaces the expensive RJ and driveshaft. Inexpensive cell-phone motors are not designed to reliably attain constant velocity of rotation and this poses several challenges in producing a robust imaging device. These challenges have been mitigated through a combination of pulse-width drive modulation, appropriate loading of the motor, and by using image-based control-loop feedback. We will describe these technical solutions in detail and will present human data from our inexpensive cell-phone motor-based TCE device. Based on our results with this new technology, we believe that the cost of TCE devices can be lowered to the level that would be required to implement widespread OCT screening for BE.

Infrared Spectroscopy for Tissue Pathology Imaging: Importance of Location and Resolution of Measurements.

Michael Walsh, Vishal Varma, Hari Sreedhar, Shaiju Nazeer, David Martinez, Christie Massie, Suman Setty, Grace Guzman

Infrared (IR) spectroscopic Imaging is an approach that has the potential to revolutionize pathology by adding new biochemical markers for disease diagnosis. Exploiting these biochemical markers has been shown to enable accurate cell typing, disease classification and prediction of patient outcome. The development of both the instrumentation and computation of IR imaging requires careful consideration when applying to different pathological questions. Considerations such as the location from the tissue that spectral signatures are derived and the resolution (both spatial and spectral) are critical to ensure accurate classification along with amenability to clinical implementation. We have demonstrated that cell types in the liver that are not directly involved in diseases processes can be interrogated and accurate diagnoses made, which highlights the importance of looking in unexpected locations of the tissue for biomarkers. We have also demonstrated that a low spatial resolution and low spectral resolution when imaging the thyroid and kidneys can still achieve successful disease classification and thus significantly reduce the speed of data acquisition and the requirement for pathology expertise.

Temporal diabetes-mediated biochemical changes in distinctive mouse retinal layers

Ebrahim Aboualizadeh, Christine Sorenson, Miriam Unger, Nader Sheibani , Carol Hirschmugl

Diabetic retinopathy (DR) is a microvascular complication of diabetes and a leading cause of adult vision loss. To discover the mechanisms underlying the development and progression of diabetes changes in the retina, a more comprehensive understanding of the bio-molecular processes in individual retinal cells subjected to hyperglycemia is required. Animal models provide a suitable model for detecting the underlying pathophysiological mechanisms of

DR, which is not fully attainable in human studies. In the present study, we aimed to determine the nature of diabetes-induced, highly localized biochemical changes in the retinal tissue from Ins2Akita/+ (Akita/+; a model of Type I diabetes) male mice with different duration of diabetes. Employing label-free spatially resolved Fourier transform infrared (FT-IR) imaging engaged with chemometric tools enabled us to identify temporal-dependent reproducible biomarkers of the diabetic retinal tissue from mice with 6 or 12 weeks, and 6 or 10 months of diabetes. We report, for the first time, the origin of molecular changes in the biochemistry of individual retinal layers with different duration of diabetes. A robust classification between distinctive retinal layers namely photoreceptor layer (PRL), outer plexiform layer (OPL), inner nuclear layer (INL), and inner plexiform layer (IPL), and associated temporal-dependent spectral biomarkers were delineated. Spatially-resolved super resolution chemical images revealed oxidative stress-induced structural and morphological alterations within the nucleus of the photoreceptors. Comparison among the PRL, OPL, INL, and IPL suggested that the photoreceptor layer is the most susceptible layer to the oxidative stress with short-duration of diabetes. Moreover, for the first time, we present the temporal-dependent molecular alterations for the PRL, OPL, INL, and IPL from Akita/+ mice, as diabetes progresses. These findings are potentially important that may be of particular benefit in understanding the molecular and biological activity of retinal cells during oxidative stress in diabetes, which is still unclear. Our integrating paradigm provides a new conceptual framework and a significant rationale for a better understanding of the molecular and cellular mechanisms underlying the development and progression of DR. This approach may yield alternative and potentially complementary methods for the assessment of diabetes changes. **Acknowledgements:** US NSF under awards CHE-1112433 and CHE-1508240.

Overview of LIBS analysis of soft and hard tissues

Pavel Po?ízka, Prochazka David, Jakub Klus, Pavlína Škarková, Jan Novotný, Jozef Kaiser

In this talk, past and recent experiments targeting elemental analysis of soft and hard tissues using Laser-Induced Breakdown Spectroscopy (LIBS) will be reviewed. Presented research topics reflect the long-term experience of LIBS group at CEITEC in Brno. The main emphasis is given to the detection of heavy metals nanoparticles (NPs) administered into soft tissues. Benefits of LIBS technique will be confronted with Laser Ablation Inductively Coupled Plasma techniques. Last but not least, analysis of the degradation of hard tissues using LIBS and the in-house production of NPs will be given.

Bio-LIBS and the Role of Trace Metals When Laser-Induced Breakdown Spectroscopy is Used to Study Biological or Biomedical Systems

Steven Rehse, Alexandra Paulick, Dylan Malenfant, Vlora Riberdy, Siddharth Doshi

In recent years, it has been shown that laser-induced breakdown spectroscopy (LIBS) can provide a sensitive elemental assay of both prokaryotic (without a nucleus, e.g. bacterial) and eukaryotic (containing a nucleus, e.g. human) cells. While these cells are primarily composed of water (hydrogen, oxygen) and carbon, the dominant atomic species that are observed in the LIBS emission spectra from such ablated cells are often the trace metals that are present in the cells such as calcium, magnesium, sodium, and potassium. On the basis of this assay, numerous biomedical diagnostic tests have recently been proposed. Our group has demonstrated that sensitive and specific discrimination between bacterial specimens at both the species and strain levels is possible and that assays of in vivo human fingernails for a determination of human nutriture are feasible. In this talk I will review the methodology that has been used to obtain high signal-to-noise LIBS spectra (from human fingernails and bacterial specimens) and summarize the applications of the technique in the field of biomedical diagnostics. I will present our group's recent work to understand the impact of the elemental content of the growth environment on the LIBS spectrum obtained from bacterial cells with a focus on experiments to intentionally dope *E. coli* cells with zinc and magnesium. These studies are relevant for LIBS measurements on cells obtained from medical specimens for infection or health diagnosis as well as on cells obtained from an environmental setting for ecological surveillance.

Strategies for obtaining adequate signal to noise in bacterial LIBS spectra by isolating and concentrating bacteria from fluid suspensions will be discussed. Finally, our efforts to quantify the effect of sample preparation on the testing of clipped human fingernails and the ramifications for in vivo testing will also be presented.

LIBS and XRF: Complimentary Solid State Analysis Techniques in the Pharmaceutical Lab

Lydia Breckenridge, Sharla Wood

Laser-Induced Breakdown Spectroscopy (LIBS) and X-Ray Fluorescence (XRF) have been leveraged as complimentary analytical techniques for the purpose of pharmaceutical analysis. Both techniques offer fast analysis with elemental specificity and sensitivity and with little to no sample preparation. The small sample size requirement, spatial resolution, and detection of light elements allows LIBS to be employed for applications which may not be amenable to XRF analysis. On the other hand, the non-destructive behavior, ease of use, as well as stable and reproducible analysis due to a controlled x-ray beam, allow XRF to fulfill applications where LIBS would not be ideal. In many cases, both techniques may be employed to gather more comprehensive data about a sample or to serve as mutually orthogonal confirmatory techniques. As sample matrices in the pharmaceutical industry evolve in complexity, so does the growing need for solid analysis techniques that do not require liquid sample preparation, including dissolution or digestion. Analytical techniques that can be performed on solids are highly desirable owing to the reduced introduction of error or contamination from sample preparation, reduced consumption of solvents and reagents and a rapid analysis times. For many types of analyses routinely performed in the pharmaceutical industry, including foreign matter investigations, synthetic chemistry process optimization support and elemental component characterization, the advantages of solid analysis techniques far outweigh the disadvantages (namely, decreased accuracy and sensitivity) relative to more conventional liquid based techniques such as ICP-MS or ICP-AES. Case studies will be presented where both LIBS and XRF were leveraged in order to produce a more complete picture of the unknown elemental composition of a variety of different sample types. Results will demonstrate that each technique, with its unique advantages, served to either confirm the results obtained by the other technique or to provide additional complimentary data. In combination, LIBS and XRF have proven to be vital analytical tools for the analysis of solid pharmaceutical samples.

Quantitative Laser-Induced Breakdown Spectroscopy (LIBS) for early detection of calcification in aortic valvular interstitial cells (VICs)

Seyyed Ali Davari, Shirin Masjedi, Zannatul Ferdous, Dibyendu Mukherjee

Calcific aortic valve disease (CAVD) is a major cardiovascular disorder caused by osteogenic differentiation of valvular interstitial cells (VICs) within aortic valves. Early-stage detection of CAVD is elusive and development of CAVD at later stages leaves the patients with the only viable treatment of aortic valve transplant surgery. Early diagnosis of CAVD demands fundamental research that can enable laboratory-based rapid screening and analyses of early calcification in VICs to understand the initiation and origin of the disease. This requires systematic investigations into the progress of VICs calcifications over time. Direct evaluation of aortic valve calcification has led to the development of various techniques to measure calcium depositions in VICs or valve tissue sections on a laboratory platform. And yet, conventional methods like colorimetric assays and histology fail to detect small calcium depositions during in-vitro VIC cultures. Laser-induced breakdown spectroscopy (LIBS) is a robust analytical tool that has been effectively used for inorganic materials characterizations, but relatively new to biomedical applications. Here, for the first time, we use LIBS for quantitative in-vitro detection of calcium depositions in VICs at various osteogenic differentiation stages. VICs isolated from porcine aortic valves were cultured in osteogenic media over various days. The calcium depositions within VICs are first measured using colorimetric calcium assays based on arsenazo dye and Von Kossa staining. Moreover, Ca I (422.67 nm) atomic emission lines collected from LIBS measurements on lyophilized VIC samples are used for estimating calcium depositions. Our results indicate excellent linear correlation between the calcium assay and our LIBS measurements. Furthermore, in contrast to the assay results, the LIBS results could resolve calcium signals from cell samples with as early as 2 days of osteogenic culture. Finally, the quantitatively LIBS measurements establish the detection limit for calcium contents in VICs to be ~0.17 ± 0.04?g indicating a 5-fold improvement over calcium assay

Feasibility of LIBS for prediction of minerals in powdered infant formula

Xavier Cama-Moncunill, Maria Markiewicz-Keszycka, Yash Dixit, Raquel Cama-Moncunill, Maria Casado-Gavalda, Patrick J. Cullen, Carl Sullivan

Infant Formula (IF) is an industrially produced food intended to substitute human milk during infancy, which must satisfy by itself all the infants' nutritional requirements. Since IF may be the only source of nutrients for infants, it is of crucial importance that IF provides the adequate amounts of carbohydrates, fat, and proteins, but also those of minerals, trace elements and vitamins, which also play an indispensable role on infant nutrition. To ensure the nutritional profile, quality and safety of IF, manufacturers need efficient and reliable methods. However, conventional methods used for the analysis of minerals in IF are often time-consuming, laborious and difficult to operate. Laser-Induced Breakdown Spectroscopy (LIBS) is a promising atomic emission spectroscopic technique described as a real-time method with little to no sample preparation, which also holds the possibility of spatial distribution analysis. In this study, a Q-switched LIBSCAN-150 system (Applied Photonics, UK) equipped with a 150 mJ Nd:YAG laser operating at 1064 nm was employed to predict calcium content in powdered infant formula samples. The aim of this work was to evaluate the feasibility of LIBS as an at-line tool for the prediction of minerals in IF. To this end, IF mixtures with varying content of calcium were prepared over a range (approx. 1.5–7 mg/g of calcium) selected to be in conformity with the guidelines provided by the Codex Alimentarius Commission. Atomic Absorption Spectroscopy (AAS) was used as the reference method for mineral content determination in IF. Partial least squares regression (PLSR) was applied to the LIBS spectral data to develop a calibration model for calcium content quantification. The model was build using the cross-validation leave-one-out method. Validation of the model was then carried out to investigate its robustness, which showed a good fit with a R2 of prediction value of
0.85. Low values for root mean square errors of cross-validation (RMSECV) and prediction (RMSEP) were obtained. Furthermore, this study also illustrated the possibility of LIBS to provide mineral prediction maps as a useful tool for testing sample homogeneity.

Developing ion mobility - hydrogen deuterium exchange -mass spectrometry techniques for rapid analysis of 'omics mixtures

Hossein Maleki, Megan Maurer, Stephen Valentine

A significant challenge in small-molecule 'omics investigations is the identification of constituent molecules in complex mixtures. Even with advanced condensed-phase separation and mass spectrometry techniques, verification of compound assignment is required often using an isotopically -abeled standard. This not only significantly increases the cost of the overall measurement but also decreases the overall throughput of the methods. Here, we describe a new approach for mixture component identification that is based on rapid ion mobility spectrometry and gas-phase hydrogen-deuterium exchange measurements. The method provides unique signatures for a wide-variety of compounds that would be encountered in 'omics mixtures. Additionally, the approach is well suited for the development of predictive methods for identifying compounds for which no previous measurements are available for comparison (e.g., newly emerging neuropeptides and metabolites of new drugs). The robust nature of the approach is demonstrated and real world examples of its usage for aiding compound identification are presented.

Ion Mobility and the Omics: The Challenge of Separating Isomeric Systems
James Dodds, Jody May, John McLean

Among many recent innovations in the "Omics" sciences, mass spectrometry (MS) techniques made important strides initially in the realm of proteomics and genomics, however more recent studies have focused on the application of MS strategies targeting small molecules, including studies related to metabolomics, lipidomics, and glycomics. Modern mass spectrometers are highly selective, often able to assign an intrinsic molecular formula for small molecule analytes with mass accuracy of less than 2 ppm. However, complex biological samples may include isomeric species of interest, or compounds with identical chemical formula but differing in overall skeletal structure, which can complicate structural identification and downstream pathway analysis. In light of the challenges associated with increasing structural identification for isomeric small molecules, this work highlights recent applications of isomer separations by ion mobility-mass spectrometry (IM-MS), wherein the ion mobility provides a structural separation of analytes based on size, shape and charge in the gas phase in complementary fashion to mass selection. Ion mobility as a separation technique spans various analytical separation methods, instrument platforms, and varying degrees of difficulty in separation for specific isomeric compounds. In order to relate ion mobility separations across various instrument platforms and isomer species, all analytes in this study are evaluated based on the degree of separation difficulty in terms of percent difference in collision cross section (%?CCS), resolving power (CCS/?CCS) as a separation metric, and overall two-peak resolution. By using this technique, we have evaluated the separation capabilities of various mobility platforms, and can in turn describe the possible applications of each device in terms of specific "omics" separations.

The Evolution of FTMS Based Trapped Ion Mobility Separation
Michael Easterling, Melvin Park, Christopher Thompson, Mark Ridgeway

Out of the many implementations of ion mobility based separation that have been hyphenated on to mass spectrometry platforms, the trapped ion mobility (TIMS) approach is the most spatially efficient and offers resolution metrics that exceed other schemes. Contrasted to traditional mobility arrangements that utilize drift tube dynamics, TIMS relies on a hybrid tunnel-funnel geometry built into a multiple funnel atmospheric pressure ionization source and separates ions against a potential gradient countercurrent to the viscous gas flow. By programming the voltage associated with the gradient electric field, ions of a given mobility can be ejected out of the linear RF trap along the ion optics pathway towards the detector. While commercialized mobility arrangements were traditionally implemented on TOF geometries lending to their detection speed, FTMS mass detectors generally have much higher duty cycles requiring modifications to the implementation, workflow, and overall experimental expectations. This presentation will cover the nature and results of those optimizations contrasted to original implementations and the consequential effects on desired workflows. Several examples of FTMS hybridized mobility experiments will be illustrated to highlight their complementary nature to information that can be gathered with TOF systems.

Tuning Mobility Separation Factors for Metabolomics via Selective Ion-Neutral Clustering
Brian Clowers, Pearl Kwantwi-Barima, Zhihao Yu, Christopher Hogan

The remarkable performance and accessibility of modern mass spectrometry provides researchers across disciplines with tools to measure mass-to-charge ratios with exquisite resolution and mass measurement accuracy. These increases in performance also prompt researchers to pursue additional modes of separations that capture molecular parameters (e.g. stereochemistry) that transcend m/z. For select chemical systems ion mobility measurements can fulfill that experimental need with a separation dimension that is highly complementary to m/z measurements. By modifying the traditional ion mobility experiment using neutral gas vapors selective gas-phase interactions and can be leveraged to realize additional dimensions of separation. Viewed from a chromatographic perspective, the drift gas in a mobility separation is analogous to a stationary phase, however, the range of drift gasses has been traditionally limited to a narrow range of analytes. Using standardized plasma (NIST 1950) as a demonstration case, we illustrate how separation factors between different analytes may be adjusted and sorted based upon the degree of their mobility shift. By modulating the introduction of the drift gas modifier an additional degree of discrimination is introduced. In addition to the practical separation benefits we also provide a quantitative model to describe the degree of chemical shifts occurring target analytes. While enhanced separation factors are not observed for all chemical classes, and the exact mechanism of the selective interaction remains under investigation, we anticipate that continued exploration will aid researchers using ion mobility and mass spectrometry as a tool to probe chemical structure by providing yet another dimension of separation.

Lipidomics Analysis of Antimicrobial-Resistant Bacteria by HILIC-Ion Mobility-Mass Spectrometry

Kelly Hines, Brian Werth, Libin Xu

Ion mobility-mass spectrometry (IM-MS) is a rapid gas-phase separation technique based on structure and mass, and is orthogonal to conventional liquid chromatography techniques. We have developed a method for untargeted lipidomics that combines hydrophilic interaction liquid chromatography (HILIC) with IM-MS. Lipid species are separated by headgroup polarity and fatty acid composition in the HILIC dimension, and by backbone and headgroup structure in the IM-MS dimension. Structural information can be obtained from the IM dimension in the form of collision cross-sections (CCS) by calibration with lipid-specific CCS calibrants on a traveling-wave IM instrument. The combination of HILIC with IM-MS provides detailed molecular information in the form of lipid class, accurate mass, collision cross-section, and fragmentation pattern from untargeted MS/MS to support identification of unknown lipid species. Using this HILIC-IM-MS approach, we have characterized lipidomic alterations in three Gram-positive bacterial species that were associated with the development of resistance to daptomycin, a critically important antibiotic for the treatment of vancomycinresistant enterococci (VRE) and methicillin-resistant S. aureus (MRSA). Daptomycin is a membrane-targeting lipopeptide antibiotic, and recent studies have found a strong correlation between daptomycin resistance and the phospholipid content of the bacterial cell membrane. We evaluated daptomycin resistance in clinically-derived isogenic strain pairs of C. striatum (MIC of 0.125 μg/mL and > 256 μg/mL), and vancomycin-resistant
E. faecalis (MIC of 1 μg/mL and 12 μg/mL), and a series of three in vitro-derived MRSA strains with decreasing daptomycin susceptibility (MIC of 0.125 μg/mL, 1 μg/mL, and 8 μg/mL, respectively). Our HILIC-IM-MS lipidomics method resolved the predominant lipid species of these bacteria, including glycerolipids, glycerophospholipids and glycolipids. In all three species of bacteria, the major alteration associated with daptomycin resistance was a significant reduction in membrane phosphatidylglycerols (PG). Several glycolipid species were also altered in each of the daptomycin resistant strains, including a glycolipid unique to C. striatum. These results from the HILIC-IM-MS analysis identified both common and unique lipid pathways among several species of daptomycin-resistant bacteria, which may reveal new mechanistic insights into the development of resistance or inspire new diagnostics for resistance detection.

New Strategies for Surface-Enhanced Sensing: Carbenes as Thiol Replacements and Hyper-Raman Based Detection
Jon Camden

SERS as a sensing platform is well establish, finding application in a diverse set of fields from in vivo imaging to trace analyte detection. In this talk I present recent work to extend the application of SERS-based schemes to higher-order spectroscopies such as hyper-Raman scattering and also address the issue of surface modification, which plays a central role in the success of surface-enhanced spectroscopy.

Doped Lanthanum Hafnates as Scintillating Materials for High-Energy Photon Detection
Yuanbing Mao, Kareem Wahid, Madhab Pokhrel

Recent years have seen the emergence of nanocrystalline complex oxide scintillators for use in X-ray and gamma-ray detection. In this study, we investigate the structural and optical properties of La2Hf2O7 nanoparticles doped with varying levels of Eu3+ or Ce3+ by use of X-ray diffraction, Raman spectroscopy, scanning electron microscopy, transmission electron microscopy, and optical photoluminescence. In addition, scintillation response under X-ray and gamma-exposure is reported.

Mechanically deforming nanoparticles and impressing deformation patterns
Jeffrey Anker, Fathima Ameer, Meenakshi Ranasinghe, Shilpa Varahagiri, Daniel Willet, Yimei Wen, George Chumanov

The shape of a nanoparticle can determine its optical, magnetic, chemical, and biological characteristics. However, it is challenging to control nanoparticle shape, especially while maintaining volume. Herein, we describe a simple method to modify the shape of malleable nanoparticles post-synthesis using mechanical deformation. For example, a glass rolling pin can flatten gold and silver nanospheres deposited on a glass slide into disks at chosen locations and with controllable forces. Flattening these plasmonic nanoparticles causes the nanoparticle scattering spectrum to dramatically red-shift due to shape and orientation-dependent longitudinal localized surface plasmon resonance modes. The shape change also exposes new surface area. In addition, polymer particles can be mechanically deformed into roll and pancake shapes and embedded with plasmonic, fluorescent, and magnetic nanomaterials. Micro and nano-particle deformation patterns can be generated using a stencil to impress a pattern, or using a glass rod to apply force at specific locations one at a time. We studied changes in particle shape and spectral properties using electron microscopy, dark field microscopy & spectroscopy, and surface enhanced Raman spectroscopy (SERS).

SERS headspace sampling for a polysulfide cyanide antidote
David Thompson, Md Nure Alam, Reece Thompson, Joie Games

Dimethyl trisulfide (DMTS) is a promising candidate for treating cyanide intoxication. To better understand its biochemistry, rapid sensitive detection methods are needed for determining DMTS and its reaction products. This talk will present progress in the development of a surface enhanced Raman spectroscopy (SERS) method to detect DMTS and products in the headspace above DMTS containing solutions. Headspace sampling is used to improve specificity and to reduce sensor fouling. Problems in the reproducibility of signals in the headspace sampling led to the observation of condensation on the sample cell windows and sensor. This motivated us to work toward differential control of the sample, sampling window, and SERS sensor temperatures. An initial model to inform the setting of temperatures will be presented, along with experiments showing that local heating removed condensation from the sampling window without observably altering the temperature of the liquid solution. Complimentary experiments focused on detection rate. In these experiments, DMTS was reduced to form more volatile products, and sub-minute detection times were achieved.

Planar Array Substrate Based Surface Enhanced Raman Spectroscopy: Effect of Metal Types and Geometries on Enhancement and Thermodynamics of Binding
Ashish Tripathi, Erik D. Emmons, Augustus W. Fountain III, Jason A. Guicheteau

Surface enhanced Raman spectroscopy (SERS) is widely reported as a possible tool for trace chemical detection. Planar SERS substrates are formed when nano-geometrical structures coated with a metal film and the combination of geometry and metal type provides for plasmonic enhancement with a particular laser excitation wavelength. The type of metal used to form the planar SERS substrates (typically gold or silver) will have varying binding affinity with different classes of chemicals. To probe these influences, we are studying the formation of mono-molecular layers for many nitrogen and sulfur containing aromatic chemicals on planar SERS substrates of varying geometries and noble metal coating. Our studies reveal the combined influence of metals and geometries on the effectiveness of planar SERS substrates as a tool for trace detection.

Approach to Multivariate Model Lifecycle Management

Daniel Hill

The evolution of predictive models supporting biologic processes have sped up in recent years. While the use of models has always added process understanding value, the use of models in the control strategy and batch release process has raised the robustness and scrutiny requirements. This talk with provide an overview of the best practices of the Biogen model development and Validation platform.

Towards a turnkey process Raman spectroscopy analyzer in upstream bioprocessing operations

Karen Esmonde-White, Maryann Cuellar, Alexander Pitters, Sean Gilliam, David Strachan, Herve Lucas, Bruno Lenain, Ian Lewis

The past 20 years has brought more interest in process Raman spectroscopy, as scientists and engineers have realized benefits in polymer, petrochemical, pharmaceutical and bioprocessing applications. In many instances, customers and vendors work closely to optimize not only the hardware but also the statistical method for analyzing data. Instrumentation reliability and usability has become more important as an increased number of Raman spectroscopy users do not have expertise in spectroscopy, data analysis, or statistical model development. Reliability and usability are especially important features in a process environment because of the need to support 24/7 operations and the range of users' spectroscopy knowledge. Understanding the needs of all users affects technological, ergonomic, safety and logistical aspects of analyzer and sampling probes. In this paper, we describe how consideration of these aspects affected the development of Raman spectroscopy hardware and software for upstream bioprocessing applications, enabling in situ simultaneous measurement of multiple biochemical parameters within a cell culture or fermentation bioprocess. We describe developments in Raman spectroscopy for real-time understanding and monitoring in bioprocesses. Exciting Raman-based control of glucose and lactate in cell culture bioprocesses demonstrates improved understanding for process robustness and realize optimal product quality.

Model System Based Comparison of NIR and Raman Spectroscopy by the Prediction of the Glucose Concentration of CHO Cell Cultivations

Bence Kozma, László Párta, Szilveszter Gergely, András Salgó

NIR spectroscopy is an established spectroscopic technique for decades that is capable of on-line measurements of bacterial, yeast, fungi fermentations and mammalian cell cultivations in harmony with the PAT guidance of FDA. Raman spectroscopy is also applied recently for bioprocess monitoring and it is a very promising technique for the biopharmaceutical industry because, in contrast with NIR spectroscopy, it is insensitive to water. Glucose is the primary carbon source of the cells thus on-line monitoring and controlling of its concentration is crucial in bioprocesses. In this industry focused study, the capabilities of NIR and Raman spectroscopy were compared together by predicting the glucose concentration on-line, in the same CHO cell cultivations. By monitoring and controlling glucose concentration, the cell physiology is balanced that is positively correlates with produced protein quantity and quality. A shake flask model system was used to generate on-line NIR and Raman spectra of cell cultivations to test the glucose predicting ability of the two methods and to develop a working model prior to bioreactor cultivations. Shake flask cultivations are faster and cheaper than bioreactor runs but can also be conducted in fed-batch mode to mimic a bioreactor process. A unique model building workflow was applied to test the effectiveness of different pre-processing and variable selection methods. Spectra of different shake flasks were divided to a calibration and a validation part and PLS regression models were developed using the calibration part. The glucose prediction ability of the models was tested on the validation part subsequently. Both NIR and Raman based PLS models could predict glucose similarly with an RMSEP under 3 mM. The models with the lowest RMSEP were applied to spectra that were obtained during a 10L and a 100L scale bioreactor run thus testing the models scalability for both techniques. NIR could predict the trend of the glucose concentration during cultivations but the accuracy was low. However, the two different scales were equally accurately predicted by Raman with an RMSEP of 4.0 mM therefore it proved to be accurate and scalable as well. In conclusion, the shake flask model system is capable of generating data for fast calibration that is also scalable but only for Raman spectroscopy. Therefore, an accurate working model for monitoring the glucose concentration in a bioreactor during cultivation is available for the first bioreactor cultivation before any other bioreactor runs were conducted. This model could be developed to model predictive control with a combination of bioreactor data.

Detailed Characterization of Culture Media Through the Use of Imaging Technology Leading to Enhanced Understanding and Control of Bioprocesses

Mark Kemper, Rudy Hofmeister, Scott Tandy

Imaging technology provides strategic detailed physical and chemical information that enables manufacturers to better understand and control their processes. Culture media represent seminal components of bioprocesses. The quality of these media and the understanding of the mechanisms of their contribution to their respective processes are critical. However, because of their complexity, they are very difficult to characterize and control. By using a combination of UV/Vis-fluorescence and Raman imaging, chemical media can be characterized in detail. Over 90% of chemical components can be identified, lot-to-lot and vendor-to-vendor differences can be understood, impurities and foreign bodies can be detected and particle size information can be obtained. One practical use for this is incoming quality control to prevent sub-par lots from being used in production. However, a second, more fundamental application is the investigation of spent media to gain insight into the way in which cells interact with the constituents, thus leading to enhanced process understanding. These goals are not easily achieved by any other means and both will be described in this talk. Examples of real-world cases will illustrate the insights gained by this approach.

How to develop and implement Raman glucose control for biomanufacturing mammalian cell culture processes
John Paul Smelko

Raman Spectroscopy has long been discussed as a novel instrument to measure cell culture analytes in near real time due to its unique attributes such as specificity and limited water interference. While the use of bolus glucose daily shots has been the traditional method of control, Raman spectroscopy provides a more strategic and accurate control paradigm opening new opportunities for process optimization. This talk will outline the key steps in model development, control strategy definition, and system integration when driving a technology from concept to reality at 15,000L bioreactor scale to support a PPQ campaign.

Lighting up non-fluorescent molecules with Stimulated Raman Scattering microscopy
Dan Fu

Cells are the fundamental building blocks of life. A major challenge in biology is to understand how individual cells process information and respond to perturbations. Much of our knowledge comes from ensemble measurement, where intrinsic cell heterogeneity is ignored. Yet, it is now widely accepted that cell heterogeneity plays critical roles in many pathophysiological processes, especially cancer. We are interested in developing novel optical imaging tools that are capable of interrogating biological systems noninvasively at single cell or subcellular resolution. In particular, we use stimulated Raman scattering (SRS) microscopy – an emerging vibrational imaging technique - to perform in situ precision chemical measurements. SRS microscopy combines the molecular fingerprinting capability of Raman spectroscopy with high speed optical imaging to enable chemical imaging of living biological systems. In this talk, I will present an overview of the current status of SRS microscopy and our own efforts in applying SRS microscopy to study the transport and metabolic dynamics of lipids and small molecule drugs in living cells, tissues, and organisms.

Optimized methods for spontaneous Raman-imaging of immune cells
Nicholas Smith, Alison Hobro, Nicolas Pavillon

Spontaneous Raman imaging is an all-optical method that uses light scattering to determine aspects of the molecular composition of a sample. In a scanning mode, it can be used to map intracellular features in live cells, and thereby provide a view of how the cells differ from each other, or over time following a change in the cell state. Important examples of this include measuring how macrophage cells respond to bacterial triggers, as well as the use to discern characteristic features that define different types of cells. With a wide range of applications, the utility of Raman imaging is limited by the relatively low amounts of Raman scattered light from the sample. To achieve the most useful information from Raman measurement of a sample, we are usually constrained by laser power, and sample lifetime. Efficient sampling is therefore important, and reducing the sampling rate is often necessary to optimize the imaging. This presentation will show Raman measurement of immune cells, and with an aim of higher throughput, demonstrate compressed sensing, as well as use multimodal approaches, which can significantly improve the Raman imaging speed. Compressed sensing is a method of reducing the total amount of sampling in order to reduce the redundancy of measurements. It then employs a reconstruction model, the accuracy of which depends on both the sample characteristics and the measurement methods. We will show that for typical Raman imaging conditions, and improvement in speed by between 3 and 10 times is not unrealistic. In terms of sample preparation, while live cells are ideal in many ways, for some measurements, fixed cells can be useful, but without a way of measuring the changes that can occur during fixation, fixation may introduce unknown influences on the measurements. We therefore also investigate the effects of various fixation methods on the Raman spectra, and show that depending on the features of interest, several fixation processes are acceptable for Raman imaging.

Confocal Raman Microscopy for Investigating the Internal Surface Chemistry of Porous Particles
Joel Harris, Jay Kitt, David Bryce

Functional supports for practical applications of interfacial chemistry, such as heterogeneous catalysis or chemical separations, are most typically porous materials. This situation arises because high surface area is needed to provide molecular capacity for separations or catalytic throughput. Materials such as silica xerogels or porous polymer supports have very high specific surface areas (>200 m2/g) providing high capacity for these applications. The challenge for probing the interfacial chemistry of these materials, in order to understand their structure and functioning, is that nearly all of their active surface is within the *interior* of the material, while most surface-sensitive spectroscopies are not capable of probing buried interfaces. In this work, confocal Raman microscopy has been adapted to examining the interior surfaces of *individual* porous chromatographic silica particles, reporting information on the structure of the interface and its interactions with solute molecules. The poor sensitivity of Raman spectroscopy arising from the small scattering cross sections, which limits its application for studying planar interfaces, is overcome in this case by the high surface area of the porous substrate, where a monolayer of a small-molecule adsorbate generates a local concentration within a particle of ~0.5 M. Because silica is highly porous, one might anticipate significant loss of image fidelity due to scattering from the pores; however, the integrity of a focused excitation beam and the scattered radiation is well maintained within a particle because the 10-nm pores are much smaller (1/60th) than the wavelength of light; the resulting spatial resolution of the well-defined confocal volume allows selective in-situ probing of the interior structure and composition of an individual particle. We apply this technique to kinetic studies of solute exchange measured under high-retention conditions. We also investigate the structure and properties lipid bilayers deposited on bare silica surfaces and hybrid-lipid monolayers assembled onto C18-modified silica surfaces; these materials can be used for assessing lipid membrane affiliation of analytes from their retention on these model lipid-membrane phases. Raman spectroscopy provides insight into conformations of the acyl chains of the lipid, and how these conformations change with lipid deposition conditions and temperature.

3D Raman Imaging: A Method to Study the Effects of Lubrication on the Microstructure of Tablets

Shashwat Gupta, Savitha Paniakr, Fernando Muzzio

When it comes to pharmaceutical tablets, there are known knowledge gaps in the current manufacturing processes, particularly regarding the relationship between material properties of ingredients, processing conditions, and resulting tablet attributes. These three aspects can be linked through a common feature of understanding the microstructure of a sample. A better understanding of tablet microstructure has the potential to unlock the black box between process parameters and performance attributes. Product structure understanding of this nature can aid in prediction of product performance, process development as well as deviation investigations. Understanding the microstructure of a tablet requires knowing the spatial distribution of the different ingredients and void space, which can be enabled by Chemical Imaging (CI). Raman imaging is a powerful tool for studying tablet structure, with superior chemical specificity than NIR and is also less affected by physical characteristics of the sample. This study focuses on using 3D Raman Imaging system (H2Optx), as an off-line measurement system, to bridge this gap for tablets made using a Direct Compression (DC) continuous line, for the lubrication process. For pharmaceutical operations, lubrication is essential in order to reduce the friction between the surfaces of manufacturing equipment and that of organic solids as well as to ensure the continuation of an operation. The most common pharmaceutical lubricant used is Magnesium Stearate (MgSt). Till now, extensive structural studies have not been conducted to study the effects of lubricant(s) on tablet properties. In this work, we investigate the particle size distribution, spatial variability (Variograms) of MgSt, using the 3D Raman Images. These parameters are then correlated with product performance (tablet hardness and dissolution profile) as well as formulation (amount of MgSt) and process parameters (Total exposure to shear and Residence Time), with the help of Principle Component Analysis.

Advances in approaches and techniques for the acquisition, analysis and post-processing of three-dimensional Raman imaging data

Joachim Koenen, Ute Schmidt, Wei Liu

Raman imaging has become a widely-used technique, with the number of papers featuring the method doubling in the last five years. Developments in hardware operation and data evaluation have enabled this trend. Though more researchers now use Raman imaging, compromises in performance have at times been made to further accessibility. This submission details advancements in Raman imaging systems that allow the acquisition of data from rough and inclined surfaces and its quick evaluation without any compromise in throughput or resolution. A new operating concept has reconciled high-performance and user-friendliness. The positioning of the sample, selection of an objective, illumination control and auto-focus have all been integrated and placed under the control of a tactile and intuitive user interface. This facilitates the location of a sample's area of interest and enables easy access to the full capabilities of microscopes such as the alpha300 or apyron series. Microscopes with high confocality, on the order of <800nm FWHM, strongly reject out-of-focus light and therefore rough surfaces can pose a challenge for them. A variation in sample height of only several hundred nm can result in a drastic change in signal intensity. The TrueSurface option was introduced in 2010 to remedy this [1]. It conducted a non-contact measurement of the surface topography with an optical profilometer, which then guided a subsequent Raman imaging measurement, allowing measurements that were previously impossible [2, 3]. The drawbacks of the two-step procedure were that it took longer and no real-time compensation could be made for changes due to humidity or thermal expansion. The new generation TrueSurface eliminates these limitations with one-pass simultaneous operation and closed-loop feedback. The results of measurements performed without and with the new TrueSurface are shown in Fig. 1. Additional examples will be presented to showcase the technology's advantages. Many algorithms exist for evaluating Raman imaging data sets, including cluster analysis and non-negative matrix factorization. They are effective but not necessarily intuitive or easy to use. A new approach, called True Component Analysis extracts useful information from data sets while guiding the user through the process.

Trends and Potentials of Raman Spectroscopy in biological safety and security

Juergen Popp

Raman spectroscopy has emerged in the last years as an extremely powerful analytical method. Especially the field of bio Raman spectroscopy is expanding rapidly and a variety of Raman techniques are now available to enhance the signals of the intrinsically weak Raman scattering effect and to quench the interfering fluorescence. The most prominent approaches is surface enhanced Raman scattering (SERS). The first part of this presentation reports about the application of surface enhanced Raman spectroscopy to detect contaminants. It will be shown that SERS using metal colloids as SERS active substrate in combination with a microfluidic lab-on-a-chip (LOC) device enables high throughput and reproducible measurements with highest sensitivity and specificity. The application of such a LOC-SERS approach for the detection of e.g. antibiotics in waste and surface waters will be presented. Furthermore, we will introduce innovative bottom-up and top-down strategies to prepare SERS-active substrates like e.g. nanostructures coated with a lipophilic sensor layer as one-time use SERS substrates for specific food analysis (e.g. quantitative detection of toxic food colorants). The last part of this presentation introduces Raman biochip approaches for a fast, reliable and on-site sensing of microbial pathogens for water-, air- and soil monitoring (e.g. identification of potential bioterrorism agents, identification of food-born microorganisms etc.). It will be shown that Raman microspectroscopy permits the rapid identification of single bacteria cells without the need of time-consuming cultivation steps. The Raman spectra of the individual cells allow for quickly identifying pathogens based on their characteristic Raman spectra by using innovative chemometric approaches.
Acknowledgements Financial support of the EU, the "Thüringer Ministerium für Wirtschaft, Wissenschaft und Digitale Gesellschaft", the "Thüringer Aufbaubank", the Federal Ministry of Education and Research, Germany (BMBF), the German Science Foundation, the Fonds der Chemischen Industrie and the Carl-Zeiss Foundation are greatly acknowledged.

Development of Standoff Deep UV Resonance Raman Determination of Trace Explosives
Sanford Asher, Sergei Bykov, Katie Gares, Kyle Hufziger

We are developing deep UV resonance Raman standoff spectrometers to detect trace explosives. For this effort we are developing novel compact lasers, ultrahigh efficient spectrometers and imaging spectrometers. We will discuss both rastering as well as an imaging instrument that utilizes a novel photonic crystal narrow wavelength band optic. We are developing methodologies to detect explosives at standoff distances greater than 2 m. We are exploring the spectroscopic signatures of explosives as well as their photochemistries, which gives rise to specific dynamic spectral signatures. Lastly, we will discuss the deep UV Raman spectral signatures of $NaNO_3$, NH_4NO_3, TNT, RDX, PETN and HMX.

Novel through-barrier detection of Explosives, Narcotics and their Precursors

Matthew Bloomfield, Robert Stokes

Hand-held detectors play a critical part in the role of first responders and a variety of analytical techniques are exploited to achieve *in situ* identification. Raman spectroscopy has become a significant part of the 'tool-kit' since being introduced as a portable detector some 10 years ago. This vibrational spectroscopic technique has unparalleled specificity and wide applicability to the compounds of interest, but has been limited, historically, to the analysis of materials with direct, 'line of sight' to the detector. The development of a novel variant, Spatially Offset Raman Spectroscopy (SORS), has extended the range of materials to those contained within a variety of common containers. This presentation will include examples from paper, fabrics, glass, opaque and coloured plastics. SORS leverages the differences in spectra acquired at different positions to separate the characteristic signal arising from the sub layers of a sample from the spectrum at the surface with no *a priori* knowledge of the surface. In the context of security, this method has the advantage that the operator is able to obtain a positive identification without opening or disturbing the container to take a sample for laboratory analysis, potentially improving the safety and efficiency of operating procedures in explosive ordinance disposal and first response.

Universal Detection of Body Fluid Traces In Situ with Raman Hyperspectroscopy for Forensic Purposes

Marisia Fikiet, Gregory McLaughlin, Masahiro Ando, Hiro-o Hamaguchi, Igor Lednev

Raman Spectroscopy has been a boon to the field of forensic science. One area that this technique shows exceptional promise is in body fluid identification and characterization, but substrate interference remains a major impediment to its practical implementation. Here, we present an approach for the universal detection of body fluids regardless of the substrate. This approach, which is based on Raman hyperspectroscopy and Multivariate Curve Resolution (MCR), was applied to datasets representing simulated semen evidence. In every instance, the signal of the fluid was extracted and matched to a reference of semen. This approach has immediate application for body fluid detection and allows for a universal, automatic, nondestructive, on-field method for confirmatory identification of body fluid traces at a crime scene. This approach is applicable where an analyte is either a minor contributor or spatially distributed on a strongly interfering substrate.

Raman spectroscopy technique in Toxicology: detecting and quantifying cocaine in seized drug samples from an Amazon State of Brazil.

Landulfo Silveira, Ciro Penido, Marcos Pacheco, Renato Zângaro, Igor Lednev

Identification of cocaine and subsequent quantification immediately after seizure are problems for the Police in developing countries such as Brazil. Also, overdose of cocaine accounts for a significant number of hospital admissions, and identification and quantification of the drug by the Emergency units is crucial for the quick medical intervention or determination of the cause of death. In this study we evaluated the possibilities of Raman spectroscopy as a method to identify and quantify cocaine and adulterants employed to produce crack and adulterants commonly used to increase volume, and also possible degradation products in samples seized by the Police. Seized samples of freebase cocaine, cocaine hydrochloride, cocaine paste and crack were obtained from the Police Evidence File of Amapá State, Brazil, as well as samples of pure adulterants such as caffeine, lidocaine and sodium carbonate. Ternary mixtures of crack with caffeine/lidocaine and carbonate were prepared in different concentrations using the simplex centroid design. Raman spectra at 830 nm excitation were obtained from these samples and both qualitative and quantitative analyses were performed. Freebase cocaine, cocaine hydrochloride and crack can be distinguished by the Raman spectral, revealing differences in their chemical structure. Most of the cocaine samples, including paste and crack, showed characteristic peaks of degradation products such as benzoylecgonine and benzoic acid, and some presented evidence of adulteration with aluminum sulfate and sodium carbonate. Multivariate regression models based on partial least squares (PLS) were used to obtain the concentrations of the compounds of the ternary mixtures, showing high correlation coefficients ($r > 0.98$) and small cross-validation prediction errors ($< 6\%$) for cocaine. It has been demonstrated that Raman spectroscopy is suitable for both qualitative and quantitative analysis of seized samples, determining the type of the drug, presence of adulterants (that may indicate drug's origin and traffic route) and degradation products, and the concentration of the drug and the adulterants in ternary mixtures, therefore becoming a technique of choice in Forensic Toxicology, and for the Police/Emergency in cases of intoxication or overdose, aiming a quick medical intervention or determination of the cause of death.

Super-resolution chemical imaging with SERS and STORM
Nathan Lindquist, Aeli Olson, Kelsey Spies, Anna Browning, Paula Soneral

Applications of Surface Enhanced Raman Spectroscopy (SERS) for chemical imaging are limited in spatial resolution by diffraction. To overcome this limit, surface plasmon resonances in metallic nanostructures can be used to manipulate optical energy within sub-wavelength dimensions. These subwavelength regions of intense, confined light are often called "hotspots" and can give chemical and spectral information of proximate molecules via SERS. These hotspots can also give rise to a "blinking" SERS effect that can be processed using Stochastic Optical Reconstruction Microscopy (STORM) algorithms for super-resolved imaging. Furthermore, by imaging SERS blinking through an optical transmission grating, STORM algorithms can be modified to also extract a full SERS spectrum, thereby capturing spectral as well as spatial content simultaneously. This presentation describes a chemical imaging strategy that combines SERS and STORM for super-resolved chemical imaging. Images taken with this technique of biological samples, e.g. such as gram-positive vs. gram-negative bacteria, show excellent agreement with scanning electron microscope (SEM) images, very high spatial resolution on the <50 nm scale, as well as spectral content obtained from SERS that can be correlated to different regions of the image.

Advanced Microscopies for Nanophotonics
Matthew Sheldon

The cost of solar energy is a primary factor that limits more widespread adoption, and improved conversion efficiency is the most effective strategy for lowering cost. Although commercial solar cells today generally provide less than 20% conversion efficiency, thermodynamic analyses indicate a maximum efficiency of 95% is possible. In this talk I will show how nanofabricated optoelectronic and plasmonic materials provide a route to achieve this maximum efficiency, and describe some of the cutting edge microscopy techniques that can be used to study materials design for this purpose. By properly optimizing light-matter interactions, nanostructured materials can enable systematic control of the thermodynamic parameters governing optical power conversion. I will discuss several studies of nanoscale structuring that exemplify these opportunities, as well as powerful microscopic and spectroscopic characterization techniques for understanding applications in solar energy conversion.

Spatially Resolving Vibrational Couplings for Structure Identification with Hyperspectral 2D Infrared Microscopy
Joshua Ostrander, Martin Zanni

Two-dimensional infrared spectroscopy (2D IR) has been invaluable for the studies of liquid structure, protein biophysics and materials science. In this talk, we report the first wide-field microscope that can simultaneously collect over 16 000 2D IR spectra to spatially resolve molecular structure and femtosecond dynamics. Enabled by a dual Acousto-Optic Modulator (AOM) pulse shaper and an IR Focal Plane Array (FPA) detector, the 2D IR microscope exhibits diffraction-limited spatial resolution and enhanced image contrast compared to conventional FTIR microscopy. The dual-AOM enables rapid scanning of time delays and phase cycling, which suppresses background signals and exhibits improved signal-to-noise. We first demonstrate the principles of 2D IR microscopy by imaging polystyrene beads with transition metal carbonyl labels. We specifically highlight how the extra spectral dimension, relative to linear FTIR microscopy, can be used to improve molecular discrimination and eliminate noise. Cross peaks present in 2D IR spectroscopy offer the possibility of spatially resolving differences in molecular structures. We will present data from ongoing experiments that use 2D IR microscopy to resolve these differences in scientifically interesting samples, such as the heterogeneous distribution of amyloid protein structures. This ongoing work shows that 2D IR microscopy is poised to impact the study of amyloid diseases in tissues, including the importance of amyloid polymorphism in fibril formation.

Multimodal nonlinear imaging for sensitive and specific analysis of multiple solid state forms and their changes on pharmaceutical tablets

Clare Strachan, Dunja Novakovic, Jukka Saarinen, Antti Isomäki, Sara Fraser-Miller, Leena Peltonen, Timo Laaksonen

Two complimentary nonlinear imaging modalities, coherent anti-Stokes Raman scattering (CARS) and sum-frequency generation (SFG), were combined and investigated for sensitive, specific, rapid, and nondestructive imaging of multiple solid-state forms of a drug and changes on tablet surfaces. CARS variants included both hyperspectral imaging, combined with multivariate analysis, and narrowband imaging. SFG included detection of both the second harmonic generation (SHG) signals from each of two laser wavelengths, as well as the combined SFG signal from the two wavelengths. A fully-integrated Leica TCS SP8 CARS microscope was used for the CARS and SFG imaging, with complimentary scanning electron microscopy, ATR FTIR and Raman microscopy serving as comparative analysis methods. Indomethacin, which exists as the amorphous form and multiple crystalline polymorphs, was used as the model drug. In the first part of the study, tablets containing mixtures of the amorphous form and two crystalline polymorphs, alpha and gamma, were prepared and their surfaces imaged. The three different forms of the drug exhibited different CARS spectra that allowed both hyperspectral and narrowband imaging of the three forms on the (sub)micron level. SFG imaging allowed differentiation of the SFG active (noncentrosymmetric) alpha form from the non SFG-active gamma and amorphous forms, and was complimentary to the CARS data. Narrowband CARS and SFG imaging was faster and simpler, while hyperspectral CARS and SFG would allow selective imaging of more complex samples. In the second part of the study, amorphous indomethacin tablets were prepared and stored at 30 C/23% RH and 30 C/75% RH. Very early stage crystallisation was detectable and the resulting polymorphs could be identified. The surfaces crystallized mainly, but not exclusively, to the gamma form at 23% RH and the alpha form at 75% RH. The nonlinear imaging was more sensitive and specific than any of the reference analytical methods. Overall, this study suggests that multi-modal nonlinear imaging is a viable approach for highly sensitive and specific analysis of solid-state changes in pharmaceuticals. Since it appears to be more sensitive than most established analysis methods, it may be useful for understanding and optimizing diverse drug formulations, including amorphous drug formulations.

Spatially Encoded Polarization-Dependent SHG Microscopy of Pharmaceutical Materials

Garth Simpson, Changqin Ding, James R. W. Ulcickas, Fengyuan Deng, Ellen J. Gualieri

Polarization-dependent imaging provides rich structural information in numerous optical microscopy measurements, including bright-field, fluorescence, and Raman microscopy. In second harmonic generation (SHG) imaging, polarization analysis is particularly advantageous, given the high sensitivity of SHG to molecular-scale structure and orientation. In the present work, introduction of a depolarizer into the appropriate location in the beam path has the opposite effects of its intended purpose, providing a simple yet precise method for modulating the polarization across the field of view. The depolarizer consists of a liquid crystal thin film, which induces a different rotation of the polarization state of the beam as a function of position. Ad designed, this spatial variation in polarization results in a depolarized plane-wave beam following passage of a plane wave through the optic. However, placing this same optic in the rear conjugate plane of a microscope with a polarized source instead produces well-defined discrete polarization states that vary smoothly with location within the field of view. This simple approach was integrated into a beam-scanning microscope for polarization-dependent analysis in second harmonic generation (SHG) microscopy, with the mechanics of beam-scanning allowing for polarization modulation at ~60 kHz. Spatially encoded polarization-dependent SHG microscopy have been used to deduce molecular arrangements of collagen microfribrils relative to the primary fiber axis for diagnostics of tissue samples, and for rapid identification of pharmaceutical crystal form targeting high-throughput screening applications.

Race Differentiation by Raman Spectroscopy of a Bloodstain for Forensic Purposes
Ewelina Mistek, Lenka Halámková, Kyle Doty, Claire Muro, Igor Lednev

Bearing in mind forensic purposes, a nondestructive and rapid method was developed for race differentiation of peripheral blood donors. Blood is an extremely valuable form of evidence in forensic investigations so proper analysis is critical. Because potentially miniscule amounts of blood traces can be found at a crime scene the ideal method is nondestructive while providing a substantial amount of information about the sample. In this study Raman spectroscopy was applied with advanced statistical analysis to discriminate between Caucasian (CA) and African American (AA) donors based on dried peripheral blood traces. Spectra were collected from 20 donors varying in sex and age. Support vector machines-discriminant analysis (SVM-DA) was used for differentiation between the two races. An outer loop subject-wise cross-validation
(CV) method served to evaluate the performance of the SVM classifier for each individual donor from the training data set. The performance of SVM-DA, evaluated by the area under the curve (AUC) metric, showed 83% probability of correct classification for both races, and a specificity and sensitivity of 80%. This preliminary study shows promise for distinguishing between different race donors of human blood. The method provides rapid and reliable results without any preparation, destruction, or consumption of the sample— thus making it an ideal method for use at real life crime scenes.

Detection and identification of illicit substances in articles of mail by a shortwave infrared (SWIR) hyperspectral imager
Oksana Olkhovyk, Nathaniel Gomer, Robert Schweitzer, Jeffrey Beckstead, Matthew Nelson

Shortwave infrared hyperspectral imaging (SWIR HSI) has been proven to be a valuable analytical tool. In particular, it can be used for a variety of applications to detect and identify hazardous narcotic, explosive, chemical and biological threats based on the materials' molecular composition. This technology is especially valuable for wide-field, non-contact verification of illicit substances that may be concealed within regular articles of mail and sent to correctional facilities. Tainted mail items pose a threat to the mail inspectors, prison guards and prisoners, as well as provide grounds for criminal prosecution of the parties involved. Due to the proliferation of drugs being smuggled into prison in various forms (e.g., mail soaked in drug solutions, drugs hidden within envelope seams or under stamps, or drugs mixed with crayons, paints, markers, etc.), a need exists for a sensor capable of detecting and identifying illicit materials beyond what an inspector is capable of detecting through visual and manual inspection. The development of the mail screener system, equipped with a SWIR multi-conjugate liquid crystal tunable filter (MCF), will be discussed. The MCF system enables wavelength-dependent detection of dozens of illicit substances and cutting agents by providing spectrally and spatially-resolved contrast of the materials of interest within a complex background in the form of an image. Both reflection and transmission spectral properties of the suspected drug may be probed in the spectral range from 850 to 1700 nm. Presumptive identification is then possible by obtaining and processing spectral information of the detection and its surrounding background using advanced chemometric image processing tools. Advantages of nondestructive HSI analysis of mail articles tampered with illicit substances will be presented and compared to other methods of illicit drug identification in complex systems. The limits of detection and identification will be presented for the most commonly interdicted illicit drug substances. Detector performance analysis will be demonstrated using receiver operating characteristic (ROC) curves.

Gas Composition Measurements during One-Dimensional Time to Explosion Experiments

Greg Klunder, Paul Spackman, Fowzia Zaka, Nick Muetterties, Evan Kahl, Peter Hsu

One-Dimensional Time to Explosion (ODTX) experiments measure the time to explosion under isothermal conditions. Explosive material is placed in a sealed spherical chamber where the geometry provides well-defined boundary conditions and the time and temperature data are used for modeling the kinetics of the material decomposition. In recent developments, a pressure transducer has been incorporated into the experiment (P-ODTX) to determine gas concentrations during thermal ramp boundary conditions. The pressure measurement is useful in quantifying extent of reaction for modeling, but P-ODTX experiments still do not provide chemical speciation information. Extracting evolved gases for analysis with a mass spectrometer can alter the experiment as many materials have autocatalyzing and pressure dependent reactions in gas-solid and gas-gas phases. In-situ spectroscopy could provide chemical speciation with minimal alteration to the experimental configuration. Given that each experiment is designed to run to explosion, the near-infrared (NIR) spectral range offers a more cost-effective option for detection than the mid-infrared (MIR). Thus, NIR spectroscopy has been implemented to measure the dominant gases being produced during the ODTX experiments. The pressures monitored during the P-ODTX experiments indicate that concentrations of gases should be detectable in a short pathlength cell in this wavelength range. This presentation will describe ODTX experiments, discuss the engineering required to detect evolved gases in-situ, and present results from early experiments. This work was performed under the auspices of the United States Department of Energy by Lawrence Livermore National Laboratory under Contract DE-AC52-07NA27344.

Calibration of the Likelihood Ratio for the Evaluation of Forensic Glass Evidence
Ruthmara Corzo, Daniel Ramos, Jose Almirall

The continuous likelihood ratio provides an objective, quantitative measure for the evaluation of forensic evidence. This approach offers an improvement over the use of a subjective verbal scale to describe the significance of evidence or the use of a fixed match criterion that can lead to the "fall off the cliff" effect. A two-level model previously proposed by Aitken, Zadora, and Lucy has been used in forensic casework but the output of this model yields unreasonably high (or unreasonably low) values for the LRs. Therefore, a post-hoc calibration step is necessary in order to improve the accuracy of the likelihood ratio and limit the magnitude of the LRs to correlate with the size of the available databases. In the present study, elemental concentration data collected with LA-ICP-MS was used to generate two glass databases: a casework database consisting of 398 samples and a vehicle windshield database consisting of 420 samples. The two glass databases were then used to evaluate different "calibration" methods for the likelihood ratio calculation. The first calibration approach involves the use of the Pool Adjacent Violators (PAV) algorithm, while the second approach involves the use of density models followed by an empirical upper and lower bound method. The performance of the two methods is compared in terms of their rates of misleading evidence (false exclusions and false inclusions) and the cost log likelihood ratio (Cllr). The likelihood ratios determined by both calibration methods present an objective and quantitative assessment that communicates reasonable significance values for glass evidence.

Monitoring Sulfonate Ester Potential Genotoxic Impurities in Pharmaceuticals by GCMS and LCMS
Joseph Snodgrass, Shunyan Mo, David Moon, Ricardo Borjas

The requirements for monitoring alkyl esters of alkyl or aryl sulfonates as potential genotoxic impurities in pharmaceuticals remain stringent, despite progress in understanding the conditions and kinetics leading to their formation in synthetic processes. A quantitation limit of 0.1 – 1 ppm is often required. Analytical approaches most often employ GCMS (headspace or direct injection) and LCMS (positive ion via ammonium adducts or negative ion via stable anions formed by dissociative electron attachment). Sample preparation may include derivitization or extraction; internal standards are sometimes used. Monitoring in the final drug substance and/or drug product is always required. Measurements of in-process solution samples, starting materials and process intermediates are also required if Quality by Design and Spike/Purge studies are conducted. Because of the variety of approaches, an overview is useful for navigating the analytical options. Analytical monitoring is most commonly needed when a sulfonate salt is the chosen salt form for the active pharmaceutical ingredient. Monitoring may also be required if aryl or alkyl sulfonic acids are present in the synthetic route, or are incorporated as a structural moiety in the final drug substance structure. In this case, monitoring free sulfonic acid in process steps may also be of interest. We describe the development of a novel method for monitoring alkyl benzenesulfonates in an active pharmaceutical ingredient containing a sulfonyl substructure. This presented significant selectivity challenges for GCMS analysis due to potential degradation of the drug substance in the injection port and on-column. We desired a direct injection method, without internal standards, derivitization, extraction or complex sample preparation to provide flexibility for in-process samples and intermediates. We discuss how a variety of alternative GCMS and LCMS methods may be useful during development. An adjunct, negative ion LCMS method for benzenesulfonic acid is also described.

Detailed molecular composition analysis of middle distillates using GC×GC-TOFMS; A tool to upgrade low value refinery streams

Kalicharan Chattopadhyay, Anil Yadav, Dheer Singh, Anju Chopra, J Christopher , G. S. Kapur

Detailed molecular composition analysis of middle distillates using GC×GC-TOFMS; A tool to upgrade low value refinery streams Kalicharan Chattopadhyay, Anil Yadav, Dheer Singh, Anju Chopra, J. Christopher and G.S. Kapur Research & Development Centre, Indian Oil Corporation Limited, Sector-13, Faridabad-121007, Haryana (India) Email: kalicharanc@indianoil.in Corresponding author; Tel.: +91 129 2294737; fax: +91 129 2286221. Abstract Strategy for the up gradation of residues and byproducts is emerging and could be achieved only if the unique chemical value of residues and byproducts can be tracked out. This accomplishment needs advanced analytical capabilities to have detailed molecular composition of refinery stream having low value. A two dimensional gas chromatography coupled with time-of-?ight mass spectrometry (GC×GC–TOFMS) method has been developed for simultaneous estimation of benzene, toluene, ethylbenzene & xylenes (BTEX) and detailed hydrocarbon class composition in various middle distillates like LCO, RCB which could be used as value-added products in portfolio of petrochemical industry. Rapid expansion of aromatics markets and the search for increased amounts of BTEX feedstocks from LCO is promising due to its depressed value for upcoming diesel specifications. On the other hand BTEX is not recommended in Raf?nate column bottom (RCB), which generates as a byproduct from linear alkyl benzene (LAB) plants and proposed to be used as less hazardous petroleum products. Therefore, estimation of BTEX in middle ditilates from percentage to ppb level is important and disclosed in the developed method. Molecular compositional analysis in turn of hydrocarbon class composition (PIONA) is included in the method as well for the stated middle distillates. Usually number of compounds and their isomers increases exponentially with the boiling point; consequently, conventional GC becomes rather ineffective for the analysis of middle distillates. GC×GC–TOFMS was found to be only approach where compounds get separation based on boiling point as well as polarity. However, quanti?cation in TOF–MS needs response factor for every compound present in sample, which is not possible because of a lack of available standards. The limitation of non-uniformity of compounds' response in TOF–MS has been dealt effectively through plotting response factors against carbon numbers for each class of compound. The evaluation of a compound's response versus carbon number was carried out using the gravimetric mixture of 46 compounds belonging to paraf?n, isoparrafin, naphthene and aromatics classes. The outcomes of hydrocarbon class composition analysis have been validated with the HPLC-based ASTM D6591 method to check the appropriateness of the developed method, and good correlation was found.

Charge Transfer Dissociation (CTD) Mass Spectrometry of Sulfated Oligosaccharides

Zachary Sasiene, Praneeth Mendis, Glen Jackson

Tandem mass spectrometry has been a vital tool in the fields of proteomics, peptidomics, metabolomics, and other areas of biomolecule analysis. Some of the current fragmentation techniques being used to analyze biomolecules include collision induced dissociation (CID), electron transfer dissociation (ETD), and electron capture dissociation (ECD). These techniques work well but they have their limitations. CID does not always provide full fragmentation of a peptide backbone and often fails to identify the position of labile post-translational modifications, like phosphates and sulfates. ECD and ETD typically only fragment precursor ions that have a charge state of 2 or greater (z ? +/-2). These limitations have prompted the development of a variety of high energy and radical-initiated fragmentation techniques, including charge transfer dissociation (CTD). CTD involves the placement of a saddle field fast ion source above a 3D ion trap. A variable leak valve controls the amount of helium flowing through the source. An arbitrary function generator is used to generate a square wave with 0-5 V, which is amplified by a high voltage amplifier to generate helium cations with kinetic energies up to 10 keV, which are timed to line up with the fragmentation portion of the scan function. CTD of oligosaccharides has been shown to yield similar results as ultraviolet photodissociation (UVPD) with 1+ and 2+ precursor ions, including very valuable cross-ring cleavages. This current project extends the application of CTD to extensively sulfated oligosaccharides, which are prevalent in pectins, algae, and cartilage. The ability to preserve these labile modifications while producing extensive cross-ring cleavages provides a valuable tool to assist the characterization of highly-sulfated complex oligosaccharides.

Measurement of Material Properties from Levitated Micron Sized Particles

Matthew Hart, Vasanthi Sivaprakasam, Jozsef Czege, Jay Eversole

The linear electrodynamic quadrupole trap (LEQ) provides an excellent mechanism to study single or many particles in a "touchless" environment. Our experimental approach provides the opportunity to investigate liquid or solid particles in controlled environments over relatively long periods of time. Particles are two dimensionally constrained along the axis of the trap where they are free to move along that axis. By applying airflow and electrostatics we control the movement and positioning of the particles for various types of optical interrogation. We have developed a system where we employ electrostatics, light scattering and spectroscopy to measure physical properties with high sensitivity from materials, as a free aerosol, with sizes ranging from
0.25 ?m to 100 ?m in diameter. These properties include surface charge, diffusion coefficients, particle sizing accurate to within 10 nm, vapor pressure lower than 10-5 Pa, fluorescence, and Raman spectra from picograms of material. Here we present an overview of the LEQ capabilities and examples of aerosol particle measurements as well as possible future applications.

Identification of Streptococcus gallolyticus subsp. gallolyticus TX20005 (biotype I) competence stimulating peptide pheromone

Anthony Harrington

Quorum sensing is a communication mechanism bacteria utilize to facilitate physiological group behavior such as biofilm formation, virulence, motility, sporulation, antibiotic production and competence.[1] *Streptococcus gallolyticus subsp. gallolyticus* (biotype I), previously known as *Streptococcus bovis* biotype I, is an emerging opportunist pathogen normally found in the gut of humans and is responsible for causing bacteremia, infective endocarditis (IE) and colorectal cancer (CRC).[2-3] In the last decade, the genomes of a few isolates of *S. gallolyticus subsp. gallolyticus* have been sequenced.[4-5] Using bioinformatics analysis of the known genome sequences along with isolation and analysis of supernatant contents by mass spectrometry, we have found preliminary evidence suggesting that *S. gallolyticus subsp. gallolyticus* TX20005 produces a 21-amino acid competence stimulating peptide that activates competence through the ComCDE pathway. Determining the correct signaling molecule is vital to better understand pathogenicity in this organism. 1. Bassler, B. L., How bacteria talk to each other: regulation of gene expression by quorum sensing. Current opinion in microbiology 1999, 2 (6), 582-587. 2. Beck, M.; Frodl, R.; Funke, G., Comprehensive study of strains previously designated Streptococcus bovis consecutively isolated from human blood cultures and emended description of Streptococcus gallolyticus and Streptococcus infantarius subsp. coli. Journal of clinical microbiology 2008, 46 (9), 2966-2972. 3. Kumar, R.; Herold, J. L.; Schady, D.; Davis, J.; Kopetz, S.; Martinez-Moczygemba, M.; Murray, B. E.; Han, F.; Li, Y.; Callaway, E., Streptococcus gallolyticus subsp. gallolyticus promotes colorectal tumor development. PLoS pathogens 2017, 13 (7), e1006440. 4. Lin, I.-H.; Liu, T.-T.; Teng, Y.-T.; Wu, H.-L.; Liu, Y.-M.; Wu, K.-M.; Chang, C.-H.; Hsu, M.-T., Sequencing and comparative genome analysis of two pathogenic Streptococcus gallolyticus subspecies: genome plasticity, adaptation and virulence. PLoS One 2011, 6 (5), e20519. 5. Rusniok, C.; Couvé, E.; Da Cunha, V.; El Gana, R.; Zidane, N.; Bouchier, C.; Poyart, C.; Leclercq, R.; Trieu-Cuot, P.; Glaser, P., Genome sequence of Streptococcus gallolyticus: insights into its adaptation to the bovine rumen and its ability to cause endocarditis. Journal of bacteriology 2010, 192 (8), 2266-2276.

Microwave-Assisted Electrospray Ionization (μAESI)
Maria Rivera, Jaime Orejas-Ibanez, Andrew Schwartz, Steven Ray

Electrospray Ionization (ESI) is one of the most extensively used ionization techniques for mass spectrometry, capable of generating gas phase ions from a variety of small and large molecules direct from solution phase. During the ESI process, molecules are released from solution as charged ions from highly-charged droplets. These droplets are formed from by an electrostatic, liquid free-jet expansion (Taylor cone), and while ion formation is thought to involve aspects of coulombic repulsion/explosion, all hypothesized ionization mechanisms involve some degree of solvent evaporation. Therefore, it is common for ESI sources to implement heating interfaces or gas streams in order to promote desolvation. Recently, our laboratory has developed a novel method to study the effects of microwave heating and desolvation upon the ESI process. Microwaves, at a frequency of 2.45 GHz, are focused near the tip of an ESI emitter by means of a novel waveguide structure. The coaxial waveguide focuses microwave energy in a very small and controlled volume. When the microwave radiation is absorbed, it induces very rapid evaporation through dielectric heating, modifying the features of the Taylor cone and affecting processes such as droplet generation, ionization efficiency, jet stability, and ion fragmentation. Here, we examine microwave-assisted electrospray ionization (μAESI) as a means to explore the mechanisms of ion generation, searching for methods to expand the capabilities of ESI through alterations in the fragmentation mechanisms and ionization processes.

The use of a capillary dielectric barrier discharge ionization (DBDI) source for spatially-resolved measurements of cholesterol in mouse brains
Mercede Erickson, Isabella James , Richard Carson, John C Price, Paul B Farnsworth

Abnormalities in cholesterol metabolism have been linked to several neurological disorders, including Alzheimer's disease. In this presentation we will describe our efforts to develop an efficient ambient ionization source that can be used to map concentrations and turnover rates of cholesterol in tissue samples. Desorption electrospray ionization (DESI) has proven to be an effective tool for mass spectrometric imaging of some lipids, but it does not efficiently ionize cholesterol. A miniature dielectric barrier discharge (DBD), contained in a fused silica capillary, samples a small spot on a sample surface and generates a significant (M+H-H2O)+ signal for cholesterol from samples deposited on a glass slide in solution, and also from thaw-mounted tissue samples. The remaining challenges are to characterize and optimize the spatial resolution achievable with the discharge source and to find a discharge configuration that is compatible with raster scanning and chemical image generation. In this poster we will describe our latest results with the DBD source and our progress toward addressing these challenges.

Use of Nonlinear Optics for Materials State Awareness

James Patterson

As a material component is used for its intended purpose, it can experience a variety of mechanical, environmental and chemical stresses. These stresses can cause changes in the properties of the component, which may affect its ability to perform its intended function. Optical methods have significant advantages over more invasive approaches in determining the state of a material and its likelihood of failure. We are applying nonlinear optics, specifically second harmonic generation (SHG) and vibrational sum-frequency generation (VSFG), to the investigation of metals and polymers to determine the response of these materials to a variety of stresses. We have observed correlations between the SHG response and the degree of mechanical extension of 2024 (aerospace grade) Al alloys under tensile load. We have also observed changes in the SHG response of 5456 (naval grade) Al alloys due to heat treatment. This heat treatment leads to formation of beta phase (Mg_2Al_3), which also makes the alloy much more sensitive to corrosion. For polymeric materials, VSFG has
great potential to determine structural changes imposed by tensile loading, as seen in stretched high density polyethylene

(HDPE) samples. In all cases, these nonlinear optical approaches are noninvasive and thus may serve as useful platforms for nondestructive evaluation (NDE) and materials state awareness (MSA).

Quantitative Spectroscopic Chemical Imaging Enables Product Purity Assessment Resulting from Operational Changes in Industrial Processing
David Wetzel, Mark Boatwright

Quantitation of product purity for the refinement of natural products has been a longstanding analytical challenge. The high volume, low margin nature of this industry means that even relatively small increases in product purity can provide a large multiplier. Thus, a sensitive method is necessary to monitor the effect of process changes. Spectroscopic chemical imaging with array detection enables objective determination of the chemical identity of each pixel within an x, y field of view. Summation of the results from thousands of individual pixel yields a robust quantitative result. Traditionally, a low quantity of impurity was used as a measurement of product quality and processing efficiency in the industrial process of wheat milling. For a group of product streams, weighted summation of the cumulative contribution of mineral residue (ash) resulting from the ignition of organic matter or off color of the product have been used to rate the overall process result. In this manner, the informed operations manager determines the economic penalty of his current operation and target operational parameters requiring adjustment for optimization. In contrast, quantitative chemical imaging provides a direct, objective molecular structural basis with enhanced contrast to determine purity. The utility of modern quantitative spectroscopic chemical imaging is demonstrated by its application to the determination of purity for 44 product streams from an industrial milling process before and after making significant operational adjustments. Applying this advanced analytical technique enabled the observation of selective product purity shifts among different streams within the processing flow. The results illustrate the potential for further optimization at multiple stages of the process.

Visible Light Tomography to Define Edges in FTIR Tomographic Reconstructions
Nicholas Walter, Carol Hirschmugl, Sugato Ray, Ghazal Azarfar, Alex Schofield

Fourier transform infrared (FTIR) spectro-microtomography, allows visualization of how chemical signatures are distributed throughout a biological or materials sample. In FTIR tomography, computational tomography algorithms are applied to a series of two-dimensional (2D) FTIR absorption images, taken while a sample is rotated, to create three-dimensional (3D) reconstructions of the sample at each wavelength. It can be problematic to visually display these reconstructions for weak chemical bands and spectral regions with high levels of noise, as distinguishing the edge of the sample from the background noise may be difficult. To correct for this, 3D masks are created by implementing visible light tomography. By rotating a sample while a series of visible images are taken, visible light tomography is implemented using the same experimental setup as FTIR tomography. Applying the same computational tomography algorithms used to create FTIR tomographic reconstructions, visible light reconstructions are created. From these reconstructions, 3D masks are made. In recent years, a line of cameras has been produced by FLIR Systems, Inc., which record both visible and IR images. While collecting IR images, the camera displays the visible light images concurrently with the IR images. This creates hard edges in the IR images, which help to identify sources of infrared radiation. By using visible light tomography, a similar overlay of images will be demonstrated for FTIR tomography. Visible light tomography will be used to create 3D mesh reconstructions, which will be superimposed with the FTIR tomographic reconstructions. These meshes will define the hard edges and outline the boundaries of the sample, allowing a better understanding of how chemical signatures are spatially distributed within the sample. This work was supported by the UWM Physics department and NSF grants CHE-1508240 and CHE-1112433.

Multi-modal Super-Resolution Microscopy through Super-Resolution Radial Fluctuations (SRRF)
Justin Cooper, Mark Browne, Hugh Gribben, Martin Catney, Colin Coates, Geraint Wilde, Ricardo Henriques

Super-resolution radial fluctuations (SRRF) is a synthesis of temporal fluctuation analysis and localization microscopy. One of the key differences between SRRF and other super-resolution methods is its applicability to live-cell dynamics because it functions across a very wide range of fluorophore densities and excitation powers. SRRF is applied to data from imaging modes which include widefield, TIRF and confocal, where short frame bursts (e.g. 50 frames) can be processed to deliver spatial resolution enhancements similar to or better than structured illumination microscopy (SIM). On the other hand, with sparse data e.g. STORM, SRRF can deliver resolution similar to Gaussian fitting localization methods. Thus SRRF could provide a route to super-resolution without the need for specialized optical hardware, exotic probes or very high power densities. We present a fast GPU-based SRRF algorithm and apply it to imagery from an iXon EMCCD coupled to a multi-modal imaging platform, Dragonfly. The new implementation is >300 times faster than the current CPU version running on an Intel Xeon 3.5GHz 4 core processor, and > 20 times faster than the NanoJ GPU implementation, while also being integrated with acquisition for real time use. We explore the image resolution and quality with sCMOS cameras and various fluorophores including fluorescent proteins and organic dyes.

True Resolution Enhancement for Optical Spectroscopy
Jeffrey Oleske, Justin Cooper, Hugh Gribben, Martin Catney, Colin Coates, Geraint Wilde, Ricardo Henriques

Resolving spectrally adjacent peaks is important in techniques such as tracking small shifts in Raman or fluorescence spectra, quantifying pharmaceutical polymorph ratios, or molecular orientation studies. Thus, spectral resolution is a vital consideration when designing most spectroscopic systems. However, most of the parameters that influence spectral resolution are fixed for a given system (spectrometer length, grating groove density, excitation source, CCD pixel size, etc). This is non-problematic if the spectrometer is dedicated for a single purpose; however, these specifications cannot be optimized for different applications with wider range resolution requirements. Data processing techniques including peak fitting, PLS or PCA are typically used to achieve sub-optical resolution information. These techniques can be plagued by spectral artifacts introduced by post-processing as well as the subjective implementation of statistical parameters. TrueRes™, from Andor Technology, uses an innovative optical means to greatly improve and expand the range of spectral resolutions accessible on a single setup. This increases the versatility of the spectrometer and allows one to fine tune the spectral resolution without sacrificing spectral bandpass. Data presented here shows a true spectral resolution enhancement of >30% without mathematical spectral alteration, data-processing or spectrometer component changes, on the Raman spectra of a characteristic analyte with spectrally-adjacent peaks The total throughput change and improvement in hyperspectral imaging quality are also characterized. We show that TruRes™ allows imaging of several individual fibers on both sides of the CCD detector. The enhanced resolution enables the spectrograph to produce a much higher hyperspectral track density by sacrificing some throughput.

Label-free Imaging of Amphotericin B Interacting with Live Cells Using Transient Absorption Microscopy

Kevin Higgins, Tessa Calhoun

Amphotericin B (AmB) is an effective, polyene antibiotic that has been in clinical use for five decades now. It has displayed limited microbial resistance compared to multiple other antifungal drugs, although it does have multiple side effects. Despite being highly studied, the mechanism of action for AmB attacking fungal cells is still debated as shown by multiple conflicting proposed mechanisms. Much of the difficulty in studying AmB comes from its poor quantum yield, limiting fluorescent imaging of the without additions or manipulations to the molecule. In this study, we use ultrafast, transient absorption microscopy (TAM) to directly image AmB in its native state, interacting with live fungal cells for the first time. TAM is a nonlinear point scanning technique which involves two laser pulses. The first pulse (pump) arrives at the sample promoting it to an excited electronic state, followed by a second pulse (probe), after some time delay, and the difference in transmitted probe light is measured generating the TA signal. Images obtained using TAM have revealed contradictory results compared to that of the previously proposed mechanisms of AmB. Our images show that AmB is internalized into the yeast cell in large clusters leading to new insight towards the mechanism of action for AmB. The results obtained within this study will lead to a more thorough understanding of AmB and how it is still able to remain effective against severe fungal infections.

Two Dimensional silver nanoparticle polymer composites for organic vapor sensing

Yimei Wen, George Chumanov

A sensitive volatile organic compound (VOC) sensor based on the coherent plasmon coupling in two dimensional (2D) silver nanoparticle (AgNP) arrays is presented here. Silver nanoparticles were synthesized by the hydrogen reduction of silver (I) oxide in water. The 2D AgNP arrays were fabricated by the self-assembly of silver nanoparticles on poly (4-vinylpyridine) modified glass slides. Different polymers including but not limited to polystyrene, poly (methyl methacrylate), polybutyl methacrylate, polyvinyl acetate, poly (4-vinylpyridine), polyvinylpyrridone were spin-coated onto the 2D AgNP arrays to form the nano-composites Upon exposing the nano-composites to organic vapor, such as acetone, ethanol, toluene and formaldehyde, the sharp resonance due to coherent plasmon coupling changed as was monitored by UV/Vis absorption. The response was reversible when removing the vapor source. The sensor exhibited high sensitivity that was achieved by combing the sharp resonance of the AgNP arrays and the swell and shrinkage ability of polymer matrix upon the exposure to organic vapors. This work demonstrated the proof-of-concept for the new type of sensors to be able to use at room temperature, working in ambient environment with reasonable reproducibility and sensitivity.

Interactive web-based visualization tool for monitoring optical properties of nanoparticles during synthesis reactions

Bryan Calderón-Jiménez, Gabriel Sarmanho, Karen E. Murphy, Antonio R. Montoro-Bustos, José R. Vega-Baudrit

Engineered nanoparticles (NPs) are being used for a broad array of high technology applications including sensing, imaging, targeted drug delivery, bio-diagnostics, catalysis, optoelectronics, film growth seeding, and other purposes. The enhanced optical, electrical, and catalytic properties of metal NPs are strongly correlated to control of their size, shape, and structure. As such, physicochemical characterization of NPs is critically important and ultraviolet/visible spectroscopy (UV/VIS) is one of the most widely used methods for measuring the optical properties and electronic structures of NPs [1]. We have developed an interactive web application to analyze multiple UV-VIS absorbance spectra as a function of time to create different graphical visualizations in two and three dimensions including, spectrum, surface, and contour plots. This tool evaluates important parameters of the absorption bands of the NPs. Specifically, it provides the maximum optical absorbance, the surface plasmon resonance (SPR) peak and full width at half maximum (FWHM). These parameters are closely related to the diameter, shape, and polydispersity of metal and semiconductor NPs. Therefore, this application can be used to monitor the processes occurring during the manufacture of NPs, to understand NP stability under different conditions and mediums, and to establish optical properties of NPs for studies that use spectrochemical analysis (UV-VIS) as a technique to characterize NPs. [1] Calderón-Jiménez B, et al. (2017). NanoUV-VIS: an interactive visualization tool for monitoring the evolution of optical properties of nanoparticles throughout synthesis reactions. Journal of Research of NIST, (in review process).

Algae-Biotemplated Water-Splitting Copper Oxide Nanocatalysts for Hydrogen Production
Paloma Salazar, Sakr Elsaidi, Marina Avram, Daniel Nde, Wei Zhao

Water-splitting hydrogen fuel production using carbon-free sources of energy, such as solar and wind energy, is a promising renewable and sustainable solution to solve the world problem of carbon pollution and global climate change associated with fossil fuels. The water-splitting reaction is composed of two half reactions, hydrogen evolution reaction (HER) and oxygen evolution reaction (OER). Low-cost, low over potential, and earth-abundant HER and OER catalysts are under intense investigation in order to enhance and speed up the reaction for practical applications on a large scale. One of the best strategies currently under development is using low-cost approaches to develop the catalysts at the nanoscale (<10 nm), and enhance their dispersion and electrical conductivity on conductive supports with a large number of active sites on porous surfaces. Here, we used carbonized Nannochloropsis oculata cells as three-dimensional scaffolds for the synthesis of copper oxides, which have been previously shown to have HER electrocatalytic properties. Oculata algal cells were first converted to carbonized cells (cCells) with a biotemplate approach. CuOx nanoparticles were then grown on the cCells by a facile hydrothermal reaction, forming a CuOx-cCell composite (CuOx@cCells). The structure of the composite was analyzed by scanning electron microscopy and the HER electrocatalytic properties were studied by cyclic voltammetry and linear sweep voltammetry. The enhanced HER performance and stability of CuOx@cCells will be discussed in detail, in comparison with Pt/C benchmark HER catalyst.

Calibration-Free Quantification of Trace Crystallinity within Amorphous Solid Dispersions by SHG Microscopy
Garth J. Simpson, Casey J. Smith, Janny Dinh, Paul Schmitt, Ellen J. Gualtieri

A calibration-free method for quantifying trace crystallinity within nominally amorphous solid dispersions provides allows quantitation over five orders of magnitude in crystallinity into the parts per million regime without the need for calibration standards. Trace crystallinity can potentially greatly impact both the stability of amorphous solid dispersions and the subsequent dissolution kinetics by acting as seeds for crystal formation. SHG microscopy provides contrast that is highly selective for chiral crystals, but with large variance in intensity as a function of particle composition, size, orientation, and crystal form. In the limit of low crystallinity, much of this measurement variance is shown to be removed by particle counting approaches combined with shape estimates to determine single-particle volumes. Statistical approaches to correct for particle-particle overlap greatly extends the dynamic range of particle counting from the ppm regime to ~1% crystallinity. Because the volume fraction is based on particle cross-sectional areas rather than SHG intensities, no prior knowledge of the SHG activity is required for quantification of crystallinity. Furthermore, determination of the volume fraction independent of the SHG intensity allows determination of the average SHG intensity per unit volume, which is equivalent to the slope of a conventional calibration curve. This recovered SHG activity in turn allows for the determination of crystallinity in the high-crystallinity regime (>5%), in which particle/particle overlap complicates volume estimation by particle counting, all without the need for an external reference standard.

Multi-bounce ATR FTIR measurement of live cells in response to Anticancer Drugs

ALI ALTHARAWI

Fourier transform infrared (FTIR) is a non-destructive and label-free method that have been found to be a promising technique in the study of drug in live cells. The spectrum reflects their chemical compositions, which can be exploited to detect the cellular changes of live cells in response to chemotherapy. Previously, we tested the ATR FTIR measurements of live cells approach on three different cell lines (HeLa, PC3, CaCo2) exposed to Doxorubicin. The study has shown than that the ATR method is highly suitable to detect the subtle changes in the chemical compositions of living cells which is related to the inhibition of the DNA synthesis due to the DNA-intercalating effect of doxorubicin.1 In this current study, we are investigating the response of MDA-MB-231 breast cancer cells to Tamoxifen (Selective Estrogen Receptor Modulators-SERMs) and Imatinib (Tyrosine Kinase Inhibitor-TKI). Principal component analysis (PCA) was applied to highlighted the subtle chemical changes in the living cells spectra after exposed to tamoxifen and imatinib. Changes in the spectral regions 1200-900 cm-1 assigned to nucleic acids, phosphates and carbohydrates were observed. These changes could be related to changes in phosphatases and kinases activity due to the cytotoxicity induced by tamoxifen and imatinib to MDA-MB-231 cells. Furthermore, we applied clustering analysis to the data collected. The analysis has successfully shown three separated clusters for untreated cells, cells treated with tamoxifen and cells treated with imatinib demonstrating the potential of the FTIR method in the detection of chemical pathways of anticancer drugs and eventually its application as a tool in early screening of chemotherapy. [1] P.V. Fale, A. I. Altharawi and K.L.A. Chan, BBA-Mol. Cell Res. 1853, 2640

Quantification of solid-state co-amorphous mixtures of crystalline carbamazepine by Terahertz spectroscopy
Yi Li, James K. Drennen, III, Carl A. Anderson

The development of co-amorphous pharmaceutical formulations is a viable approach to improve the physiochemical properties of solubility-limited BCS II class drugs. Quantitative determination of the crystalline and amorphous fractions of active pharmaceutical ingredients(API) in co-amorphous formulations facilitates the physical stability evaluation and thus is critical to the design of such novel formulations. The objective of this study is to determine the phase compositions of an API in tablets using Terahertz spectroscopy. Terahertz time-domain spectroscopy is sensitive to the crystalline lattice structure and phonon vibrations. It is a powerful tool that can identify and quantify crystalline APIs. Partial least squares models were developed from transmission Terahertz spectra measured on tablets composed of crystalline carbamazepine, crystalline tryptophan, co-amorphous components and mannitol. The root mean square errors of prediction were between 1.7% to 3.5% for different constituents. In addition to the establishment of quantitative models for phase quantification, a separate model was calibrated to predict tablet density based on refractive index. Tablet Yield pressure was calculated from Heckel plots; they were shown to decrease with the amorphous contents in tablets. In a summary, Terahertz spectroscopy is an effective technique for bulk quantification of API crystallinity in a formulation involving a co-amorphous form.

Benefits of spectroscopic analysis over standard analytic techniques within the laboratory

Keely Bergqvist

Titration methods have traditionally been used to measure water content within solvents and reagents during pharmaceutical development and manufacture. However, it is also possible to determine water content using spectroscopic techniques such as Near Infra-red. Spectroscopic analysis offers a number of advantages such as minimal or no sample preparation, rapid analysis times and recovery of an intact sample. This presentation will look to compare the different techniques with emphasis on efficiency, productivity and safety. The presentation will also outline the method development and validation aspects, as well as cover the time, cost benefits, and return on investment, of this 'Green Chemistry' approach.

Applications of UPLC-MSMS in pharmaceutical impurity profiling
Lynn X. Zhang, Heather Fleming, Jason Batchelor, Rachel Rensing

The current pharmaceutical regulations require high purity drug substances, along with the knowledge of the impurities present in the manufactured drug substances. This includes not only the identification/structural information, but also the formation process of such impurities and quantitative determination of impurities. The understanding of such impurities will greatly improve the process chemistry and the drug safety, thus the impurity profiling after forced degradation procedures have become a very important step in the pharmaceutical research and development. Liquid Chromatography-Mass spectrometry (LC-MS) has been a developing technique in the application of the modern pharmaceutical analytical chemistry field. UPLC-MSMS combines the fast separation/purification and identification/structural information at the same time. Presented here is the application of UPLC-MSMS in the studies of pharmaceutical impurity profiling. Advantages and current challenges of UPLC-MSMS in comparison of other commonly utilized techniques in the pharmaceutical industry are also discussed.

Qualifying ATR Accessories for Use in Pharmaceutical Applications
Steve Lowry, Garry Ritter

Single Bounce ATR accessories for FT-IR instruments have become one of the most popular mid-infrared techniques for characterizing materials. High quality spectra can be obtained from milligrams of sample in less than a minute with minimal sample preparation or cleaning. While many of the international Pharmacopoeia organizations have provided clear guidance in qualifying FT-IR instruments for transmission analysis, there is presently very little discussion on how to qualify the instrument with an ATR accessory. In fact the general section of the US Pharmacopeia describes the problems with qualifying an ATR measurement. There is also some debate whether the ATR accessory falls under Instrument Qualification or System Suitability since the choice of ATR internal reflectance element (IRE), angle of incidence and number of reflections can significantly change the resulting infrared spectrum. In this presentation we will discuss some of the causes of variance and why the criteria set for transmittance spectroscopy are not necessarily appropriate for ATR spectroscopy. We will suggest some metrics that might be included in a SOP for qualifying a specific ATR configuration on an FT-IR instrument.

Beyond Elemental Impurities Analyses in Pharmaceutical Research: LC-ICP-MS Applications
Brittany Kassim, Lanfang Zou, Qiang Tu, Xiaodong Bu, Yun Mao

Control of elemental impurities including residual metals is required by regulatory agencies to ensure the quality of active pharmaceutical ingredients (API) and formulated drug products. Elemental analyses are an integral part of modern pharmaceutical research and development, as elemental impurities have been regulated for decades. Elemental impurities in API and drug products originate from reagents, catalyst, excipients, and manufacturing equipment. Alternatively, metals and metalloids can be incorporated into the drug design to enhance the efficacy, such as platinum based chemotherapy agents. The role of elemental analyses in pharmaceutical research has expanded over the past several years, with the introduction of analytical techniques such as inductively coupled plasma mass spectrometry (ICP-MS) and high resolution ICP-MS. While these advanced techniques allow for rapid and accurate quantitation of residual metals, additional analytical techniques of separations via liquid chromatography (LC) expand the ability of ICP-MS to differentiate metal species. Utilization of LC-ICP-MS in the field of pharmaceutical elemental analysis has a tremendous impact on drug discovery and development by delivering tools that provide insights on reaction mechanisms, catalyst degradation, and metal removal to enable chemistry development. This presentation will highlight three applications of LC-ICP-MS in pharmaceutical research. Specifically, the role of iron speciation for the understanding of iron induced polymerization in drug formulation. Secondly, the investigation into oligomerization characterization via sulfur and silicon analysis, utilizing high resolution ICP-MS. As well as, the advantages of a highly sensitive and selective zinc analysis to bridge the understanding of enzyme deactivation. With these highlighted applications it is the aim to describe the utility of the metals laboratory in pharmaceutical research beyond the elemental impurities testing.

Bioecological Study of Sargassum sp. and its Extract Bioactivity as Anti-MDR bacteria
Anggara Mahardika, Rini Pramesty, Wilis A. Setyani, Muhamad Zainuddin, AB Susanto

Sargassum sp. was collected from five difference locations, i.e. Teluk Awur, Panjang Island, Bandengan, Ujung Piring and Bondo. There were seven species of Sargassum identified from the sampling locations, i.e. *S. duplicatum, S. polycystum, S. echinocarpum, S. cinerium, S. crassifolium, S. plagyophillum* and *S. binderi*. Diversity index at Panjang Island, Bandengan, and Ujung Piring were categorized as medium with H'; 1.24,
1.09 and 1.09, respectively. Evenness index at Teluk Awur, Panjang Island, Bandengan and Ujung Piring categorized as high showed "e" values ranging from 0.91 to 0.99. There were no dominant species found in Teluk Awur, Panjang Island, Bandengan, and Ujung Piring. The principal component analysis (F1xF2) exhibited the main components that affected the *Sargassum* at Teluk Awur, Panjang Island, and Ujung Piring, were phosphate, depth, and salinity, respectively. As well as in Bondo, the main components were temperature, dissolved oxygen (DO) and pH. Clustering analysis revealed there was character similarity between Panjang Island and Bandengan (99.73%), meanwhile, at Teluk Awur, Ujung Piring, and Bondo, the similarity was 99.62%, 99.60%, and 99.52%, respectively. Based on Pearson's analysis, the correlation of DO with abundance at Teluk Awur was positive (99.45%). There was a positive correlation between salinity and phosphate at Panjang Island (97.88%), negative correlation with DO at Bandengan (99.94%), depth at Ujung Piring (96.22%) and Bondo (99.73%). Multivariable regression analysis between water quality and abundance resulted an equation having the similar component, which was phosphate. Antibacterial activity of Sargassum sp. demonstrated a positive activity against the MDR bacteria, Staphylococcus aureus (6,28±0,04 mm), Escherichia coli (6,96±0,19 mm), and S. epidermidis (12,65±0,23 mm). Clustering analysis of extracts with bactericidal characteristic having a broad spectrum was performed and grouped into three classes with similarity of class 1, 2 and 3 at 72.33%, 76.80% and 58.50% respectively. Sargassum sp. categorized in class 1 had the best antibacterial activity and the highest abundance value.

Raw Material Identification: Method Driven Raman for Increased Speed and Accuracy
Adam Hopkins

Handheld Raman spectroscopy holds the promise of eliminating sample quarantine and statistical sampling for raw material identification (RMID) applications and enabling 100% container verification. However, the variety of packaging materials and types of analytes can cause difficulty for many handheld instruments. Some of the difficulties are very long acquisition or library building times, and sample or packaging material degradation. In this work, we examine common excipients and active pharmaceutical ingredients (APIs) inside a variety of common packaging materials with the MIRA M-3. We show that these difficulties can be eliminated through complete control over the data collection and analysis method combined with orbital raster scanning.

Development of Vapor Diffusion Chamber for in situ and High Throughput X-Ray Diffraction Analysis
Kathleen Sokolowsky, Andrew Brunskill, Alexander Chin, Matthew Hagan, Timothy Rhodes

Hydrated and solvated crystalline phases are very common for pharmaceutical solids and require thorough investigation during drug development. A common technique is the exposure to variable relative humidity or organic solvents followed by analysis by X-ray powder diffraction (XRPD). Hydrates and solvates will commonly dehydrate/desolvate when removed from the environment in which they were generated. Static humidity chambers can be used keep the sample in the desired environment, but it must be removed for analysis; humidity systems for XRPD instruments are available but tie up the instrument for the long experimental time window. In order to enable characterization of hydrates and solvates, we have developed 3D printed XRPD holders which can expose single samples to either variable relative humidity or organic solvent. However, these experiments were still conducted one sample at a time and each require a long equilibration time. To improve on the throughput of analysis, we have developed a sealed 384-well plate which allows for solids to be exposed to up to 192 different environmental conditions. The plate is allowed to equilibrate, and *in situ* XRPD is collected at any time points of interest. Using the 3D printed holders, conversion of both Zetia (Ezetimibe) and creatine between anhydrous and hydrated phases by exposure to water vapor was monitored *in situ*. Proof of concept studies with the 384-well plates and nine relative humidity conditions correctly report the Zetia anhydrous to hydrate phase change.

A New Approach to Characterize Crystallization in Complex Transdermal Delivery System (TDS) by Raman Mapping and Modeling

Teng Xu, Daniel Willett, Sam Raney, Caroline Strasinger, Jason Rodriguez, David Keire, Anna Wokovich

Transdermal delivery systems (TDS) are designed to deliver a therapeutically effective amount of drug across a patient's skin and into the systemic circulation. To enhance drug delivery through the skin, and to facilitate a relatively sustained rate of drug delivery, the concentration of the active pharmaceutical ingredient in the TDS formulation is often close to the saturation state, or even in a supersaturated state. Drug crystallization can occur over time due to the thermodynamically metastable nature of the supersaturated state. Crystallization (or lack thereof) is a critical quality attribute for TDS because crystals can lead to altered drug delivery rate, adhesion failure and thus result in product recall or even discontinuation. In this study, confocal Raman spectroscopy together with chemometric modeling were used to study the crystallization of four drug-inadhesive Fentanyl TDS, 25 μg/hr products sold on the U.S. market by different manufacturers (FTDS-1, FTDS-2, FTDS-3, and FTDS-4), before and after aging. The result showed that no fentanyl crystal was identified in FTDS-1 and FTDS-2 both before and after aging. Raman mapping together with principal component analysis (PCA) suggested a relatively uniform fentanyl distribution in the adhesive matrix of FTDS-1/2. By contrast, fentanyl crystallization was identified in the FTDS-3 adhesive matrix both before and after aging. Raman mapping and PCA were used to characterize crystal distributions in the complex TDS matrices. For FTDS-4, no crystals were identified by Raman microscopy before aging. However, large elongated fentanyl crystals about 20 μm in width and up to 5 mm in length were observed and identified by Raman microscopy at the edge of die cutting after aging. This work demonstrates a new approach to characterize complex TDS by confocal Raman microscopy mapping together with chemometric modeling.

Prediction of in vitro drug release of a multparticulate dosage form by process analytical techniques (PAT)

Hanzhou Feng, Shikhar Mohan, James Drennen III, Carl Anderson

In vitro drug release testing is one of the most common practices in the pharmaceutical industry to ensure the performance of final products. This is currently performed using a dissolution apparatus and often requires substantial time for analyses. The food and drug administration (FDA) has encouraged the industry to implement lower-cost and more efficient surrogate methods to predict drug release. Vibrational spectroscopic methods, such as Near-Infrared (NIR) spectroscopy, offer a potential solution to these concerns. Near-Infrared spectroscopic methods have advantages such as rapid and non-destructive measurement. It requires minimal (if any) sample preparation. The NIR spectrum is generally sensitive to chemical and physical information from pharmaceutical products. The critical quality attributes that have significant effects on drug release performances are often well-correlated with NIR spectra. This enables the development of a non-destructive method for testing drug release by NIR. This project aims to predict the final drug release behaviors of pharmaceutical products by NIR spectroscopy. The feasibility was investigated on an extended release multiparticulate dosage form. The proposed method holds the promise for improving the efficiency in product development and reducing the financial cost, by decreasing the dependence of quality control on in vitro drug release testing.

APPLICATIONS OF A MULTISPECTRAL VISION SYSTEM TO SUPPORT DEVELOPMENT OF PHARMACEUTICAL PRODUCTS

Brian Marks, Rich Steinbeiser, Simon Hamilton, Megan Mackey, W Peter Wuelfing

A multispectral vision system, covering UV-Vis-NIR, can be used for non-destructive, fast and accurate analysis of color, texture, gloss, shape, size and chemical composition of surfaces. The VideometerLab uses 19 different LEDs covering 370nm – 970nm to captures digital images, which can be combined into a single high-resolution, multispectral image. Each pixel of this image contains a NIST calibrated spectrum with a resolution of 45um. Advanced digital image analysis and statistical software provides quantitative measurements of visual and chemical properties of surfaces. Multispectral vision analysis was evaluated for applications that support the development of pharmaceutical products. For tablets, the technique was successfully used to (1) estimate drug loading, (2) provide false-color images to visualize moisture uptake, (3) estimate thickness of a transparent film coating, and (4) evaluate active and placebo tablets for effective clinical study blinding under several simulated lighting conditions. Additionally, multispectral vision analysis was used as a microbiology colony counter, for visualization of non-visible extraneous matter in drug substance and for evaluation of inks used in counterfeit packaging.

Methods for Approaching Substrate Interference in Raman Spectroscopy for Forensic Science

Marisia Fikiet

Raman spectroscopic mapping has become more and more popular as an analytical method. One drawback of Raman analysis is interference from substrates. Our lab has evaluated several approaches to overcome this problem. First, we varied the excitation wavelength, but no one wavelength worked well for all the common substrates. Using different excitation wavelength for different substrates was impractical. Then background subtraction was attempted and works very well for homogeneous and non-fluorescent substrates. For heterogeneous and fluorescent substrates a more complicated method was used. First a partial least squares discriminant model was used to separate spectra based on how much the spectra was dominated by either the body fluid signal or the substrate signal. In the spectra where the body fluid dominated manual subtraction was used. Using a combination of the above techniques a body fluid signal was able to be extracted from every substrate tried, but no one method worked for every substrate. Since no one previously tried method worked for all the substrates tried an advanced statistical program based on MCR was used. This program was trained on the target body fluid, so the identity of the substrate is not needed. The program was used for substrates that are fluorescent and heterogeneous and was able to extract the body fluid signal in all cases.

Confocal Raman Microscopy Investigation of Self-Assembly of Hybrid Supported Phospholipid Bilayers within Individual Porous Silica Chromatographic Particles

Jay Kitt, David Bryce, Joel Harris

Hybrid supported phospholipid bilayers (HSBs), comprising an upper-leaflet of phospholipid and a lower leaflet of surface-bound hydrocarbon acyl-chains, are a commonly employed model for studying interactions which normally occur at cell membranes. Recently HSBs have been adapted from gold interfaces, where they were used for surface-enhanced Raman spectroscopy and surface plasmon resonance experiments, to alkylsilane-modified porous chromatographic silica particles for use in chromatographic experiments. It was recently shown, however, that HSBs formed within C18-chromatographic silica are structurally different from alkanethiol HSBs and vesicle bilayers; Raman spectra of within-particle HSBs indicate an interdigitated structure where phospholipid acyl-chains extend into the surface-bound C18 chains. Interestingly, within-particle HSBs in chromatography are typically formed by self-assembly from phospholipid-vesicle suspensions containing a fraction of alcohol modifier. This is in contrast to non-interdigitated HSBs on gold which are typically formed by vesicle fusion from buffer and vesicle membranes formed by hydrating phospholipid films in water. The difference in assembly conditions, along with the observed interdigitation, motivate investigation of HSB formation within chromatographic silica. In this work, we employ confocal Raman microscopy within individual C_{18}-chromatographic particles to investigate time-dependent self-assembly of 1,2-dimyristoyl sn-glycero-3-phosphocholine HSBs from aqueous 15% isopropanol. Self-modeling curve resolution is employed to resolve spectral and composition vectors describing time-dependent evolution of the Raman spectra revealing a four-component process: disordered phospholipid is initially adsorbed to the C_{18} interface followed by an ordering step where phospholipid and C_{18} acyl chains adopt trans conformations with increasing lipid coverage. As solution is switched from isopropanol-water to pure water further ordering is observed, indicating isopropanol significantly alters bilayer acyl-chain structure. To further address the impact of isopropanol on formation of HSBs in C_{18} silica, HSBs were formed where no isopropanol was included in the assembly solution. Raman spectra of these HSBs showed two-fold greater intensity of phospholipid vibrational modes indicating phospholipid density corresponding to a phospholipid cross-sectional area of $60\pm4Å^2$ in close agreement with cross-sectional areas in vesicle membranes. These results suggest that during HSB formation isopropanol acts to lower the surface tension of the C_{18} interface between phospholipid molecules resulting in lower acyl-chain density leading to interdigitation.

Raman Chemical Imaging of Adsorption and Reactions on Surfaces

Erik Emmons, Ashish Tripathi, Gregory Mogilevsky, Chris Karwacki

Our group is investigating the use of Raman chemical imaging as a technique for imaging adsorption as well as chemical reactions on surfaces. There is currently a lack of techniques that provide spatial and temporal resolution of molecular interactions on complex materials such as filters and textiles. Since Raman spectroscopy probes molecular identity via observation of chemical bonds, it has the capability to map transport of chemicals into surfaces as well as their reactions. We are examining model systems including adsorption of aromatic molecules on silica as well as transport and adsorption of chemicals on fabrics containing activated carbon. Results from imaging and time-resolved studies will be presented. This technique has the potential to be useful for studying adsorption and decontamination of chemical warfare agents on suits and filter materials.

Swept-Wavelength Raman Spectroscopy for the Detection of Controlled Substances

Pratima Kunapareddy, Calvin Zulick, Jacob Grun

The recent increase in drug related fatalities has created an urgent need for an analytical technique that can rapidly and accurately identify controlled substances in seized evidence in the field or in a laboratory setting. The goal of this work is to examine the utility of multi-wavelength resonance Raman spectroscopy for the detection and identification of controlled substances. The Swept-Wavelength Optical resonance-Raman Device (SWOrRD) is an NRL patented spectroscopic method that combines the advantage of resonance-Raman spectroscopy with multi-wavelength operation. This system is capable of illuminating samples at multiple wavelengths between 210 - 2000 nm. SWOrRD generates a two-dimensional (2D) Raman signature with the excitation wavelength and the Raman frequency shift as the two axes. We have used SWOrRD successfully characterize biological materials, chemicals, explosives, and isotopes. We have obtained two-dimensional Raman signatures of controlled substances like heroin, methamphetamine, fentanyl, and cocaine and cutting agents using deep-UV illumination wavelengths. We will discuss resonance behavior of these substances in the UV region and present preliminary data on detection limits. Controlled substances are rarely encountered in their pure form except at the supply point. They are often mixed with other excipients. We will discuss identification algorithms that can identify target substances in realistic mixtures. The techniques developed in this research will enable the development of devices that can be used to process seized drug evidence in a rapid and accurate manner.

Quantitative Analysis of Saccharides in Kappaphycus alvarezii Using Raman Imaging
ANGGARA MAHARDIKA, AB Susanto, Rini Pramesti, Yusuke Matsuda, Hidetoshi Sato

Purpose: Development of a Raman technique to study enzymatic reaction and quantitative analysis of saccharides in cell size using Raman Spectroscopy. Enzymes are widely used in many industries for biocatalyst to accelerate chemical reactions. Many enzymes are produced from microorganisms i.e. bacteria, fungi, archaea, etc. In the natural environment, interaction inter organisms are involve enzymatic reaction, such as the process of fruits decay by bacteria. To understand the biological process of enzymatic activity within cell size, a special tool that can observe in cell size is need to employed. In this study, we applied Raman imaging Spectroscopy to monitoring enzymatic reaction within cell size in intact algae. To analyze sugar concentration in algae, we have developed analytical model base on partial least square regression (PLSR). Commercial Alpha amylase converts starch into maltose and glucose was used in this study. Initially, Raman spectra of starch, maltose, and glucose were measured to get the fingerprint of each component. Raman spectra under 600 cm^{-1} was selected; 479 cm^{-1} as starch band, 519 and 554 cm^{-1} as maltose band, and 519 and 541 cm^{-1} as glucose band. Standard concentration model of starch, maltose and glucose were calculated by PLSR. Regression square of each component was calculated up to 0.9, and it is reliable enough for estimate the enzymatic component concentration. For the results, Raman spectroscopy with PLS model can estimate the concentrations of each enzymatic components and it is show the same result comparing with 3,5-Dinitrosalicylic acid (DNSA) reagent. On the other hand, Raman imaging spectroscopy is reliable to figure out saccharide in the intact algae *K.alvarezii.*

Spectroscopy and DFT Studies of Uranyl Carbonate, UO2CO3: A Model for Uranium Transport, Carbon Dioxide Sequestration, and Seawater Species
N. Kalashnyk, D. L. Perry, F. Massuyeau, E. Faulques

Rutherfordine, UO2CO3, is a member of a set of minerals that consist of a uranium(VI) center surrounded by oxygen atoms through which are coordinated ligands such as carbonates and phosphates. The components of the molecular structure include UO8 hexagonal bipyramids in neutral sheets in which the UO8 units are involved in edge- and corner sharing, along with CO3 triangles. The research described in the present work focuses on Raman and optical spectral studies of rutherfordine and their use to provide spectral sets of data that exhibit a unique signature for the mineral. The data are used to address and clarify several issues of the bonding and composition of the parent rutherfordine structure, correlation of exact structural details with the collected spectral data, and an overall comparison with previously known spectral data. As in the case of uranyl phosphates previously studied by us and other authors, it is shown that micro-Raman spectroscopy, reflectance, and photoluminescence (PL) techniques provide good spectral markers that can be used to differentiate rutherfordine from other uranyl carbonates. Additionally, intense steady-state PL spectral and reflectance data of natural rutherfordine are, to our knowledge, reported for the first time in the present investigation. The PL spectrum analysis provides information that can be linked with Raman data, while excited vibrational states are probed with reflectance spectroscopy. Periodic DFT calculations have been carried out using PBE and PBEsol functionals with ultrasoft pseudopotentials and were compared to experimental data. Optimized lattice parameters and U-O bond length are increased by more than 2%. Band structure calculations show an electronic gap of ~ 3 eV, not too far from that obtained experimentally. Calculated vibrational frequencies are in good agreement with Raman data and allow precise assignment of the vibrations. Finally, the electron localization function ELF has been evaluated and is useful to probe covalency contribution of the different types of bonding in the structure.

Tip-Enhanced Raman Spectroscopy with Plasmon-Resonance Thin-Film Waveguide Probe
Kaifeng Zhang, Takehiro Tachizaki, Ryota Matsumoto, Toshihiro Okamoto, Masanobu Haraguchi, Shin-ichi Taniguchi

Tip-enhanced Raman spectroscopy (TERS) is a powerful technique for inspecting materials (e.g. carbon nanotubes, graphene, DNA, etc.) with nanometer resolution. There are growing needs for nondestructive *in situ* TERS. However, it is still challenging to get a strong tip-enhanced Raman signal without direct illumination of the end of the tip or without the use of resonant substrates (e.g. gold or silver) to achieve excitation of the gap-mode plasmons. To satisfy these needs, we have developed a TERS system that does not require direct illumination or resonant substrates. We theoretically calculated and designed a new type of probe with a plasmon-resonance thin-film waveguide composed of gold film with an optimized thickness (46.5 nm) on a silicon thin-wedge. Far-?eld excitation light of 660 nm wavelength was focused on the interface of silicon and gold across the silicon film. An excitation beam with a designated incident angle generated surface plasmons on the gold surface. The surface plasmons transferred the electromagnetic energy of the optical frequency to the apex of the tip and then generated a near-field spot at the end of the tip. The electromagnetic energy carried by the surface plasmons was focused during the propagation of the surface plasmons. This enhanced near-field spot gave a strong Raman signal without any resonant substrates. To evaluate our technique, we demonstrated Raman spectroscopy of bulk 4'-(pyridin-4-yl)-biphenyl-4-yl) methanethiol (4-PBT) with a waveguide probe (L. Yang *et al. Nanoscale*, 2015, **7**, 18225–18231). Characteristic Raman signals from 4-PBT were observed only when the tip was engaged. When comparing our results with the microscopic spectroscopy performed with the same sample by a conventional optical microscopy, the enhancement factor of our waveguide probe was estimated as -10^5. This result falls within the range of the well-known TERS enhancement factor 10^3-10^9 (W. Zhang *et al. J. Phys. Chem. C*, 2007, **111**, 1733-1738; J. Steidtner *et al. Phys. Rev. Lett*, 2008, **100**, 236101). In this presentation, we will explain the principle of our waveguide probe using the finite-difference time-domain simulation results. We will give the details of our experimental system and demonstrate the measurement results.

Confocal-Raman Microscopy Monitor of Temperature-Dependent Pyrene Partitioning and Hybrid-Bilayer Structure within C18-Modified Silica Particles

Xin Wen, David Bryce, Jay Kitt, Joel Harris

An ongoing challenge in chromatography is separating compounds of similar molecular weights and hydrophobicity. Polyaromatic hydrocarbons (PAHs) are one example, where many analytes exhibit similar retention under traditional reversed phase chromatographic conditions. Planarity is a characteristic which has been previously used as a basis for separation of chemically-similar PAHs. Here, tightly packed C_{18} stationary

phases, where the acyl chains adopt a more ordered conformation, have been observed to more strongly retain planar vs. non-planar compounds. It has been suggested that this shape selectivity is due to the partitioning of planar compounds into 'slots' between adjacent trans-conforming acyl chains in the C_{18} stationary phase.

Recently, hybrid bilayers have been formed within C_{18}-modified silica particles for use in biomimetic separations. Characterization of these materials has revealed that the acyl chains from the upper phospholipid leaflet interdigitate into the lower leaflet of C_{18} chains leading to an all-trans acyl chain structure, and the first observation of a brush-like phase which would be optimal for shape-selective separations. Additionally, this

structure shows temperature dependent melting, which eliminates trans character in the bilayer acyl chains making temperature dependent shape selectivity a possibility. In this work, interdigitated hybrid bilayers were formed by adsorbing 1,2-dimyristoyl-sn-glycero-3-phospho-choline (DMPC) onto the interior surface of the pores of C_{18}-chromatographic particles. To test for temperature-dependent shape selectivity, these DMPC

hybrid-bilayer particles were equilibrated with an aqueous 1-?M solution of pyrene (a planar PAH) in 5% methanol. After equilibration, temperature-dependent confocal Raman microscopy was used to simultaneously monitor pyrene partitioning and hybrid-bilayer structure as the temperature was increased

from 18°C through the hybrid-bilayer phase transition to 46°C. Here, pyrene partitioning was observed to decrease with loss of trans character in the acyl-as the bilayer melted, where pyrene was more strongly retained in the all-trans gel-phase bilayer. This shows that an ordered acyl-chain structure directly impacts pyrene partitioning, consistent with the previously proposed 'slot' model. These results support the use of hybrid bilayers as shape-selective stationary phases for chromatographic separation of planar vs. non-planar PAHs where the selectivity for planar vs. non-planar molecules could be controlled by adjusting the proportion of trans-conformers with temperature.

Bioanalytical Methods for Investigating Dynamic Behavior of Pancreatic Cells
Michael Roper

Islets of Langerhans are the endocrine portion of the pancreas responsible for maintaining glucose homeostasis through the regulated secretion of numerous hormones, most notably insulin. With highly sensitive analytical methods pioneered almost 25 years ago, the automated measurement of insulin release with a time resolution of seconds from a single islet can be performed. These single endocrine units display complex hormone secretion patterns consisting of rapidly changing phases and oscillations. Despite the presence of hundreds of thousands of these islets in vivo, these complex patterns are also manifested in the blood stream and are critical for proper glucose utilization. In this talk, various analytical approaches that enable monitoring of insulin secretion from single islets with a time resolution appropriate to discern these various patterns will be discussed. The ability to make intracellular Ca^{2+} measurements simultaneously with insulin measurements for investigating stimulus-secretion coupling will also be shown. Finally, a mechanism of how populations of individual islets can synchronize their hormonal pulses to produce coordinated oscillations, similar to those found in vivo, will be proposed.

Sheath-Flow SERS for Online Chemical Detection
Zachary Schultz, Emily Peters, Anh Nguyen, Rafael Masitas

The increased sensitivity of surface enhanced Raman scattering provides an efficient method to increase the sensitivity and utility of Raman detection for bioanalysis. Efficient coupling to separation techniques, such as liquid chromatography and capillary electrophoresis can enable complementary detection to the more commonly used mass spectrometric techniques. Sheath flow SERS is a recent development that has enabled online detection of multiple analytes in complex mixtures. Comparative studies show that the SERS signal exhibits a narrowing of bands associated with the chemical separation that is not observed in other detection methods. The narrow features and times associated with these features suggest that only a small number of molecules are sampled by SERS detection. Here we present results that show the evolution of SERS detection in flow. Incorporating electrodes for quantitative analysis of analytes provides new insight and further supports the importance of molecular interactions with the surface. Additionally, incorporation microfluidics to better control fluid dynamics also provides additional information. Based on our results, we continue to develop approaches that combine chemical separations with SERS detections. These new designs, which maximize the interaction of molecules with the SERS substrate, are shown to be important for next generation SERS assays.

Microchip Electrophoresis-based Methods for Measuring Oxidative Stress
Susan Lunte

Microchip electrophoresis (ME) is a powerful tool for the analysis of biological samples. In particular, its ability to perform fast, efficient separations makes it possible to monitor several compounds in a single run with high temporal resolution. The extremely small sample volume requirements, as well as the micron size dimensions of the separation channels, make it possible to analyze the contents of single cells. When ME is coupled on-line with microdialysis sampling, continuous near real time monitoring of endogenous biomarkers in the brain and other tissues is possible. In this paper, the use of microchip electrophoresis (ME) for the determination of reactive oxygen and nitrogen species (RNOS) and other oxidative stress biomarkers will be described. A method has been developed to simultaneously measure superoxide and nitric oxide using MitoSox and DAF-FM and microchip electrophoresis with fluorescence detection in macrophage lysates. Microchip electrophoresis makes it possible to separate the different fluorescent RNOS products from interferences and side products. This ME-FL method provides an integrated amount of RNOS produced during the incubation time. In order to determine the instantaneous concentrations of RNOS, microchip electrophoresis with electrochemical detection is being evaluated. With this direct detection method, the instantaneous concentrations of nitrite, peroxynitrite, and nitric oxide can be determined. Current progress concerning ME-EC methods for the detection of RNOS will be described. This includes the development of a dual electrochemical/fluorescence detection system using a biopolar electrode. Lastly, the use of the on-line microdialysis-microchip electrophoresis for monitoring biomarkers of oxidative stress *in vivo* will also be discussed.

Metabolomic Analysis of Diabetic Complications using Microfluidics
Jim Edwards

Abstract: Small intracellular metabolites play a critical role in cellular viability and function. Investigating the structural heterogeneity and broad dynamic range of the metabolic state requires the development of high sensitivity analytical systems. Using capillary liquid chromatography-nanospray ionization mass spectrometry based platforms signal enhancement of 100 fold and detection of > 50,000 unique metabolite features is achieved. To further boost selectivity, detection and quantitation of specific classes of metabolites, novel isotopic chemical tags with specific functional group reactivity have been developed. Here, these tools are applied using a semi-targeted approach in conjunction with untargeted nanoLC-MS analyses to investigate metabolic defects of diabetic complications in murine aorta. Previously unidentified metabolic pathway changes involved in diabetic complications will be discussed. Development of a mass spectrometry-compatible microfluidic human vascular mimetic system will be introduced. Our discoveries support the theory that diabetic vascular complications arise from the interplay of a myriad of metabolic pathways in conjunction with oxidative and carbonyl stress, which may provide not only new and much needed biomarkers but also insights into novel therapeutic targets. Novel tagging reagents which boost MS sensitivity and improve fragmentation will be discussed.

Microfluidics Made Easy: 15 Years of Collaboration with the Martin Group

Dana Spence

While microfluidic systems first appeared in the literature nearly 40 years ago, with follow-up work performed in the 1980's by Ruzicka and Hansen, the field grew at a very fast rate in the 1990's. Over the past 2 decades, microfluidic systems have been etched, developed, and now printed; detection has occurred on-chip, off-chip, in-channel, and end-channel; valves have been integrated along with pumps, packing materials, modified surfaces, etc. In fact, if there has been a constant during the development of microfluidic systems, it has been the consistent added complexity, and resultant difficulty in adapting, of individual devices for one's own needs. Throughout this period, one group who has consistently tried to make devices that are adaptable by all groups and do not require million-dollar infrastructure to complete is the Martin research group at Saint Louis University. In this talk, I will discuss some of the early work performed by the Martin/Spence collaboration that began nearly 20 years ago while Scott was still a post-doc at Kansas with Sue Lunte. Included in this talk will be work involving one of the earliest reports of measuring secretions from a monolayer of adhered cultured cells in a microfluidic chip channel and a very early version of membrane integration into a microfluidic device. I will also highlight recent advances from our collaboration that include integration of electrodes into 3D-printed devices and finish with recent technologies from the Martin group that includes a revolutionary method of printing fibers inside device channels for culturing and capturing cells with incredibly high fidelity.

Using Advanced Statistics to Increase the Information Content and Usefulness of Radio Frequency Spectroscopy
Matthew Augustine

The study of industrial processes and consumable products is difficult as the samples of interest are typically shielded from most forms of spectroscopic analysis e.g. material flowing in a pipe, products stored in sealed containers, etc. Typically these materials or products are characterized before and after they enter the inaccessible region and details are inferred from these measurements. Radio frequency spectroscopy e.g. nuclear magnetic resonance (NMR), magnetic susceptibility, and electric susceptibility is an attractive characterization method as low frequency electromagnetic radiation can penetrate most container materials including metal. In most cases a direct relation between measured low resolution response and molecular parameters is not possible nor in many cases needed as samples of interest are complicated mixtures that change in response to an external variation such as spoilage or contamination. Here statistical methods such as principle component analysis, multiple linear regression (MLR), and partial least squares (PLS) are used to bridge the measured low resolution response to macroscopic material variations. Three experimental examples that rely heavily on these statistical methods will be provided. Principle component analysis is used to highlight the usefulness of radiofrequency electric and magnetic susceptibility based wine authentication, MLR and PLS are used to discriminate wine based on the multi-color luminescence afterglow created with a fast ultraviolet light pulse applied to a frozen sample, and PLS is used to identify cancer patients based on the NMR spectra of human urine and blood serum.

Spectroscopic and Chemometrics Applications in Synthetic Drug Substance and Drug Product Development
Wencan Chen, Yong Xie

Spectroscopic technique and chemometrics tools are used both by academic researchers and industry groups. In this presentation, a few representative examples are included to demonstrate the applications within Amgen Attribute Sciences to support synthetic drug substance (DS) and drug product (DP) development. Two off-line applications with NIR and FT-Raman will be introduced for monitoring the polymorphic form purity in DS and DP. The NIR method was qualified and transferred from Amgen to CRO to control the DS polymorphic impurity. The Raman method was developed with using solid-state NMR as an orthogonal method for polymorphic purity of DS in DP. The result was used to investigate the chemical stability of DP. On-line applications using Raman and infrared probes were also used to support synthetic DS process development. For one case, Raman probe was used in monitoring an ozonolysis reaction; in another, a quantitative model was developed to provide reaction endpoint and reaction kinetics for a H2O2 mediated reaction. IR probe was also used as a PAT tool to safe guard an exothermic Grignard reaction. The consumption of Grignard reagent and generation of reaction product can be provided at the same time. Through these examples, we hope to show how the spectroscopic and chemometrics applications fit into our day-to-day fast pace project work, what the challenges are, what we can do and what are needed from the instrument vendor and academicals researchers.

Automated Multivariate Calibration & Classification in Portable/Handheld Instruments with Systemic Modeling Approaches
Christopher Brown

The application of spectroscopic techniques for increasingly broad/wide applications by non-experts with portable instruments has substantially challenged the techniques of traditional chemometrics. The usual approaches reliant on stationarity assumption (e.g. PCR, PLS, DA) are deeply problematic due to the scale of the problems of greatest import. Alternative approaches that drive the modeling approaches from first-principals offer numerous advantages for these very large scale problems. Examples from Raman, FTIR, and MS / GCMS will be used to discuss these approaches and opportunities in both qualitative and quantitative applications.

Calibration updating by sample and feature augmentation
Erik Andries, John Kalivas, Anit Gurung

In spectroscopy, calibration updating (transfer and/or maintenance) without using a standardization set has historically been implemented using a simple but effective technique: in addition to the primary samples, include a small number of secondary samples and re-weight them. However, it would be ideal if we could only include unlabeled secondary samples, i.e., secondary samples with unknown reference measurements. We examine two alternative calibration updating scenarios involving unlabeled secondary samples: the sample augmentation and feature augmentation of labeled and/or unlabeled secondary samples to a pool of samples consisting of many primary samples. The unlabeled secondary samples are then used to construct an "undesirable" subspace from which the model vector of regression coefficients are projected away from. The feature augmentation approach is a multi-component model vector framework whereby the final model vector used to predict novel secondary samples now consists of a dedicated primary model vector plus a perturbation vector that can accommodate secondary samples.

Forensic Science R&D Funding Program at the National Institute of Justice
Gregory Dutton, Minh Nguyen

The National Institute of Justice (NIJ) is the research, development and evaluation agency of the U.S. Department of Justice. NIJ maintains a program of external funding for research and development in the forensic sciences which is a leading federal funder in this mission space. The portfolio spans a broad range, from fundamental research with the potential for application to forensic science, to the development of prototype devices, to the evaluation of novel instruments and methods. Since 2009, NIJ has funded over $179M of external R&D to support the advancement of accuracy, validity and efficiency in the forensic sciences. Strengthening forensic sciences through R&D ensures that perpetrators of crimes can be identified quickly, increasing public safety. Forensic science is a collection of applied disciplines that draws from all branches of science. Nevertheless, practicing forensic scientists most often tend to be concerned with the detection, collection, separation, and analysis of biological and chemical samples. Due to the unique circumstances forensic evidence, there is an ongoing need for these analyses to be done on ever smaller, degraded, compromised or mixed samples. Additionally, increased backlogs in operational forensic science laboratories create a need to increase the speed and decrease the cost of analyses. These needs drive NIJ's continuing R&D investments in analytical chemistry and bioanalytical science. Advances in mass spectrometry, electrophoresis, applied spectroscopies, microscopy and microfluidics, among other tools, have yielded or show promise for successful application to forensics. NIJ anticipates continued interest in advancing these technologies, as well as emerging analytical methods, for forensic application. In this effort, NIJ strives to engage the research community to bring novel perspectives to solving forensic problems. An overview of NIJ's research and development portfolio will be presented, highlighting relevant examples in Trace Evidence (fibers, glass, paint, dust, microbes, etc.); Controlled Substances and Toxicology; and Forensic Biology and DNA. Information on the funding cycle and general information about research and development funding opportunities at NIJ will be presented.

Trace blood detection with infrared spectroscopy, infrared imaging, and latent heat thermography in the context of achievable detection limits
Stephen Morgan, Michael Myrick, Raymond Belliveau, Brianna Cassidy, Zhenyu Lu, Stephanie DeJong

We have investigated the potential for detecting trace blood amounts using various forms of infrared imaging for several years. We initially confirmed the ability of IR imaging to detect trace blood down to levels of 100 times diluted. The standard for comparison has been the capability of luminol to detect blood. However, published literature on detection limits span several orders of magnitude, and depend intimately on the control of experimental variation during the calibration process. To acquire a knowledge of realistic detection limits that can be achieved with luminol, we have performed controlled experiments to optimize detection of blood with luminol. Using a video camera to record the luminscent response, the estimated detection limit for blood with luminol is in the range of 200,000 x dilute blood. More recently, we have employed steam thermography to image ridge patterns of blood-transferred artificial fingerprints on acrylic, polyester and cotton fabrics. Microscale thermal imaging confirms that the ridge patter transfer occurs on the tops of the most prominent threads in weave. Prints are readily imaged on the more hydrophobic fabrics in both heat-up and cool-down phases of vapor exposure. On the more hydrophilic cotton fabric, ridge detail is obscured by the strong background response of the fabric to moisture. Microscale thermography reveals that loose cotton strands with transferred blood may respond differentially to moisture relativeto uncoated strands. Ridge detail in blood transfers can be resolved in at least some cases and suggest a possible method for detecting transferred material on cotton despite its strong inherent response to moisture, based on observations of isolated cotton fibers occurring as a natural feature of the fabric.

On-Site GC/MS Analysis of Drugs: Reasoning, Reliability and Return on Investment
Glen Jackson

Instrumental methods of chemical analysis provide objective and largely irrefutable circumstantial evidence in the support of criminal prosecutions. Chemical measurements can also be archived for future validation or cross-examination by third-party expert witnesses, and they are widely accepted as amongst the most reliable of forensic techniques. It is for these reasons that chemical measurements received almost no criticism in the highly critical 2009 National Academy of Sciences (NAS) report on the Strengthening Forensic Science in the United States: A Path Forward. More recently, several recent advancements have been made in different areas of instrumental analysis that have resulted in a shift towards on-site and in situ chemical measurements. The reliability of portable instruments relative to bench-top instruments is not well documented, so the forensic community must demonstrate the validity and cost-benefit of portable instrumentation towards forensic casework. The first part of this presentation will evaluate the pros and cons of on-site or in situ measurements in forensic casework. This evaluation will include a return-on-investment analysis, which attempts to incorporate the entire criminal justice system, not just the crime laboratory system. The talk will also discuss the performance characteristics of portable instruments relative to bench-top instruments, in and whether or not collected data is likely to be admissible in court. This talk is intended to help researchers and instrument vendors understand and overcome the barriers to implementation in the forensic community.

Sampling and Analysis of Breath Components for Cannabis Detection Using Capillary Microextraction of Volatiles (CMV)
William MacCrehan, D'Nisha Hamblin, Bruce Benner, Michelle Schantz, Jose Almirall , Mimy Young, Sigalit Gura

The noninvasive collection of exhaled breath makes for an attractive sample for field detection of drugs from suspected impaired drivers. Previous studies characterizing the breath of individuals who have smoked cannabis used filters as sample collection devices resulting in low collection efficiencies. This study proposes the evaluation of a new technology, capillary microextraction of volatiles (CMV), for its utility in the collection of breath aerosols and volatiles from exhaled breath. The CMV preconcentrates breath components using a mini-capillary tube filled with PDMS-coated glass filter strips. The CMV offers dynamic sampling of volatile organic compounds (VOCs) with a simple coupling to the inlet of a gas chromatograph (GC) for analysis, avoiding expensive thermal desorption instrumentation needed for bulk sorbent type collection devices. CMV offers a 5,000-fold increase in surface area and an improved collection capacity over the static, solid-phase microextraction (SPME) fiber. In addition, the CMV sampling is dynamic, allowing for a relatively large volume (> 2L) of lung air to be sampled in less than one minute. The collection efficiency and analysis of VOCs associated with normal breath and from VOCs associated with tobacco smoking were studied and the CMV coupled to GC-MS was capable of distinguishing between non-smokers and smokers. The sampling of headspace VOCs from cannabis plants also provide potential target markers. The comparison of the direct thermal desorption and online supercritical CO2 extraction of a CMV into a GC/mass spectrometer (GC/MS) inlet was used to determine the recovery profiles of the two recovery approaches. The chemical characterization of breath components provides a foundation and better understanding of breath collection with CMV. Other forensic applications (VOC sampling for the detection of hidden drugs and the analysis of gunshot residue on the hands of a suspect) are also shown. Finally, the incorporation of a cryosampling method that can be used to enhance the recovery of VOCs is also presented.

Analysis of Blood Traces by Attenuated Total Reflection (ATR) Fourier Transform-Infrared (FT-IR) Spectroscopy for Forensic Purposes
Ewelina Mistek, Igor Lednev

Blood is one of the most common and informative types of biological evidence found at a crime scene. In forensic investigations it is crucial to identify the origin of a blood stain. However, current standard methods employed for the analysis of blood samples are destructive and time-consuming. In this study, attenuated total reflection (ATR) Fourier transform-infrared (FT-IR) spectroscopy was used as a confirmatory, nondestructive, and rapid method to distinguish between human and animal (nonhuman) blood. Partial least squares-discriminant analysis (PLS-DA) models demonstrated complete separation between human and animal donors. The models also distinguished between three separate species, namely human, cat, and dog. Classification predictions of unknown blood donors performed by the model resulted in 100 % accuracy. This study demonstrates ATR FT-IR spectroscopy's great potential for blood stain analysis and species discrimination. Furthermore, the commercial availability of portable ATR FT-IR instruments affirms the potential for the implementation of such blood stain analyses both at the crime scene as well as in the lab.

Raman applications in orthopaedics: from gout to arthritis
Ozan Akkus, Mustafa Unal, Bolan Li

Raman spectroscopy is a powerful analytical tool for analysis of tissues and biofluids, particularly due to its low demand for sample handling and preparation. Furthermore, Raman spectroscopy is ideal for the analysis of hydrated specimens. This presentation is going to focus on two relatively recent biomedical applications of Raman spectroscopy. The first application is geared toward diagnosis of crystals which lead to arthritis like symptoms in the joint space. Specifically, monosodium urate and calcium pyrophosphate dihydrate crystals accumulate in human joints and result in significant pain. Their symptoms can be confused with other forms of arthritis. Conclusive diagnosis requires analysis of aspirated synovial fluids which are currently inspected microscopically by certified and trained operators in clinical lab setting. In this presentation we will be showing a microfluidic crystal isolation kit along with a point of care Raman device for automated and unbiased detection of crystals. Clinical data from a patient set (N= 174) indicated that diagnosis by Raman was not significantly different from that of trained pathologists. The second application will be a novel shortware infrared Raman system that is specifically optimized for studying the high wavenumber range of 2900-4000 cm-1. High fidelity acquisition of Raman spectra in this range allowed us to measure the hydration of cartilage and bone. Systematic drying of unbound and bound water coupled with deutration of bound water enables us to identify different compartments of water in these tissues. Results indicated that the peaks that are identified had significant and meaningful correlations with the mechanical properties of cartilage and bone. These efforts bring Raman spectroscopy one step closer to creating translational niche for a number of applications in analysis of the musculoskeletal system.

Resonance Raman spectroscopy based label-free approach for HbA1c detection
Rishikesh Pandey

Glycated hemoglobin (HbA1c) is an important biomarker that reveals average value of blood glucose over preceding three months, and HbA1c value is routinely quantified in clinical laboratory for long-term glycemic control and recently gained considerable traction in screening of pre-diabetic. Here we present label-free platform for classifying normal, mild-diabetic and chronic diabetic populations based on resonance Raman spectral signal in both hemolysate and whole blood samples. Our resonance Raman spectroscopic results in conjunction with multivariate chemometric analyses reveal that the glycation results in subtle and reproducible changes in Raman spectrum that subsequently enables high degree of classification accuracy. We envisage that the substantive advantages of Raman spectroscopy especially label-free approach and its ability to couple to a fiber-probe will enable its ready translation to a point-of-care setting.

Integration of Nanopillar SERS Substrates in a Microfluidic Platform for Analyte Separation and Quantitative Sensing

Michael S. Schmidt, Onur Durucan, Kaiyu Wu, Marlitt Viehrig, Oleksii Ilchenko, Lidia Morelli, Tommy S. Alstrøm, Tomas Rindzevicius, Anja Boisen

We have previously developed a surface enhanced Raman spectroscopy (SERS) substrate that in addition to a very high enhancement also has a very uniform signal across its entire surface. The substrate consists of free-standing silicon nanopillars capped with gold. In this contribution we show how we have utilized this unique nanostructure in a centrifugal microfluidic platform to enable quantitative measurements of various analytes in complex matrices. The key feature of this SERS substrate is its ability to act as a chromatographic media which due to capillary forces will essentially "suck" the analyte up when immersed into a liquid. Due to the pillar structure it will effectively separate/filter a mobile analyte from larger particles in the matrice. This wicking effect has proven to be very effective in the well known case of melamine detection in milk where separating melamine from the proteins, which will otherwise block the plasmonic structures, is essential in order to obtain a signal. To precisely control the movement of the analyte on the SERS substrate the chip is integrated into a centrifugal microfluidic platform. This enables the analyte to be introduced onto the liquid without the use of pumps and ensures a uniform coverage each and every time. This becomes important when one wants to obtain quantitative measurements. Furthermore, in the case of melamine in milk we obtain a 70 times larger signal when using the microfluidic platform as opposed to merely dipping the substrate into the analyte solution. The enhancement uniformity and the capillary filtration effect of the substrate combined with precise sample handling in the centrifugal microfluidic platform is the basis for what we can comfortably describe as quantitative SERS measurements. In the case of p-coumaric acid detection we have corroborated our observed Raman intensities with high performance liquid chromatography (HPLC) thus illustrating that SERS can be used in a quantitative capacity. We find that automated analyte delivery to a uniform SERS substrate is a feasible pathway to using SERS as a quantitative too in analytical chemistry applications.

Monitor reactions by in situ IR and Raman spectroscopy

Xiaoyun (Shawn) Chen

We will describe recent application examples of in situ IR and Raman spectroscopy for reaction monitoring in Dow Chemical.

Surface-enhanced Raman Scattering from Synergistic Contribution of Graphene and Semiconductor in Graphene-TiO2 Assembly for Stem Cell Related Bioanalysis

Tingting Zheng, Enduo Feng, Yang Tian

2D layered materials have been recognized as active platforms for plasmon-free surface-enhanced Raman spectroscopy (SERS). Herein, several novel EG-TiO2 SERS platforms were specially designed through electrochemically deposition of graphene (EG) onto TiO2 nanoarrays. The proposed EG-TiO2 SERS platforms possess excellent SERS activity with an enhancement factor of ~ 48 while using copper phthalocyanine (CuPc) as probe molecules, which is 2.4–fold Raman enhancement compared with pristine graphene. Both experimental and theoretical calculation results demonstrated that the enhancement was attributed to TiO2-induced shift of the Fermi level of graphene and consequently change the match degree of the energy band between graphene and the molecules. Taking advantage of the specific SERS response of CuPc molecule on the surface of EG-TiO2 platform as well as the CuPc deposition induced by multiple telomeric G-quadruplex, the telomerase activity in the proliferation and differentiation process of stem cells was tactfully and ultrasensitively monitored for the first time. The unique structural and SERS properties of the EG-TiO2 platform have made the detection limit of telomerase activity down to 1 stem cell. Remarkably, the self-cleaning characteristic of TiO2 successfully realized the recycling of the plasmon-free EG-TiO2 substrate. Our work is important for gaining fundamental insights into the chemical mechanism (CM) in Raman enhancement, and will open a new way in the investigation of the long-term dynamic of stem cell differentiation and clinical drug screening.

On-Line Monitoring of Materials by Chemometric Methods Applied to LIBS
Arne Bengtson, Jonas Petersson, Bertrand Noharet, Baptiste Ottino, Mattias Åslund, Tania Irebo Schwartz

Laser-induced breakdown spectroscopy (LIBS) is a method capable of monitoring the chemical composition of materials without the need for sample preparation. The analysis can be made at high speed and at a distance, making the method highly suitable for on-line analysis with fast response. LIBS is therefore a highly useful analysis tool in many steps of metal processing, including monitoring of the input materials e.g. raw materials and recycled metal scrap. At Swerea KIMAB we have research projects in close collaboration with industry. A critical aspect is to bring the technique from the clean lab-environment to the harsh conditions in industrial on-line analysis by developing robust LIBS instruments. This means designing equipment appropriate for the specific applications by selecting different lasers, detectors and system layouts. In all cases the output will be spectra that consist of a multitude of emission-lines, which makes up a wealth of information about the material. To efficiently use all this information is a challenging field of research. The focus of the presentation is on recent results on classification of materials (scrap sorting) using different types of chemometric methods.

Effect of laser parameters on the analysis of precious metals in minerals by Laser-Induced Breakdown Spectroscopy (LIBS)
Daniel Diaz, David Hahn, Alejandro Molina

Surrogate samples containing synthetic Au and actual ore samples containing Au and Ag were analyzed by LIBS using 355 mm and 1064 nm lasers at various laser irradiances. The LIBS net, background and signalto-background signals were analyzed for both wavelengths. At higher laser irradiances, large LIBS emission signals were registered for synthetic Au and Ag-ore analytes. The calibration curve metrics for Au in surrogate and Ag in ore samples improved at higher laser irradiances, but that did not translate into better limits of detection. Calibration curves for Au in surrogate samples were statistically invariant at 355 nm and 1064 nm. Contrary, the Ag-ore analyte showed higher emission intensity for the 1064 nm ablation. However, the signalto-background normalization reduced the effect of laser wavelength on Ag calibration curves. Plasma diagnostics by the graphic Boltzmann and Stark broadening methods showed that the plasma temperature and electron density did not follow a specific trend as the wavelength changed. Finally, it was not possible to quantify natural-occurring Au in ore samples with neither 355 nor 1064 nm lasers in the explored range of laser irradiances (0.11 to 15.3 E9 W/cm2).

Testing of Carbon and Silicon in Low-Alloy Carbon Steels by Field-Portable Handheld LIBS
Brendan Connors, David Day

Field-portable handheld LIBS analyzers have been available commercially for several years, and have proven themselves well for alloy sorting and other metals applications. The ability of LIBS to analyze light elements has been of particular interest for metal alloy analysis, since field-portable techniques to analyze light elements are limited. In this paper we will present data to show that a field-portable LIBS analyzer can quantify carbon (C) in steels at levels as low as 0.1%. The level of precision allows for effective separation of many common grades of carbon steel. This testing makes use of several features recently available in portable LIBS analyzers including inert gas (argon) purge, variable detector time gating, and two-dimensional laser rastering. Data showing effective screening of silicon (Si) in carbon steels will also be presented. Screening of silicon content of carbon steels is useful in the petrochemical industry as a protection against corrosion. Fieldportable handheld LIBS can effectively quantify Si as low as 0.02% with short acquisition times. Data from laboratory testing as well as in-service components at a refinery will be discussed.

Analysis of glass pharmaceutical containers with a handheld LIBS spectrometer
Katherine Bakeev, Qun Li, Dan Liu, Jing Li, Sean Wang

Glass containers are among the primary packaging material that has found use in the pharmaceutical industries. Glass containers are classified into Type-I (borosilicate), Type-II (treated soda-lime), and Type-III (soda-lime) glasses based on their degree of chemical/hydrolytic resistance. A critical step in assuring the biological safety and suitability of a glass container is the ability to characterize the composition and type of the glass container. In this paper, we demonstrate the capability of a handheld LIBS spectrometer for analyzing glass pharmaceutical containers. The handheld LIBS spectrometer employs a micro diode-pumped Q-switched solid-state laser with a high repetition rate of well above 1 kHz in comparison to 1-10 Hz repetition rate lasers as used in a traditional LIBS instrument. The laser beam is fast scanned over a pre-defined area, hence generating several thousands of micro-plasma emissions per second at different locations. Synchronized miniature CCD array spectrometer modules measure these thousands of micro-plasma emissions to produce a LIBS spectrum. By integrating the intensity of this plurality of micro-plasma emissions, the signal-to-noise ratio (SNR) as well as the stability and repeatability of the obtained LIBS spectrum is greatly improved. This unique feature of the high repetition rate laser based LIBS spectrometer allows it to measure elements at trace levels, hence reducing the limit of detection (LOD). In addition, the energy of the individual laser pulse can be reduced in comparison to traditional LIBS system to obtain the same signal level, making the laser pulse less invasive to the subject under measurement. With the handheld LIBS spectrometer, we measured and analyzed the LIBS spectra of a variety of glass pharmaceutical containers. The results demonstrate that the handheld LIBS spectrometer is capable of characterizing the composition and type of the glass pharmaceutical containers within a few seconds. The handheld LIBS spectrometer is compact and easily portable for use at point of need.

Subsurface Mineral Exploration Using LIBS
Pablo Sobron, Kris Zacny

Honeybee Robotics and Impossible Sensing have developed and successfully deployed a downhole wire-line LIBS sensing system for investigating the subsurface within a pre-drilled borehole and obtaining detailed elemental composition as a function of depth. The robotic probe integrates a laser, an array of spectrometers, and data and power interfaces. The probe is tethered to a remote control unit that sits on the surface. We demonstrated the functionality of a LIBS sensor in a downhole environment. Specifically, we showed that we can obtain quantitative estimates of the concentration of multiple elements as a function of depth in a borehole. We are currently developing a novel achitecture that integrates (1) a demonstrated rotary-percussive drill with a hollow drill stem adapted to any type of material; (2) a novel miniature (<5mm diameter; <50mm length) optical probe to perform in-situ LIBS; (3) automated chemical and mineralogical quantitative analyses routines, combining hardware designs and high-performance signal and data processing algorithms to enable unprecedented measurements: rapid, depth-resolved downhole chemical and mineralogical identification and quantitation of subsurface materials while drilling. Our innovative approach consists of combining, for the first time, drilling and LIBS to build an integrated tool that responds to critical challenges at the scientific/engineering boundaries of drilling and sensing; in particular, the challenges involved in characterizing subsurface materials in-situ and in real-time. Conventional methods involve drilling and coring, and the analysis of cores in off-site laboratories. Not only is this approach laborious, time consuming, and dangerous for human operators, but the results are not uniformly reliable and typically not available for weeks. Our new LIBS system has applications in the Earth, environmental, and planetary sciences, e.g. geological prospecting; environmental monitoring/assessment; agricultural soil quality monitoring; oil & gas exploration and development; and homeland security.

Development of a CE-MS/MS Platform for Sequencing Glycosaminoglycans
Jon Amster, Patience Sanderson, Morgan Stickney, Franklin Leach, James Xia, Yanlei Yu, Fuming Zhang, Robert Linhardt

Sulfated glycosaminoglycan (GAG) carbohydrates are linear, acidic polysaccharide chains abundant on the surface of mammalian cells, and which effect a number of biological processes through protein-binding interactions. Tandem mass spectrometry methods have been developed for the structural analysis of purified oligomers, but the analysis of mixtures remains a significant challenge. Using capillary electrophoresis-mass spectrometry (CE-MS) with a novel interface, we can separate and analyze GAG mixtures containing chains up to dodecamers. This advance enables the determination of binding motifs responsible for the interaction of GAG chains with proteins as these sequences typically range from at least tetrasaccharides to octasaccharide and greater in length with the most notable occurring as the pentasaccharide sequence responsible for the heparin-antihrombin binding relationship.

Gas-Phase Ion-Electron Reactions for Carbohydrate and Glycopeptide Structural Characterization
Kristina Hakansson

Protein glycosylation serves a number of biological functions, including improved protein stability and solubility as well as proper folding and interaction with other molecules. Thus, for complete characterization of such biomacromolecules, both the carbohydrate structure and linkage site(s) need to be determined. The latter information can be achieved on a glycopeptide level whereas the former typically requires glycan release and separate analysis. However, due to their non-linear structure and lower ionization efficiency, carbohydrates and glycopeptides are more challenging to structurally characterize by mass spectrometry compared with, e.g., unmodified peptides. For carbohydrate structural determination, extensive glycosidic and cross-ring cleavages are needed in tandem mass spectrometry (MS/MS). However, cross-ring fragments, crucial for determining linkage and branching, are scarce in conventional collision activated dissociation (CAD) and, thus, alternative MS/MS activation strategies are needed. Similarly, CAD is inefficient in characterizing glycopeptides as glycosidic bond cleavages dominate and thus preclude determination of the peptide sequence and carbohydrate attachment site. Alternative electron-based, radical driven MS/MS activation strategies that can alleviate some of these problems will be discussed, including di-and trivalent metal-assisted electron capture dissociation (ECD)/electron transfer dissociation (ETD), electron induced dissociation (EID), electron detachment dissociation (EDD), and negative ion ECD (niECD). ECD and ETD are limited to multiply charged cations whereas EDD is limited to multiply charged anions and also shows low fragmentation efficiency. EID has the benefit of being compatible with singly charged ions of both polarities and niECD is optimum for larger singly charged anions. For carbohydrates, metal adduction is necessary to generate the multiply charged cations necessary for ECD/ETD. This strategy has shown promise for glycan isomer differentiation. However, for acidic glycans/glycopeptides, including sialylated and sulfated species, negative ion mode is preferred with particular compatibility between glycopeptide analysis and niECD.

LC-MS/MS Analysis of Glycan and Glycopeptide Isomers
Yehia Mechref

Glycosylation, as one of the most common post-translational modification (PTM), plays critical roles in various biological processes. Development of quantitative glycomics/glycoproteomics profiling methods is essential for understanding the biological attributes of glycans. Although high-resolution mass spectrometry facilitates accurate sequential identification of glycans, identification of glycan isomers is not readily attainable without LC separation. Moreover, the interpretation of isomeric microheterogeneity associated with the glycosylation sites of proteins is one of the major technical challenges of glycoprotein and glycopeptide analysis. Here, isomeric separation of released N-glycans and glycopeptides on the porous graphitic column at elevated temperatures will be described and discussed. Permethylation of glycans enhances ionization efficiency, stabilize sialic acid and eliminates fucose rearrangement. However, the isomeric separation of permethylated glycans is always not satisfactory due to the increased intramolecular interaction after permethylation. In this study, we have achieved efficient isomeric separation of permethylated glycans by using PGC-LC at 75oC. Similarly, isomeric separation of glycopeptides was also achieved using PGC column at elevated temperatures. Base peak isomeric separation of glycopeptide mixture was attained at 75oC. We are reporting here for the first time, the ability to attain isomeric separation of released N-glycans and glycopeptides on PGC columns at elevated temperatures. Accordingly, comprehensive characterization of the isomeric microheterogeneity of the glycosylation sites of proteins is readily achieved by interfacing PGC columns to mass spectrometry. We are reporting here for the first time the sensitive and efficient isomeric separation of glycans and glycopeptides on PGC column at temperatures higher than ambient. Accordingly, comprehensive characterization of the isomeric micro heterogeneities of the glycosylation sites of proteins is readily achieved by LC-MS/MS analysis on PGC column at elevated temperature.

A Multidimensional Strategy to Resolve Carbohydrate Isomerism in the Gas Phase

Eric D. Dodds, Katherine N. Schumacher, Jessica L. Minnick, Richard L. Backhus

The rapid and unambiguous discrimination of carbohydrate isomers is among the major challenges in contemporary bioanalysis. The scope of this challenge is matched by the structural diversity and biological centrality of carbohydrates. Ion mobility spectrometry (IMS) and tandem mass spectrometry (MS/MS) have both shown promise with respect to distinguishing carbohydrate isomers, although each of these individual methods remains somewhat limited in their ability to resolve carbohydrate isomers in mixtures. Previous work in our group has shown that metal cationized carbohydrates can assume isomer-distinguishing conformations, as inferred by IMS measurements, depending on the identity of the bound cation. Furthermore, we have found that gas-phase ion-ion reactions can produce charge-reduced carbohydrate / metal ion complexes that result in isomers taking on a wide range of IMS behaviors, while also providing strikingly unique MS/MS spectra upon vibrational activation. Our current efforts are aimed at establishing a fundamental understanding of these phenomena, while also leveraging these insights in concert to provide for rapid analysis of realistic mixtures of isomers. This presentation will focus on three major themes: (1) systematic study of how carbohydrate fine structures impact the IMS behaviors of isomers across multiple metal adduct species; (2) elucidation of charge carrier dependent oligosaccharide fragmentation pathways that yield unique MS/MS signatures for isomers;
(3) combination of the IMS and MS/MS dimensions to deconvolute partially resolved IMS profiles based on unique MS/MS fragment ions. In aggregate, these approaches are shown to enable analytical disentanglement of carbohydrate mixtures on the millisecond timescale.

Mapping the Glycoproteome with Activated Ion Electron Transfer Dissociation

Nicholas M. Riley, Alexander S. Hebert, Nicholas W. Kwiecien, Michael S. Westphall, Joshua J. Coon

Protein glycosylation is a prevalent, chemically complex, and biologically diverse post-translational modification (PTM) involved in a wide array of intra- and inter-cellular functions. Approximately half of all expressed proteins undergo glycosylation, and this heterogeneous modification accounts for the greatest proteome diversity over any other PTM. Changes in protein glycosylation are associated with cellular proliferation, inter-cellular communication, and metabolic processes, making the characterization of the cellular landscape of protein glycosylation integral to advancing our understanding of cell biology. Glycan microheterogeneity, i.e., different glycans modifying the same glycosite, makes glycan identity at a given site crucial to the biological context of the modification. This unique feature of glycosylation makes analysis of intact glycopeptides imperative for glycoproteome characterization, but current analytical tools are ill-suited for this task. Tandem mass spectrometry (MS) is an ideal platform to advance glycoproteomic technology, but current dissociation methods are often suitable only for characterization of either peptide or glycan moieties. This mandates multiple analyses of the same precursor ions that limit throughput and challenge data interpretation. We have developed a tandem MS dissociation method called activated ion-electron transfer dissociation (AI-ETD) that addresses several of challenges of intact glycopeptide analysis. Through the use of concurrent ion-ion reactions and infrared photoactivation, AI-ETD can access glycan and peptide information from intact glycopeptides in a single MS/MS scan, and with this AI-ETD provides (1) improved product ion generation for peptide backbone sequencing, (2) higher MS/MS success rates to sequence more glycopeptides per experiment, and (3) valuable fragmentation for glycan composition determination. Here we showcase AI-ETD for large-scale intact glycopeptide characterization on the newest generation of Orbitrap instruments. We show that >2000 localized N-glycosites can be confidently identified from approximately 30,000 localized N-linked glycopeptide spectral matches (>6,500 unique) in mouse brain tissue. This represents a more than 3-fold increase over recent studies in the number of glycopeptides and glycosites that can be identified via intact glycopeptide analyses, and it rivals, if not outmatches, glycoproteomic studies that analyze only deglycosylated peptides. Furthermore, we extend this global survey of the glycoproteome to multiple mouse tissues, mapping thousands of glycosites in a tissue-specific atlas.

Dual emissive gold nanoparticles for ratiometric pH sensing

Jie Zheng

Luminescent gold nanoparticles have found broad biomedical applications such as cancer detection and kidney functional imaging. However, since many diseases are involved in pH changes of local microenvironment, rendering these luminescent gold nanoparticles a new capability for reporting local pH will further broaden their biomedical applications. In this talk, I will present two strategies for developing dual-emissive gold nanoparticles and demonstrate their applications in ratiometric pH sensing and imaging. The first strategy is to fine tune binding geometries of surface ligands on the gold nanoparticles, which allows two emissive centers to be incorporated into one gold nanoparticle. As a result, these two emissive centers can be coupled and give a ratiometric response to pH changes. The second strategy is to take advantage of dimerization of pH-insensitive dyes on ultrasmall gold nanoparticles. The coupling between organic dyes and local surface ligands will also render the hybrid nanoparticles a capability that can fluorescently report local pH in a ratiometric way. These two strategies open up a new pathway to design ultrasmall fluorescent ratiometric nanoindicators with tunable wavelengths for in vivo pH imaging.

Tailoring Luminescent Nanoparticles in Biology
Gang Han

Functional luminescent nanoparticles are promising materials for in vitro and in vivo optical imaging and therapy due to their unique optical and chemical properties. In I will present two new types of biocompatible luminescence nanoparticles. The first type of materials is upconversion nanoparticles (UCNPs). I will present new developments regarding engineering UCNPs towards deep tissue imaging, photodynamic therapy, optogenetic applications in neuroscience and immunotherapy. The second type of nanoparticles is persistent luminescence nanoparticles (PLNPs). They are bioluminescence-like and possess unprecedented in vivo deep tissue energy rechargeability, outstanding signal-to-noise-ratio with no need for an excitation resource (light) during imaging, and they can be directly detected with existing imaging systems. These nanoparticles continue to emit light for minutes or hours and, in some cases, days, after turning off the excitation source. These long-lasting, light-emitting nanocrystals can provide noninvasive imaging technology for evaluating structural and functional biological processes in living animals and patients.

Photoelectrochemistry of colloidal TiO2 nanoparticles: from aggregates to single crystals

Mario Alpuche-Aviles, Krishna Barakoti, Pushpa Chhetri, Rezvan Kazemi, Nelum Karunathilake, Ganesh Rana

We present our measurements of photocatalytic activity of suspended nanoparticles (NPs) of anatase TiO2 nanoparticles. Our particles are suspended in MeOH under constant illumination. When the NPs collide with an ultramicroelectrode, that is, an electrode of a few micrometers in diameter, a current is detected. The current transient can show increases that are step by step or it can also be a current spike. The different transients are assigned to the cases of an irreversible adsorption of individual NPs and the current spikes are assigned to agglomerates that interact transiently with the electrode surface. The stochastic photocurrent transients that we measure are correlated with the colloidal properties. We have studied the products of MeOH photooxidation for the colloidal nanoparticles and we compare the colloidal conditions to the stochastic electrochemical measurements. The colloidal properties were studied by dynamic light scattering.

Single Nanoparticle Plasmonic Spectroscopy for Biomedical Applications

X. Nancy Xu, Pavan K. Cherukuri, Preeyaporn Songkiatisak, Asia Poudel, Andrea Korell

Plasmonic optical properties of single noble metal nanoparticle (NPs) highly depend upon their sizes, shapes, chemical and surface properties. We have developed single NP plasmonic spectroscopy to study heterogeneity of nanomaterials and to detect and characterize plasmonic properties of individual rare NPs in a highly heterogeneous population of NPs in situ in real time for better understanding and design of physicochemical dependent plasmonic properties of NPs. Further, single NP plasmonic spectroscopy enables single noble metal NPs (e.g., Ag NPs) to serve as distinctive optical spectroscopic imaging probes and sensors to achieve multiplex high-throughput sensing and super-resolution imaging of individual molecules of interest on/in single live cells in real time at nanometer resolution. We have demonstrated that single plasmonic NPs possess superior photostability, and they can serve as photostable imaging probes to quantitatively and continuously track and image individual molecules of interest over a desired long period of time. We have developed a set of powerful new nano tools surrounding single NP plasmonic spectroscopy for a wide range of biomedical applications, from probing single membrane transporters to understand molecular mechanisms of multidrug resistance of single live cells, mapping signaling transduction pathways of single live cells, investigating the effects of NPs on cellular functions, to detecting and sensing of rare subsets of single cancer stem cells (CSCs) cells in situ in real time for effective diagnosis and therapy. The detailed experimental design and updated findings will be discussed in this presentation. The work is supported in part by NSF (CBET 0507036 & 1450936) and NIH (5R01 NIGMS076440 & 119116).

Near Infrared ElectroChemiLuminescence of Gold Nanoclusters for Redox and Metal Ion Sensing

Gangli Wang

Electrochemiluminescence (ECL) refers to the photoemission activated by electrode driven electron transfer (ET) reactions. ECL has found broad applications in i.e. assays, sensing, imaging and bioanalysis. A major challenge is the side reactions of highly reactive radical intermediates in the ECL generation pathways which limit the efficiency and stability of ECL under ambient or physiological conditions. This talk will discuss our recent progresses in the enhancement of the near infrared ECL from thiolate-protected Au nanoclusters. ECL is less explored compare to the photoluminescence in the near infrared spectrum window that is highly advantageous in bioimaging and sensing with reduced background interferences. Redox activation in ECL eliminates the need of photo excitation that further reduces possible interferences and simplifies the hardware requirements. Two strategies to enhance near IR ECL from different types of Au nanocluster under ambient conditions will be presented. The first is an intra-cluster charge transfer mechanism by aligning the related electronic and molecular structures. In another route, EDTA as representative chelators is adopted as coreactants for ECL enhancement under physiological pH. Sensing of metal ions is demonstrated by the modulation of the ECL signals. Generalizable sensing and prototype cell imaging studies will be discussed based on the luminescence activated by heterogeneous electron transfer, ECL, as well as the well-known photon-excitation/emission.

Innovative Applications of Raman Microscopy
Peng Wang, Thomas Tague, Juergen Sawatzki, Sergey Shilov

Raman spectroscopy is one of the most commonly used techniques for materials identification and characterization. By measuring vibrational as well as lattice and other low frequency modes in a system, it provides unique information for molecular compositions as well as backbone and crystalline structures. Therefore, Raman spectroscopy finds extensive application to various fields including but not limited to pharmaceuticals, polymers, forensics, cultural heritage, life science, mineralogy, materials science, and more. State-of-the-art Raman microscopy enhances the investigation of materials by providing non-invasive microscopic analysis featuring minimal sample preparation, diffraction limited lateral resolution and subsurface accessibility due to its confocal design. It can not only be used to analyze pure compounds but also be applied to mixture analysis, composite structure analysis, orientation and crystallinity analysis and so on. In this paper, some recent applications of Raman microscopy are presented and discussed.

2D and 3D Micro-Raman Imaging of Crystal Transformation due to Indentation of ZrO2 Ceramic and ZrO2 Containing Glass Ceramic Materials
Galan Moore, Charlene Smith, Benjamin Hanson, Sara Cole

Raman is well known to be able to detect and identify crystalline materials; however, little work has been done to observe how crystals (or crystals in glass) respond to pressure, in particular, due to micro-indentation. ZrO_2 is known to undergo a phase transition from tetragonal to monoclinic due to applied pressure at room temperature. However, observing this phase transformation within and surrounding an indent can be challenging due to the strict requirements of spatial resolution. New 2D and 3D Raman imaging methods have been developed to visualize and quantify the degree of this crystal transformation due to the indentation process as a function of applied force and location with respect to the indent center. The methods allow for the direct comparison and visual interpretation of this crystal transformation both inside the indent and outside the indent. The crystal transformation zone, observed outside of the indents, due to the distribution of applied force from to the indentation process can now be observed with spatial resolution down to a 1?m^2.

A novel wide-field Raman system for fast chemical imaging
Haithem Mustafa, Ozan Akkus

Raman spectroscopy is a fundamental tool used in materials science, biology, energy, medicine, chemistry and physics. Embedded in the Raman spectrum is the information on the identities and amounts of chemical species, the conformation of the bonds and molecular morphology. Traditionally, Raman spectroscopy has been used to generate chemical images by point by point mapping over the sample space. As useful and dramatic as these images are, it takes hours, sometimes days to generate such images when high resolution is needed. Furthermore, point by point scanning does not allow studying the entire field of view simultaneously which is a limitation in studying dynamic systems. Utilization of 2D-CCDs help address the limitations of point by point scanning; however, global imaging as such require uniform illumination and creative wavenumber filtering schemes to generate Raman images. We demonstrate an advanced Raman Imaging System which integrates a digital micro-mirror device (DMD) as an adaptive optical beam former to attain uniform intensity illumination on the sample space. A tunable spatial filter is designed to select the desired wavenumber and then collected by a 2D-CCD camera with a high resolution (1024 x 102 pixels) and sensitivity. When discrete wavenumber images are generated the spatial filters are fixed, expediting the imaging rate. In hyper spectral imaging applications, filters kinematically scan the wavenumber range of interest to obtain 4D data. Based on this proposed setup system, fast and higher spatial resolution (image details versus spatial resolution) Global Raman Imaging is experimentally demonstrated. In addition, spatially offset Raman Spectroscopy (SORS) technique for subsurface imaging is also feasible using this setup.

Raman Chemical Imaging Based Cell Cytometry: Differentiation and Quantification of viable and Gamma Deactivated B. Anthracis Sterne Spore

Jason Guicheteau, Ashish Tripathi, Erik Emmons, Michael Kim, Phillip Wilcox

In 2015, a private company notified the CDC that inactivated Bacillus anthracis (BA) spores in their possession were viable. Upon an exhaustive review of the situation it was determined that the material in question originated from a DoD lab and was further distributed to other areas around the country and world. Since that time a major focus by the CDC and DoD has been to re-assess and better characterize all inactivation procedures currently being performed across all capable laboratories. ECBC has been a crucial part of this process and has been tasked to provide the Chemical Biological Defense Program with a methodology to properly inactivate BA spores while maintaining molecular and immunological targets for assay development, performance verification, and proficiency testing. To that end, along with more robust inactivation methods needing to be developed, analysis techniques to rapidly detect and determine viable and non-viable biological material is a priority. Raman Chemical Imaging (RCIM) has demonstrated that bacterial species discrimination via unique spectral signatures is achievable. The source of that discrimination however has not been linked to specific chemical content, especially considering linkage to signals related to inactivation protocols and damaged bacteria. This work presents a study of Raman spectral features to specific biological sources and identify those vibrational chemical components that give rise to the spectral features responsible for viable, non-viable discrimination.

Dynamic sampling in Raman microscopy

Garth Simpson, Shijie Zhang, Zhengtian Song, Azhad U. Chowdhury, G. M. Dilshan P. Godaliyaddab, Dong Hye Ye, Atanu Sengupta, Gregery T. Buzzard, Charles A. Bouman

The timeframe for image generation in Raman microscopy is greatly reduced using sparse sampling strategies informed dynamically by the preceding measurements. In brief, Raman spectra acquired at an initial set of random positions inform the next most information-rich location to subsequently sample within the field of view, which in turn iteratively informs the next locations until image reconstruction is complete. Using this approach, reliable chemical images of pharmaceutical materials were obtained with 20% sampling, representing a 5-fold reduction in measurement time with negligible loss in image quality. This supervised learning approach to dynamic sampling (SLADS) has the distinct advantage of being directly compatible with standard Raman instrumentation. Furthermore, SLADS is not limited to Raman imaging, providing a time-savings in image reconstruction whenever the single-pixel measurement time is the limiting factor in image generation.

Method development for the classification of drugs with identical API content using Raman spectroscopy

Md. Nayeem Hossain, Carl Anderson, James Drennen

Raman spectroscopy is a promising technique for qualitative applications, e.g. pharmaceutical product authentication. This authentication method should be able to avoid misclassification of genuine samples, which reflects the sensitivity of the model. It is equally important to recognize any kind of counterfeit products and reject them by the method, which represents specificity. The degree of counterfeit product varies between 'low quality' fakes and 'high quality' fakes. 'Low quality' fakes include products with no API but excipients such as starch and calcium carbonate instead, which is comparatively easy to identify by using spectral correlation coefficient method. However, 'high quality' fakes refer to the products produced with a recipe that is common to the genuine product with modern manufacturing equipment. Rejecting such identical sample is challenging. Multivariate classification methods (PCA/SIMCA) can be one option to resolve this problem. However, rigorous sample representation for calibration and validation set is required to develop robust method which has not been discussed in the current literature. This work demonstrated how different sources of variability have effect on raman qualitative method performance and addressed the required steps for method development. The calibration set included acetaminophen samples from a genuine manufacturer. To test the specificity against 'high quality' fakes, it is wise to use the legitimate analogs of the calibration set that are manufactured by various producers. Such drug should contain identical API. Acetaminophen tablets manufactured by other manufacturers were considered for this type of test sample. Sensitivity was tested by analyzing a sample set from the same manufacturer. Long-term sensitivity was tested by preparing samples in different seasons. During the study, relative humidity changes. Although water has weak raman scattering, relative humidity variability had an indirect effect on qualitative model performance. Because many pharmaceutical components have intrinsic fluorescent characteristics (with broad spectral features), moisture variability caused spectral artifacts in the form of baseline variation associated with fluorescence quenching. Study found that spectral variabilities due to water-induced fluorescence quenching caused rejecting of genuine sample. It was proposed to consider water variation for robust raman classification method development.

Low-frequency Raman spectroscopy can improve quantitative solid-state analysis of solid-state form mixtures of pharmaceuticals

Tiina Lipiäinen, Sara J. Fraser-Miller, Keith C. Gordon, Clare J. Strachan

Low-frequency Raman spectroscopy has much potential for solid-state pharmaceutical analysis. Unlike conventional mid-frequency Raman spectroscopy, it probes low-energy lattice vibrations that are directly defined by the crystal structure. Compared to other solid-state characterization techniques, Raman spectroscopy offers advantages through simplicity and potential for in-line measurements. For pharmaceuticals, the control and monitoring of solid-state structure is essential because it affects many critical quality attributes, such as solubility and stability. Low-frequency Raman has been used in several solid-state studies, but direct comparison to mid-frequency Raman for quantitative analysis of crystal form mixtures, without confounding sampling and instrumental variation between the regions, is still lacking. The aim of this study was to evaluate the potential of low-frequency Raman analysis for quantification of ternary solid-state form mixtures, and to directly compare it with mid-frequency Raman analysis through simultaneous detection of both spectral regions. Piroxicam was used as a model drug. Mixtures of three solid-state forms (?, ?2 and monohydrate) were prepared to calibrate and test the quantitative models. Measurements were performed using two dispersive low-frequency Raman systems one capable of simultaneous low- and mid-frequency spectral data collection (785 nm excitation, 7 cm^{-1} resolution), the other low frequency focused (830 nm, 2 cm^{-1} resolution), and a FT-Raman spectrometer (1064 nm, 2 cm^{-1} resolution). Quantitative PLS models were built using different spectral regions after SNV transformation of the data. Both the low (8-200 cm^{-1}) and mid-frequency (700-1700 cm^{-1}) spectral regions contained distinct spectral features for each solid-state form and could be used for quantification. Results with the instrument capable of simultaneous low-and mid-frequency analysis, where confounding sampling and instrument-induced error was avoided, showed that intensities were higher and quantitative models performed better in the low-frequency region. The prediction errors were lower with low frequency analysis (2.7%, 3.1% and 3.2% RMSEP for forms ?, ?2 and monohydrate, respectively) compared to mid-frequency analysis (6.3%, 5.4%, and 4.8%). In conclusion, significant spectral differences between the solid-state forms combined with higher a signal-to-noise ratio resulted in better analytical performance with low-frequency than mid-frequency Raman analysis, suggesting an advantage in using low-frequency Raman spectroscopy in quantitative solid-state analysis of pharmaceuticals.

Monitoring Form Conversion of Pharmaceutical Drug Using In-line and At-line Raman Spectroscopy
Ming Huang, John Wasylyk, Robert Wethman

The crystalline form in the Active Pharmaceutical Ingredient (API) and pharmaceutical drug product needs to be controlled carefully as it could impact physical properties such as solubility and stability as well as the efficacy of the drug. Different crystalline forms may typically have unique Raman spectral bands which can be used to ensure that the desired form is being produced. Raman methods can also be developed to measure quantitatively the level of desired and undesired forms in the sample and can be used to monitor and control the conversion from one form to the other. Raman spectroscopy was recently used in the API drying to study the kinetics of conversion between hydrate and dehydrated form as a function of temperature, pressure and humidity. The results of which was used to develop a drying protocol to produce the desired product with the correct form. Many times the crystallized form after the reaction is different from the desired form thus a conversion in slurry to the desired form is often needed. Raman spectroscopy can be used to monitor the rate of form conversion in slurry. The form conversion can be monitored at-line by taking a small sample and determine the ratio of forms by Raman. Alternatively an immersion probe can be inserted into the reaction vessel and the form conversion can be monitored in-line enabling real-time analysis and control. Raman spectroscopy has become a preferred method for studying crystalline form in our labs. Several examples using these approach will be presented.

Monitoring API Concentration by Raman Spectroscopy
Zachary Harms, Zhenqi Shi, Rajesh Kulkarni, James Hermiller, David Myers

Reflectance Raman spectroscopy is an excellent candidate for process analytical technology (PAT) applications in continuous manufacturing of pharmaceutical tablets. Spectroscopic methods provide a real-time, non-destructive measurement of active pharmaceutical ingredient (API) concentration in order to ensure product quality and uniformity. Of particular interest are tablets with low drug loads (< 5% w/w), where Raman is shown to be a promising alternative to near infrared spectroscopy (NIR). Our Raman spectrometer is equipped with a fiber-optically coupled probe head for non-contact measurement of powder blends. A mockup of the interface between the spectrometer and powders within the feed frame of a rotary tablet press is used to simulate the movement of powder blends from the mixer to the press. A port on the feed frame allows measurement by Raman or NIR of the blends just before compression. Powder blends containing varying known concentration of API are analyzed at the feed frame. The effects of spot size, acquisition time, and paddle speed are determined. A multivariate PLS model is used to calculate concentrations of API in unknown samples and accuracy, precision, and limit of detection are compared to identical measurements made by near infrared spectroscopy. Raman is shown to be accurate across a range of API concentrations with minimal day-to-day variability. The ability of Raman to be used in the control of a process within a set specification limit will be discussed.

Quantitative Raman Assays for On-Site Analysis of Stockpiled Drugs
Daniel Willett, Jason Rodriguez

We present a rapid aqueous Raman assay for on-site analysis of stockpiled drugs using a combination of portable Raman spectroscopy and chemometrics. To demonstrate the effectiveness of this approach it was tested on Tamiflu (oseltamivir phosphate). Tamiflu is a high priority drug approved for treatment of influenza. Tamiflu is the most common antiviral included in stockpiles for use in the event of a national emergency. The rapid assays feature a straight forward sample preparation step which involves dissolving each of the capsules in 10 ml of water. Testing was performed on the three approved Tamiflu capsule concentrations which contain 30, 45 & 75 mg of oseltamivir. Three different commercially-available portable Raman instruments were used. PLS Toolbox was used to build a model on each of the instruments using an identical calibration set of samples. Raman assay values were validated using HPLC to demonstrate the comparability of this approach to the compendial techniques. The average Raman assay value was within 0.5% from the HPLC value for the 35 different capsules tested. This Raman-based method demonstrates the potential for rapid screening of stockpiled pharmaceuticals using portable Raman instrumentation.

Fabry-Pérot Photothermal Interferometry (FP-PTI) for trace gas sensing in small sample volumes
Bernhard Lendl, Jakob Hayden, Johannes Paul Waclawek

Advances in Analytical Chemistry are often linked to technological developments in neighboring disciplines. This also concerns the recent development of quantum cascade lasers, which allows realization of sensing approaches in the mid-IR range that are different and more powerful compared to classical absorption measurements based on Lambert-Beer law. This presentation will focus on trace gas sensing using Photothermal Interferometry (PTI) on the example of sulfur dioxide. In PTI modulated absorption of the analyte produces a modulated change in the refractive index of the sample gas due to sample heating. The proposed set-up uses the cavity of a NIR Fabry Pérot (FP) interferometer as gas cell. The gas sample is excited using a distributed feedback Quantum Cascade Laser operated in the 2f-WM mode. The change in the refractive index of the sample produces a change in the transmission curve of the FP interferometer which is probed using a NIR laser. A new optical layout for rugged FP-PTI will be presented for the first time. This geometry efficiently assures long term operational stability along with single digit ppb sensitivities in a gas volume of just a few cubic millimeters.

Analysis of oxides in steel slag and of trace elements in steel by laser-induced breakdown spectroscopy
Johannes D. Pedarnig, Simon Eschlböck-Fuchs, Ludwig Birklbauer, Christoph M. Ahamer, Wolfgang Gaderbauer, Hubert Duchaczek, Josef Hofstadler, Andreas Pissenberger, Roman RösslerNorbert Huber

We report on the application of laser-induced breakdown spectroscopy (LIBS) for the chemical analysis of slag and steel from industrial steel production. Plasma is induced on the sample surface by short intense laser pulses and the optical emission of plasma is detected in the UV and VIS ranges by means of Echelle spectrometers and time-gated detectors. The slags are from secondary metallurgy and consist of several major and minor oxides (CaO, Al2O3, MgO, SiO2, FeO, MnO, TiO2, Cr2O3, P2O5, ...). The nominal concentration of oxides Cn is determined by X-ray fluorescence spectrometry (XRF). The steel samples contain traces of various elements (Si, Mn, Al, Ti, Nb, Cu, Cr, Ni, ...). The reference concentration of trace elements is determined by optical emission spectroscopy using spark excitation (spark-OES). For the slags the LIBS spectra are evaluated by an automated calibration-free (CF) method. CF-LIBS allows determining the concentration of major oxides Ccf directly from measured spectra without the use of standards. We present the results of more than 230 slag samples taken during steel production and analyzed at-line (total time-toanalysis ~3 minutes per sample). Major oxides with concentration Cn ? 1 wt% are quantified; minor oxides with Cn ? 0.1 wt% are detected. The influence of sample structure and temperature is investigated and the errors are compared for slags from the production and for homogenized slags. For the steel samples the LIBS spectra are evaluated by calibration-based methods. More than 40 samples are analyzed. The emission line intensities reveal a dependence on sampling depth and stable signals are measured for sufficient depth. For higher alloyed steel, the parameters of laser-induced plasma and the emission intensities are found to depend on sampling depth and on the concentration of alloy elements (matrix effect). Calibration curves of various elements are determined. The LOD values are in the low ppm range depending on the element. Acknowledgements: Financial support by the Austrian Research Promotion Agency FFG (Projects 843546 and 838861) is gratefully acknowledged.

Investigating Li7-3xAlxLa3Zr2O12 garnets using laser based spectroscopic analysis techniques
Stefan Smetaczek, Maximilian Bonta, Andreas Wachter-Welzl, Stefanie Taibl, Reinhard Wagner, Daniel Rettenwander, Jürgen Fleig, Andreas Limbeck

Due to their high Li-ion conductivity as well as chemical and electrochemical stability, cubic $Li_7La_3Zr_2O_{12}$ (LLZO) garnets are among the most promising candidates for solid-state electrolytes to be used in future Li-ion batteries [1]. Although LLZO has been studied extensively in recent years, the reproducible synthesis of highly conductive garnets remains challenging. In order to investigate the processes involved in the synthesis as well as to study the relationship between chemical composition and conductivity behaviour, information about the exact stoichiometry of the garnets is required. In this contribution, we present a LA-ICP-OES procedure for the laterally resolved stoichiometry determination of aluminium stabilized LLZO ($Li_{7-3x}Al_xLa_3Zr_2O_{12}$). Matrix-matched standards were successfully prepared by pressing ground LLZO powders with varying Al content into pellets. For signal quantification, an internal-standard independent calibration strategy based on the normalization of the sum of all metal oxides to 100 wt% was applied [2]. Using this approach, two-dimensional elemental distribution images of whole LLZO pellets were created, which revealed significant local variations in sample stoichiometry. [1] R. Murugan, V. Thangadurai, W. Weppner (2007), Angew. Chem. Int. Ed., 46: p. 7778 7781. [2] Y. Liu et al (2008), Chemical Geology, 257: p. 34-43.

Mid-infrared spectroscopy with supercontinuum laser sources
Markus Brandstetter, Jakob Kilgus, Christoph Gasser, Bernhard Lendl, Kristina Duswald

With the recent emergence of mid-infrared (MIR) supercontinuum (SC) sources an attractive new type of fiber laser source became available for the MIR spectral region. SC sources combine a series of desirable characteristics, such as high brightness and spatial coherence typical for quantum cascade lasers as well as broadband spectral coverage typical for thermal MIR sources. For this contribution a commercial MIR SC source covering the spectral range from 1.2 to 4.6 μm was employed, targeting a spectral region with limited or no availability of quantum cascade lasers. The average output power of the source was 500 mW at a repetition rate of 2.5 MHz. In order to evaluate the potential of MIR SC radiation for spectroscopic application the source was characterized with regard to its noise performance and compared to low-noise thermal sources [1]. A suitable detection scheme was developed, using MOEMS-based Fabry-Pérot filter spectrometers, which are both compact and rugged. These properties make them ideally suited for field application. Furthermore, they allow to overcome limitations associated with source noise. Results obtained with different spectroscopic measurement schemes will be presented. Among them standoff detection of various substances, such as paints and explosive precursors, time-resolved measurements of polymer curing and evaporation processes as well as sensitive liquid phase measurements using an ATR module that was specifically designed for that purpose. This work was supported by the strategic economic- and research program "Innovative Upper Austria 2020" of the province of Upper Austria. [1] Jakob Kilgus, Petra Müller, Peter M. Moselund and Markus Brandstetter; "Application of supercontinuum radiation for mid-infrared spectroscopy", Proc. SPIE 9899, Optical Sensing and Detection IV, 98990K (April 29, 2016); doi:10.1117/12.2225886

Recent Advances of Vibrational Spectroscopy in Phytomics
Christian Huck

The field of molecular vibrational spectroscopy applied to natural product analysis is further developing very fast. Whereas traditional separation techniques (LC, μ-LC-MS/MS, GC, CE-MS, MALD-TOF/MS) offer analytical investigations with high selectivity and sensitivity, vibrational spectroscopy benefits from the short analyses times, non-invasiveness and the possibility to screen for chemical and physical properties simultaneously. Chemometrical univariate and multivariate data treatment enables efficient spectral interpretation and the establishment of sufficient calibration and validation models. Quantum chemical approaches can further support the challenge of band assignment, especially in case of complex overlapping peak patterns. Near-infrared (NIR, 4.000-10.000 cm^{-1}), attenuated total reflection (ATR, 400-4.000 cm^{-1}) and Raman spectroscopy have been demonstrated being very efficient for even qualitative and quantitative attempts in combination with selected reference analytical methods. Qualitative attempts comprise analyzing, e.g., species and in some cases also origin, quantitative analysing chemical and physical parameters (e.g., potent ingredients and anti-oxidative potential). Two-dimensional correlation spectroscopy (2D-COS) has been developed towards a powerful analysis tool for monitoring the dynamics of a system considering temperature, pressure changes, e.g. during selective extraction procedures. The miniaturization of spectrometers is a highly demanding field, enabling to carry out investigations at any independent place. In this context, the critical evaluation of miniaturized spectrometers performance compared to powerful benchtop spectrometers is of high importance. Linear variable tunable filter (LVTF) or micro-electromechanical systems (MEMS) belong to the most frequently used techniques. Imaging and mapping spectroscopic attempts (MIR, NIR, Raman) enable high-resolution analysis of potent ingredients down to 1 μm. In the present contribution, the latest technical developments will be discussed followed by several selected applications in medicinal plant [1] analysis. Their limits and advantages over traditional methods will be critically evaluated to point out the future trends. [1] C.W. Huck, Phytochem. Lett. 2015, 11, 384-393.

Probing Cancer by Exploiting Spontaneous and Stimulated Raman Scattering
Ji-Xin Cheng, Chien Sheng Liao

Most of our knowledge about the chemical contents in cancer cells or cancerous tissue is obtained from in vitro analysis of fixed cells and tissues or homogenates. These in vitro assays identify the presence of molecules but do not provide information about their spatial distribution or temporal dynamics. A detailed mapping of molecules in live cells or intact tissues would pave the way for disease-specific drug developments or better marker-based diagnosis. This presentation will cover our latest advances in mapping cancer cells and tissues through Raman spectroscopic imaging. We realized that spontaneous Raman spectroscopy and coherent Raman microscopy, when integrated, can be a powerful analytical tool that enables discovery of hidden signatures in cells or tissues. Using this hybrid Raman system, we discovered cholesteryl ester as a metabolic marker of aggressive prostate cancer [Cell Metabolism, 2014, 19: 393]. More recently, we found lipid desaturation as a metabolic marker and therapeutic target of ovarian cancer stem cells [Cell Stem Cell, 2017, 20: 303]. To enable in vivo cancer detection in the clinic, we have developed a hand-held stimulated Raman scattering imaging system, at the speed of 8 frames per second, for chemical imaging of human subjects under ambient light. We showing applications of this system to in situ detection of brain tumor.

Effective Light Directed Assembly of Building Blocks with Microscale Control
Chia-Hung Chen

Light-directed forces have been widely used to pattern micro/nanoscale objects with precise control, forming functional assemblies. However, a substantial laser intensity is required to generate sufficient optical gradient forces to move a small object in a certain direction, causing limited throughput for applications. In this study, we demonstrated a high-throughput light-directed assembly as a printing technology by introducing gold nanorods to induce thermal convection flows that move microparticles (diameter = 40 μm to several hundreds of micrometers) to specific light-guided locations, forming desired patterns. An automatic stage integrated with a light source was developed as a printing system to design the patterns with programmable manner. With the advantage of effective light-directed assembly, the microfluidic-fabricated monodispersed biocompatible microparticles were used as building blocks to construct a structured assembly (~10 cm scale) in ~2 min. The control with microscale precision was approached by changing the size of the laser light spot. After crosslinking assembly of building blocks, a novel soft material with wanted pattern was approached. To demonstrate its application, the mesenchymal stem-cell-seeded hydrogel microparticles were prepared as functional building blocks to construct scaffold-free tissues with desired structures. This light-directed fabrication method can be applied to integrate different building units, enabling the bottom-up formation of materials with precise control over their internal structure for bio-printing, tissue engineering, and advanced manufacturing.

Optical Reflection and Waveguiding of Sound in Free Space
Daniel Kazal, Ellen Holthoff, Brian Cullum

Control and manipulation of sound waves in various media plays a key role in many scientific and engineering fields, generally requiring the use of solid, rigid physical structures for precise acoustic control. This requirement for solid, rigid structures to reflect and guide incident acoustic waves can be highly limiting in terms of environments and applications in which such structures can be employed. In this work, we describe the ability to optically reflect and guide acoustic waves in free space (i.e., air) efficiently. Abrupt, air density barriers are generated by photothermally depleting the air in a laser beam's path, generating a sharp difference in compressibility that results in significant acoustic reflection. Using this technique, reflection of incident acoustic signals as loud as 70 dB have been achieved with as much as 25% reflection efficiency. By altering the shape of the optical excitation beam, control of the direction of the reflected acoustic wave can be easily achieved. Furthermore, by employing multiple optical beam paths to generate a series of acoustic barriers, it is possible to completely suppress the transmission of incident acoustic waves. In addition to optically induced acoustic reflection and suppression, optical waveguiding of acoustic waves can also be achieved using hollow cylindrical (donut shaped) laser beams. By trapping the acoustic signal inside the center of the cylindrical beam, dramatic improvements in the distance over which acoustic waves can be transmitted with minimal amplitude loss can be achieved. Optical waveguiding of acoustic waves offers a new paradigm in the manipulation of sound over extended distances, providing potentially significant improvements to photoacoustic sensing, encrypted communications, and many other applications. In addition to describing this novel phenomenon, this talk will also discuss its application to standoff photoacoustic sensing.

Dried Blood Spheroids: A Versatile Paper-based Biofluid Sample Collection Platform for Improved Analyte Stability
Abraham Badu-Tawiah, Deidre Damon

A new paper-based biofluid sample collection platform is described that is based on three-dimensional dried blood spheroids as opposed to the traditional two-dimensional dried blood spot (DBS) sample collection procedure. This new technique uses functional hydrophobic paper substrates (prepared in-house) to overcome three major challenges associated with the traditional DBS procedure: (i) blood samples applied on the hydrophobic paper forms a spherical drop because of mismatch in surface energies, which dries to yield a dried blood spheroid. Experiments have shown that analytes (e.g., hydrolytically labile drugs such as cocaine and diazepam) trapped in the 3D dried blood spheroid are stabilized compared with storage done under the porous DBS conditions where a major portion of the sample becomes susceptible to oxidative stress from atmospheric air. (ii) Chromatographic and volcanic effects are eliminated on the hydrophobic paper, enabling easy validation without the need for chemical tracers or complex geometrical calculations to estimate sample volume in blood spot punch. It was experimentally determined that dried blood spheroids are confined to a reproducible area (e.g., 0.017 ± 0.003 cm2) compared to an area 0.13 ± 0.05 cm2 on untreated filter paper, with >2X the relative standard deviation. (iii) The hydrophobic paper strips also provide a direct mass spectrometric detection platform through a novel electrostatic spray ionization for sensitive analyte quantification. In-situ extraction of drugs (cocaine, benzoylecgonine, amphetamine and methamphetamine) from the dried blood spheroids resulted in sub-ng/mL limit of detections using ethyl acetate spray solvent. This solvent provides stable spray and enables efficient analyte extraction, limited only by the evaporation of the organic solvent. This presentation will focus on the preparation of the functional hydrophobic paper strips, the characterization of surface energies of the resultant paper strips using a novel electrostatic method, and the application of the dried blood spheroid sample collection approach for analyte stabilization and subsequent mass spectrometric detection. Mechanisms of various spray modes derived from the hydrophobic paper substrate will also be discussed.

The Science of Debunking Misconceptions
Panayiota Kendeou

Massive digital misinformation is one of the main threats to our society. Once adopted by an individual, misinformation is highly resistant to correction and influences subsequent learning and decision making. At the same time, the communication of science is caught up in political battles and mistrust in science is at a record high. In this talk, I will discuss the main mechanisms that cause misinformation to gain acceptance and approaches that have the potential to reduce and/or eliminate its impact.

The New 3R's: Risk, Reward, and Regulation
Fred LaPlant

In the current business and political climate, government regulations are a popular target for scorn from all sides. It's as easy to find ridiculous government regulations as it is to find bloated Defense spending, pork-barrel infrastructure appropriations, or inexplicable scientific research funding. But while all human activities are subject to excess and avarice, the need for regulation – and defense, and infrastructure, and research – is no less now than it has ever been. The challenge is promoting sensible, timely, equitable regulation that balances the needs of all stakeholders, including business, consumers, workers, and the regulatory agencies themselves. This talk will discuss the current state of affairs across a range of industries and ways to maintain oversight without putting our economy – or our safety – at risk.

AUTHOR INDEX

AUTHOR INDEX

AUTHOR INDEX

AUTHOR INDEX

AUTHOR INDEX

AUTHOR INDEX

AUTHOR INDEX

AUTHOR INDEX

AUTHOR INDEX

AUTHOR INDEX

AUTHOR INDEX

AUTHOR INDEX

AUTHOR INDEX

AUTHOR INDEX

AUTHOR INDEX

AUTHOR INDEX

AUTHOR INDEX

AUTHOR INDEX

AUTHOR INDEX

AUTHOR INDEX

Federation of Analytical Chemistry and Spectroscopy Societies (FACSS)
PO Box 24379
Santa Fe, NM 87502
USA

ISBN 978-1-5108-5214-3

CURRAN ASSOCIATES INC.
proceedings.com